MULTI-CARRIER SPREAD-SPECTRUM

Multi-Carrier Spread-Spectrum

For Future Generation Wireless Systems, Fourth International Workshop, Germany, September 17–19, 2003

Edited by

Khaled Fazel

Marconi Communications GmbH

and

Stefan Kaiser

German Aerospace Center

KLUWER ACADEMIC PUBLISHERS

BOSTON / DORDRECHT / LONDON

A C.I.P. Catalogue record for this book is available from the Library of Congress.

ISBN 1-4020-1837-1

Published by Kluwer Academic Publishers,
P.O. Box 17, 3300 AA Dordrecht, The Netherlands.

Sold and distributed in North, Central and South America
by Kluwer Academic Publishers,
101 Philip Drive, Norwell, MA 02061, U.S.A.

In all other countries, sold and distributed
by Kluwer Academic Publishers,
P.O. Box 322, 3300 AH Dordrecht, The Netherlands.

Printed on acid-free paper

Printed in the Netherlands.

TABLE OF CONTENTS

***Invited paper**

EDITORIAL INTRODUCTION

Khaled Fazel

Radio System Engineering
Marconi Communications
D-71522 Backnang, Germany

Stefan Kaiser

German Aerospace Center (DLR)
Institute for Communications and Navigation
D-82234 Wessling, Germany

In the last decade the technique of ***multi-carrier spread-spectrum*** (MC-SS) for wireless broadband multimedia applications has been receiving wide interests [1]. Since 1993 various combinations of the multi-carrier (MC) modulation and the spread spectrum (SS) technique have been introduced. Today, the field of MC-SS communications is considered to be an independent and important research topic with increasing activities. Several deep system analysis and comparisons of Multi-Carrier CDMA and Multi-Carrier DS-CDMA with DS-CDMA have been performed that show the superiority of MC-SS systems. New application fields have been proposed such as high rate cellular mobile (4G), high rate wireless indoor and fixed wireless access (FWA). In addition to system level analysis, a multitude of research activities has been addressed to develop appropriate strategies on detection, interference cancellation, channel coding, modulation, synchronization and low cost implementation design.

Offering a trade-off between coverage, data rate and mobility with a generic air interface architecture will be the primary goal of the next generation wireless systems. Users having no mobility and the lowest coverage distance (pico cells) with an ideal channel condition shall be able to receive the highest data rate, where on the other hand the subscriber with the highest mobility conditions and highest coverage area (macro cells) shall be able to receive the necessary data rate to establish the required communication link. Besides the introduction of new technologies to cover the need of higher data rates and new services, the *integration* of the existing technologies in a common platform as it is illustrated in Figure-1 will be an important objective of the next generation wireless systems beyond 3G.

In other words, the design of a *generic* multiple access scheme for the incoming 4G wireless systems will be challenging. This new multiple access scheme shall enable *i)* the integration of existing technologies, *ii)* to provide higher data rates in a given spectrum, i.e., maximizing the spectral efficiency, *iii)* to support different cell configuration and automatic adaptation to the channel conditions, *iv)* simple protocol and air interface layers, and finally, *v)* a seamless adaptation of new standards and technologies in the future.

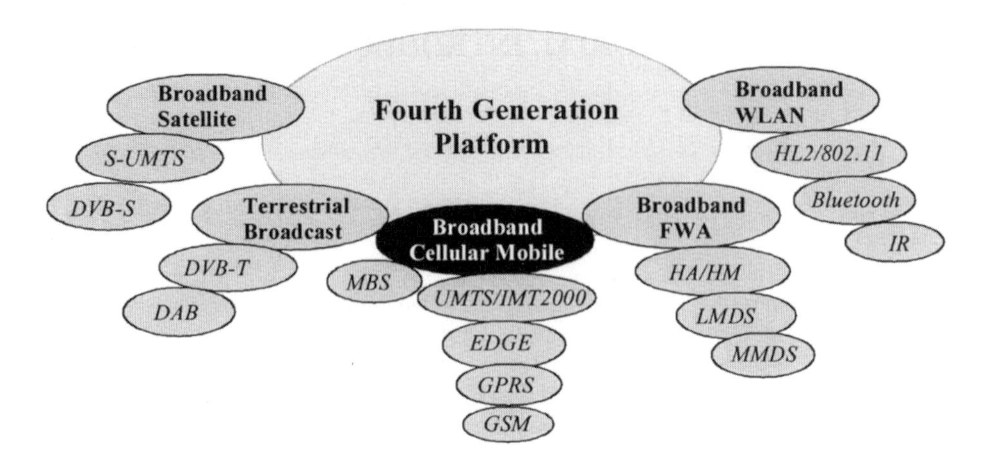

Figure-1 **Beyond 3G: *Integrated perspectives***

Here certainly ***multi-carrier spread-spectrum*** (MC-SS) with its generic air interface and adaptive technologies will be considered as a potential candidate to fulfill the above mentioned requirements of 4G [1].

SCOPE OF THIS ISSUE

The aim of this issue, consisting of six parts is to edit the ensemble of contributions presented during three days of the fourth international workshop on *multi-carrier spread-spectrum (MC-SS)*, held from September 17-19, 2003 in Oberpfaffenhofen, Germany.

The first part is devoted to the ***general issues*** of MC-SS. First, Atarashi and Sawahashi give an overview of multiple-access techniques based on MC-SS approaches for 4G System. Their concept is based on a variable two dimensional spreading in frequency and time domain. Then, Svensson *et al* analyze an alternative candidate for 4G. This paper proposes for the downlink the use of a pure OFDM technology. Based on their already presented concept of an overlay of different multiple access schemes Vanhaverbeke and Moeneclaey present a channel overloading in CDMA with scalable signature sets using an iterative Turbo detection strategy. Hara's paper deals with an overview of the main principle and applications of the virtual sub-carrier assignment (VISA) concept, where the impact of intelligent antenna in a frequency selective fading channel is considered. Analysis of the effect of cellular interference in MC-CDMA and its impact on channel estimation has been treated by Auer *et al.* A new two dimensional spreading code for MC-DS-CDMA based on orthogonal complete complementary codes is introduced by Farkas and Turcsany. Li, and Latva-aho analyze the performance of MC-CDMA multirate transmission schemes, needed especially for 4G systems. A concept for priority swapping sub-carrier and user allocation techniques to support an adaptive multi-rate system is presented by Al-Susa *et al.* Benvenuto *et al* discuss the strategy for sub-carrier loading for a pre-equalized uplink MC-CDMA system. A study on sub-carrier level power control for an OFDM transmission scheme with diversity and data spreading combination is shown by Kanamor *et al.* Novak and Krzymien present several techniques of packet re-transmission for a SS-OFDM system using frequency and time allocation. Finally, a detailed comparison of several iterative detection schemes for MIMO systems for MC-CDMA is presented by Yacoub *et al.*

The second part of this issue is devoted to ***coding and modulation***. First, the performance of Turbo product codes for an orthogonal MC-DS-CDMA system is analyzed by Guimaraes and Portugheis. Then, Fujii *et al* evaluate the joint performance of diversity and coding gain in an OFDM transmission system. The strategy of an adaptive coding in MC-CDMA/FDM systems with adaptive sub-carrier allocation is detailed by Trifonov *et al.* Xing and Renfors give a deep study of MC-CDMA systems with differential modulation. The choice of spreading codes to reduce the peak to average power ratio of an MC-CDMA system in the presence of nonlinear amplifier is analyzed by Saito *et al.* Raos *et al* make an analysis of linear receivers for MC-CDMA with digital prolate functions. Finally, a performance analysis of different spreading sequences for MC-CDMA in WLAN environments is made by Garcia-Armanda *et al.*

The ***synchronization and channel estimation*** aspects for MC and MC-SS transmission systems are discussed in the third part of this issue. An overview of analytical performance evaluation of synchronization sensitivity for uplink and downlink of a MC-DS-CDMA is presented by Steendam and Moeneclaey. Zhang *et al* study the performance of symbol synchronization in MC-CDMA systems. An efficient iterative channel estimation technique for space-frequency coded OFDM systems with transmitter diversity is proposed by Cirpan *et al.* Feuersänger *et al* make a detailed comparison of different pilot multiplexing schemes for maximum likelihood channel estimation in coded MC-CDMA systems. Honglei and Juntti study the performance of a data aided MMSE channel estimation technique in an OFDM system. The performance of channel estimation in case of multi-user transmissions for OFDM is detailed by Chenu-Tournier *et al.* An adaptive pilot assisted channel estimation technique for OFDM systems is discussed by Sand *et al.* In case of multi-user OFDM transmission in mobile radio channel the benefits of exploiting *a-priori* information for channel estimation is studied by Maniatis *et al.* The problem of timing of the FFT-window in SC/FDE systems is deeply analyzed by Koppler *et al.* A joined channel estimation and equalization for OFDM systems in mobile radio is presented by Gligorevic *et al.* The performance analysis of the downlink and uplink of MC-CDMA in the presence of carrier frequency offset is studied by Zhang *et al.* The effects of a not perfect, i.e. real channel estimation in case of space-time multi-user detection for MC-CDMA systems is analyzed by Sanguinetti and Morelli. Finally, the synchronization and power control processes for an uplink on an MC-CDMA system is detailed by Legouable *et al.*

The fourth part is devoted to ***MIMO, diversity and space time/frequency coding schemes***. The first paper, presented by Lee and Nakagawa discusses a new handoff technique using space time transmit diversity (STTD) technique for an MC-CDMA system in a frequency selective fading channel. An array antenna assisted Doppler spread compensator with vehicle speed estimator for OFDM receiver is presented by Nagai *et al.* Silva, and Gameiro analyze several pre-filtering techniques for the downlink of a TDD based MC-CDMA system using antenna arrays. Furthermore, for the downlink of an MC-CDMA scheme, Sälzer and Mottier study several strategies using antenna arrays for interference mitigation. Witschnig *et al* present for a single carrier transmission scheme with frequency domain equalization an overall analysis of several antenna diversity techniques. A performance comparison between MC-CDMA and OFDM transmission schemes in Rayleigh fading channels is made by Stitz *et al.* Several efficient diversity techniques using linear pre-coding and STBC for Multi-carrier systems is presented by Le Nir *et al.* Auffray *et al* evaluate the performance of MMSE STBC for an MC-CDMA scheme in a mobile radio channel. A deep comparison between space-time block coding and eigen beamforming in TDD MIMO-OFDM downlink with partial CSI knowledge is made by Codreanu and Latva-aho. Jankiraman and Prasad study the performance of a space-time block code for OFDM-MIMO systems, proposed as a candidate for fourth generation systems. Space-frequency coding and detection techniques for the downlink of a MIMO MC-CDMA scheme is presented by Vehkapera *et al.* Del Cacho *et al* propose an adaptive V-BLAST architecture based on broadband MIMO systems. Finally, Liu and Bourdoux propose a new design of a low complexity Turbo MIMO receiver for WLAN applications.

The fifth part assembles all issues related to ***detection, multiplexing and interference cancellation*** techniques. Here first Thomas presents a distributed multiplexing scheme for multi-carrier wireless networks application. An inter-cell interference suppression technique using virtual sub-carrier assignment (VISA) for MC-CDMA uplink is analyzed by Tsumura *et al.* The effect of synchronisation loss on the signal detection performance at the base station using an OFDM-CDMA system is evaluated by Bader and Zazo. A minimum BER multi-user transmission for spread-spectrum systems in frequency selective fading channels is analyzed by Irmer *et al.* Cosovic *et al* study a combined pre- and post-equalization techniques for uplink of a TDD-MC-CDMA in fading channels. A time varying equalization technique for MC-CDMA via Fourier basis functions is analyzed by Zemen *et al.* Finally, the impact of channel variation on a code multiplexed pilot in multi-carrier transmission systems is evaluated by Krauss and Baum.

The last part of this book is devoted to the ***realization and implementation*** aspects. First a joint compensation of IQ imbalance, frequency offset and phase noise in multi-carrier systems is presented by Tubbax *et al.* Some practical issues of a parallel interference cancellation scheme in MC-CDMA systems is analyzed by *Duan* *et al.* The impact of oscillator imperfection on performance of MC-DS-CDMA and MC-CDMA systems is evaluated by Hicheri *et al.* Wu and Bar-Ness present a new phase noise mitigation method in OFDM systems with simultaneous common phase estimation and inter-channel interference correction. Finally, the performance comparison of OFDM transmission affected by phase noise with and without PLL is done by Rave *et al.*

In conclusions, we wish to thank all of the authors who have contributed to this issue, and all those in general who responded enthusiastically to the call. We also hope that this edited book may serve to promote further research in this new area and especially for the success of the next generation wireless technology beyond 3 G.

REFERENCE

[1] Fazel K, Kaiser S., "Multi-Carrier and Spread Spectrum Systems", John Wiley and Sons Ltd., Sept. 2003.

ACKNOWLEDGMENTS

The editors wish to express their sincere thanks for the support of the chairmen of the different sessions of the workshop namely, Dr. H. Atarashi from NTT-DoCoMo, Prof. S. Hara from University of Osaka, Prof. W. Koch from University of Erlangen, Prof. W. A. Krzymien from University of Alberta / TRLabs, Prof. J. Lindner from University of Ulm, Prof. U. Mengali from University of Pisa, Prof. M. Nakagawa from University of Keio and Prof. E. Panayicri from ISIK University. Many thanks to our invited spreaker Prof. G. Fettweis from University of Dresden and all authors that through their contributions made the workshop successful. Furthermore, many thanks to Ms. J. Uelner from DLR for her active support for the local organization of the workshop.

This fourth international workshop on Multi-Carrier Spread-Spectrum could not be successfully happened without the

- assistance of the ***TPC members***:

P. W. Baier (Germany)	S. Kaiser (Germany)	M. Nakagawa (Japan)
Y. Bar-Ness (USA)	K.-D. Kammeyer (Germany)	S. Pasupathy (Canada)
K. Fazel (Germany)	W. A. Krzymien (Canada)	R. Prasad (Denmark)
G. Fettweis (Germany)	J. Lindner (Germany)	M. Renfors (Finland)
G. B. Giannakis (USA)	U. Mengali (Italy)	H. Rohling (Germany)
J. Hagenauer (Germany)	L. B. Milstein (USA)	H. Sari (France)
S. Hara (Japan)	M. Moeneclaey (Belgium)	M. Sawahashi (Japan)
H. Imai (Japan)	W. Mohr (Germany)	R. Z. Ziemer (USA)

- technical and financial support of:

German Aerospace Center (DLR)
DoCoMo Eurolabs
Marconi Communications

- and technical support of

IEEE Communication Society, German Section
Information Technology Society (ITG) within VDE
ITG Fachausschuss 5.1 (Information and System Theory)

ACKNOWLEDGEMENTS

[illegible] the [illegible] of the workshop [illegible] Prof. S. [illegible] from University of Osaka, Prof. S. [illegible] University of [illegible], Prof. [illegible] from University of Alberta [illegible] from University [illegible] from University [illegible] Prof. [illegible] from [illegible] and Prof. [illegible] from [illegible] from University [illegible] of the workshop [illegible]

Section I

GENERAL ISSUES

HIROYUKI ATARASHI AND MAMORU SAWAHASHI

BROADBAND WIRELESS ACCESS BASED ON VSF-OFCDM AND VSCRF-CDMA AND ITS EXPERIMENTS

Abstract. This paper presents broadband packet wireless access schemes based on Variable Spreading Factor (VSF)-Orthogonal Frequency and Code Division Multiplexing (OFCDM) in the forward link and Variable Spreading and Chip Repetition Factors (VSCRF)-CDMA in the reverse link for the systems beyond IMT-2000. In our design concept for wireless access in both links, radio parameters such as the spreading factor are optimally controlled so that the system capacity is maximized according to the cell configuration, channel load, and radio channel conditions, based on the tradeoff between efficient suppression of other-cell interference and the capacity increase in the target cell by exploiting orthogonality in the time and frequency domains. We demonstrate that the peak throughput of greater than 100 Mbps and 20 Mbps is achieved by the implemented base station and mobile station transceivers using the 100-MHz and 40-MHz bandwidths in the forward and reverse links, respectively. Moreover, the simulation results elucidate the possibility of the peak throughput of approximately 1 Gbps for short-range area application using the 100-MHz bandwidth OFCDM forward link by applying four-branch Multiple Input Multiple Output (MIMO) multiplexing with 16QAM data modulation and punctured turbo coding.

1. INTRODUCTION

Specifications of the High-Speed Downlink Packet Access (HSDPA) based on the W-CDMA air interface are almost complete with the aim to establish much higher-speed packet data services than 2 Mbps in the forward link. In HSDPA, key techniques, such as adaptive modulation and channel coding (AMC), hybrid automatic repeat request (ARQ) with packet combining, and fast packet scheduling, are employed [1]. However, anticipating the current and future tremendous increases in the amount of data traffic, new broadband wireless access schemes for the systems beyond IMT-2000 must establish broadband packet transmission with a maximum data rate above 100 Mbps in the forward link using an approximate 50-to-100-MHz bandwidth [2]-[4] (note that the target data rate corresponds to approximately ten fold higher than that achievable in HSDPA with a 5-MHz bandwidth). Furthermore, this broadband wireless access scheme must flexibly support both isolated-cell environments such as hot-spot areas and indoor offices as well as cellular systems from the standpoint of further reducing the cost of radio access networks (RANs).

To develop a broadband wireless access scheme, we elucidated that Orthogonal Frequency and Code Division Multiplexing (OFCDM), which is originally based on multi-carrier CDMA (MC-CDMA) [5],[6], or Orthogonal Frequency Division Multiplexing (OFDM) exhibits better performance than conventional DS-CDMA wireless access [2]-[4]. This is because OFCDM and OFDM mitigate the degradation caused by severe multipath interference (MPI) in a broadband channel owing to a low symbol rate associated with many sub-carriers. In OFCDM wireless

K. Fazel and S. Kaiser (eds.), Multi-Carrier Spread-Spectrum, 3–14.

access, we proposed introducing the variable spreading factor (VSF) concept, (hereafter VSF-OFCDM) [7], which changes the spreading factor in both the time and frequency domains of OFCDM corresponding to the cell structure, channel load, propagation channel conditions, and major radio link parameters (e.g., data modulation and channel coding rate). Through VSF-OFCDM, the seamless and flexible deployment of the same wireless access method is possible both in cellular systems and isolated-cell environments.

Meanwhile in the reverse link, we elucidated that the DS-CDMA based wireless access achieves a higher link capacity using coherent Rake combining with a dedicated pilot channel than does using numerous sub-carriers, such as in the case of MC-CDMA and OFDM [2],[4]. The DS-CDMA approach is also advantageous in the application to a mobile terminal owing to lower power consumption for its inherently much lower peak-to-average power ratio (PAPR) feature compared to MC-CDMA and OFDM. Furthermore, in order to increase the link capacity of CDMA wireless access in isolated-cell environments, we proposed using Variable Spreading and Chip Repetition Factors (VSCRF) based CDMA (VSCRF-CDMA hereafter) [8] by applying a symbol repetition principle [9]. In the proposed scheme, the spreading and chip repetition factors are adaptively changed in accordance to the cell structure, the number of simultaneous accessing users and the propagation channel conditions. The conventional DS-CDMA feature employing only spreading is applied to achieve easily one-cell frequency reuse exploiting the cell (or user)-specific scrambling code in a cellular system with a multi-cell configuration. Meanwhile, by increasing the chip repetition factor by more than one in an isolated-cell environment, multiple access interference (MAI) is suppressed by the orthogonality in the frequency domain. Therefore, VSCRF-CDMA also supports the seamless and flexible deployment of the same wireless access method both in cellular systems and isolated-cell environments.

By unifying our evaluations on the constituent techniques, this paper presents broadband wireless access schemes employing VSF-OFCDM in the forward link and VSCRF-CDMA in the reverse link as a promising wireless access candidate for the system beyond IMT-2000. Furthermore, we show the operating principle of VSF-OFCDM and VSCRF-CDMA, which enable the flexible and seamless support of both cellular systems and isolated-cell environments using the same air interface. In the rest of the paper, we first describe the proposed design concept of the broadband wireless access schemes and the details of VSF-OFCDM and VSCRF-CDMA in Section 2. In Section 3, we show the laboratory experimental results utilizing the implemented testbed based on VSF-OFCDM and VSF-CDMA, in which multipath fading simulators are used, to demonstrate the achievement of throughput exceeding 100 and 20 Mbps in the forward and reverse links, respectively. Furthermore, in Section 4, we discuss the appropriate multiple antenna transmission and reception techniques for VSF-OFCDM to achieve much higher spectrum efficiency such as 10 bps/Hz in the forward link, considering the application to hot-spot areas and indoor office environments.

2. PROPOSED BROADBAND WIRELESS ACCESS

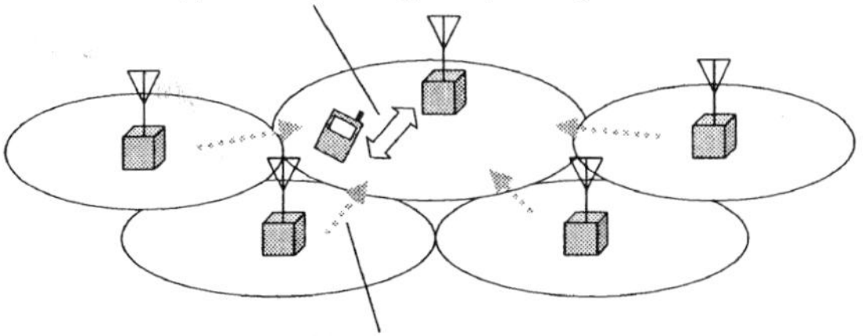

Figure 1. Proposed concept for 4G wireless access.

2.1. Proposed Concept for Broadband Wireless Access

In future RANs, a further decrease in the network cost is a very important requirement for offering rich multimedia services to customers via wireless communications. The higher system capacity, i.e., higher frequency efficiency accommodating a large number of simultaneous users, will definitely contribute to the further reduction of RAN cost. In a cellular system with a multi-cell configuration, one-cell frequency reuse is essential to increasing the system capacity. On the other hand, focusing on one target cell with low-level interference from the surrounding cells such as in isolated-cell and indoor environments, the approach exploiting orthogonality in the time or frequency domains achieves higher capacity than does the spreading approach (i.e., the use of code domain) in frequency-selective (multipath) fading channels. This is because the orthogonality among simultaneously coded channels is destroyed by increasing the number of multipaths in a broadband multipath fading channel. Therefore, in the proposed concept, radio parameters such as the spreading factor (*SF*) and chip repetition factor (*CRF*) are optimally controlled so that the system capacity is maximized according to the cell configurations whether multi-cell or isolated-cell and according to the radio channel conditions such as other-cell interference both in the forward and reverse links. As shown in Fig. 1, optimization is achieved by choosing the optimum balance for the tradeoff relationship between efficient suppression of other-cell interference and the capacity increase in the target cell by exploiting orthogonal channels in the time or frequency domains. Consequently, we aim to achieve the maximum system capacity in the respective cell configurations and radio channel conditions using the same air interface, thereby, the same broadband wireless access scheme with the same air interface (i.e., the same carrier frequency, frequency bandwidth, and radio frame format).

2.2. VSF-OFCDM in Forward Link

Figure 2 shows the principle of the proposed VSF-OFCDM employing two-dimensional spreading, where the spreading factors in the time and frequency domains, i.e., SF_{Time} and SF_{Freq}, are adaptively controlled based on the cell structure, the cell configuration, channel load, and channel conditions such as the delay spread and fading maximum Doppler frequency, in order to achieve higher link capacity both in cellular systems and isolated-cell and indoor environments (note that the speed of updating the SF_{Time} and SF_{Freq} values is even more gradual than that of the data modulation scheme and channel coding rate in the AMC). We introduced time domain spreading [10] and two-dimensional spreading [11], [12] into our proposed VSF-OFCDM [13]. As shown in Fig. 2, VSF-OFCDM employs a total spreading factor, SF (= SF_{Time} x SF_{Freq}), of greater than 1, in a multi-cell environment to achieve higher link capacity. This is because one-cell frequency reuse is possible for $SF > 1$ by introducing a cell-specific scrambling code, and a direct increase in the radio link capacity is expected by employing sectorization. Furthermore, in two-dimensional spreading, we prioritize time domain spreading rather than frequency domain spreading. This is because, in a frequency selective fading channel, time domain spreading is superior to frequency domain spreading in general to maintain the orthogonality among the code-multiplexed channels, which is important in applying AMC employing multi-level modulation to achieve a higher data rate. Meanwhile, in a lower received signal-to-noise power ratio (SIR) region, such as the cell boundary, QPSK data modulation associated with a lower channel coding rate is effective in satisfying the required transmission quality. In this case, employing frequency domain spreading, i.e., $SF_{Freq} > 1$, along with time domain spreading is very beneficial, since the frequency diversity effect derived by frequency domain spreading and interleaving enhance the transmission quality while the impact of the inter-code interference in QPSK data modulation is slight.

On the other hand, in an isolated-cell environment, in order to avoid inter-code interference caused by the destroyed orthogonality in the frequency domain, we employ $SF_{Freq} = 1$. However, in the time domain, we apply $SF_{Time} > 1$ in order to utilize the benefits of code-domain multiplexing as described below, while still achieving the orthogonality among the code-multiplexed channels. First, by introducing time domain spreading, within the same frame timing, i.e., without incurring any additional transmission delay, the data channel is flexibly code-

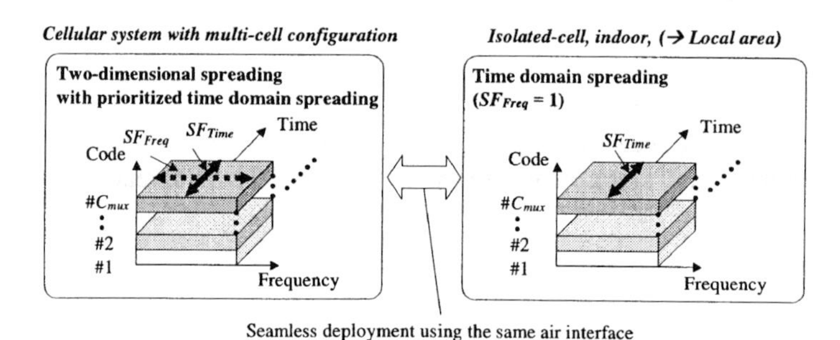

Figure 2. VSF-OFCDM wireless access scheme in forward link.

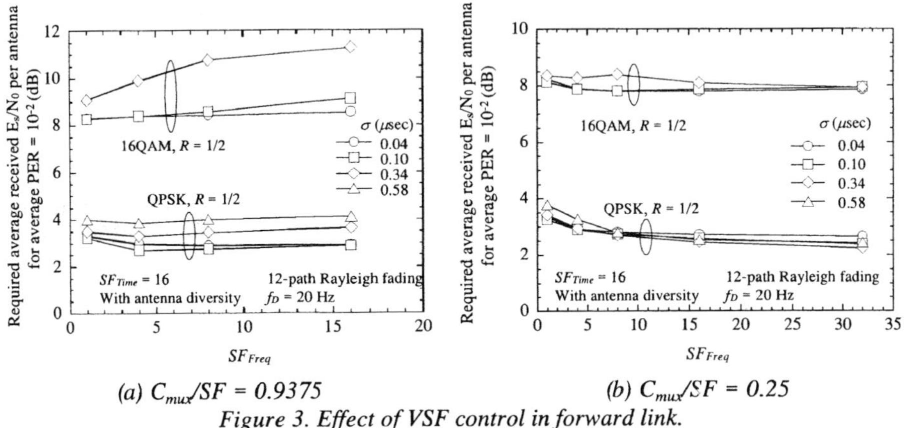

(a) C_{mux}/SF = 0.9375 (b) C_{mux}/SF = 0.25

Figure 3. Effect of VSF control in forward link.

multiplexed with the associated control channel independently at any slot by fast packet scheduling, which is a very advantageous feature in transmitting the control data when AMC and hybrid ARQ are applied in the data channel. Second, a new call, i.e., a new physical channel, is flexibly reallocated in addition to existing physical channels in the code domain. Third, multiplexing of multiple physical channels with different symbol rates is easily achieved. Fourth, a low-rate physical channel is easily supported by simply increasing the spreading factor value. Fifth, transmission powers among simultaneous physical channels are flexibly allocated in the code domain. Sixth, a code-multiplexing pilot channel is easily realized.

In addition to the VSF concept in OFCDM according to the cell structure, it should be noted that the overall received signal quality employing two-dimensional spreading through a multipath fading channel depends on the tradeoff relationship between the increasing frequency (time) diversity effect and the impairment of the code orthogonality. Furthermore, these two factors are strongly affected by the channel load, propagation channel conditions, and other-cell interference. Thus, by simply introducing two-dimensional spreading uniformly only according to the data rate, i.e., symbol rate of the physical channel, the maximum system capacity reflecting the aforementioned various conditions is not achieved. It is advantageous to employ the optimum spreading factor values taking into account the channel load, propagation channel conditions, and other-cell interference [14]. The optimum design of the spreading factor control in the frequency domain is investigated taking into account the influence of the spreading factor value in the frequency domain, SF_{Freq}. Since the code-orthogonality in the frequency domain spreading is destroyed due to the channel variation in the frequency domain, the root mean squared (rms) delay spread, σ, is varied as a parameter. Figures 3(a) and 3(b) show the required average received signal energy per symbol-to-noise power spectrum density ratio (E_s/N_0) per antenna for achieving the average packet error rate (PER) = 10^{-2} as a function of the SF_{Freq} for QPSK and 16QAM data modulation, assuming the number of multiplexed codes normalized by the spreading factor C_{mux}/SF = 0.9375 and 0.25, respectively. In the figure, we assume a 12-path Rayleigh fading channel with various rms delay spread values, σ. In the QPSK data modulation, the increase in

SF_{Freq} up to 4 for C_{mux}/SF = 0.9375 and up to 16 for C_{mux}/SF = 0.25 exhibits better performance as shown in Figs. 3(a) and 3(b), respectively. This is because the larger SF_{Freq} value brings about an improved frequency diversity effect associated with less influence of the inter-code interference with QPSK data modulation in σ = 0.04 to 0.58 μsec. Therefore, we can say that adaptive control of the spreading factor in the frequency domain is effective in QPSK data modulation according to the delay spread and channel load.

Meanwhile, in 16QAM data modulation, the required average received E_s/N_0 is degraded for σ = 0.10 and 0.34 μsec according to the increase in the SF_{Freq} value for C_{mux}/SF = 0.9375. This is because 16QAM data modulation is degraded by the inter-code interference more severely due to the efficient amplitude and phase modulation. Therefore, in 16QAM data modulation under heavier channel load conditions, the optimum SF_{Freq} is one to avoid inter-code interference in frequency domain spreading.

2.3. VSCRF-CDMA in Reverse Link

We recently elucidated that DS-CDMA based wireless access achieves a higher radio link capacity than does wireless access with numerous sub-carriers such as OFDM and MC-CDMA, since the impact of channel estimation error is greater, which is caused by the lower pilot signal energy per sub-carrier using numerous sub-carriers [2]. Furthermore, in a multi-cell environment, one-cell frequency reuse is easily achieved by the spreading gain in DS-CDMA, and thereby, a direct capacity increase is achieved by sectorization. Therefore, in a multi-cell environment, the system capacity using DS-CDMA is higher than that of Time Division Multiple Access (TDMA) together with a single carrier or multi-carriers that require cell frequency reuse of more than one. DS-CDMA is one of the most promising broadband wireless access schemes in the reverse link taking into account the application to cellular systems.

In an isolated-cell environment, however, where other-cell interference is almost negligible, DS-CDMA wireless access is disadvantageous due to the degradation by MAI within the cell. This is because the orthogonality in the code domain among the simultaneously accessing users is not maintained due to the different propagation delay times among the users. Furthermore, in an isolated-cell environment, the benefit of one-cell frequency reuse is ineffective. As a result, the system capacity per sector normalized by the spreading factor value becomes approximately 20 to 30% without voice activation, which is much lower than that of TDMA-based

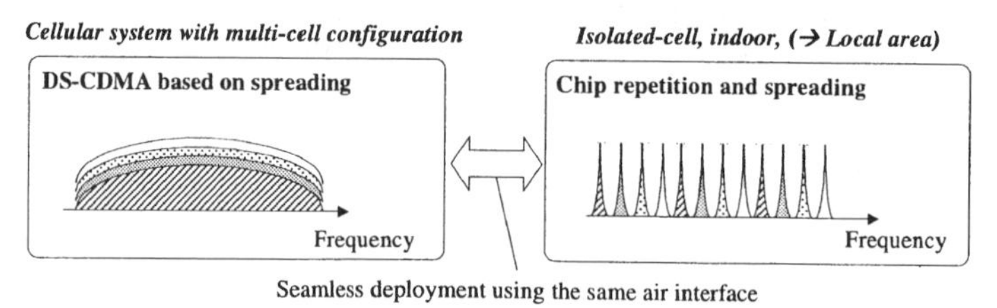

Figure 4. VSCRF-CDMA wireless access scheme in reverse link.

wireless access without spreading. Here, we aim to achieve seamless and flexible deployment of a RAN with the same air interface while supporting a higher system capacity both in multi-cell and isolated-cell environments. In order to achieve this goal, further capacity increase, i.e., improvement in the spectrum efficiency of CDMA based wireless access, in an isolated-cell environment is required by maintaining the orthogonality among the accessing users in the reverse link.

Therefore, as shown in Fig. 4, in the proposed VSCRF-CDMA [8], we implement the chip repetition principle into a part of the conventional time-domain spreading of CDMA by applying the symbol repetition principle [9] in isolated-cell environments. By setting the chip repetition factor, *CRF*, to a value greater than one, the simultaneous physical channels can be orthogonal in the frequency domain. Meanwhile, in a cellular system employing a multi-cell configuration, we basically only use the conventional time-domain spreading (i.e., *CRF* = 1), in order to suppress effectively other-cell interference.

The average received E_s/N_0 employing DS-CDMA with and without chip repetition to satisfy the average PER of 10^{-1} and 10^{-2} is plotted in Fig. 5 as a function of number of simultaneous accessing users, K_u. We clearly find that as the K_u value becomes larger, the required average received E_s/N_0 is monotonously increased due to the increasing MAI in DS-CDMA without chip repetition. However, by introducing chip repetition with *CRF* = 8, an almost constant required average received E_s/N_0 is achieved since the MAI from the simultaneous accessing users up to eight can be removed by the orthogonality in the frequency domain. When K_u is one or two, the required average received E_s/N_0 of the DS-CDMA with chip repetition is degraded. This is because in DS-CDMA with chip repetition, the MPI suppression effect is reduced since the spreading factor is small such as two. Nevertheless, we emphasize that, as indicated by the open-circle plots, the spreading factor and chip repetition factors are changed according to the number of simultaneous accessing users and the propagation channel conditions in our proposed VSCRF-CDMA. Thereby, the performance of VSCRF-CDMA with optimum *SF* and *CRF* combinations according to number of accessing users achieves the best performance.

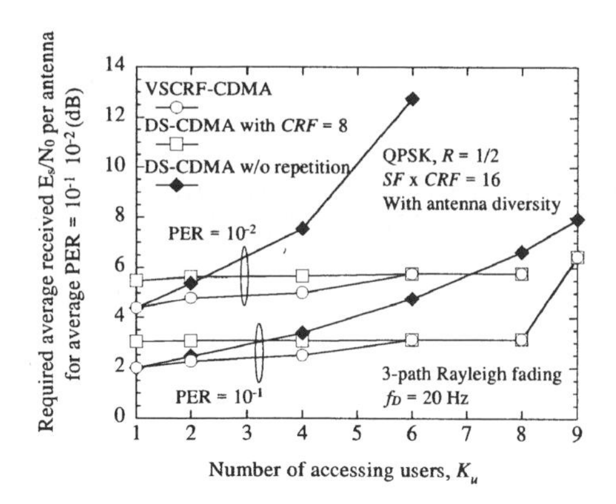

Figure 5. Performance of VSCRF-CDMA in isolated-cell environment.

3. EXPERIMENTAL EVALUATION OF PROPOSED BROADBAND WIRELESS ACCESS

3.1. Configuration of Implemented Base Station and Mobile Station Transceivers

The major radio link parameters of the implemented base station (BS) and mobile station (MS) transceivers are given in Table 1 [15]. We applied VSF-OFCDM with a 101.5-MHz bandwidth and 768 sub-carriers in the forward link. In addition, we employed VSF-CDMA, i.e., $CRF = 1$ for VSCRF-CDMA, with a 40-MHz bandwidth with two sub-carriers in the reverse link. The frame formats for the forward and reverse links are shown in Figs. 6(a) and 6(b), respectively.

In the BS transmitter, the binary information bit is first turbo encoded. After data modulation mapping, the symbol interleaving in the frequency domain is performed to randomize the burst errors due to frequency-selective fading. The interleaved sequence is mapped into the Downlink Packet Data Channel, which is spread two-dimensionally by SF_{Time} x SF_{Freq} and then, C_{mux} code channels are multiplexed. In the Downlink Packet Data Channel, AMC based on the measured SIR is applied. In addition, the Downlink Packet Control Channel is used to convey the modulation and channel coding scheme (MCS) information in the AMC. Furthermore, the pilot symbols used for channel estimation and SIR measurement for the AMC are time-multiplexed into the Downlink Packet Data Channel and Downlink Packet Control Channel. After conversion into baseband in-phase (I) and quadrature (Q) components by a D/A converter, quadrature-modulation is performed. Finally, the IF modulated signal is up-converted into an RF signal and amplified by the power amplifier. At the BS receiver, we apply two-branch antenna diversity reception. The frequency down-converted IF signal is first linearly amplified by an AGC. The received spread signal is converted into the baseband I and Q components by a quadrature detector, and the resultant signals are converted into digital format by an A/D converter, and filtered by a square-root raised cosine Nyquist filter. We generate a power delay profile in order to search the propagation paths for Rake combining. The composite signal sample sequence is despread by a matched filter and Rake-combined by using a channel estimate. Finally, the soft-decision data sequence from the coherent Rake combiner is Turbo-decoded to recover the transmitted data sequence. Max-Log-MAP decoding with eight iterations is used as the decoding algorithm.

Table 1. Major radio link parameters of testbed.

		Forward link	*Reverse link*
Bandwidth		101. 5 MHz	40 MHz
Number of sub-carriers		768 (131.836 kHz sub-carrier separation)	2
OFCDM symbol duration		7.585 μsec + GI 1.674 μsec (1024 + 226 samples)	–
Chip rate		–	16.384 Mcps / carrier
Spreading factor (Packet data channel)		SF_{Time} x SF_{Freq} = 1 x 1, ..., 16 x 1, ..., 16 x 8	SF = 4, 8, ..., 16
Modulation	Data	QPSK, 16QAM, 64QAM	QPSK, 16QAM, 64QAM
	Spreading	QPSK	HPSK
Channel coding		Turbo code (R = 1/3 - 5/6)	Turbo code (R = 1/16, 1/8, 1/3 – 5/6)

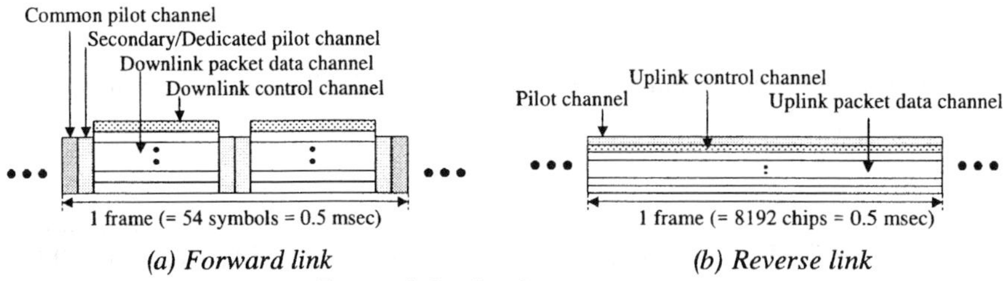

(a) Forward link *(b) Reverse link*

Figure 6. Packet frame structure.

At the MS transmitter, the binary information bit sequence in the Uplink Packet Data Channel is channel-encoded using turbo coding. After the coded bit sequence is serial-to-parallel converted into C_{mux} code channels, the data sequence of each code channel is QPSK data-modulated. Subsequently, the data modulated sequence is spread by the combination of the orthogonal variable spreading factor (OVSF) channelization code with a SF = 4-chip length and a scrambling code with the repetition period of 8192 chips. Finally, after the pilot channel comprising 32 symbols with the spreading factor of SF = 256 is code-multiplexed into a coded data channel, a square-root raised cosine Nyquist transmit filter is applied before frequency conversion into the carrier frequency per sub-carrier.

At the MS receiver, we apply two-branch antenna diversity reception. The channel gain of each frame at each sub-carrier is estimated by coherently accumulating pilot symbols within a frame. By employing the channel estimate, the spread sequence at each sub-carrier is despread in the time domain with equal gain combining and in the frequency domain with minimum mean square error (MMSE) combining. After de-interleaving in the frequency domain, turbo decoding is performed to recover the information bit sequence (Max-Log-MAP decoding with six iterations). Meanwhile, the measured received SIR information at the MS for selecting the optimum MCS is sent to the BS through the Uplink Packet Control Channel.

3.2. Throughput Performance

The measured throughput performance in VSF-OFCDM forward link when AMC with five MCSs, i.e., MCS1 (QPSK, R = 1/3), MCS2 (QPSK, R = 1/2), MCS3 (QPSK, R = 3/4), MCS4 (16QAM, R = 1/2), and MCS5 (16QAM, R = 3/4), is applied is plotted in Fig. 7(a) as a function of the average received E_s/N_0 per antenna for L = 6-path Rayleigh fading, maximum Doppler frequency f_D = 20 Hz, and σ = 0.3 μsec. We employ the threshold values for selecting the optimum MCS at the measured received SIR over one frame length. The throughput performance for the five respective MCSs is shown in the figure. We clearly find that by performing AMC based on the measured SIR per frame, nearly the maximum throughput at each average received E_s/N_0 is achieved, suggesting that the near optimum MCS is accurately selected according to the measured received SIR. We observed that the throughput of 100 and 200 Mbps using AMC is obtained at the average received E_s/N_0 of approximately 7.5 and 19 dB, respectively, when AMC is actually

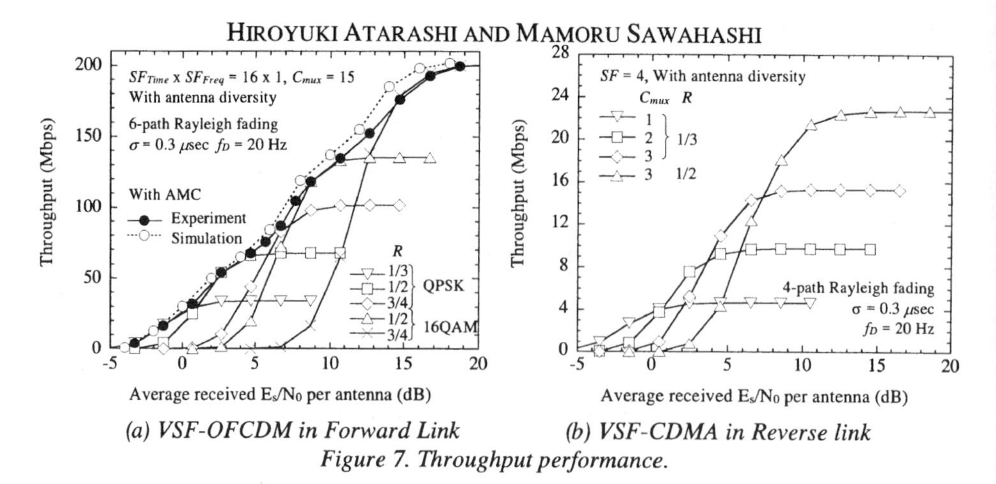

(a) VSF-OFCDM in Forward Link *(b) VSF-CDMA in Reverse link*

Figure 7. Throughput performance.

performed. Furthermore, we find that the loss in the required average received E_s/N_0 of the experimental results from that of the simulation results is within 1 dB.

The throughput performance in the VSF-CDMA reverse link using the four combinations of R and C_{mux} with QPSK data modulation is plotted in Fig. 7(b) as a function of the average received E_s/N_0 per antenna for L = 4-path Rayleigh fading, maximum Doppler frequency f_D = 20 Hz, and σ = 0.3 μsec. Figure 7(b) shows that the adaptive changes in the R and C_{mux} values according to the increase in the average received E_s/N_0 value are very beneficial in achieving the maximum throughput. Furthermore, we see that the average throughput of 20 Mbps employing a 40-MHz bandwidth is achieved at the average received E_s/N_0 per antenna of approximately 9 dB with the combination of R = 1/2 and C_{mux} = 3.

4. MULTIPLE-ANTENNA SIGNAL TRANSMISSION TECHNIQUES

Considering the application to hot-spot areas and indoor office environments where an extremely high amount of traffic is concentrated in a very small coverage area, the achievable throughput of approximately 100 Mbps as the target of future broadband wireless access schemes around 2010 will not be sufficient. Therefore, we set our target for the achievable peak throughput in these areas to approximately 1 Gbps, in contrast to that of more than 100 Mbps in a cellular system assuming a 100-MHz bandwidth employing other identical air interfaces such as carrier frequency, wireless access, and frame structure. To achieve this goal, employing a space division multiplexing technique utilizing multiple antenna transmitters and receivers is inevitable, which is represented by the Multiple Input Multiple Output (MIMO) signal transmission technique. In [16], members of our research group investigated the following three schemes assuming N_{TX} transmitter antennas and N_{RX} receiver antennas: MIMO multiplexing utilizing simultaneously parallel data transmission from a multiple-antenna transmitter [17]; MIMO diversity represented by Space Time Block Code (STBC) or Space Time Trellis Code (STTC) [18]; and the adaptive antenna array beam forming (AAA-BF) transmitter associated with antenna diversity reception employing maximal ratio combining. In order to achieve identical information bit rates in the three schemes, higher multi-levels in the data

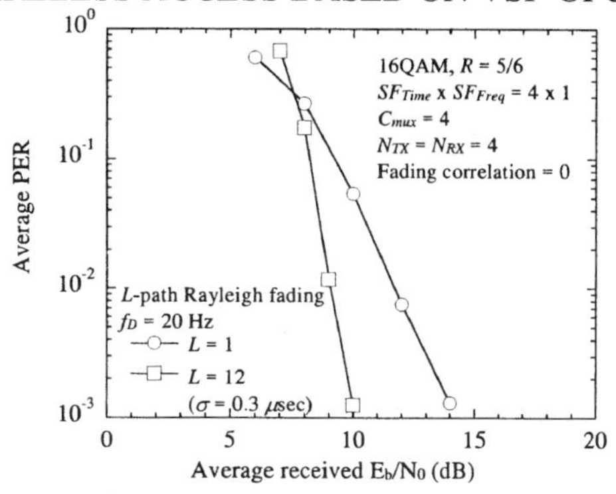

Figure 8. Average PER performance for 1-Gbps information bit rate in VSF-OFCDM.

modulation or a higher rate channel coding is needed in the MIMO diversity and AAA-BF schemes compared to the MIMO multiplexing scheme. This is because a direct increase in the achievable information bit rate is achieved in the MIMO multiplexing scheme by utilizing space-domain multiplexing. Meanwhile, the diversity gain and an increasing antenna gain by directive transmission are achieved using the MIMO diversity and AAA-BF schemes, respectively. According to the evaluation in [16], it was clarified that to achieve high frequency efficiency of greater than approximately 4 bps/Hz, MIMO multiplexing is more promising compared to the other two schemes, since the required transmission power is maintained at its lowest associated with the application of lower-level data modulation or lower channel coding to maintain the same data rate.

Figure 8 shows the average PER performance at the information bit rate of 1 Gbps based on the VSF-OFCDM air interface of approximately 100 MHz bandwidth and 16QAM with $R = 5/6$ and $N_{TX} = N_{RX} = 4$ MIMO multiplexing. The horizontal axis is the transmit signal energy per bit-to-noise power spectrum density ratio (E_b/N_0) for $L = 1$ and 12-path Rayleigh fading channel. The figure clearly shows that there is no error floor and the average PER of 10^{-2} is achieved at the average received E_b/N_0 of approximately 10 dB. Therefore, the figure indicates the possibility of the information bit rate of 1 Gbps based on the radio parameters with the bandwidth of 100 MHz (the corresponding spectrum efficiency of 10 bps/Hz). The figure also shows that the PER performance for $L = 12$ is improved compared to that for $L = 1$ owing to the increasing frequency diversity associated with the turbo coding effect.

5. CONCLUSION

This paper presented broadband packet wireless access schemes based on VSF-OFCDM in the forward link and VSCRF-CDMA in the reverse link for the systems beyond IMT-2000. In our design concept for wireless access in both links, radio parameters such as *SF* and *CRF* values are optimally changed so that the system capacity is maximized according to the cell configurations, channel load, and radio channel conditions, from the tradeoff relationship between efficient suppression of interference from the surrounding cells and the capacity increase in the target cell by exploiting orthogonality in the time and frequency domains. Through VSF-OFCDM and VSCRF-CDMA wireless access, the seamless and flexible deployment of the same wireless access method is possible both in cellular systems and isolated-cell and indoor environments. The experimental results showed that the peak throughput of greater than 100 Mbps and 20 Mbps is achieved by the implemented BS and MS transceivers using the 100-MHz and 40-MHz bandwidths in the forward and reverse links, respectively. Moreover, the simulation results elucidated the possibility of the peak throughput of approximately 1 Gbps for short-range area application in the 100-MHz band OFCDM forward link by applying four-branch MIMO multiplexing together with 16QAM data modulation and punctured turbo coding.

Wireless Laboratories, NTT DoCoMo, Inc.
3-5 Hikari-no-oka, Yokosuka-shi, Kanagawa, 239-8536 Japan

REFERENCES

[1] 3GPP, 3G TR25.858, "Physical Layer Aspects of UTRA High Speed Downlink Packet Access."
[2] S. Abeta, et al., "Performance of coherent Multi-Carrier/DS-CDMA and MC-CDMA for broadband packet wireless access," IEICE Trans. Commun., vol. E84-B, no. 3, pp. 406-414, Mar. 2001.
[3] S. Abeta, et al., "Forward link capacity of coherent DS-CDMA and MC-CDMA broadband packet wireless access in a multi-cell environment," IEEE VTC2000-Fall, pp. 2213-2218, Sept. 2000.
[4] H. Atarashi, et al., "Broadband packet wireless access appropriate for high-speed and high-capacity throughput," IEEE VTC2001-Spring, pp. 566-570, May 2001.
[5] N. Yee, et al., "Multi-Carrier CDMA in indoor wireless radio networks," PIMRC'93, pp. 109-113, Sept. 1993.
[6] K. Fazel et al., "On the performance of convolutional-coded CDMA/OFDM for mobile communication systems," PIMRC'93, pp. 468-472, Sept. 1993.
[7] H. Atarashi, et al., "Variable spreading factor orthogonal frequency and code division multiplexing (VSF-OFCDM) for broadband packet wireless access," IEICE Trans. Commun., vol. E86-B, no. 1, pp. 291-299, Jan. 2003.
[8] Y. Goto, et al., "Variable Spreading and Chip Repetition Factors (VSCRF)-CDMA in reverse link for broadband wireless access," PIMRC2003, pp.254-259, Sept. 2003.
[9] M. Schnell, et al., "A promising new wideband multiple-access scheme for future mobile communication," European Trans. on Telecommun. (ETT), vol. 10, no. 4, pp. 417-427, July/Aug. 1999.
[10] K. Miyoshi, et al., "A study on time domain spreading for OFCDM," Technical Report of IEICE, RCS2001-179, Nov. 2001 (in Japanese).
[11] A. Sumasu, et al., "An OFDM-CDMA system using combination of time and frequency domain spreading," Technical Report of IEICE, RCS2000-3, Apr. 2000 (in Japanese).
[12] A. Persson, et al., "Time-frequency localized CDMA for downlink multi-carrier systems," IEEE ISSSTA 2002, pp. 118-122, Sept. 2002.
[13] H. Atarashi, et al., "Broadband packet wireless access based on VSF-OFCDM and MC/DS-CDMA," IEEE PIMRC2002, pp.992-997, Sept. 2002.
[14] N. Maeda, et al., "Variable spreading factor-OFCDM with two dimensional spreading that prioritizes time domain spreading for forward link broadband wireless access," IEEE VTC2003-Spring, pp.127-132, Apr. 2003.
[15] N. Maeda, et al., "Experimental evaluation of throughput performance in broadband packet wireless access based on VSF-OFCDM and VSF-CDMA," PIMRC2003, pp.6-11, Sept. 2003.
[16] J. Kawamoto, et al., "Comparison of space division multiplexing schemes employing multiple antennas in OFDM forward link," to appear in IEEE VTC2003-Fall.
[17] G. J. Foschini, Jr., "Layered space-time architecture for wireless communication in a fading environment when using multi-element antennas," Bell Labs Tech. J., pp. 41-59, Autumn 1996.
[18] V. Tarokh, et al., "Space-time block coding for wireless communications: performance results," IEEE J. Select. Areas Commun., vol. 17, no. 3, pp. 451-460, Mar. 1999.

AN OFDM BASED SYSTEM PROPOSAL FOR 4G DOWNLINKS

ARNE SVENSSON[1], ANDERS AHLÉN[2], ANNA BRUNSTROM[3], TONY OTTOSSON[1], MIKAEL STERNAD[2]

[1] Dept. Signals and Systems, Chalmers Univ. of Technology, SE-41296 Göteborg, Sweden, {arne.svensson, tony.ottosson}@s2.chalmers.se

[2]Signals and Systems, Uppsala University, PO Box 528, SE-751 20 Uppsala, Sweden, {anders.ahlen, mikael.sternad}@signal.uu.se

[3]Dept. Computer Science, Karlstad University, SE-651 88 Karlstad, Sweden, anna.brunstrom@kau.se

Abstract. In this paper we describe an OFDM based 4G downlink for a wide area coverage and high mobility system. User data are multiplexed and OFDM modulated such that the user with the best predicted channel conditions are always using the channel. This user employs the linear modulation scheme that maximizes the spectral efficiency. We show that the system obtains a sector capacity that is significantly better than current 3G systems. Various combinations of OFDM and CDMA are also discussed and it is concluded that it is difficult to motivate the significantly increased complexity of such schemes. Moreover, we also doubt that these combinations can increase spectral efficiency when predicted channel information is utilized at the transmitter.

1. INTRODUCTION

Higher spectral efficiency will be a key feature of any acceptable radio interface beyond 3G. A promising approach for the downlink, is to adaptively multiplex user data onto an OFDM transmission scheme. This will minimize interference between users within a cell and efficiently allows users to share the total bandwidth. In such a system, spectral efficiency can be improved by allocating the time-frequency resources based on throughput requirements, quality of service constraints and the channel qualities of each user. A scheduler, which optimizes the resource allocation for multiple active users, becomes a key element in the system. In present CDMA systems, the spectral efficiency decreases with an increasing number of active users having conventional detectors. This is caused by intra-cell interference due to imperfect orthogonality of the downlinks. In an adaptive multiplexing and OFDM system, where orthogonal

K. Fazel and S. Kaiser (eds.), Multi-Carrier Spread-Spectrum, 15–22.

time-frequency resources are given to the user that can utilize them best, the spectral efficiency will instead *increase* with the number of active users. This *multiuser diversity* effect [1] is quantified and illustrated by analytical results in Section 3, assuming independently frequency-selective fading channels, and an adaptive joint multiplexing and modulation scheme, which is optimized in a novel way.

Designing an adaptive multiuser multiplexing and OFDM system that works also for vehicular users and wide area coverage scenarios is a challenging task [2]. In Section 2 below, we outline such a system.[1] It assumes FDD, a base station infrastructure and a tight reuse of the bandwidth. The quality of downlink channels must in such a solution be predicted by the terminals, and reported to the system. A potential problem is that the required amount of feedback information might become unreasonably large. The uplink control bit rate increases with the granularity of the resource partitioning of the downlink, i.e. with the size of time-frequency bins that may be adaptively allocated to different users. An important issue is whether resources can be partitioned into bins that are large enough to keep feedback data rates at reasonable levels, while the reduction in spectral efficiency due to channel variability *within* these bins, caused by frequency selectivity and time variation, remains acceptable. According to our results it can in fact be done up to reasonably high vehicle speed [6].

2. THE ADAPTIVE DOWNLINK

The available downlink bandwidth within a base station sector is assumed to be slotted in time. Each slot is partitioned into time-frequency bins of given bandwidth and duration. These resources are shared among K active terminals.

During each slot, each terminal predicts the signal to interference and noise ratio (SINR) for all bins, with a prediction horizon which is larger than the time delay of the transmission control loop. All terminals then signal their predicted quality estimates on an uplink control channel. They transmit the suggested appropriate modulation format to be used within the frequency bins of the predicted time slot. A scheduler that is located close to the base station then allocates these time-frequency bins exclusively to different users and broadcasts its allocation decisions. In the subsequent downlink transmission of the predicted slot, the modulation formats used are those which were suggested by the appointed users.

The bin size should be selected so that all payload symbols within a bin can be given the same modulation format, without too large reduction in spectral efficiency relative to an ideal case with a completely flat and time-invariant channel within each bin. Assuming a design vehicle speed of 100 km/h and a carrier frequency of 1900 MHz (Doppler 174 Hz), we here select a bin size of 0.667 ms times 200 kHz. A cyclic prefix of length 11 μs is introduced. It eliminates intersymbol interference if paths more than 3.3 km longer than the shortest path are insignificant. We here also select a sampling period 0.20 μs, subcarrier spacing of 10 kHz, and a symbol period of 111 μs.

---

[1]The system serves as a focus for research within the Wireless IP project [3], supported by the Swedish Foundation for Strategic Research SSF.

Thus, each time-frequency bin carries 120 symbols, with 6 symbols of length 111 μs on each of the 20 10 kHz subcarriers. Of the 120 symbols, 12 are for training and downlink control, leaving 108 payload symbols [2].

For the payload symbols, we utilize an adaptive modulation system that uses 8 uncoded modulation formats: BPSK, 4-QAM, 8-QAM, 16-QAM, 32-QAM 64-QAM, 128-QAM, and 256-QAM. This adaptive modulation system is optimized and evaluated in [6]. It is designed to maximize the spectral efficiency for each user for constant transmit power, by balancing the throughput against the loss due to erroneous packets. Variable transmit power provides only minor improvements [4] and would require a large amount of feedback.

Prediction of the whole channel with a given horizon can be performed either in the time-domain or in the frequency-domain. The best known power predictor performance on measured broadband data is obtained with the unbiased quadratic power predictor presented in [7]. Prediction over 2 ms seems attainable at 1900 MHz also for users travelling at 100 km/h for this predictor.

A more detailed description of the proposed system can be found in [2].

3. ANALYSIS AND RESULTS

We now estimate the resulting spectral efficiency for best effort services under some simplifying assumptions. The channel is assumed flat and time-invariant AWGN within bins and independent Rayleigh fading between bins[2]. All K users are assigned equal average received power[3] and channels to different users fade independently. Accurate SINR predictions and channel estimation are used for symbol detection. Finally, all users always have data to transmit, and the allocated bins are fully utilized by their designated users.

The here assumed scheduler works as a selection diversity scheme, where the user with the best predicted SINR out of all K users will transmit in a bin. In the receiver we assume *maximum ratio combining* (MRC) with L *antennas*.[4] The resulting pdf of the received SINR (γ) after MRC and multiuser selection diversity can then be calculated analytically [6]. The SINR limits for selecting the appropriate M-QAM format have been optimized to maximize the number of bits that arrives in bins which are declared correctly received by the CRC check. We believe this is a useful and novel approach to the optimization of adaptive modulation schemes. The spectral efficiency when using adaptive modulation will then be obtained by a weighted average over all the modulation formats, weighted by the probabilities that those particular formats will be utilized, and also by the corresponding frame acceptance rates. This raw spectral

[2]Such a channel is sometimes referred to as a block Rayleigh fading channel where a block in our case contains the symbols in a time-frequency bin.

[3]This assumed power control scheme is wasteful from a system capacity perspective. The capacity of the proposed adaptive downlink with a better power control strategy is evaluated by Monte-Carlo simulation for an interference-limited environment in [2].

[4]Equivalently, we could assume downlink beamforming with L transmit antennas, but this variant would require much more control information.

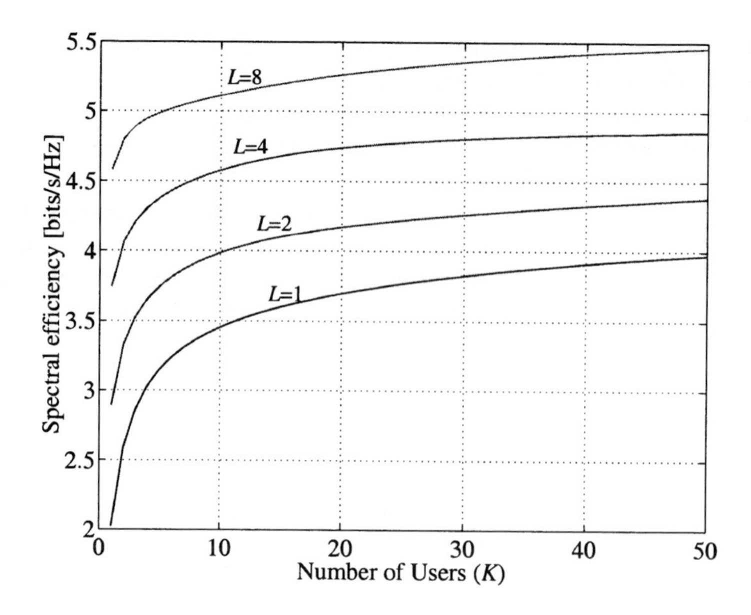

Figure 1. Payload spectral efficiency at SINR 16 dB per receiver antenna, when using adaptive multiplexing and modulation with Lth order MRC diversity in the mobile and Kth order of selection diversity between the users. Degradation due to channel variation within a bin, prediction errors, link layer overhead and frequency reuse is not taken into account.

efficiency must in our target system be multiplied by $100/111$ due to cyclic prefixes and by $108/120$ due to the 12 pilots and control symbols per bin.

This somewhat ideal sector capacity is evaluated numerically in Fig. 1 for an average symbol energy to noise ratio of $\bar{\gamma} = E(\gamma) = 16$ dB per receiver antenna, for all users. There is a notable improvement with an increasing multiuser selection diversity and of course also an increase with the number of receiver antennas. The spectral efficiency saturates for a high number of users, when most bins are occupied by users who can utilize a high modulation format. The addition of more receiver diversity branches (larger L) decreases the OFDM channel variability [5]. Here, this tends to decrease the multiuser diversity effect.

Our proposed system has also been simulated on the channels used for UMTS performance evaluation. These channel models are more realistic and takes into account correlation between time-frequency bins and channel variation within each bin. The simulations show that the degradation compared to the results in Fig. 1 is small for vehicle speeds below 120 km/h [6]. With very few users, a loss of slightly more than 10 % might appear while with 20 users the loss is less than 3 % for an average SNR of 16 dB. Other time-frequency bin sizes have also been evaluated and the results show that the proposed bin size of 0.667 μs times 200 kHz makes a good compromise between spectral efficiency and feedback information rate.

Imperfect channel predictions will also slightly reduce the spectral efficiency. In

[8] it is shown that this results in approximately 10 % degradation of spectral efficiency at an average SNR of 16 dB when all channels have to be predicted 1/3 wavelength ahead (corresponding to 2 ms at a velocity of 100 km/h at 1900 MHz). The loss is smaller for lower velocities. At the link layer, there will be some degradation in spectral efficiency due to CRC bits and sequency numbers. This loss is expected to be between 3 % (for 20 users) and 7 % (for one user). The overall system capacity also must take frequency reuse into account. Preliminary studies show that a frequency reuse of 1.73 is possible in a fully loaded system [2]. Thus in total, the numbers in Fig. 1 should be multiplied by 0.41 for one user and 0.5 for 20 users to obtain a more practical sector capacity for high vehicle speeds and system loads.

4. ON SPREADING AND OFDM IN THE DOWNLINK

Above we outlined a downlink proposal that utilizes uncoded adaptive OFDM. Among other alternative solutions, several researchers have proposed 4G downlinks based on different combinations of OFDM and CDMA. Here we discuss some of the more common spread schemes:

MC-CDMA User bits are spread to N chips, the chip sequences from different users are added and then mapped to different subcarriers of the same OFDM symbol. This is spreading over frequency. Unfortunately, MC-CDMA is also often used as a name for the whole family of spread OFDM schemes.

MC-DSCDMA User bits are spread to N chips, the chip sequences from different users are added and then mapped to different consecutive OFDM symbols on the same subcarrier. This is spreading over time.

TFL-CDMA User bits are spread to N chips, the chip sequences from different users are added and then mapped to a rectangular OFDM time-frequency bin of size N. This is spreading over both time and frequency.[5]

The systems are illustrated in Fig. 2 for the case of 4 chips per bit, 4 subcarriers and no interleaving of chips. When chips are interleaved, the situation becomes more involved. MC-CDMA is still spreading over frequency but the chips for a particular bit are now further apart in frequency such that the fading correlation between them is further reduced. MC-DSCDMA is still spreading over time but the chips corresponding to one bit is again further apart in time. TFL-CDMA probably only makes sense without interleaving since it is especially designed to reduce correlation between users' signals. In the figure we for completeness also include ordinary unspread OFDM.

The performance of the spread schemes depends both on the diversity order that can be obtained with the scheme (which limits the single user performance) and the amount of multiuser interference that is influencing the decision. These factors are quite different in the case of chip-interleaving or no chip-interleaving.

[5]TFL is an abbreviation for time-frequency localized.

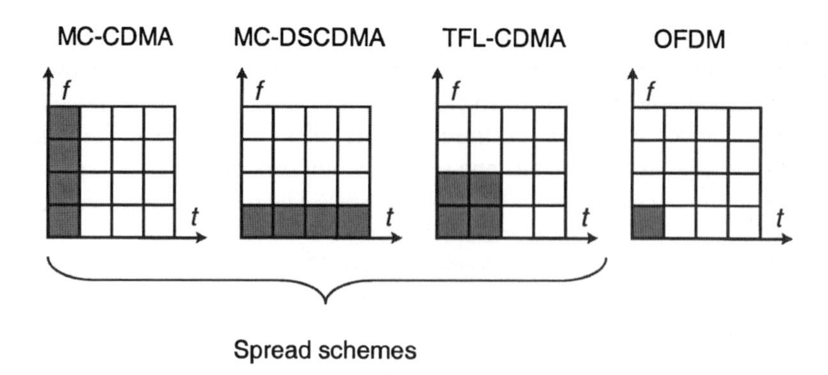

Figure 2. Various combinations of CDMA and OFDM transmission with 4 chips per bit. The red/dark areas represent the time-frequency resources used for each transmitted bit when no chip-interleaving is used. In the spread schemes, several users share these resources.

Without chip-interleaving, the diversity order depends on the frequency/time selectivity of the channel and the outer channel coding, while the multiuser interference (MUI) depends only on the selectivity of the channel (assuming ortogonal codes are used at the transmitter). The impact of selective fading on these different systems are quite different. The diversity order and the MUI of MC-CDMA depend on the coherence bandwidth of the channel. When the coherence bandwidth is large compared to the signalling bandwidth, the spreading does not contribute to the diversity order and the MUI is low. On the contrary, a small coherence bandwidth leads to large diversity orders (maximum N) but also large MUI. In MC-DSCDMA it is the coherence time that influences the diversity order and amount of MUI in the same way. In TFL-CDMA, the area that a symbol occupies can be adjusted to better fit the coherence time and bandwidth, but this area may still exceed one coherence bandwidth times one coherence time for systems with large spreading and large channel variation. Thus, the spread schemes will perform very similar to pure OFDM when the channel variation is low in frequency and/or time and/or the signalling bandwidth is low.

However in many practical systems, the spread schemes will result in significant MUI and also reasonably large diversity gains. The MUI is smaller for TFL-CDMA than for the other two schemes, but the diversity order is also smaller for TFL-CDMA. With pure OFDM, outer channel coding must be used to obtain diversity, but this scheme is free from multiuser interference. From [9] and [10], the conclusion is that in most cases the coded pure OFDM scheme has a performance advantage when the same outer code is used in all systems. With very few users or very efficient multiuser detectors, the spread schemes may performance slightly better than the pure OFDM scheme but at the expense of either much lower spectral efficiency (few users) or much higher complexity.

With chip-interleaving in the spread systems, the diversity order due to spreading increases and can be made to approach N if a long delay due to interleaving is acceptable. On the other hand, the amount of MUI also significantly increases since the spreading codes becomes random in the receiver due to the uncorrelated (with long interleavers) fading on the different chips in a bit. Thus, multiuser detection is necessary leading to much more complex receivers. In [11] it is concluded that the spread schemes with outer channel coding has some performance improvement compared to coded pure OFDM in single cell systems. If this is the case also in cellular system with more than one cell, seems to still be an open question.

Another problem with spread OFDM schemes, besides the large receiver complexity with multiuser detectors, is that it is an open question how to efficiency use adaptive modulation with them. It is straight forward when no interleaving is used and the channel variation is small in time and/or frequency. Then each bit is transmitted on a constant channel and the same principles as with pure OFDM can be used. In this case, there is however no performance advantage of the spread schemes and thus no reason to use them. With chip-interleaving and/or high variation on the channel, different chips for a given symbol will see different channel gains which means that adaptive modulation becomes more difficult to employ. As far as we know, there is no good current solution on how to employ adaptive modulation when the channel variation of each bit is large. Thus, we conclude that in systems like the one we propose, where channel state information is available at the transmitter, coded pure OFDM is the more attractive choice and also the one with the best known performance. In systems that do not utilize channel state information in the transmitter, we still find it very difficult to motivate all the additional receiver complexity of spread schemes based on the quite limited performance gains that are available. Overall, we believe a simpler system is obtained by utilizing a more powerful outer channel code in the pure OFDM system. This will increase the complexity of the receiver too, but not at all in the same way as the increased complexity due to multiuser detection. So in summary, we believe that pure OFDM has a better tradeoff between performance and complexity on downlinks in almost all cases. For the uplink, the conclusion might be different since the uplink is asynchronous, but this is not the scope of this paper.

5. SUMMARY AND CONCLUSIONS

Adaptive multiplexing and OFDM transmission based on predicted channels seems to be a very promising technique for the downlink in future high mobility and wide area coverage systems. In this paper, we show that with a reasonable number of users and at least two receiver antennas, it is possible to reach a sector capacity of about 2 bits/s/Hz, which is far better than current 3G systems. This is however obtained with a very simple scheduler that does not take quality of service and fairness into account. More advanced scheduling must be done in a practical system and this will sacrifice some capacity to improve quality of service and fairness. An issue that seems to be difficult though is to find proper criteria for optimizing the scheduler since such a criteria should depend on the business models that operators will use.

We also discuss some combinations of CDMA and OFDM. Our conclusion is that pure OFDM has a better tradeoff between performance and complexity on downlinks. In systems with channel state information available in the transmitter, it also has much higher spectral efficiency.

6. REFERENCES

[1] R. Knopp and P. A. Humblet, "Multiple-accessing over frequency-selective fading channels," *Proceedings IEEE Personal, Indoor and Mobile Radio Communications*, Toronto, Canada, Sept. 1995, pp. 1326–1330.

[2] M. Sternad, T. Ottosson, A. Ahlén, and A. Svensson, "Attaining both coverage and high spectral efficiency with adaptive OFDM downlinks", *Proceedings IEEE Vehicular Technology Conference Fall*, Orlando, Florida, USA, Oct. 2003.

[3] Online: www.signal.uu.se/Research/PCCwirelessIP.html.

[4] S. T. Chung and A. J. Goldsmith, "Degrees of freedom in adaptive modulation: A unified view," *IEEE Transactions on Communications*, Vol. 49, No. 9, Sep. 2001, pp. 1561–1571.

[5] T. H. Liew and L. Hanzo, "Space-time block coded adaptive modulation aided OFDM", *Proceedings IEEE Global Telecommunications Conference*, Vol. 1, San Antonio, Texas, USA, Nov. 2001, pp. 136–140.

[6] W. Wang, T. Ottosson, M. Sternad, A. Ahlén, and A. Svensson, "Impact of multiuser diversity and channel variability on adaptive OFDM," *Proceedings IEEE Vehicular Technology Conference Fall*, Orlando, Florida, USA, Oct. 2003.

[7] T. Ekman, M. Sternad, and A. Ahlén, "Unbiased power prediction on broadband channels," *Proceedings IEEE Vehicular Technology Conference Fall*, Vancouver, Canada, Sep. 2002.

[8] M. Sternad and S. Falahati, "Maximizing throughput with adaptive M-QAM based on imperfect channel predictions," submitted to *IEEE International Conference on Conference*, Paris, France, Jun. 2004.

[9] A. Persson, T. Ottosson, and E. Ström, "Analysis of direct-sequence CDMA for downlink OFDM systems, Part I", submitted to *IEEE Transactions on Communications*, Jun. 2003.

[10] A. Persson, T. Ottosson, and E. Ström, "Analysis of direct-sequence CDMA for downlink OFDM systems, Part II", submitted to *IEEE Transactions on Communications*, Jun. 2003.

[11] G. Auer, A. Dammann, S. Sand, and S. Kaiser, "On modelling cellular interference for multi-carrier based communication systems including a synchronization offset," *Proceedings IEEE Wireless Personal Multimedia Communications*, Yokosuka, Japan, Oct. 2003.

FREDERIK VANHAVERBEKE, MARC MOENECLAEY

CHANNEL OVERLOADING IN CDMA WITH SCALABLE SIGNATURE SETS AND TURBO DETECTION

Abstract- We compare the achievable load of different types of scalable overloaded binary signature sets with turbo detection: random spreading (PN), scrambled multiple OCDMA (s-m-O/O), overall-permuted multiple OCDMA (o-m-O/O) and PN/OCDMA (PN/O). Both with and without coding, PN/O and m-O significantly outperform PN. When coding is applied, m-O outperforms both PN/O and PN .

1. INTRODUCTION

It is well-known that in a multiple access system based on code-division multiple access (CDMA) with spreading factor N, N is the maximum number of orthogonal users. Although almost perfect time alignment of the different users is required for these schemes, synchronization can always be achieved in the downlink. Even in the uplink, some systems like for example multicarrier-CDMA can maintain orthogonality by application of an appropriate cyclic prefix and single-tap equalization [1]. If the number of users K exceeds the spreading factor N in such a system (i.e. a so-called 'oversaturated' or 'overloaded' channel), Welch-Bound Equality (WBE) sequences minimize the total squared correlation between a set of K unit-norm signatures [2,3]. Unfortunately, these WBE sequences are not scalable and have to be recomputed for every change in the number of users K, which makes them useless for any practical system with a variable number of users. At the other hand, several scalable signatures sets are proposed in the literature.

One can choose the signatures at random, i.e. random spreading [4]. This is an evident choice for the signatures, as orthogonality between the users is impossible if K > N. Moreover, for high channel loads (K/N $\rightarrow \infty$) or for high signal-to-noise ratios, random spreading incurs almost no loss in spectral efficiency as compared to optimal signature sets [4]. So, one can even raise the question whether the choice of spreading sequences in oversaturated channels is important whatsoever [5].

A second approach is to design a scalable signature set that is especially suited to be detected by means of an optimum MUltiuser Detector (MUD) [6]. Examples are the tree-structured channel overloading [7] and excess signaling [8,9,10]. Both overloading schemes allow for a restricted number of excess users (K-N) only, and the spreading sequences are not binary, which makes them rather unattractive for mobile systems.

Two types of overloaded signature sets, which are especially suited to be detected by means of interference cancellation [6], were introduced by the present authors: OCDMA/OCDMA (O/O) in [11,12,13] and PN/OCDMA (PN/O) in [13]. In these systems, the first N users (set 1 users) are assigned orthogonal sequences, while the excess users (set 2 users) are assigned other orthogonal sequences (O/O) or random sequences (PN/O). In this way, the set 1 users suffer from interference of the set 2 users only. In the O/O system, the set 2 users suffer from interference of the set 1 users only, while the set 2 users suffer also from interference of the other set 2

This work has been supported by the Interuniversity Attraction Poles Program - Belgian State - Federal Office for Scientific, Technical and Cultural Affairs.

K. Fazel and S. Kaiser (eds.), Multi-Carrier Spread-Spectrum, 23–30.

users in the PN/O system. So, at a first glance, we expect O/O to outperform PN/O with interference cancellation. Because of the special structure of these signatures, PN/O and O/O can be detected easily by means of interference cancellation, where in every stage, the set 1 users are detected first, followed by a detection of the set 2 users. In [14], O/O was extended to multiple-O (m-O), in order to be able to cope with a number of users in excess of 2N.

In any of the above mentioned oversaturated systems, the interference levels of the users can be very high, so that MUD will be required to obtain a satisfactory performance of the users. Linear MUD's, such as the decorrelator [15], the minimum mean-squared error detector [16] or linear decision directed interference cancellation [17], are devised to detect users in a nonsaturated system and are unable to cope with the high interference levels of oversaturated systems. Also Maximum Likelihood (ML) detection [18] is not an option because of its complexity that is exponential in the number of users. On the other hand, nonlinear decision-directed MUD [6], and more precisely nonlinear parallel interference cancellation (PIC) and nonlinear successive interference cancellation (SIC), are considered to have a good complexity-performance trade-off as compared to other MUD's and are the evident choice of multiuser detection in an oversaturated system. PIC has the advantage of speed over SIC, since the users can be detected in parallel at every stage for PIC, while this detection has to be performed successively in SIC. Multistage SIC (m-SIC) on the other hand, results in general in a better performance than PIC.

For coded CDMA systems, extensive research has been devoted to the so-called Turbo Detector (TD), e.g. [19,20,21]. In this turbo detector, the data of the users are determined by an iterative procedure, where MUD and decoding are performed in succession at each stage of the detection. The aim of this paper is to compare the performance with turbo detection of oversaturated systems with BPSK modulation, based on random spreading, O/O and PN/O. In section II, we summarize the three types of channel overloading. In section III, the turbo detector is explained. In section IV, the achievable channel loads are presented, and finally, in section V, the conclusions are drawn.

2. RANDOM SPREADING, m-O AND PN/O

Consider a CDMA system with spreading factor N, where each user i (i = 1, …, K) encodes a set of L information bits $\mathbf{d}_i = (d_i(1),\ldots,d_i(L))$ into a set of M code bits $\mathbf{b}_i = (b_i(1),\ldots,b_i(M))$. The code rate C = L/M and the channel load $\beta = K/N$. The obtained code bits are randomly permuted (interleaved) so that we obtain the set $\mathbf{c}_i = (c_i(1),\ldots,c_i(M))$. In a next step, we map the interleaved code bits to the BPSK constellation {+1,-1} and obtain the symbol set $\mathbf{a}_i = (a_i(1),\ldots,a_i(M))$, with $a_i(k) = 2.c_i(k)-1$ for k = 1, …, M. If the users are symbol-synchronous and block-synchronous over an AWGN channel, the received vector $\mathbf{r}(t)$ in symbol interval t (t = 1, …, M) after chip-matched filtering is given by the real-valued vector

$$\mathbf{r}(t) = \mathbf{S}(t).\mathbf{a}(t) + \mathbf{n}(t) = \sum_{i=1}^{K} a_i(t).\mathbf{s}_i(t) + \mathbf{n}(t) \tag{1}$$

In this expression,

- $\mathrm{S}(t) = [\mathbf{s}_1(t) \ldots \mathbf{s}_K(t)]$ is composed of the signatures of the respective users in symbol interval t. We restrict our attention to binary signature sets, so that $\mathbf{s}_j(t) \in \{1/\sqrt{N},-1/\sqrt{N}\}^N$ for every $(j,t) \in \{1,\ldots,K\}\mathrm{x}\{1,\ldots,M\}$.

- $\mathbf{n}(t)$ is a vector of independent Gaussian noise samples with $E[\mathbf{n}(i).\mathbf{n}(j)^T] = \sigma^2.\delta_{i-j}.\mathbf{I}_N$ for $(i,j) \in \{1, \ldots, M\}^2$, where $\mathbf{I}_N$ is the identity matrix of order N, δ_z is the discrete delta function, $\sigma^2 = (N_0/2)/(C.E_b)$ with N_0 the variance of the thermal noise and E_b the energy per bit.

The choice of the signatures determines the type of overloaded system. For PN/O, the users are split up into 'set 1 users' (the first N users) and 'set 2 users' (the (K-N) excess users), so that we can rewrite expression (1) as

$$\mathbf{r}(t) = \mathbf{S}_1(t).\mathbf{a}^{(1)}(t) + \mathbf{S}_2(t).\mathbf{a}^{(2)}(t) + \mathbf{n}(t) \qquad (2)$$

where $\mathbf{a}^{(i)}(t)$ and $\mathbf{S}_i(t)$ are the code bits and the signature matrix of set i (i = 1, 2) respectively. For m-O[1], the number of different sets is m, where the total number of users $K = (m-1).N + M$ with $0 < M \leq N$. The first (m-1).N users are split up into (m-1) sets of N users, while the last M users make up set m:

$$\mathbf{r}(t) = \sum_{j=1}^{m} \mathbf{S}_j(t).\mathbf{a}^{(j)}(t) + \mathbf{n}(t) \qquad (3)$$

2.1 Random Spreading (PN)

This is the simplest signature set: in every symbol interval, each signature is chosen independently and completely at random out of the set $\{1/\sqrt{N},-1/\sqrt{N}\}^N$.

2.2 PN/OCDMA (PN/O)

Signature matrix $\mathbf{S}_1(t)$ is the orthogonal Walsh-Hadamard matrix $\mathbf{WH}_N$ of order N [22], where each signature is overlaid by a common scrambling sequence:

$$\mathbf{S}_1(t) = \begin{bmatrix} \Lambda_1^{(1)}(t) & & \\ & \cdots & \\ & & \Lambda_N^{(1)}(t) \end{bmatrix} .\mathbf{WH}_N \qquad (4)$$

In every symbol interval, the scrambling sequence $\mathbf{\Lambda}^{(1)}(t) = \left(\Lambda_1^{(1)}(t),\ldots,\Lambda_N^{(1)}(t)\right)$ is chosen at random out of $\{+1,-1\}^N$.

The signatures of the set 2 users are selected in every symbol interval and for every user with equal probability over $\{1/\sqrt{N},-1/\sqrt{N}\}^N$.

2.3 m-OCDMA

With m-O the signatures of the set 1 users are obtained in the same way as with PN/O (see (4)). The difference between the PN/O and the m-O system lies in the fact that also the signatures of each other set are orthogonal. So, $\mathbf{S}_j(t)$ (j = 2, ..., m) is obtained by overlaying the columns of a (binary and orthogonal) Hadamard matrix $\mathbf{H}_N^{(j)}(t)$ with a scrambling sequence $\mathbf{\Lambda}^{(j)}(t)$:

$$\mathbf{S}_j(t) = \left(\begin{bmatrix} \Lambda_1^{(j)}(t) & & \\ & \cdots & \\ & & \Lambda_N^{(j)}(t) \end{bmatrix} .\mathbf{H}_N^{(j)}(t) \right)_{1..M_j} \qquad (5)$$

where $\mathbf{X}_{1..L}$ denotes the submatrix of $\mathbf{X}$, consisting of the first L columns of $\mathbf{X}$, and M_j is the number of users in set j ($M_j = N$ for j = 2, ..., m-1; $M_m = M$). Scrambling

[1] Note that for m = 2, we have O/O.

sequence $\mathbf{\Lambda}^{(j)}(t) = \left(\Lambda_1^{(j)}(t),\ldots,\Lambda_N^{(j)}(t)\right)$ is chosen in every symbol interval at random over $\{+1,-1\}^N$ and independently of all other $\mathbf{\Lambda}^{(l)}$ $(1 \neq j)$. The choice of the Hadamard matrices $\mathbf{H}_N^{(j)}(t)$ will define the type of m-O: s-m-O and o-m-O:

- s-m-O: here we take $\mathbf{H}_N^{(j)}(t) = \mathbf{WH}_N$ in every symbol interval and for all sets j.
- o-m-O: here we select as Hadamard matrix $\mathbf{H}_N^{(j)}$ in every symbol interval at random one of the matrices of set $\Phi_{o\text{-}O}$. The set $\Phi_{o\text{-}O}$ is obtained by means of matrix-operator $\mathbf{\Psi}_n(\mathbf{Z}_{n/2},\mathbf{T}_{n/2},s,t)$, which transforms matrices of order n/2 x n/2 into a matrix of order nxn:

$$\mathbf{\Psi}_n(\mathbf{Z}_{n/2},\mathbf{T}_{n/2},s,t) = \frac{1}{\sqrt{2}}\begin{bmatrix} \mathbf{Z}_{n/2} & \mathbf{Z}_{n/2}^{s,t} \\ \mathbf{T}_{n/2} & -\mathbf{T}_{n/2}^{s,t} \end{bmatrix} \tag{6}$$

In this expression, submatrix $[(\mathbf{Z}_{n/2}^{s,t})^T\ (-\mathbf{T}_{n/2}^{s,t})^T]^T$ is obtained from $\mathbf{A}_{nxn/2} = [(\mathbf{Z}_{n/2})^T\ (-\mathbf{T}_{n/2})^T]^T$ by multiplying every column of this matrix by 1 or –1, according to the binary representation of s (s = 0, …, $2^{n/2}$-1), and by permuting the columns of **A** by means of permutation t (t = 1, …, (n/2)!). Note that (6) is merely a permutation of and sign-assignment to the last n/2 columns of the matrix

$$\mathbf{\Psi}_n(\mathbf{Z}_{n/2},\mathbf{T}_{n/2},0,0) = \frac{1}{\sqrt{2}}\begin{bmatrix} \mathbf{Z}_{n/2} & \mathbf{Z}_{n/2} \\ \mathbf{T}_{n/2} & -\mathbf{T}_{n/2} \end{bmatrix} \tag{7}$$

So, if both $\mathbf{Z}_{n/2}$ and $\mathbf{T}_{n/2}$ are orthogonal (binary) matrices, $\mathbf{\Psi}_n(\mathbf{Z}_{n/2},\mathbf{T}_{n/2},s,t)$ will be orthogonal (and binary) too.
Matrix-operator $\mathbf{\Psi}_n$ enables us to construct a binary N x N orthogonal matrix **B**, starting from N/2 orthogonal binary matrices $\mathbf{K}_i$ (i = 1, …, N/2) of order 2:

$$\mathbf{B} = \mathbf{P}_N^1 = \mathbf{\Psi}_N(\mathbf{P}_{N/2}^1,\mathbf{P}_{N/2}^2,s_N^1,t_N^1) \tag{8}$$

where, for i = 1, …, N/n, $\mathbf{P}_n^i$ is defined as:

$$\mathbf{P}_n^i = \begin{cases} \mathbf{\Psi}_n(\mathbf{P}_{n/2}^{2i-1},\mathbf{P}_{n/2}^{2i},s_n^i,t_n^i) & n = 2^l \geq 4 \\ \mathbf{K}_i & n = 2 \end{cases} \tag{8-bis}$$

Now, $\Phi_{o\text{-}O}$ is the set of all matrices **B** that we obtain by starting with a random choice of the orthogonal matrices $\mathbf{K}_i$ (i = 1, …, N/2) of order 2 and a random choice of (s_n^i,t_n^i) for all (n,i).

3. TURBO DETECTION

The turbo detector is an iterative detector, where in each stage, we detect the data of the users successively or in parallel. If the data are coded, the detection of each user consists of two successive steps: interference cancellation and single user decoding. In the uncoded case, the turbo detector reduces to an m-SIC or PIC detector. For any turbo detector, the received vector **r**(t) is first transformed to the vector **y**(t):

$$\mathbf{y}(t) = \mathbf{S}(t)^T\mathbf{r}(t) = \mathbf{R}(t).\mathbf{a}(t) + \mathbf{n}'(t) \qquad t = 1,\ldots,M \tag{9}$$

where **R**(t) is the crosscorrelation matrix and **n'**(t) is the contribution of the noise.

If the users are detected successively, the users should be detected according to decreasing signal-to-(interference + noise) ratio (SINR). In the PN system, the interference level of each user is the same, because of the random generation of the signatures in the successive symbol intervals. Consequently, the users are detected in an arbitrary order, that is kept the same in every stage of the iterative detection. In PN/O, the set 1 users suffer from less interference than the set 2 users, while in the m-O system, the users of the (m-1) first sets suffer from less interference than the users of the last set. In order to limit the decision delay in the turbo detector, we propose to detect the users of each set in parallel. Taking into account the interference levels of the users, it is sensible to detect the users of the sets according to increasing set number, both for PN/O and m-O. In this way, the detection for PN/O and m-O is a hybrid version of PIC and m-SIC.

We start with the initial estimates of the codebits: $\tilde{\mathbf{a}}^{(0)}(t) = \mathbf{0}$ (t = 1, ..., M). Now, at stage i (i = 1, ..., I), the users are detected successively in the predetermined detection order, as discussed above. Let's focus on the detection of user k at stage i. It consists of the succession of interference cancellation and single user decoding. We first produce the set of observables $\mathbf{z}_k^{(i)} = (z_k^{(i)}(1),\ldots,z_k^{(i)}(M))$ by subtracting the estimated interference of the other users:

$$z_k^{(i)}(t) = y_k(t) - \left(\sum_{j \in B_k} R_{k,j}(t).\tilde{a}_j^{(i)}(t) + \sum_{j \in A_k} R_{k,j}(t).\tilde{a}_j^{(i-1)}(t) \right) \tag{10}$$

where B_k and A_k are the set of users that are detected *before*, respectively *after or at the same time as* the detection of user k in the chosen detection order. In order to obtain the estimates of the codebits, we make a distinction between the uncoded case and the coded case.

3.1 Uncoded case (L = M = 1)

Since L = M = 1, we omit the time dependence for convenience. The data estimates are obtained by means of a soft decision function with the optimal nonlinearity of [14]:

$$\tilde{a}_k^{(i)} = \begin{cases} sign(z_k^{(i)}) & \left|z_k^{(i)}\right| > \lambda \\ z_k^{(i)} / \lambda & \left|z_k^{(i)}\right| \le \lambda \end{cases} \tag{11}$$

where λ is selected so as to minimize the BER after I iterations.

3.2 Coded case

Here we obtain the estimates of the codebits from the soft decoding of user k, based on $\mathbf{z}_k^{(i)}$:

$$\tilde{a}_k^{(i)}(t) = 2.P\left[a_k(t) = +1 \mid D^k(\mathbf{z}_k^{(i)})\right] - 1 \qquad t = 1,\ldots,M \tag{12}$$

where $P[a_k(t)=+1|D^k(\mathbf{z}_k^{(i)})]$ is the A Posteriori Probability (APP) of codebit $a_k(t)$, after soft decoding of $\mathbf{z}_k^{(i)}$ (i.e. event $D^k(\mathbf{z}_k^{(i)})$).

In order to perform the soft decoding, we have to supply the decoder with the *a priori* probabilities of the codebits and the estimated variance of the interference on the codebits. Since each codebit is equally likely to be +1 or –1 *a priori*, the a priori probabilities are readily found as $P[a_k(t) = +1] = P[a_k(t) = -1] = 1/2$. In order to estimate the variance of the interference of each user k, we take a closer look at the observables for the decoding/detection of user k (with h(j,i) = i if $j \in B_k$ and h(j,i) = i-1 if $j \in A_k$):

$$z_k^{(i)}(t) = a_k(t) + \sum_{j \neq k} (\mathbf{R}(t))_{k,j} \left(a_j(t) - \tilde{a}_j^{(h(j,i))}(t)\right) + (\mathbf{n}(t)')_k = a_k(t) + N_k^{(i)}(t) \tag{13}$$

where the variance of $N_k^{(i)}(t)$ is easily obtained as

$$\sum_{j\neq k}(\mathbf{R}(t))_{k,j}^2\left\{1-\left(\tilde{a}_j^{(h(j,i))}(t)\right)^2\right\}+\sigma^2 \qquad t=1,...,M \tag{14}$$

The estimated variance $\sigma_k^2(i)$ of user k in iteration i is calculated, based on (14), as

$$\sigma_k^2(i)=\sigma^2+\frac{1}{M}\sum_{t=1}^{M}\left(\sum_{j\neq k}(\mathbf{R}(t))_{k,j}^2\left\{1-\left(\tilde{a}_j^{(h(j,i))}(t)\right)^2\right\}\right) \tag{15}$$

4. SIMULATION RESULTS

Denoting the required value for E_b/N_0 in order to have a BER $< 10^{-5}$ in a system with a single user as $(E_b/N_0)_c$, we introduce the notion of *critical load L_{max},* defined as the maximum load K_{max}/N the overloaded system can accommodate with a number of iterations I = 10, so that the average BER $< 10^{-5}$ at $(E_b/N_0)_{th} = (E_b/N_0)_c + 0.35$dB for every number of users $K \leq K_{max}$. This implies that for every number of users $K \leq K_{max}$, the loss in signal-to-noise ratio as compared to a single user system, will be lower than about 0.35dB at BER = 10^{-5}.

For the simulations, we consider three possible ways of coding:

- Uncoded: L = M = 1. Here we have $(E_b/N_0)_{th}$ = 10dB.
- Convolutional coding by means of the 21/37 Recursive Systematic Code (RSC). Decoding is performed by means of the BCJR algorithm, and with L = 1000, we have that $(E_b/N_0)_{th}$ = 5.6dB.
- Turbo coding, with a turbo code made from the 21/37 RSC. Decoding is performed by means of a turbo decoder with 5 turbo iterations. For L = 1000, we find $(E_b/N_0)_{th}$ = 1.95dB.

In table 1, the critical load is shown for PN, s-m-O, o-m-O and PN/O with spreading gains N = 16, 32, 64, 128 and 256 for uncoded transmission. All four types of overloaded systems allow for a considerably increasing number of users with increasing spreading factor. PN is inferior to any other type of channel overloading. o-m-O is clearly superior to s-m-O for spreading gains up to 128, which is in agreement with results, obtained in [23], where the superiority of o-O/O to s-O/O was illustrated with optimal multiuser detection. Surprisingly, we see that PN/O outperforms both m-O types for N ≤ 32. For higher spreading gains, PN/O has the same performance as o-m-O, although its set 2 users suffer from much higher interference levels than those of the o-m-O and s-m-O system. An explanation for this surprising result, lies in the fact that the distance properties of s-m-O are far from good. This was the motivation to make the better performing o-m-O type in [23].

Table 1: Critical loads for uncoded transmission.

	N=16	*N=32*	*N=64*	*N=128*	*N=256*
PN	6%	41%	73%	95%	109%
s-m-O	100%	103%	111%	123%	152%
o-m-O	100%	113%	131%	145%	153%
PN/O	100%	116%	131%	143%	153%

The results with coding are shown in table 2. Here, we make no distinction between s-m-O and o-m-O, since the results of both m-O types are the same. Comparing table 2 to table 1, we see that the critical load with coding is significantly higher than that without coding. This is especially pronounced for convolutional coding. Another remarkable fact is that the dependence of the critical load on the spreading gain is much lower in the coded case than in the uncoded case. We see that with coding, the advantage of m-O over PN/O becomes apparent. For convolutional coding and turbo coding, channel loads are about 20 % and 12% higher respectively than that of PN/O for N = 128.

Table 2: Critical loads for coded transmission.

Convolutional Coding						
	N=4	*N=8*	*N=16*	*N=32*	*N=64*	*N=128*
PN	125%	150%	156%	156%	158%	158%
PN/O	175%	188%	188%	194%	194%	195%
m-O	175%	200%	206%	209%	213%	214%
Turbo Coding						
	N=4	*N=8*	*N=16*	*N=32*	*N=64*	*N=128*
PN/O	125%	138%	144%	147%	148%	149%
m-O	125%	138%	150%	153%	156%	161%

5. CONCLUSION

In this paper, the performance with turbo detection of four different types of scalable binary overloaded sequence sets was compared: random spreading, s-m-O, o-m-O and PN/O. We summarise our findings:

- All systems show a considerable gain in critical load with increasing spreading gain. This gain is much more pronounced when the data are send uncoded than with coding of the data. For o-m-O, with uncoded transmission, the critical load jumps from 100% for N = 16 to 153% for N = 256. With convolutional coding and turbo coding of m-O, the load increases from 206% to 214% and from 150% to 161% respectively for N = 16 and N = 128.
- The acceptable load in a coded system is much higher than in the corresponding uncoded system. Especially with convolutional coding, considerable loads can be achieved, in excess of 200% for m-O.
- In the uncoded case, PN/O performs as good as o-m-O. Both PN and s-m-O are inferior to PN/O and o-m-O.
- The advantage of m-O over PN/O becomes apparent in the coded case. With convolutional coding, the channel can be loaded by 20% more users when m-O is applied instead of PN/O for N = 128. With turbo coding, about 12% more users are tolerable in the m-O system.
- PN is substantially inferior to any of the other considered overloaded systems, both with and without coding and for any spreading gain.

REFERENCES

[1] S. Hara, R. Prasad, " Overview of Multicarrier CDMA", IEEE Communications Magazine, Vol. 35, pp. 126-133, Dec. 1997.

[2] M. Rupf, J. L. Massey, "Optimum Sequence Multisets for Synchronous Code-Division Multiple-Access Channels," *IEEE Trans. Information Theory*, vol. 40, pp. 1261-1266, July 1994.

[3] P. Viswanath and V. Anantharam, "Optimal Sequences and Sum Capacity of Synchronous CDMA Systems," *IEEE Trans. Information Theory*, vol. 45, pp. 1984-1991, September 1999.

[4] S. Verdu, S. Shamai,"Spectral Efficiency of CDMA with Random Spreading," *IEEE Trans. Info.*, vol. 45, pp. 622-640, March 1999.

[5] J.L. Massey, "Is the Choice of Spreading Sequence Important ?," *ISSSTA 1996*, September 1996.

[6] S. Verdu, *Multiuser detection*, Cambridge University Press, New York 1998.

[7] R. E. Learned, A. S. Willisky and D. M. Boroson, "Low complexity joint detection for oversaturated multiple access communications," *IEEE Trans. Signal Processing*, vol. 45, pp. 113-122, January 1997.

[8] J. A. F. Ross and D. P. Taylor, "Vector assignment scheme for M+N users in N-dimensional global additive channel," *Electron. Lett.*, vol. 28, August 1992.

[9] J. A. F. Ross and D. P. Taylor, "Multiuser signaling in the symbol-synchronous AWGN channel," *IEEE Trans. Info. Theory*, vol. 41, July 1995.

[10] F. Vanhaverbeke, M. Moeneclaey and H. Sari, "An Excess Signaling Concept with Walsh-Hadamard Spreading and Joint Detection," *Globecom 2000*, vol. 2, pp. 906-909, November 2000, San Francisco.

[11] F. Vanhaverbeke, M. Moeneclaey and H. Sari, "DS/CDMA with Two Sets of Orthogonal Sequences and Iterative Detection," *IEEE Communication Letters*, vol. 4, pp. 289-291, Sept. 2000.

[12] Sari H, Vanhaverbeke F and M. Moeneclaey, "Multiple access using two sets of orthogonal signal waveforms," *IEEE Communication Letters*, vol. 4, pp. 4-6, January 2000.

[13] H. Sari, F. Vanhaverbeke and M. Moeneclaey, "Extending the Capacity of Multiple Access Channels," *IEEE Communications Magazine.*, pp. 74-82, Jan. 2000.

[14] F. Vanhaverbeke, M. Moeneclaey and H. Sari, "Increasing CDMA Capacity Using Multiple Orthogonal Spreading Sequence Sets and Successive Interference Cancellation," *ICC2002*, May 2002, New York.

[15] R. Lupas and S. Verdu, "Linear multiuser detectors for synchronous code-division multiple-access channels ", *IEEE Trans. Information Theory*, vol. 35, pp. 123-136, January 1989.

[16] U. Madhow and M.L. Honig, "MMSE interference suppression for direct-sequence spread spectrum CDMA", *IEEE Trans. Communications*, vol. 42, pp. 3178-3188, December 1994.

[17] L. K. Rasmussen, T. J. Lim, and A.-L. Johansson, "A Matrix-Algebraic Approach to Successive Interference Cancellation in CDMA.", *IEEE Trans. Commun.* , vol. 48, Jan. 2000.

[18] S. Verdu, *Optimum multiuser signal detection*, PhD thesis, University of Illinois at Urbana-Champaign, August 1984.

[19] X. Wang and G. V. Poor, "Iterative (Turbo) Soft Interference Cancellation and Decoding for Coded CDMA ", *IEEE Trans. Communications*, vol. 47, pp. 1046-1061, July 1999.

[20] M. Kobayashi and J. Boutros, "Successive Interference Cancellation with SISO Decoding and EM Channel Estimation ", *IEEE JSAC*, vol. 19, pp. 1450-1460, August 2001.

[21] M. C. Reed et al., "Iterative Multiuser Detection for CDMA with FEC: Near-Single User Performance ", *IEEE Trans. Communications*, vol. 46, pp. 1693-1699, December 1998.

[22] E. H. Dinan and B. Jabbari, "Spreading Codes for Direct Sequence CDMA and Wideband CDMA Cellular Networks," *IEEE Commun. Mag.*,pp 48-54, Sept. 1998.

[23] F. Vanhaverbeke and M. Moeneclaey, "Influence of Different Base-pairs on the Optimal Performance of an Oversaturated OCDMA/OCDMA System," *CIC2002*, October 2002, Seoul.

Frederik Vanhaverbeke and Marc Moeneclaey
University of Ghent, department TELIN
Sint-Pietersnieuwstraat 41, B-9000 Gent (Belgium)
fv@telin.rug.ac.be

SHINSUKE HARA

VIRTUAL SUBCARRIER ASSIGNMENT (VISA): PRINCIPLE AND APPLICATIONS

Abstract. This paper proposes a novel spatial filtering technique of orthogonal frequency division multiplexing (OFDM) signals, which is called "VIrtual Subcarrier Assignment (VISA)." Just selecting or assigning different virtual subcarrier positions for different users, VISA can selectively receive one of OFDM signals or simultaneously receive all the OFDM signals. As applications of VISA, this paper discusses coexistence of Wireless Personal Area Networks (WPANs) in a localized area and a space division multiple access (SDMA) in an isolated cell environment.

1. INTRODUCTION

Orthogonal frequency division multiplexing (OFDM) signal has a lot of freedom in the frequency axis, namely, a lot of subcarriers. Here, we call subcarriers which convey information "data subcarriers" whereas sucarriers which are not used for actual data transmission, in other words, transmit zero-information "virtual subcarriers" (they are also called "carrier holes"). When an OFDM system assigns different *data* subcarriers to different users, they freely access the system. This is well known as "frequency division multiple access (FDMA)."

Someone may imagine that nothing will happen when assigning different *virtual* subcarriers to different users. This is true, without the use of array antenna. However, something effective and interesting does happen if using array antennas. This is because array antenna works as a spatial filter, and with the use of array antenna, we can make a receiver accept or reject only signals with a certain spectral (or temporal) characteristic. In this paper, we propose a novel spatial filtering technique of OFDM signals called "VIrtual Subcarrier Assignment (VISA)."

2. PRINCIPLE OF VISA

In wireless local area network (WLAN) standards such as IEEE 802.11a, HIPERLAN/2 and MMAC[1], there is a pre-determined virtual subcarrier position around the center of the transmitted subcarrier arrangement. In [2],[3], we have proposed a pre-fast Fourier transform (FFT) type OFDM adaptive array antenna for interference suppression. If interfering signals have some signal components at the frequency of the virtual subcarrier, the array antenna can easily steer nulls toward the interfering signals, by controlling the array weights so as to force the virtual subcarrier output to be zero. In this case, we can make the array antenna accept only the signals which have no signal component at the pre-determined virtual subcarrier position.

K. Fazel and S. Kaiser (eds.), Multi-Carrier Spread-Spectrum, 31–38.

The interfering signal from the viewpoint of the array antenna can be a different OFDM signal which has a signal component at the pre-determined virtual subcarrier position. In this case, the array antenna rejects the OFDM signal as an interfering signal. This implies that, *if we color OFDM signals with different virtual subcarrier positions, we can selectively receive one of OFDM signals, or we can simultaneously receive all the OFDM signals.*

Figure 1 shows the principle of VISA, where there is a receiver with array antenna. There are two types of OFDM signals A and B, which have different virtual subcarrier positions A and B, respectively. Assume that the receiver knows their virtual subcarrier positions. When receiving the OFDM signal A, the receiver controls the array weights so as to force the output of the virtual subcarrier A to be zero. Even if the signals from A and B simultaneously arrive at the receiver, the receiver can *spatially filter out* only the OFDM signal B that has a signal component at the virtual subcarrier position A. On the other hand, the receiver is successful in reception of only the signal B, by forcing the output of the virtual subcarrier B.

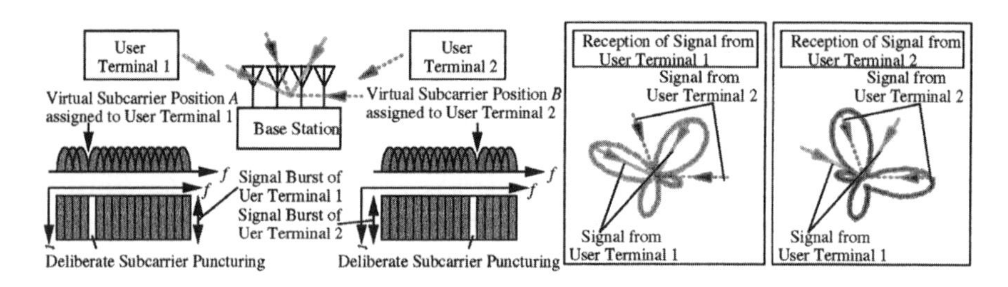

Figure 1. Principle of VISA

3. TWO TYPES OF VISAS

There are two types of how to make virtual subcarriers: mapping VISA and puncturing VISA.

One way is to map information data over the remaining subcarriers after virtual subcarrier positions are selected or assigned (Mapping-VISA). Assuming the number of total usable subcarriers=M and the number of virtual subcarriers=L, Mapping-VISA reduces the transmission rate by the factor of $(M - L)/M$ without any other performance degradation.

The other way is to deliberately puncture (decimate) data subcarriers at the selected or assigned positions (Puncturing-VISA) after mapping information data over all the M subcarriers . Puncturing-VISA keeps the transmission rate (but introduces some performance degradation). Another advantage of the subcarrier puncturing is that the demodulator at the receiver does not care about the punctured data, namely, can deal with them as those that happen to be close to zero by multipath fading. Therefore, Puncturing-VISA is introducible in already-existing OFDM-based standards without any significant change to transmitter/receiver structures, just inserting a puncturer between a serial-to-parallel converter (S/P) and an FFT processor at the transmitter, and adding an array antenna to the receiver.

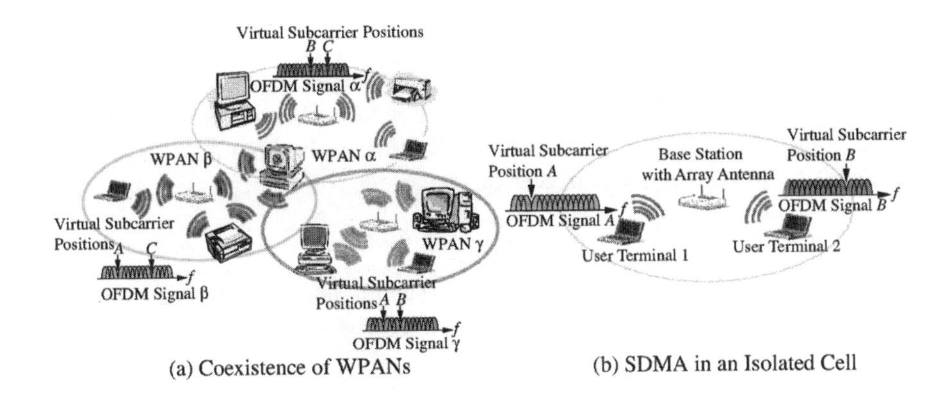

Figure 2. Applications of VISA

4. APPLICATIONS OF VISA

4.1 Coexistence of WPANs in a Localized Area

Wireless personal area network (WPAN) means grouping electronic devices such as note personal computers (PCs), lap-top PCs, printers and access points in a private area with wireless connection. In office environments, different persons will have different WPANs, all of which may use the same OFDM-based standard. In this case, coexistence of different WPANs in a localized area is a problem.

Figure 2(a) shows such a situation where three different WPANs: α, β and γ want to coexist in a localized area. Here, assume that all the terminals are quipped with array antennas and, for instance, data transmission is based on carrier sense multiple access with collision avoidance with request-to-send and clear-to-send (CSMA/CA with RTS/CTS).

Assume that there are three pre-determined candidates in virtual subcarrier positions: *A*, *B* and *C*. According to VISA, in each WPAN, a master terminal first selects two positions of virtual subcarriers from the three candidates, and then broadcasts the information on the positions to all the terminals. Note that the pattern of the virtual subcarrier positions should be different in different WPANs, namely, a virtual subcarrier position which is not selected in one WPAN should be a virtual subcarrier position in the other two WPANs.

In each WPAN, data transmission is made with an identifier (ID) to identify its own WPAN, namely, whenever a terminal belonging to a WPAN transmits a packet, it must use the selected virtual subcarrier positions of the WPAN. Communication starts with *carrier sense of the virtual subcarrier position which was not selected in its own WPAN.* This is because all the terminals in a WPAN do not have to care about interfering signals from neighboring WPANs, which are automatically rejected by VISA. For instance, in WPAN α, when a terminal wants to transmit a packet, it senses the virtual subcarrier position of A. Even when there are signals from WPANs β and γ at the sensing timing, it cannot sense the existence of the signals, because they do not have any signal component at the virtual subcarrier position of A. No signal sensed at the virtual subcarrier position of A means no signal transmission only in WPAN α, so the terminal can attempt to transmit the packet with RTS/CTS handshake. In this way, VISA supports coexistence of WPANs in a localized area.

4.2 SDMA in an Isolated Cell

Figure 2(b) shows an application of VISA for an uplink in an isolated cell, where each user terminal with no array antenna transmits a signal to a base station with an array antenna.

Assume that the two user terminals 1 and 2 want to transmit their packets to the base station. According to VISA, when the base station receives request packets for data transmission from the two user terminals, it first assigns the virtual subcarrier position A to the user terminal 1 and the virtual subcarrier position B to the user terminal 2 (of course, their positions are different). In the data transmission at the two user terminals, they transmit their packets to the base station with VISA. The base station can simultaneously receive both of the OFDM signals: for the OFDM signal A by forcing the output of the virtual subcarrier position A to be zero whereas for the OFDM signal B by forcing the output of the virtual subcarrier position B to be zero. In this way, VISA supports an SDMA where *the position of a virtual subcarrier assigned is an ID to identify an individual user terminal.*

5. SYSTEM MODEL

VISA does not require any special pilot signal, so it is applicable for already-existing OFDM-based WLAN systems. Let us use the signal burst formant and subcarrier arrangement in IEEE802.11a[1].

A signal burst (packet) is composed of a payload and a preamble. The payload is composed of 10 OFDM data symbols, where in one OFDM symbol, the guard interval is N_G samples long and the useful symbol is N_U samples long. The preamble can be commonly used among the users. The first half of the preamble is 2 OFDM symbols long and is used for automatic gain control and FFT window timing synchronization, so we do not pay attention to the first half any more. The second half of the preamble is also 2 OFDM symbols long and used for array weight control and subcarrier recovery. The first N_G samples are unusable to avoid inter-symbol interference due to multipath fading, and at least the last N_U samples are required to perform the FFT for subcarrier recovery. Therefore, the remaining $(N_G + N_U)$ samples can be used for array weight control. Let us call the part "the control pilot signal" and set the first sampling instant in the control pilot symbol at the base station to m=0.

Figure 3 shows a pre-FFT type OFDM adaptive array antenna with N antenna elements. The array output at sampling instant of m is written as

$$y(m) = \mathbf{w}^H(m)\mathbf{r}(m) \tag{1}$$

where H denotes Hermitian transpose. In Eq.(1), $\mathbf{r}(m)$ and $\mathbf{w}(m)$ are the received signal vector ($N \times 1$) and the weight vector ($N \times 1$), respectively, which are given by

$$\mathbf{r}(m) = [r_1(m), \cdots, r_n(m), \cdots, r_N(m)]^T \tag{2}$$

$$\mathbf{w}(m) = [w_1(m), \cdots, w_n(m), \cdots, w_N(m)]^T \tag{3}$$

where T denotes transpose.

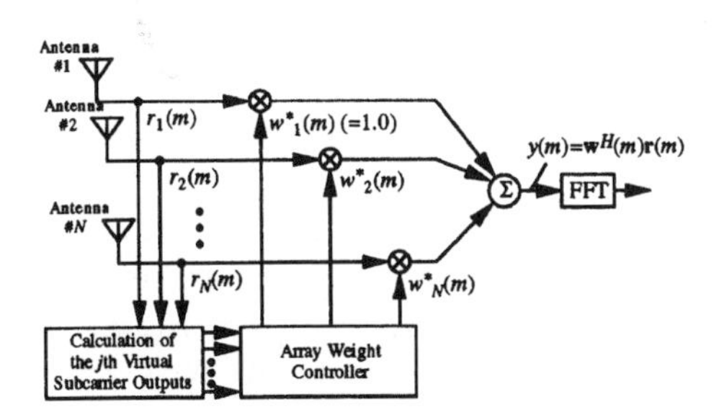

Figure 3. A Pre-FFT type OFDM Adaptive Array Antenna

Assume that the jth virtual subcarrier is assigned. Here, we focus our attention only on a single subcarrier puncturing. To derive the array weight control algorithm, define the jth virtual subcarrier output vector ($N \times 1$) as

$$\mathbf{v}_j(m) = [v_{j1}(m), \cdots, v_{jn}(m), \cdots, v_{j(N)}(m)]^T \tag{4}$$

where $v_{jn}(m)$ is the jth virtual subcarrier output of the n-th antenna element at the m-th sampling instant. With the following complex-valued sinusoidal function, $v_{jn}(m)$ is written as

$$v_{jn}(m) = \sum_{q=0}^{N_U-1} b_j(q) r_n(m-q), \quad b_j(q) = e^{-j2\pi \frac{f_{vj} q}{N_U}} \tag{5}$$

where f_{vj} is the frequency corresponding to the jth virtual subcarrier. *Note that calculation of the virtual subcarrier output means down-conversion of the received signal with the frequency of the virtual subcarrier. Therefore, it takes N_U samples to give the first output.* This means that the array weight algorithm can have only N_G iterations.

Next, define the exponentially weighted cost function as[4]

$$|\varepsilon_R(m)|^2 = \sum_{p=N_U-1}^{m} \lambda^{m-p} |e_j(p)|^2 \tag{6}$$

where λ is a forgetting factor and $e_j(p)$ is the error between the desired response and the jth virtual subcarrier output given by

$$e_j(p) = 0 - \mathbf{w}^H(p)\mathbf{v}_j(p). \tag{7}$$

The recursive least square (RLS) algorithm is applicable for the minimization problem of Eq.(6). However, it is important to note that Eq.(6) has a solution of $\mathbf{w}(p)$=$\mathbf{0}$. Therefore, a constraint is required to avoid the trivial solution, for instance,

$$subject\ to \quad w_1(p) = 1. \tag{8}$$

6. PEFORMANCE EVALUATION AND DISCUSSIONS

We evaluate the potential of VISA by computer simulation. For an antenna configuration, we assume an 8 element-circular array antenna with element spacing of half wavelength.

In a spatial-temporal (frequency selective fading) channel model, Each path that has a Rayleigh-distributed amplitude with the same power arrives forming a cluster composed of 5 waves with Root Mean Square (RMS) angle spread of 5 deg. The arrival time of each path is uniformly distributed in [0, N_G]. The number of paths for a desired user is set to 3, and the direction of arrival for each path is randomly chosen out of [0, 360 deg]. In addition, taking into consideration the speed of man walking, we assume that the temporal variation of the channel fading is slow enough not to give any significant change to the channel impulse response over one signal burst.

An OFDM symbol is generated with the 64 point-FFT, a 16 sample-long guard interval is inserted in each OFDM symbol, and the number of data subcarriers including 4 pilot subcarriers is 52 (similar to IEEE 802.11a standard[1]). For a different user, a different virtual subcarrier position is selected and assigned out of the 48 data subcarrier positions. For modulation/demodulation and channel coding formats, we assume a coherent QPSK and a half-rate convolutional code with constraint length of 7 (133, 171). In addition, we set the depth of interleaving to 8×12. Furthermore, we set the forgetting factor of the RLS-based array weight control algorithm to 1.0, and also set the number of oversampling (N_{os}) to 4, so the number of iterations for the VISA-based null-steering is 16×4.

Let us first discuss the effect of the subcarrier puncturing. Figure 4 shows the bit error rate (BER) versus the average E_b/N_0 for the deliberate subcarrier puncturing. Imagine that the operation is like "putting the OFDM signal into a bit severe multipath fading channel." The channel coding/interleaving is employed among the subcarriers, so no significant errors occur. Even when puncturing 1 subcarrier, the power loss is around 0.2 dB at the BER of 1.0×10^{-3} in both an additive white Gaussian noise channel and the frequency selective Rayleigh fading channel. In addition, there was no dependency observed between the position of the virtual subcarrier selected and the BER because of the bit interleaving. In the following, we discuss the performance of Puncturing-VISA.

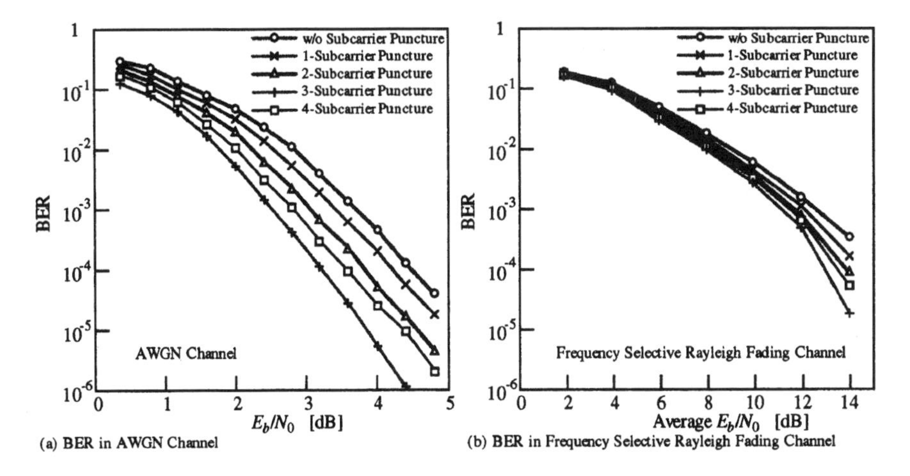

Figure 4. Deliberate Subcarrier Puncturing

Let us next discuss the BER, which is typical in the coexistence problem of 2 WPANs, as shown in Fig.2(a). Figure 5(a) shows the BER versus the average E_b/N_0 per arrival path. Without array antenna, a good BER is obtained only for the case of no interfering path. For VISA, as expected, the BER becomes worse as the number of interfering paths increases, however, it remains a relatively low up to the number of interfering paths=6. This is because VISA has the degree of freedom of array antenna=6 (7-1, note that the number of adjustable array weights is 7 in the null-steering).

Finally, let us discuss the BER in the SDMA of Fig.2(b). Figure 5(b) shows the BER versus the number of users. Here, we assume the number of paths per user is 3. Also in the figure, the BER of a beam-steering is shown for comparison purpose, which tries to catch only the first desired paths. Note that the beam-steering requires a special preamble to distinguish each user, so unlike VISA, it is un-introducable in already-existing systems. The VISA-based system always outperforms the beam-steering system. The degree of freedom of the null-steering array antenna is 6, so it can correctly steer nulls toward 6 arrival paths from 2 multiple access users. Therefore, for the system/channel parameter setting, the VISA-based OFDM system can simultaneously accommodate 3 users. This means that, if introducing VISA into an OFDM system, it can *triple* the capacity of the OFDM system. On the other hand, the degree of freedom of the beam-steering array antenna is 7 (8-1), but a part of the degree is dedicated for suppression of not only the arrival paths of interfering multiple access users but also the second and third arrival paths of desired users. Therefore, it can correctly steer nulls toward 3 arrival paths from 1 multiple access user. This means that the beam-steering can double the capacity of the OFDM system. When the number of users reaches 4, even with the array antenna, the BER becomes poor. This is because the number of total arrival paths becomes more than the degree of freedom of the array antenna.

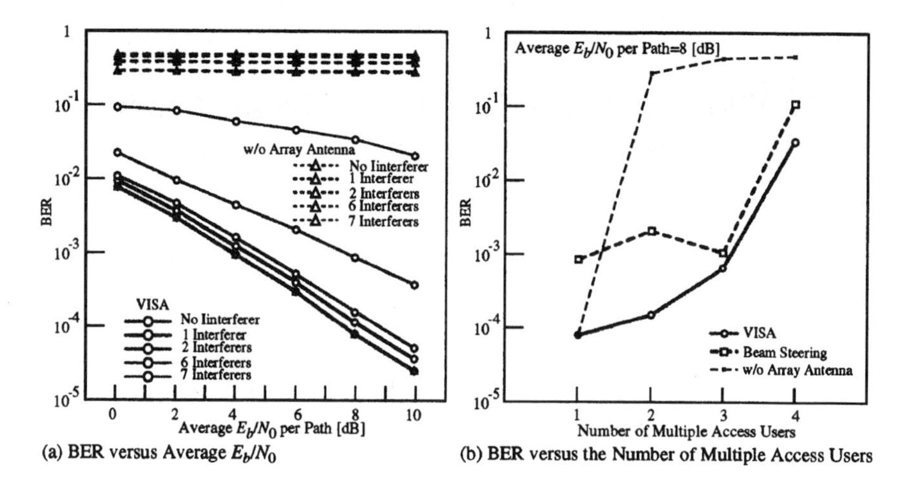

(a) BER versus Average E_b/N_0 (b) BER versus the Number of Multiple Access Users

Figure 5. BER evaluations

7. COCLUSIONS

This paper has shown the principle of a novel spatial filtering technique for OFDM scheme called "VISA," and has discussed some computer simulation results on the basic performance. Furthermore, this paper has proposed "a deliberate subcarrier puncturing," which means zeroing data-mapped subcarriers.

ACKNOWLEDGEMENT

This work was supported by the R&D support scheme for funding selected IT proposals 2003, top priority research and development to be focused (frequency resources development) of the Ministry of Public Management, Home Affairs, Posts and Telecommunications of Japan.

References

[1] IEEE Std. 802.11a, "Wireless Medium Access Control (MAC) and Physical Layer (PHY) Specifications: High-speed Physical Layer Extension in the 5 GHz Band," IEEE, 1999.

[2] S. Hara, A. Nishikawa, and Y. Hara, "A Novel OFDM Adaptive Antenna Array for Delayed Signal and Doppler-Shifted Signal Suppression," *Proc. IEEE ICC 2001*, pp.2302-2306, Helsinki, Finland, 11-14 June 2001.

[3] S. Hara, S. Hane and Y. Jia "A Simple Null-Steering Adaptive Array Antenna in OFDM-Based WPAN/WLAN," *Proc. IEEE VTC 2003-Spring*, in CD-ROM, Jeju, Korea, 22-25 Apr. 2003.

[4] S.Haykin, *Adaptive Filter Theory, Third Edition*, Prentice-Hall, 1996.

Shinsuke Hara is with Graduate School of Engineering, Osaka University, Osaka, Japan.

G. AUER, A. DAMMANN, S. SAND, S. KAISER

On Pilot-Symbol Aided Channel Estimation for MC-CDMA in the Presence of Cellular Interference

Abstract. We address the downlink of a cellular multi-carrier CDMA (MC-CDMA) system taking into account channel estimation. The system performance in presence of a synchronization mismatch between two interfering base stations (BS) is analysed in the way that a mobile terminal receives the perfectly synchronized signal from the desired BS as well as the signal from one interfering BS with a synchronization offset. The cellular interference not only corrupts the transmitted data but also the multiplexed pilot symbols which are used for channel estimation. The effects of the cellular interference on the channel estimator in presence of a synchronization offset is analyzed in this paper.

1. INTRODUCTION

Multi-carrier modulation, in particular orthogonal frequency division multiplexing (OFDM) [1], has been successfully applied to various digital communications systems. OFDM can be efficiently implemented by using the discrete Fourier transform (DFT). Furthermore, for the transmission of high data rates its robustness in transmission through dispersive channels is a major advantage. For multi-carrier CDMA (MC-CDMA), spreading in frequency and/or time direction is introduced in addition to the OFDM modulation [2–4]. MC-CDMA has been deemed a promising candidate for the downlink of future mobile communications systems [5, 6].

In order to coherently detect the received signal, accurate channel estimation is essential. For pilot-symbol aided channel estimation (PACE) pilot symbols are periodically inserted in the time-frequency grid. The channel response of data symbols at an arbitrary position can be reconstructed by exploiting the correlation of the multi-carrier based signal in time and frequency [7]. Two dimensional (2D) filtering algorithms have been proposed for PACE, based on 2D Wiener filter interpolation. Unfortunately, such a 2D estimator structure may be too complex for practical implementation. To reduce the complexity, two cascaded 1D estimators in time and frequency may be used instead, termed two times one-dimensional (2×1D) PACE [7].

If a MC-CDMA is to be employed as a cellular system with high frequency reuse factor, the effects of the cellular interference on the system performance including channel estimation need to be examined. Cellular interference not only corrupts the transmitted data but also the pilot symbols used for channel estimation. Unlike, data symbols which can be protected by means of spreading and/or channel coding, the pilots cannot be protected in such a way. One way to mitigate this problem is to use a pilot reuse factor being larger than the frequency reuse factor for the data symbols [8]. Such a system, however, requires full synchronization within all BS of the cellular systems, which may be difficult to realize in practice. Therefore, we compare the performance of a cellular MC-CDMA system with a pilot and data reuse factor of one, with a sytem having a data reuse of one and a pilot reuse of three. To this end,

K. Fazel and S. Kaiser (eds.), Multi-Carrier Spread-Spectrum, 39–48.

(a.)

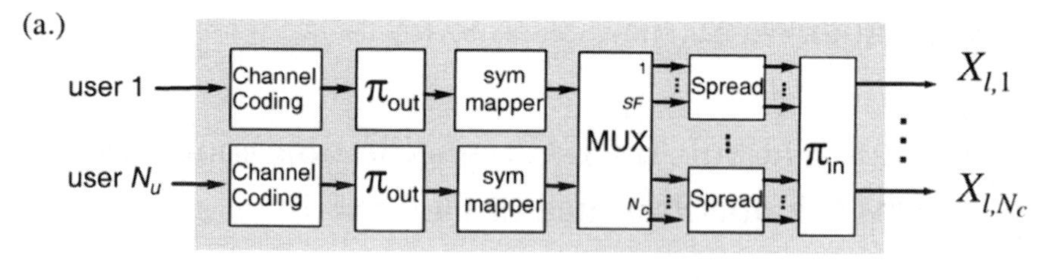

(b.)

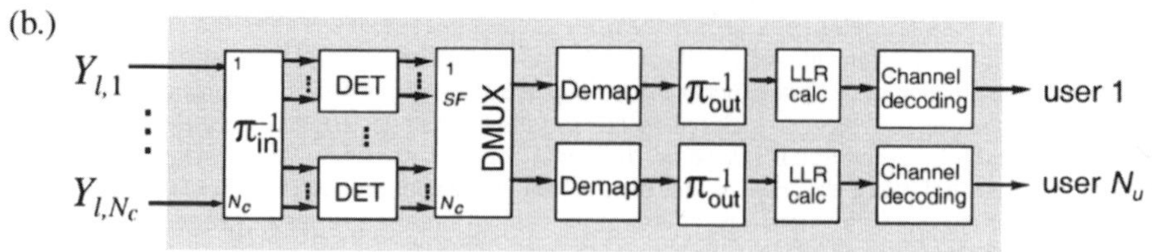

Fig. 1. Block diagram of the MC-CDMA system, (a.) transmitter, (b.) receiver.

the system performance of both approaches is investigated in a two cell scenario, dependent on the synchronization offset between two interfering BS.

2. SYSTEM & CHANNEL MODEL

Fig. 1.a shows the block diagram of a MC-CDMA transmitter with N_c subcarriers for N_u users. The bit stream for each user is encoded with a convolutional code, bit interleaved by the outer interleaver π_{out}, and fed to the symbol mapper. The symbol mapper assigns the bits to complex-valued data symbols according to different alphabets, like PSK or QAM with the chosen cardinality. A serial-to-parallel converter allocates the modulated signals to $N_d = N_c/L$ data symbols per user. Each of the N_d data symbols is spread with a Walsh-Hadamard sequence of length $L \geq N_u$. Given the vector, $\mathbf{d}_k = [d_k^{(1)}, \cdots, d_k^{(N_u)}]^T$, consisting of the kth symbol of all users, the spreading operation results in

$$\mathbf{s}_k = \mathbf{C}_L \mathbf{d}_k \ , \ \in \mathbb{C}^L \ , \ 1 \leq k < N_d \tag{1}$$

where $\mathbf{C}_L$ represents a $L \times N_u$ spreading matrix. The system load of the MC-CDMA system is N_u/L, and can be adjusted between 1 and $1/L$. Subsequently, the block $\mathbf{s} = [\mathbf{s}_1, \cdots, \mathbf{s}_{N_d}]^T$ is frequency interleaved by the inner interleaver π_{in} over one OFDM symbol to maximize the diversity gain. We choose a random interleaver for π_{in}. The interleaved symbol of the ℓth OFDM symbol, at subcarrier i, is denoted by $X_{\ell,i}$. One frame consists of N_{frame} OFDM symbols, each having N_c subcarriers. In order to distinguish signals from different BS and to further randomize the transmitted signal, $X_{\ell,i}$ is scrambled by a cell specific random sequence, $p_{\ell,i}^{(0)}$, to yield $\check{X}_{\ell,i}^{(0)} = p_{\ell,i}^{(0)} X_{\ell,i}$. The scrambler has caridinality M_s, and the M_s discrete singal points are chosen according to a PSK constellation. An inverse DFT with $N_{FFT} \geq N_c$ points is performed on each block, to yield the time domain signal $x_{\ell,n}^{(0)} = \text{IDFT}\{\check{X}_{\ell,i}^{(0)}\}$, and subsequently a guard interval (GI) having N_{GI} samples is inserted, in the form of a cyclic prefix.

A block diagram of a MC-CDMA receiver is depicted in Fig. 1.b. After D/A conversion, the signal $x^{(0)}(t)$ is transmitted over a mobile radio channel to yield the received signal $y(t)$. For the moment we only consider the signal from a single transmitter, i.e.

cellular interference is neglegted. After sampling with rate $1/T_{spl}$, the guard interval is removed and a DFT on the received block of N_{FFT} signal samples is performed, to obtain the output of the OFDM demodulation $\breve{Y}_{\ell,i} = \mathrm{DFT}\{y_{\ell,n}\}$. The last $N_{FFT} - N_c$ DFT outputs of $\breve{Y}_{\ell,i}$ contain zero subcarriers which are dismissed. Subsequently, the cell specific scrambling sequence is removed, $Y_{\ell,i} = p_{\ell,i}^{(0)*}\breve{Y}_{\ell,i}$. We assume the guard interval to be longer than τ_{max} including a possible timing synchronization offset. Assuming perfect synchronization for the moment, the received signal after OFDM demodulation of OFDM symbol ℓ at subcarrier i is in the form

$$Y_{\ell,i} = X_{\ell,i} H_{\ell,i} + N_{\ell,i} \tag{2}$$

where $X_{\ell,i}$, $H_{\ell,i}$, and $N_{\ell,i}$, denote the transmitted symbol, the channel transfer function (CTF), and AWGN with zero mean and variance N_0.

2.1. Data detection

After deinterleaving with the inner interleaver π_{in} the vector $\mathbf{r}_k = [r_{k,1}, \cdots, r_{k,L}]^T$ is obtained, containing the kth spread symbols of the N_u users. The detector comprises a linear MMSE one tap equalizer and the despreader, yielding the received data vector

$$\widehat{\mathbf{d}}_k = \mathbf{C}_L^H \mathbf{G}\, \mathbf{r}_k \tag{3}$$

The MMSE equalizer $\mathbf{G}$ is a diagonal matrix with entries [4]

$$G_{i,i} = \frac{H_{\ell,i}}{|H_{\ell,i}|^2 + \frac{1}{\gamma_c}} \tag{4}$$

where γ_c denotes the singal to noise ratio (SNR) per subcarrier, which is $\gamma_c = \frac{N_u}{L}\frac{E_s}{N_0}$ for the single transmitter scenario.

Subsequently, all data symbols of the desired user, $\widehat{d}_k^{(1)}$, are combined to a serial data stream. The symbol demapper maps the data symbols into bits, by also calculating the Log-Likelihood-ratio (LLR) for each bit based on the selected alphabet. The codebits are deinterleaved and finally decoded using softdecision algorithms [4].

2.2. Channel model

We consider a time-variant, frequency selective, Rayleigh fading channel, modeled by a tapped delay line with Q_0 non-zero taps [9]. The channel impulse response (CIR) is described by

$$h(t,\tau) = \sum_{q=1}^{Q_0} h_q(t) \cdot \delta(\tau - \tau_q) \tag{5}$$

where $h_q(t)$ and τ_q are the complex amplitude and delay of the qth channel tap. It is assumed that the Q_0 channel taps are mutually uncorrelated. Due to the motion of the moblile $h_q(t)$ will be time-variant caused by the Doppler effect, being band-limited by the maximum Doppler frequency ν_{max}. However, the channel impulse response (CIR) needs to be approximately constant during one OFDM symbol, so $h_{\ell,q} = h_q(\ell T_{sym})$. The channel of the qth tap, $h_{\ell,q}$, impinging with time delay τ_q, is a wide sense stationary (WSS), complex Gaussian random variable with zero mean.

The CTF of (2), is the Fourier transform of the channel impulse response. Sampling the result at time $t = \ell T_{sym}$ and frequency $f = i/T$, the CTF becomes

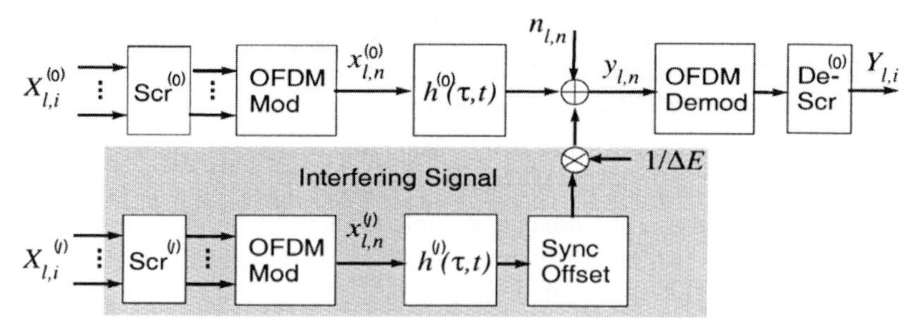

Fig. 2. Block diagram of the cellular MC-CDMA system simulator.

$$H_{\ell,i} = H(\ell T_{sym}, i/T) = \sum_{q=1}^{Q_0} h_{q,\ell}\, e^{-j2\pi\, \tau_q\, i/T} \tag{6}$$

where $T_{sym} = (N_{FFT} + N_{GI})T_{spl}$ and $T = N_{FFT} T_{spl}$ represents the OFDM symbol duration with and without the guard interval.

2.3. Synchornization offset

For synchronization the following parameters cause disturbances in the receiver [10]:

- The transmitter carrier frequency oscillators may be mistuned, resulting in a *carrier frequency offset*, Δf, that can be modeled as a time-variant phase offset $\theta(t) = \Delta f\, t$. A carrier frequency offset will cause ICI, i.e. the orthogonality between subcarriers is lost.
- The transmitter time scale is unknown to the receiver. Therefore, the receiver OFDM symbol window controling the removal of the guard interval will usually be offset from its ideal setting by a time ΔT, termed symbol timing offset. As long as the guard interval is longer than τ_{max} plus the symbol timing offset, the orthogonality of the OFDM modulated signal is maintained. Thus, requirements for the symbol time offset are rather relaxed. Otherwise, inter-symbol interference (ISI) as well as inter-carrier interference (ICI) will be observed.

Generally, the received signal having a synchronization mismatch between transmitter and receiver, sampled at time instants $t = nT_{spl} + \ell T_{sym}$, can be described by

$$y_{\ell,n} = e^{j2\pi\, t\, \Delta f} \int_0^{\tau_{max}} x(t - \tau - \Delta T) \cdot h(t,\tau)\, d\tau + n(t) \Bigg|_{t = nT_{spl} + \ell T_{sym}} \tag{7}$$

The interval $[0, \tau_{max}]$ denotes the delays τ where the CIR is non-zero, i.e. τ_{max} defines the maximum delay of the channel.

2.4. Assessing the effects of cellular interference

A block diagram of how the cellular interference is modeled is shown in Fig. 2. It is assumed that the mobile terminal is perfectly synchronized with the BS transmitting the desired signal, which is received with an energy per symbol of E_s. The signal from the interfering BS is received at the mobile with energy per symbol of $E_s/\Delta E$, having a timing offset ΔT and a frequency offset Δf. So, ΔE accounts for the difference in received signal power between the two interfering BS. After sampling the received signal at the mobile terminal is in the form

$$y_{\ell,n} = z^{(0)}_{\ell,n} + \frac{1}{\sqrt{\Delta E}}\, e^{j2\pi\,\Delta f\, T_{spl} n}\, z^{(I)}_{\ell,n-\Delta n} + n_{\ell,n} \tag{8}$$

where $\Delta n = \lfloor \Delta T / T_{spl} \rfloor$ denotes the symbol timing offset normalized to the sampling duration. Furthermore, $z^{(0)}[\cdot]$ and $z^{(I)}[\cdot]$ represent the received signal of the desired and interfering BS without noise.

After OFDM demodulation the carrier to interference ratio C/I is given by

$$\gamma_c = \frac{1}{\frac{1}{\Delta E} + \frac{N_0}{E_s}\frac{L}{N_u}} \tag{9}$$

In an celluar environment the SNR in (4) should be adjusted accordingly. The nature of the interference is not independent of Δf and ΔT. For large synchronization offsets ICI is the major source of interference, which can be approximated as white Gaussian noise. For small synchronization offsets most interference stems from only one subcarrier, so the resulting interference is non-Gaussian. The effects a cellular MC-CDMA system faces in case of a synchronization offset was analysed in [11] if perfect channel state information (CSI) is available. In this paper the effects of celluar interference are studied if channel estimation is taken into account.

In case the mobile is near the cell boundary ΔE will be close to one, so the carrier to intference ratio γ_c also approach one. In order to maintain a reliable connection the system may be operate not fully loaded, so $N_u < L$. Furthermore, we employ N_R receive antennas in order to exploit spatial diversity. We assume the receive antennas to be uncorrelated, the receiver combines the N_R signals with maximum ratio combining.

3. Pilot-Symbol Aided Channel estimation

Pilot-symbol aided channel estimation (PACE) is based on periodically inserting known symbols, termed *pilot symbols*, in the transmitted data sequence. PACE was first introduced for single carrier systems and required a flat-fading channel [12]. If the spacing of the pilots is sufficiently close to satisfy the sampling theorem, channel estimation and interpolation for the entire data sequence is possible. When extending the idea of PACE to multi-carrier systems, it must be taken into account that the fading fluctuations are in two dimensions, in time and frequency. In order to satisfy the 2D sampling theorem, the pilot symbols are scattered throughout the time-frequency grid, yielding a 2D pilot grid.

To describe pilot symbol-assisted channel estimation it is useful to define a subset of the received signal sequence containing only the pilots,[1] $\{\tilde{X}^{(\mu)}_{\tilde{\ell},\tilde{i}}\} = \{X^{(\mu)}_{\ell,i}\}$, with $\ell = \tilde{\ell} D_t$ and $i = \tilde{i} D_f$. The quantities D_f and D_t denote the pilot spacing in frequency and time, respectively.

For $2 \times$ 1D-PACE the correlation function of the channel can be factored into a time and frequency correlation function, which enables a cascaded channel estimator, consisting of two 1D estimators. The basic idea of $2\times$1D-PACE is illustrated in Fig. 3.

[1] As a general convention, variables describing pilot symbols will be marked with a ˜ in the following.

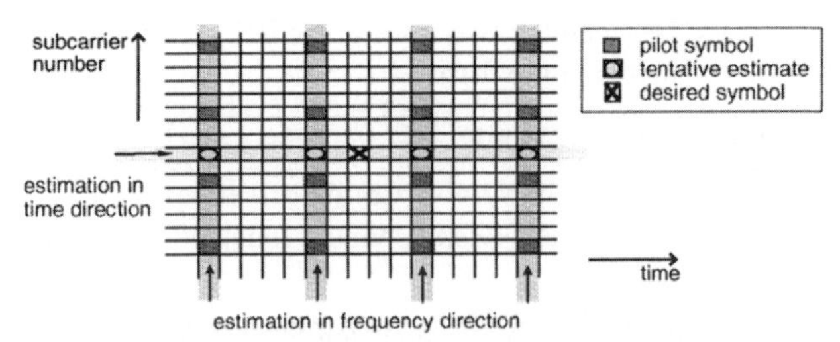

Fig. 3. Principle of $2 \times 1D$ pilot aided channel estimation (PACE).

First, channel estimation is performed in frequency direction, at OFDM symbols $\ell = \tilde{\ell} D_t$, yielding tentative estimates for all subcarriers of that OFDM symbol. The second step is to use these tentative estimates as new pilots, in order to estimate the channel for the entire frame [7]. It was demonstrated in [7], that 2×1D-PACE is significantly less complex to implement with respect to optimum 2D channel estimation, while there is little degradation in performance.

Generally, it is of great computational complexity to use all available pilots. Instead a 2D window of size $M_f \times M_t$ can be slid over the whole time-frequency grid, with $M_f < N_c/D_f$ and $M_t < N_{\text{frame}}/D_t$.

Either channel estimation in frequency direction or time direction may be performed first. The case that frequency direction is performed first corresponds to Fig. 3.

The estimator for 2×1D-PACE can be expressed as

$$\widehat{H}_{\ell,i} = \sum_{n=1}^{M_t} W''_n(\Delta\ell) \sum_{m=1}^{M_f} W'_m(\Delta i) \cdot \tilde{Y}_{\tilde{\ell}+n,\tilde{i}+m} \tag{10}$$

where $\mathbf{W}''(\Delta\ell) = [W''_1(\Delta\ell), \cdots, W''_{M_t}(\Delta\ell)]$ represents the FIR interpolation filter in time direction with filter delay $\Delta\ell = D_t\tilde{\ell} - \ell$. The filter in frequency direction $\mathbf{W}'(\Delta i) = [W'_1(\Delta i), \cdots, W'_{M_f}(\Delta i)]$ depends on the location of the subcarrier to be estimated i, relative to the pilot positions, $\Delta i = D_f\tilde{i} - i$.

The estimators $\mathbf{W}'(\Delta i)$ and $\mathbf{W}''(\Delta\ell)$ are obtained by solving the Wiener-Hopf equation in frequency and time direction, respectively [7].

3.1. Mismatched estimator

For the Wiener filter in time and frequency described above, the auto and cross-correlation matrices at the receiver need to be estimated. Alternatively, a robust estimator with a model mismatch may be chosen [7]. That is to assume a uniform power delay profile with maximum delay, $\tau_{\max_w}$, and a rectangular Doppler power spectrum with the maximum Doppler frequency, $\nu_{\max_w}$, which are to be expected in a certain transmission scenario, i.e. worst case propagation delays and maximum expected velocity of the mobile user. In order to determine the channel estimator only $\tau_{\max_w}$, $\nu_{\max_w}$, and the highest expected SNR γ_w are required.

4. System Scenarios

Cellular interference not only corrupts the transmitted data but also the pilot symbols used for channel estimation. If a high frequency reuse factor r_d is to be employed,

Bandwidth	B	101.5 MHz
# subcarriers	N_c	769
FFT length	N_{FFT}	1024
Guard interval (GI) length	N_{GI}	268
Sample duration	T_{spl}	7.4 ns
Frame length	N_{frame}	64
Spreading Factor	L	16
Modulation	QPSK	
Channel coding rate	r	1/2
Parameters for mismatched channel estimator		
Pilot spacing freq. & time	$\{D_f, D_t\}$	$\{3, 9\}$
Filter dimension freq & time.	$\{M_f, M_t\}$	$\{13, 4\}$

Table 1. MC-CDMA system parameters

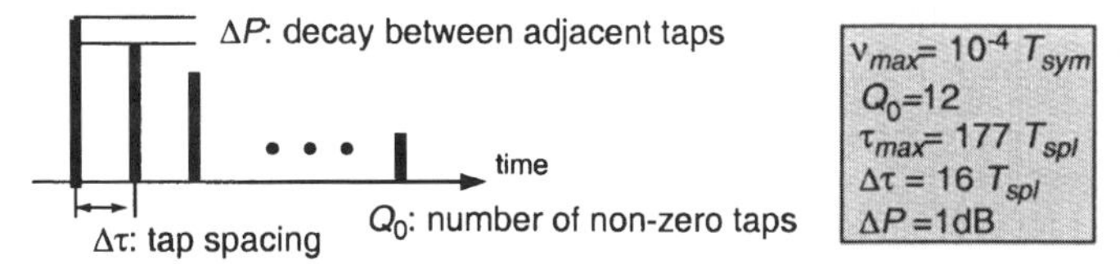

Fig. 4. The power delay profile of the used channel model.

the cellular interference near the cell boundaries is significant. While data symbols which can be protected by means of a processing gain and/or channel coding, the pilots cannot be protected in such a way. One way to mitigate this problem is to use a pilot reuse factor being larger than the frequency reuse factor for the data symbols [8], $r_p > r_d$. However, for interference free reception of the pilots, such a system requires full synchronization within all BS of the cellular systems, which may be difficult to realize in practice.

We compare two system scenarios: first, a cellular system with a pilot and data reuse factor of one, $r_p = r_d = 1$ (system A); second, a with a data reuse of one, $r_d = 1$, and a pilot reuse of three, $r_p = 3$ (system B). The advantages and drawbacks of both sytems are analysed. While the performance of System A will degrade near the cell boundaries due to strong cellular interfence, System B will be more sensitive to a synchronization offset.

5. SIMULATION RESULTS

The bit error rate (BER) performance of the cellular MC-CDMA system is evaluated by computer simulations. The system parameters of the MC-CDMA system and of the channel model were taken from [13], and are shown in Table 1. All BS are using exactly the same system parameters, i.e. the same spreading length L, number of active users, N_u, etc. This implies that if N_u decreases, the cellular interference also decreases. However, the difference in received signal power ΔE does remain constant. The channel is modeled by a tap delay line model with $Q_0 = 12$ taps, a tap spacing of $\Delta\tau = 16 \cdot T_{spl}$, with an exponential decaying power delay profile, illustrated in Fig. 4. The independent fading taps are generated using Jakes model having a U-shape Doppler power spectrum [14]. The maximum Doppler frequency of each tap was set to $\nu_{max} = 10^{-4} \cdot T_{sym}$, with T_{sym} defined in (6), corresponding to a mobile velocity of about 3 km/h @5 GHz carrier frequency. The parameters of the mismatched channel estimator are also depicted in Table 1. For the parameters of the robust channel estimator, also depicted in Table 1, we assume that the maximum delay of the channel, the maximum Doppler frequency and the average SNR is known to the receiver. A pilot spacing of $D_f = 3$ in frequency and $D_t = 9$ in time was used throughout. This induces a overhead of $r_p/26$ due to pilot symbols. Thus, System A with $r_p = 1$ has a pilot overhead of about 4% compared to an overhead of 12% for System B with $r_p = 3$.

Fig. 5.a shows the BER vs the difference in received signal power between two interfering BS, ΔE, for the considered MC-CDMA system with $N_R = 1$ and 2 receive antennas. It is seen that the difference between scenario A and B for $N_R = 1$ receive antenna is insignificant. However, for $N_R = 2$ receive antennas there is about 1.5 dB gain for scenario B where the pilots are received without interference.

Fig. 5.b shows the BER vs ΔE for a MC-CDMA system with a sync offset between the interfering BS and the mobile of $\Delta f^{(I)} = 0.5T$ and $\Delta n^{(I)} = 300$. With a synchronization offset the performance of scenario B performs somewhat better for $N_R = 1$ and 2 receive antennas. It appears that the interference levels at pilot positions of scenario B are still significantly lower than for scenario A. Thus, the synchronization requirements between two interfering BS are not as strict as between the transmitting BS and the mobile receiver. For instance on the link level the frequency offset between transmitter and receiver should not exeed 2–5% of the subcarrier spacing [15]. On the other hand, the frequency offset between two interfering BS can be significantly higher than that.

It is also interesting to note that the performance of the fully synchronized system in Fig. 5.a is similar to the performance of the system with a sync offset in Fig. 5.b. A difference occurs if pilot symbols are transmitted at a larger power than the data symbols. Then the performance degrades for the unsynchronized system. This is due to the fact that the system is not fully loaded, while the pilots are transmitted with full power. This causes the interference level to increase at data subcarriers in case of a sync offset.

6. Conclusions

The performance of a celluar MC-CDMA system on the downlink with 2×1-D PACE has been analysed. Particularly, the impact of a synchronization offset between the interfering BS and mobile receiver was taken into account. A synchronization offset between an interfering base station and the mobile has little effect on the system performance. Cellular interference causes degradation on the performance of the robust channel estimator if the pilots are not protected against the cellular interference. The degradation becomes larger as the number of receive antennas increases. While a pilot reuse factor larger than one can mitigate the interference of the pilot symbols, synchronization requirements of the cellular system are more stringent. However, the synchronization requirements between two interfering BS are not too stringent.

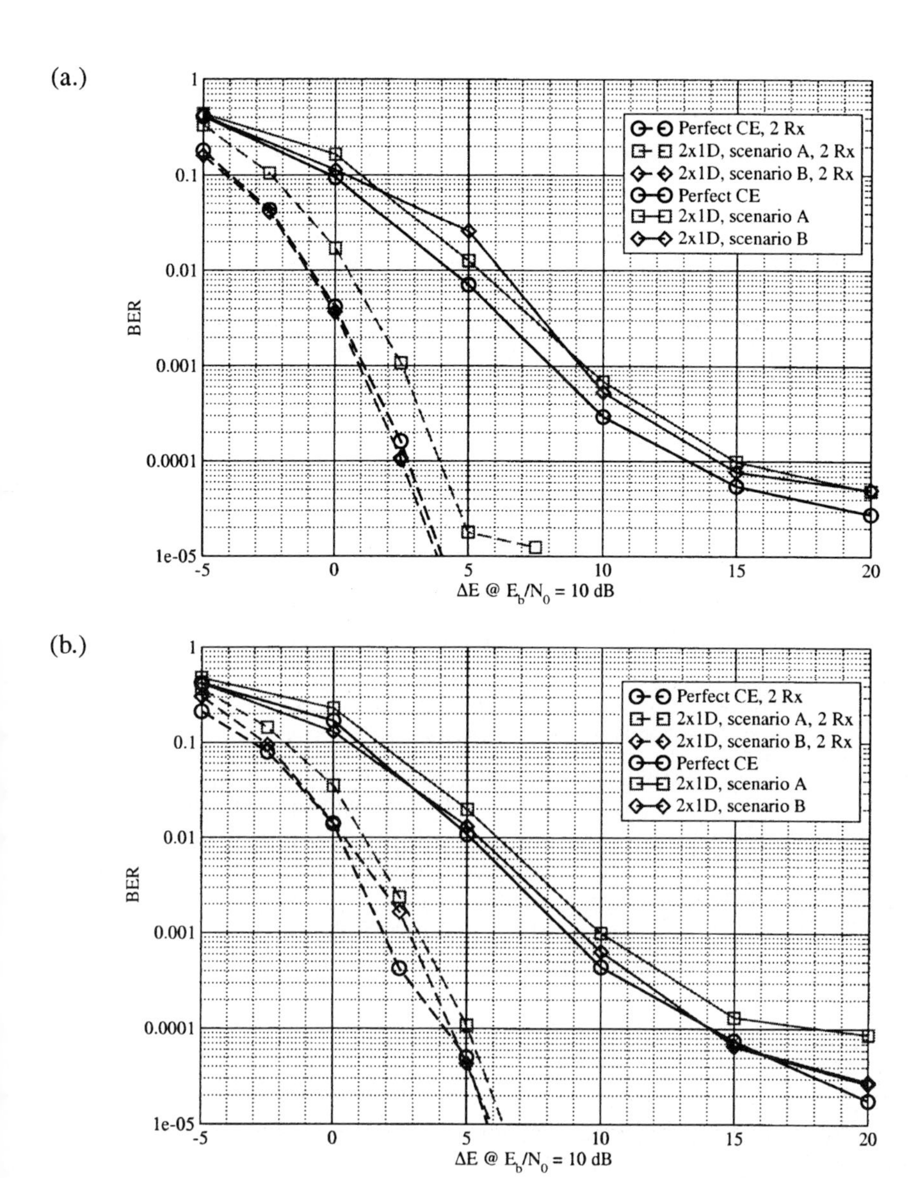

Fig. 5. *BER vs* ΔE $E_b/N_0 = 10\,dB$ *for MC-CDMA system with* $N_R = 1$ *and 2 receive antennas. The perfomance for* 2×1*-D PACE for scenario A and B are compared; results for perfect CSI are included as a lower bound. Part (a.): perfect synchronized system; part (b.): large sync offset (*$\Delta f^{(I)} = 0.5T$*,* $\Delta n^{(I)} = 300$*) between interfering BS and receiver.*

7. ACKNOLEDGEMENT

The authors would like to thank Eleftherios Karipidis from the Technical University in Munich (TUM) for his support in implementing the simulation platform.

8. AFFILIATION

Gunther Auer
DoCoMo Euro-Labs,
Landsberger Straße 312,
80687 München, Germany.
Email: auer@docomolab-euro.com

Armin Dammann, Stephan Sand
and Stefan Kaiser
German Aerospace Center (DLR),
Institute of Communications & Navigation,
82234 Wessling, Germany.

REFERENCES

[1] S. Weinstein and P. Ebert, "Data Transmission by Frequency Division Multiplexing Using the Discrete Fourier Transform," *IEEE Trans. Commun. Technol.*, vol. 19, pp. 628–634, Oct. 1971.

[2] K. Fazel and L. Papke, "On the Perfomrance of Convolutionally-Coded CDMA/OFDM for Mobile Communication Systems," in *Proc. IEEE Int. Symp. Personal, Indoor and Mobile Radio Commun. (PIMRC'93), Yokohama, Japan*, pp. 468–472, Sep. 1993.

[3] S. Hara and R. Prasad, "Overview of Multicarrier CDMA," *IEEE Commun. Mag.*, pp. 126–133, Dec. 1997.

[4] S. Kaiser, *Multi-Carrier CDMA Mobile Radio systems — Analysis and Optimisation of Detection, Decoding, and Channel Estimation.* PhD thesis, German Aerospace Center (DLR), Oberpfaffenhofen, Germany, Jan. 1998.

[5] S. Abeta, H. Atarashi, and M. Sawahashi, "Performance of Coherent Multi-Carrier/DS-CDMA and MC-CDMA for Broadband Packet Wireless Access," *IEICE Transactions on Communications*, vol. E84-B, pp. 406–414, Mar. 2001.

[6] H. Atarashi, S. Abeta, and M. Sawahashi, "Broadband Packet Wireless Access Appropriate for High-Speed and High-Capacity Throughput," in *Proc. IEEE Vehic. Technol. Conf. (VTC'2001-Spring), Rhodes, Greece*, May 2001.

[7] P. Hoeher, S. Kaiser, and P. Robertson, "Pilot-Symbol-Aided Channel Estimation in Time and Frequency," in *Proc. Communication Theory Mini-Conf. (CTMC) within IEEE Global Telecommun. Conf. (GLOBECOM'97), Phoenix, USA*, pp. 90–96, 1997.

[8] Z. Wang and R. Stirling-Gallacher, "Frequency Reuse Scheme for Cellular OFDM Systems," *IEE Electronics Letters*, vol. 38, pp. 387–388, Apr. 2002.

[9] J. G. Proakis, *Digital Communications.* New York: McGraw-Hill, NY, USA, 3rd ed., 1995.

[10] M. Speth, S. Fechtel, G. Fock, and H. Meyr, "Optimum Receiver Design for Wireless Broad-Band Systems Using OFDM—Part I," *IEEE Trans. Commun.*, vol. COM-47, pp. 1668–1677, Nov. 1999.

[11] G. Auer, A. Dammann, S. Sand, and S. Kaiser, "On Modelling Cellular Interference for Multi-Carrier based Communication Systems Including a Synchronization Offset," in *Proc. Int. Symp. Wireless Personal Multimedia Commun. (WPMC'2003), Yokosuka, Japan*, Oct. 2003.

[12] J. K. Cavers, "An Analysis of Pilot Symbol Assisted Modulation for Rayleigh Fading Channels," *IEEE Trans. Vehic. Technol.*, vol. VT-40, pp. 686–693, Nov. 1991.

[13] H. Atarashi, N. Maeda, S. Abeta, and M. Sawahashi, "Broadband Packet Wireless Access Based on VSF-OFCDM and MC/DS-CDMA," in *Proc. IEEE Int. Symp. Personal, Indoor and Mobile Radio Commun. (PIMRC 2002), Lisbon , Portugal*, pp. 992–996, Sep. 2002.

[14] W. C. Jakes, *Microwave Mobile Communications.* Wiley, NY, 1974.

[15] P. Moose, "A Technique for Orthogonal Frequency Division Multiplexing Frequency Offset Correction," *IEEE Trans. Commun.*, vol. COM-42, pp. 2908–2914, Oct. 1194.

MATÚŠ TURCSÁNY, PETER FARKAŠ

NEW 2D-MC-DS-SS-CDMA TECHNIQUES BASED ON TWO-DIMENSIONAL ORTHOGONAL COMPLETE COMPLEMENTARY CODES

Abstract. In this paper we present some new concepts of 2D-MC-DS-SS-CDMA techniques based on two-dimensional orthogonal complete complementary codes (2D-OCCC). General benefits of 2D spreading in the time-frequency space include lower power spectral density, thus better low probability of detection (LPD) and low probability of intercept (LPI) property, higher jam resistance, etc. Additional benefits of CCC include features like offset stacked spreading, MAI free system and a high spectral efficiency. 2D-OCCC provides a wider range of freedom in positioning the code elements in a channel when compared to 1D-OCCC. The proposed systems and techniques combine the advantages of 2D spreading with the advantages of 2D-OCCC. Because 2D-OCCC has special requirements on transmission channels, we explored several possibilities of 2D channel definitions. According to these, new transmission strategies are proposed. Also partial solutions of the limited user count problem using 2D-OCCC are described.

1. INTRODUCTION

In [1] a new CDMA architecture for systems beyond 3G was proposed. It is based on orthogonal complete complementary codes (CCC) [2]. The orthogonality of CCC is based not on single sequences, like it is the case for example for Walsh sequences, but instead there is a signature of sequences termed elements, which preserves that different signatures are orthogonal to each other. Every user in a system gets assigned a different signature (in [1] termed flock). In order to ensure orthogonality between users, every user has to transmit all signature elements via different channels.

It is obvious, that multirate transmission can be easily achieved just by delaying the transmission of the next bit by more than one chip. Therefore there is no need for solving the well known rate matching problem (selection of sequences of proper length for a given rate). It is also clear, that the processing gain remains constant, independent on the bit rate.

Simultaneously, there is a lot of development progress in the area of multicarrier CDMA (MC-CDMA) or OFDM-CDMA. The advantages of MC-CDMA over DS-CDMA in fading channels were proofed for example in [3], while DS-CDMA has a better anti-MAI capability [4]. When both approaches are combined, the resulting architecture has an enhanced multiple access capability and due to frequency diversity also performs well in fading channels. Such a system was proposed in [4]. For spreading in the time domain as well as in the frequency domain Walsh sequences were used, what makes this system suitable only for synchronized downlink, other way the orthogonality property is lost.

K. Fazel and S. Kaiser (eds.), Multi-Carrier Spread-Spectrum, 49–56.

This paper presents a new 2D-MC-DS-SS-CDMA architecture based on 2D-OCCC [5, 6]. The architecture proposed in [1] could be viewed as a pseudo multicarrier spread spectrum one, because the spreading is done only in the time domain in two or more parallel frequency channels. We extended this concept into a 2D space by introducing the new 2D-OCCC. Now, the spreading could be done in 2D channel in time and frequency domain simultaneously. This is realizable only by a multicarrier direct sequences spread spectrum system similar to [4]. In comparison with [4], the techniques proposed in this paper exploit all features of 2D-OCCC such as offset stacked spreading leading to higher spectral efficiency, elimination of the well known rate-matching problem, excellent correlation properties even in asynchronous environments [7] (by analyses based on modified criteria as proposed in [11]), etc. In comparison with [1], the new architecture is a true multicarrier-CDMA with all the benefits mentioned earlier.

The organization of this paper is following. In Section 2, there is a brief overview of 2D-OCCC and their properties. In Section some 2D channel examples are presented. In Section 4 the basic transmission strategies are described. In Section 6, some partial solutions to the problem of limited number of users in multiple access system are overviewed. Final conclusions are presented in Section 7.

2. TWO-DIMENSIONAL ORTHOGONAL COMPLETE COMPLEMENTARY CODES

Implementation of the 2D-MC-DS-SS-CDMA system proposed would not be possible without the 2D spreading sequences. Now we will give one example of 2D-OCCC. The synthesis of this 2D-OCCC is based on 1D-OCCC [2].

Basic property of OCCC is that the sum of autocorrelation functions of all elements within a signature is equal to zero for all nonzero shifts and the sum of cross correlation functions for all corresponding elements from every two distinct signatures is equal to zero for all shifts. This feature ensures a MAI-free system in synchronous environments and it enables offset stacked spreading. The idea behind it is that the transmission of the next sequence starts immediately after transmission of the first chip of the previous sequence. Therefore the transmitter output is a multilevel signal and the transmission is realizable only by a multilevel digital modulation scheme. It is obvious, that such overlapping increases the spectral efficiency. The rate matching problem is solved easily. If a lower rate is required, more chips are left between two consecutive sequences (there is no need to select a different sequence of a suitable length). The processing gain remains constant (rate independent) when the information rate changes.

The orthogonality of CCC is based not on single sequences like it is the case for example for Walsh sequences. Instead there is a signature composed of more sequences termed elements. In multiple access system every user gets assigned a different signature (in [1] termed flock). In order to ensure orthogonality between users, every user has to transmit all signature elements via different channels. The nature of the construction in this example requires, that if we increase the number of signatures (users) we also need to increase the number of sequences in each signature. This property implies three obvious disadvantages. The first problem is that the system complexity increases. The second problem is that the spectral efficiency in bit/chip for one user is decreasing. The third is, that the number of prototypes in the multilevel scheme increases as well because of offset stacked spreading, the transmitter output is a multilevel signal with $(N^2 + 1)$ possible levels, where N is the maximum user count. This has a great impact on the modulation scheme used. Two schemes seem to be usable: MASK [8] or M-ary QAM/PSK. Considered N being an integer power of 2, there is a need for up-mapping or down-mapping of the signal levels [9]. The process of up-mapping results in a less robust modulation scheme more susceptible to errors, while the application of down-mapping causes some loss of orthogonality property.

3. CHANNEL DEFINITIONS

To take all the advantage offered by the new classes of 2D-CCC, a new radio channel design has to be considered. It is obvious, that if a two-dimensional spreading code is used, the channel must be also considered as two-dimensional.

For simplicity let us assume, that these 2 dimensions are time and frequency. Considered the orthogonality of CCC is based on elements, which have to be transmitted via different channels, we face the problem of defining fully separated transmission channels. We solved the need for dividing the time-frequency space into clearly separated planes – channels in two fundamental ways. The first solution is the division of the time axis - time planes are the result and the second solution divides the frequency axis - frequency planes originate. More detailed descriptions of both solutions are following in next subsections.

3.1. Macro-TDM

Let N be the number of elements in each flock. The time axis would be divided into fixed width slots, where N slots would comprehend one frame. There is a direct mapping between time slots and channels in every frame – one slot represents one channel. The same position of one slot in different frames is always assumed as one channel. The corresponding elements of each flock would be assigned to one channel. The slot width should be defined allowing to transmit more than one information bit. Let us assume that L bits can be transmitted in one interval. The system would work as follows. In the first slot, every user with transmission ready data will transmit L bits spread by the first element of his flock. In the second slot, the same L bits would be spread by the second element of the corresponding flock. In the last, N-th slot, the same bits L would again be spread by the last, N-th, element of the corresponding flock. The receiver would get the complete L bits not until the last spreaded bit (spreaded with the last element) in the last slot would be received and despread. Then every user would transmit his next L bits using the same procedure.

The advantage of this approach is that all users can make use of the whole available bandwidth in every channel – the possible frequency shifts would be limited only by the system bandwidth. The disadvantage is obvious, the information rate decreases by the factor N.

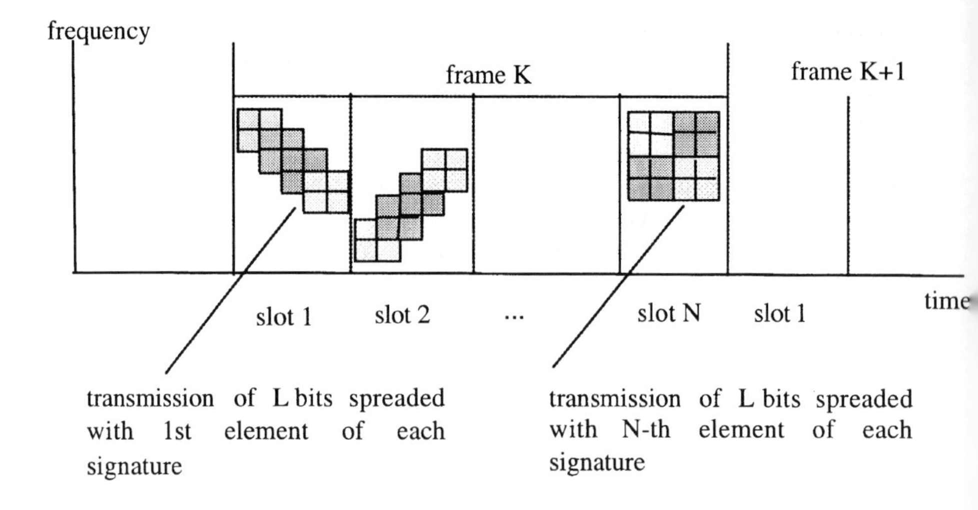

Figure 1. Macro-TDM principle.

The advantage of this approach is that all users can make use of the whole available bandwidth in every channel – the possible frequency shifts would be limited only by the system bandwidth. The disadvantage is obvious, the information rate decreases by the factor N. There is also a loss of efficiency on boundaries between slots, because the elements of signatures cannot overlap across these boundaries. However, these losses could be decreased at the expense of delay by widening the time slots.

3.2. Macro-FDM

Plane definition in the continuos time-frequency space is based unlike the first approach on dividing the frequency axis. The elements would be transmitted parallel and continuously in time, each assigned a fixed width frequency plane wide enough to allow for some frequency shifts of the transmitted matrices within. The advantage lies in the parallel and continuos transmission, the drawback is the limited spectrum available for every channel, thus the limited number of possible shifts in the frequency domain. This architecture is the generalized scheme of the multicarrier CDMA system described in [1].

4. BASIC SYSTEM DESCRIPTION

The introduction of second dimension greatly increased the possibilities of system design. While in [1], the positioning of the individual elements was limited along the time axis, we have the ability to position them anywhere in a 2D time-frequency space.

The transmitter has to be able to perform simultaneous spreading of every bit in both dimensions and then, according to the selected transmission strategy, combine the individual chips of the relevant bits and their elements. This process could be viewed as a matrix operation, where at first every element matrix of an assigned signature (every user has a different one) is multiplied by the information bit. This results in a number of matrices, each intended to be transmitted via different channel. The first column of every matrix would then be transmitted (one corresponding chip on given set of frequencies). Depending of the available amount of spectrum, various transmission strategies can be now employed. The individual matrices could overlap in any dimension and area. For example, parallel time transmis sion could be used to increase the information rate.

If no offset stacked spreading would be used, then the transmitted signal (*s*-th spreaded bit *D*) $u_m^i(t)$ of *m*-th user ($m = 1, 2, ..., N$) with his *i*-th element is defined as:

$$u_m^i(t) = D_s \sum_{k=1}^{N} \left[\sum_{l=1}^{N} c_{i,(k,l)}^m \, p_{Tc}(t - l \cdot T_c) \right] e^{j 2\pi \frac{1}{Tc} tk} \tag{1}$$

where $c_{i,(k,l)}^m$ is the chip in *k*-th row and *l*-th column in the *i*-th element of *m*-th signature $p_{Tc}(t)$ is the pulse shaping waveform and T_c denotes the chip period.

For maximum spectrum utilization, we now consider a parallel time transmission of *K* bits with maximum offset stacked spreading (one chip shift). A 2D-OCCC of order *N* is used. There are *N* signatures (users), each containing *N* elements (matrices) of size *N* x *N*. If *K* parallel bits are transmitted in uplink using maximum offset stacked spreading, then $N + (K-1)$ different frequencies are required for every sub channel (transmission of every element). Combiner block in the transmitter performs the combining of relevant chips for every element into a column vector $\mathbf{v}_p$, where all elements of that vector ($v_{p,i}$) are transmitted parallel in time, each on a different frequency. There have to be *N* parallel units in a combiner block so that *N* different vectors $\mathbf{v}_p$ can be transmitted simultaneously.

Synthesis of vector $\mathbf{v}_p$ is given by:

$$\mathbf{v}_p = \begin{bmatrix} v_{p,1} \\ v_{p,2} \\ \vdots \\ v_{p,N+(K-1)} \end{bmatrix} \tag{2}$$

$$v_{p,i} = \sum_{j=0}^{i-1} w_{j,i-j}^{p} \tag{3}$$

where $v_{p,i}$ is the i-th element of vector $\mathbf{v}_p$, i = 1, 2, ... N + (K-1) and $w_{x,y}^{p}$ is defined as:

$$w_{x,y}^{p} = c(r+x)_{y,N}^{p} + c(r+x+K)_{y,N-1}^{p} + c(r+x+2K)_{y,N-2}^{p} + \ldots + c(r+x+[N-1]K)_{y,1}^{p} \tag{4}$$

where $c(q)_{a,b}^{p}$ denotes the chip in the a-th row and b-th column of the p-th element matrix multiplied by q-th information bit, $c(q)_{a,b}^{p} = \{\pm 1\}$ if binary 2D-OCCC is used, $r = nK + 1$, where n is a non negative integer.
If $(x = N)$ *or* $(y > N)$ then $w_{x,y}^{p} = 0$ by definition.

Let **V** be $[N+(K-1)]$ x N matrix:

$$\mathbf{V} = [\mathbf{v}_1, \ \mathbf{v}_2, \ \ldots, \ \mathbf{v}_N] \tag{5}$$

We will call this matrix a combiner transmission matrix. Every Tc seconds (Tc denotes the chip time) matrix **V** is updated and each column, vector $\mathbf{v}_p$, is transmitted via different subchannel.

The above described procedure maximizes the 2D space utilization but it does apply to all transmission strategies in general (only the definition of $w_{x,y}^{p}$ changes).

The following is a more general view on 2D spreading by considering the 2D space neither as time-frequency or space-time nor any real space, but as an imaginary 2D spreading space. The earlier mentioned combiner block performs spreading in this spreading space independent on the physical implementation. As was shown, the combiner block output is only one matrix, what greatly simplifies the system design. We can now approach the implementation problem as a simple MC-DS-CDMA system. For example, every vector is parallel to serial converted and then transmitted on a distinct frequency/time slot. To satisfy the condition of chip time equality Tc, transmission duration of one vector element, $v_{p,i}$, would be reduced to $\frac{Tc}{N+K-1}$ in case of micro-FDM (see below) or $\frac{Tc}{N}$ in case of micro-TDM (see below). The problem of 2D channel definitions is also no more relevant. Because the whole functionality is hidden in the combiner block, changes in transmission strategy can be therefore easily and adaptively executed only by a software modification of the combiner block algorithm.

We will call a scheme MICRO-TDM if every row of the transmission matrix is transmitted on a different frequency. The most obvious problem of micro-TDM is time synchronization between the users. Sub channels are defined by a time-slot and to ensure orthogonality between users, the position of the different user's time slots must match (only the corresponding vectors can overlap).

We will call a scheme MICRO-FDM if every column of the transmission matrix is transmitted on a different frequency. This approach has no synchronization requirements in terms of sub channel matching – there is no problem in synchronizing frequencies.

5. INCREASING THE NUMBER OF USERS

One of the basic ideas which will help to increase the number of communicating parties is a modification of that presented in [10] for MC-SS system. Simply it could be expressed as using identical signatures for some or all users in a CDMA system from the sets given by 2D-OCCC.

Other possibility to increase the maximum user count is the introduction of quasi orthogonal 2D-CCC (2D-QOCCC) [12]. The 2D-QOCCC solves the problem in trading some orthogonality property for several times reduced number of elements (channels) while enlarging the signature count (maximum user count) by the same factor. For some vertical shifts within the zero horizontal shift there will be some nonzero values for both auto and cross correlation, but these shifts could be simply forbidden by protocol.

6. CONCLUSION

In this paper, some new 2D-MC-DS-SS-CDMA techniques based on two-dimensional orthogonal complete complementary codes were presented. There are a number of miscellaneous transmission strategies where some of them are suitable only for specific architectures while others are universally applicable. Similarly, different advantages apply for various techniques making these attractive for a larger scale of communication systems. Some architectures and strategies are suitable for highly hostile channels or for use in asynchronous environments. This was confirmed by analyses based on modified criteria as proposed in [11].

It can be expected, that some of these techniques could find their application in 4G mobile networks, particularly in PANs or secure spread spectrum communication systems.

7. REFERENCES

[1] Chen H. H., Yeh J. F., Suehiro N., "A multicarrier CDMA architecture based on orthogonal complete complementary codes for new generations of wideband wireless communications," IEEE Communications magazine, vol. 39, October 2001, pp. 126-134.

[2] Suehiro N., Hatori M.,"N-shift cross-orthogonal sequences," IEEE: Trans. Info. Theory, vol. IT-34, no. 1, Jan. 1988, pp. 143-146.

[3] Kaiser, S., "OFDM-CDMA versus DS-CDMA: Performance Evaluation for Fading Channels," Proc. IEEE Intern. Conf. on Commun. (ICC'95), Seattle, USA, pp. 1722-1726, June 1995.

[4] Xiao L., Liang Q., "A novel MC-2D-CDMA communication system and its detection methods," Proc. IEEE ICC, 2000, June 2000, pp. 1223-1227.
[5] Farkaš P., Turcsány M., "Invention Reports on 2DCCC", SIEMENS, 12. 3. 2003.
[6] Farkaš P., Turcsány M., "Two-dimensional Orthogonal Complete Complementary Codes," accepted for presentation in SympoTIC 2003, October 2003.
[7] Turcsány M., Farkaš P., "On Some Properties of Two-dimensional Orthogonal Complete Complementary Codes", submitted to DSPCS 2003, August 2003.
[8] Chen H. H., Suehiro N., Yeh J. F., Kuroyanagi N., Nakamura M., "Simulation of a Parallel Transmission System for Multipath Property to Estimate Pilot Signals and Additional Chip-Shifted Information-Transmission Signals",
[9] Ong K.B., Darnell M., Boussakta S., "A New Approach to Combining Complementary Sequences for Multi-Carrier CDMA Systems", pp. 293-298.
[10] Aue V., Fettweis G., "Multi-carrier spread spectrum modulation with reduced dynamic range," Proc. VTC'96, Atlanta, Apr./May 1996, pp. 914-917.
[11] Popovic B. M., "Spreading sequences for multicarrier CDMA system," IEEE Trans. On Commun. ,vol. 47, no. 6, June 1999, pp. 918-926.
[12] Turcsány M., Farkaš P., "Two-dimensional Quasi Orthogonal Complete Complementary Codes", accepted for presentation in SympoTIC 2003, October 2003.

8. AFFILIATIONS

Peter Farkaš (p.farkas@ieee.org) and Matúš Turcsány (turcsany@pobox.sk) are with Dept. of Telecommunications, Faculty of Electrical Engineering and Information Technology, Slovak University of Technology, Ilkovicova 3, 812 19 Bratislava, Slovakia and the second author is also with SIEMENS PSE, Slovakia, Stromová 9, 830 07 Bratislava, Slovakia.

ZEXIAN LI AND MATTI LATVA-AHO

PERFORMANCE OF MULTIRATE TRANSMISSION SCHEMES FOR MC-CDMA SYSTEMS

Abstract. Multicarrier code-division multiple-access (MC-CDMA) is a potentially attractive multiple access technique for future wireless communication systems. A unified multirate MC-CDMA system model able to fit different multirate transmission schemes is presented. The performance of the multicode system with the same processing gain and the variable spreading factor (VSF) system is compared. Based on the characteristic function (CF) of a complex Gaussian random vector (CGRV), we present a bit error rate (BER) analysis method. The accuracy of the method is verified by computer simulations in frequency selective Rayleigh fading channels. Different subcarrier interleaving schemes for VSF MC-CDMA are also investigated. It is shown that if the gain resulting from the subcarrier interleaving is taken into account, VSF MC-CDMA and multicode MC-CDMA offer quite similar performance.

1. INTRODUCTION

Multicarrier code-division multiple-access (MC-CDMA) has been viewed as a promising technique for future wireless communication services due to its advantages such as insensitivity to frequency selective channels, efficient utilization of bandwidth and flexibility to generate different data rates [1]- [3]. The interest for high data rate wireless services such as data, image and video means that future generation wireless communication systems must be able to cope efficiently with heterogeneous traffic. To serve sources with inherently different information rate in a system, it is desirable to develop systems that operate with multiple data rates.

As in direct-sequence CDMA (DS-CDMA), there are several multiplexing strategies to design a *multirate* multicarrier communication system [4]. In an MC-CDMA system, probably the most straightforward way is to design a standard MC-CDMA system and allocate several parallel communication channels, i.e., several spreading sequences, to higher rate users (multicode MC-CDMA). Another option is to use the variable spreading factor (VSF MC-CDMA) technique. Different from [5], here in VSF MC-CDMA all users are allocated the same bandwidth and the same number of subcarriers, but the processing gain of high rate users would be smaller than that of lower rate users so that the high rate users can transmit more symbols in a given time. This method implies that part of the high rate symbols with small processing gain will be orthogonal to each other in frequency domain.

In this paper, both the multicode scheme and the VSF scheme with different subcarrier interleavers are considered for a downlink MC-CDMA system with minimum mean squared error (MMSE) receiver. The bit error rate (BER) is obtained by analysis and simulations in a correlated Rayleigh fading channel. The analytical BER is derived by invoking the Gaussian approximation and the characteristic function (CF) of a

K. Fazel and S. Kaiser (eds.), Multi-Carrier Spread-Spectrum, 57–64.

complex Gaussian random vector (CGRV). It is shown that the two different multirate schemes achieve close performance if the interleaving gain is taken into account.

Notations: Column vectors (matrices) are denoted by the boldface lower (upper) case letters. The superscripts $(\cdot)^T$, $(\cdot)^*$, $(\cdot)^H$, and $(\cdot)^\dagger$ stand for transpose, complex conjugate, complex conjugate transpose and matrix inverse, respectively. $E\{\cdot\}$ denotes the statistical expectation, and $\mathbf{I}_M$ the $M \times M$ identity matrix.

2. SYSTEM MODEL AND RECEIVER DESIGN

Consider a downlink multirate MC-CDMA system with BPSK modulation and constant chip duration T_c. An MC-CDMA system transmits identical information symbols through different subcarriers. The total number of subcarriers is N. Suppose that the multirate users are classified into G groups with different data rates and let the number of users and the transmission rate of a user in the gth group be denoted by K_g and R_g, $g = 1, 2, \cdots, G$, respectively. The spreading factor for the users in the gth group is SF_g which equals to N_g ($SF_1 = N$), the number of subcarriers through which the same data symbol is transmitted. Assume that R_g is a multiple of R_1, i.e., $R_g = L_g R_1$, with the restriction that $1 = L_1 < L_2 \cdots < L_G$. Users are indexed by two variables: g indicate the group number and k indicate the user number in the group. There are total $K = K_1 + K_2 + \cdots + K_G$ users which are grouped into G groups according to different data rates.

We consider correlated, frequency selective Rayleigh fading channels for each user [6]. It is reasonable to assume that the narrowband signal transmitted through each subcarrier experiences a frequency flat fading channel. The transfer function of the fading associated with the nth ($n = 1, \cdots, N$) subcarrier can be expressed as

$$H_n = \beta_n e^{j\theta_n} \tag{1}$$

where H_n is zero mean complex Gaussian random variable (r.v.). The amplitude β_n is Rayleigh distributed with $E\left[(\beta_n)^2\right] = 2\sigma^2 = 1$ and the random phase θ_n is uniformly distributed over $[0, 2\pi)$. The amount of correlation between the fading at different subcarriers depends on the separation of subcarriers frequencies [7, Chap. 1]. The covariance matrix $\mathbf{C}_h$ of the complex Gaussian r.v.s of N subcarriers can be obtained with the (l, n)th element as [8,9]

$$\rho(l, n) = \frac{2\sigma^2 \left(1 + jh\Delta\omega T_d\right)}{1 + \left(h\Delta\omega T_d\right)^2} \qquad h = n - l, \quad (l, n = 1, \cdots, N) \tag{2}$$

in which T_d is the delay spread of the fading channel and $\Delta\omega$ is the angular frequency separation between adjacent subcarriers, i.e., $\Delta\omega_{l,n} = \Delta\omega$ if $|l - n| = 1$.

2.1 Received Signal Model

In the multirate MC-CDMA, a single user in the gth group can be viewed as L_g *virtual* users at the lowest rate. Let us consider the received vector at time interval m $\mathbf{y}(m) = [y_1(m), \cdots, y_N(m)]^T \in \mathbb{C}^N$. Let $\mathbf{n}(m) \in \mathbb{C}^N$ be the noise vector. At the receiving antenna, sampling the received frequency domain signal at chip rate yields the received vector

$$\mathbf{y}(m) = \sum_{g=1}^{G} \sum_{k_g=1}^{K_g} \mathbf{r}_{k_g}(m) + \mathbf{n}(m) \tag{3}$$

with $\mathbf{r}_{k_g}(m) = \mathcal{H}_{k_g} \mathbf{b}_{k_g}(m)$ and $\mathbf{b}_{k_g}(m)$ is the mth modulated symbol vector ($\in \mathbb{Z}^{L_g}$) with elements $\{b_{k_g}^{i_g}(m)\}$ from the k_gth user, it is assumed that the information sequence $\{\mathbf{b}_{k_g}(m)\}$ is a collection of independent equiprobable random variables (± 1) with $E\left[b_{k_g}^{i_g}(m)\right] = 0$ and $E\left[|b_{k_g}^{i_g}(m)|^2\right] = 1$.

The definition of $\mathcal{H}_{k_g}$ is different for multicode MC-CDMA and VSF MC-CDMA. In multicode systems, multiple data streams are sent in parallel by high rate users. $\mathcal{H}_{k_g}$ can be obtained as

$$\begin{aligned} \mathcal{H}_{k_g} &= \left[\mathcal{H}_{k_g}^1, \mathcal{H}_{k_g}^2, \cdots, \mathcal{H}_{k_g}^{i_g}, \cdots, \mathcal{H}_{k_g}^{L_g}\right] \\ \mathcal{H}_{k_g}^{i_g} &= \left[\mathbf{C}_{k_g}^{i_g} \mathbf{h}_{k_g}^{(i_g)}\right] \qquad i_g = 1, \cdots, L_g \end{aligned} \tag{4}$$

in which $\mathbf{h}_{k_g}^{(i_g)}$ is the fading vector and $\mathbf{C}_{k_g}^{i_g}$ is a matrix with normalized spreading code of the i_g symbol from the k_gth user on its diagonal, i.e., $\mathbf{C}_{k_g}^{i_g} = (1/\sqrt{SF_g})\text{diag}$ $(c_{k_g}^{i_g}(1), \cdots, c_{k_g}^{i_g}(SF_g))$, $c_{k_g}^{i_g}(n) \in \{\pm 1\}$, $SF_g = SF_1 = N$ in a multicode MC-CDMA. However, in the case of VSF MC-CDMA, $SF_g = SF_1/L_g$. $\mathcal{H}_{k_g}$ can be derived as

$$\begin{aligned} \mathcal{H}_{k_g} &= \left[\mathcal{H}_{k_g}^1, \mathcal{H}_{k_g}^2, \cdots, \mathcal{H}_{k_g}^{i_g}, \cdots, \mathcal{H}_{k_g}^{L_g}\right] \\ \mathcal{H}_{k_g}^{i_g} &= \left[\underbrace{\mathbf{0}}_{(i_g-1)\times SF_g}, \underbrace{\left(\mathbf{C}_{k_g}^{i_g} \mathbf{h}_{k_g}^{i_g}\right)^T}_{1\times SF_g} \underbrace{\mathbf{0}}_{(L_g-i_g)\times SF_g}\right]^T \end{aligned} \tag{5}$$

In (5) it is assumed that the same data symbol occupies the adjacent subcarriers. However, in VSF MC-CDMA, since multiple data symbols are transmitted in one OFDM symbol, different subcarrier allocation schemes exist for high rate users. The transmitter block diagram of VSF MC-CDMA with subcarrier interleaver is depicted in Fig. 1 from which we can see that the subcarrier interleaver is employed to determine the allocation schemes. Compared to the conventional MC-CDMA transmitter, the only difference is the subcarrier interleaver. From this figure it is clear that $\mathcal{H}_{k_g}^{i_g}$ is determined by the subcarrier interleaver.

Based on (3)-(5), it is worth noting that the multirate system is equivalent to a single rate system with K_{equ} ($K_{equ} = K_1 + L_2 K_2 + \cdots + L_G K_G$) *virtual* users and K_{equ} equals the total number of data symbols transmitted at the same time in the system. Eq. (3) can be written in a matrix form as

$$\mathbf{y}(m) = \mathcal{H}\mathbf{b}(m) + \mathbf{n}(m) \tag{6}$$

with $\mathcal{H} = \left[\mathcal{H}_{1_1}, \mathcal{H}_{1_2}, \cdots, \mathcal{H}_{k_g}^{i_g}, \cdots, \mathcal{H}_{K_G}^{L_G}\right]$ and $\mathbf{b}(m) = \left[\mathbf{b}_{1_1}^T(m), \cdots, \mathbf{b}_{K_G}^T(m)\right]^T$. Hence, the multirate MC-CDMA system can be written in a unified expression as (6). For brevity, the time index m is omitted in the following.

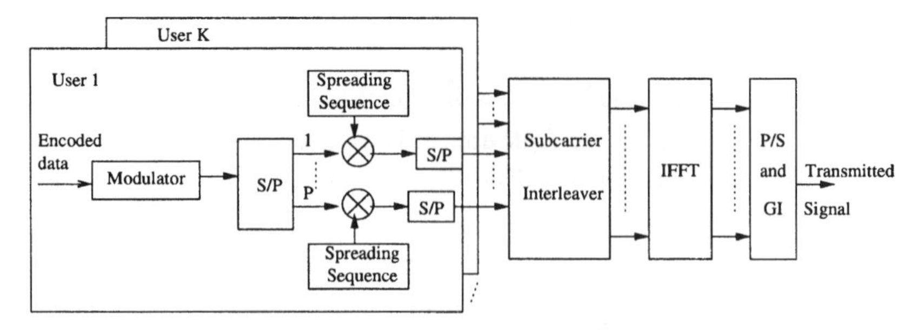

Figure 1. Block diagram of VSF MC-CDMA transmitter.

2.2 *Receiver Design*

Minimum mean squared error (MMSE) criterion is adopted to obtain the estimates of the data symbols. The tap weights are chosen to minimize the elements of the mean squared error (MSE) vector $\mathbf{J} = E\left[\left|\mathbf{b} - \hat{\mathbf{b}}\right|^2\right]$ where $\hat{\mathbf{b}} = \text{sgn}\left[\text{Re}\left(\mathbf{W}^H\mathbf{y}\right)\right]$ and $\mathbf{W}$ denotes the weight matrix for different users at different subcarriers. It is well known that the tap weight which minimizes the MSE is given by the Wiener solution [10]

$$\mathbf{W} = \mathbf{R}^{\dagger}\mathbf{P} = \left(E\left[\mathbf{y}\mathbf{y}^H\right]\right)^{\dagger} E\left[\mathbf{y}\mathbf{b}^H\right] \tag{7}$$

where the crosscorrelation matrix $\mathbf{P}$ between the received signal and the desired response and the autocorrelation matrix of $\mathbf{y}$ are, respectively, defined as

$$\begin{aligned} \mathbf{P} &= E\left\{\mathbf{y}\mathbf{b}^H\right\} = \mathcal{H} \\ \mathbf{R} &= E\left\{\mathbf{y}\mathbf{y}^H\right\} = \mathcal{H}\mathcal{H}^H + \sigma^2\mathbf{I}_N \end{aligned} \tag{8}$$

The assumption that the data and the noise are independent was used when deriving (8).

3. PERFORMANCE ANALYSIS

Without loss of generality, the data symbol b_{1_1} from the first user of the first group ($SF_1 = N, L_1 = 1$) is taken as an example for analysis. The decision variable is given by

$$z = \mathbf{w}_{1_1}^H \mathcal{H}_{1_1} b_{1_1} + \mathbf{w}_{1_1}^H \tilde{\mathbf{r}} = v_{1_1} b_{1_1} + \tilde{n} \tag{9}$$

where $\tilde{\mathbf{r}}$ represents the received interference plus noise and $\tilde{n}$ denotes the residual interference plus noise at the output of the MMSE filter. By using the Gaussian approximation [11, 12], the variance of $\tilde{n}$ can be written as $\tilde{\sigma}^2 = \mathbf{w}_{1_1}^H E\left[\tilde{\mathbf{r}}\tilde{\mathbf{r}}^H\right]\mathbf{w}_{1_1} =$

$\mathcal{H}_{1_1}^H \mathbf{R}^\dagger \tilde{\mathbf{R}} \mathbf{R}^\dagger \mathcal{H}_{1_1}$. The instantaneous signal to noise plus interference (SINR) can be obtained as [13]

$$\gamma = \frac{v_{1_1}^2}{\tilde{\sigma}^2} = \mathcal{H}_{1_1}^H \tilde{\mathbf{R}}^\dagger \mathcal{H}_{1_1} \tag{10}$$

By invoking the Gaussian approximation, z follows Gaussian distribution conditioned on the fading channel, spreading sequences and information bits of all users. The conditional BER with BPSK modulation is simply given by [14]

$$Pr(\text{error}) = \frac{1}{2}\text{erfc}\left(\sqrt{\frac{v_{1_1}^2}{2\tilde{\sigma}^2}}\right) \tag{11}$$

In order to get the average BER, it is necessary to average the conditional BER over the probability density function (PDF) of instantaneous SINR. We resort to the characteristic function (CF) of the conditional SINR to solve the problem.

Define $\mathbf{h} = [H_1, H_2, \cdots, H_N]^T$, then $\mathcal{H}_{1_1} = \mathbf{C}_{1_1}\mathbf{h}$ and the instantaneous SINR can be written as

$$\gamma = (\mathbf{C}_{1_1}\mathbf{h})^H \tilde{\mathbf{R}}^\dagger (\mathbf{C}_{1_1}\mathbf{h}) = \mathbf{h}^H \mathbf{Q} \mathbf{h} \tag{12}$$

where the $N \times N$ Hermitian matrix $\mathbf{Q} = \mathbf{C}_{1_1}^H \tilde{\mathbf{R}}^\dagger \mathbf{C}_{1_1}$ and actually the instantaneous SINR is expressed as a quadratic form of a zero mean complex Gaussian random vector (CGRV) $\mathbf{h} \in \mathbb{C}^N$. From the results in Appendix B of [15], the CF of γ is given by

$$\phi_\gamma(t) = E\left[\exp(j\gamma t)\right] = \frac{1}{\det\left(\mathbf{I}_N - jt\mathbf{C}_h\mathbf{Q}\right)} \tag{13}$$

Letting $-jt = s$, the CF of γ is

$$\phi_\gamma(s) = \det\left(\mathbf{I} + s\mathbf{C}_h\mathbf{Q}\right)^{-1} = \prod_{n=1}^{N} (1 + s\lambda_n)^{-1} \tag{14}$$

where $\{\lambda_n\}$ stand for the eigenvalues of $\mathbf{C}_h\mathbf{Q}$. It is important to note that the matrix $\mathbf{C}_h\mathbf{Q}$ includes all information about MAI and correlated channels. For correlated branch outputs, we assume, that all these eigenvalues are distinct. The SINR has the following PDF corresponding to the CF of (14) [14]

$$f_\gamma = \sum_{n=1}^{N} \frac{\alpha_n}{\lambda_n} \exp(-\frac{\gamma_n}{\lambda_n}) \tag{15}$$

in which $\alpha_n = \prod_{l=1, l\neq n}^{N} \left(\frac{\lambda_n}{\lambda_n - \lambda_l}\right)$. By invoking the results from [14, Chap. 14], we can get the average BER for the kth user as

$$P_e(k) = \frac{1}{2}\sum_{n=1}^{N} \alpha_n \left(1 - \sqrt{\frac{\lambda_n}{1 + \lambda_n}}\right) \tag{16}$$

From (16), it is obvious that the BER for the MMSE receiver can be determined only by the eigenvalues of the matrix $\mathbf{C}_h\mathbf{Q}$ and the diversity order. Since $\mathbf{Q}$ is a random matrix, Monte Carlo simulation is employed to get the final average BER.

4. NUMERICAL AND SIMULATION RESULTS

Walsh-Hadamard codes are employed as spreading sequences and all users are assumed to have the same mean power. It is also assumed that the delay spread is $T_d = 1\mu$s and the channel is assumed constant over one OFDM symbol and changes independently from one symbol to another. The subcarrier separation is assumed as $\triangle f = 100$ kHz. The correlated Rayleigh fading channel coefficients are generated according to the technique presented in [9].

The comparison between the analytical results and the computer simulation results for a single rate system is shown in Fig. 2 with $N = SF = 16$. It can be observed that a good agreement between the analytical results and simulation results can be achieved which shows that for MMSE receiver, the Gaussian approximation is pretty accurate.

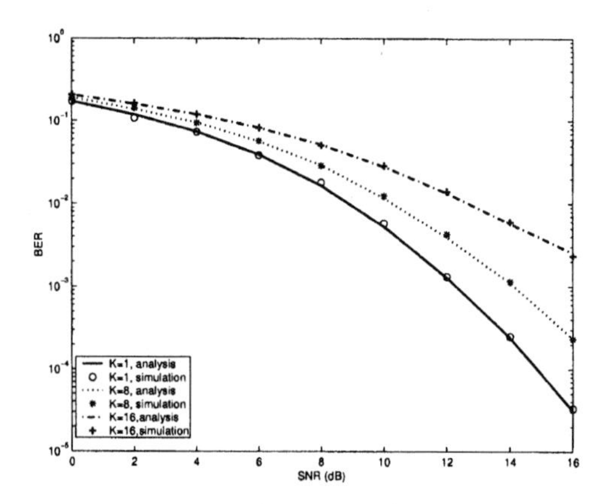

Figure 2. Comparison of analytical and simulated BER for the MMSE receiver of the downlink MC-CDMA with correlated Rayleigh fading channels ($N = 16$).

A dual-rate system with 32 subcarriers is considered in Fig. 3. In Fig. 3 (a) 12 users are active, 8 at low rate and 4 at high rate. The rate ratio is 1:2. The equivalent user $K_{equ} = 16$, that is, the system is half loaded. The spreading factor for the low rate users $SF_L = 32$. For high rate users, in order to guarantee the two data symbols transmitted simultaneously in one OFDM symbol experience the same kind of correlated fading, two schemes for placing the chips of the same data symbol are investigated, i.e., at adjacent subcarrier (without interleaving) and maximizing their space (with interleaving). It is indicated that the high rate users can benefit from the subcarrier interleaving, around 0.4 dB at BER=10^{-2} and 1 dB at BER=10^{-3}. In Fig. 3 (b) the number of users is 10, 8 at low rate and 2 at high rate with the rate ratio 1:4, $SF_L = 32$, $SF_H = 8$. As in Fig. 3 (a) the system is half loaded. For high rate user, we investigate three interleaving schemes with different subcarrier spacing $\triangle fD$ for the same data symbol. Compared with no interleaving ($D = 1$), more gain can be obtained from the best subcarrier interleaver ($D = 4$), around 1.3 dB at BER $= 10^{-2}$. From Fig. 3, we can conclude that if the interleaving gain is taken into account, lower complexity VSF MC-CDMA offers similar performance as multicode MC-CDMA, which is an important finding for the downlink system studied in this paper.

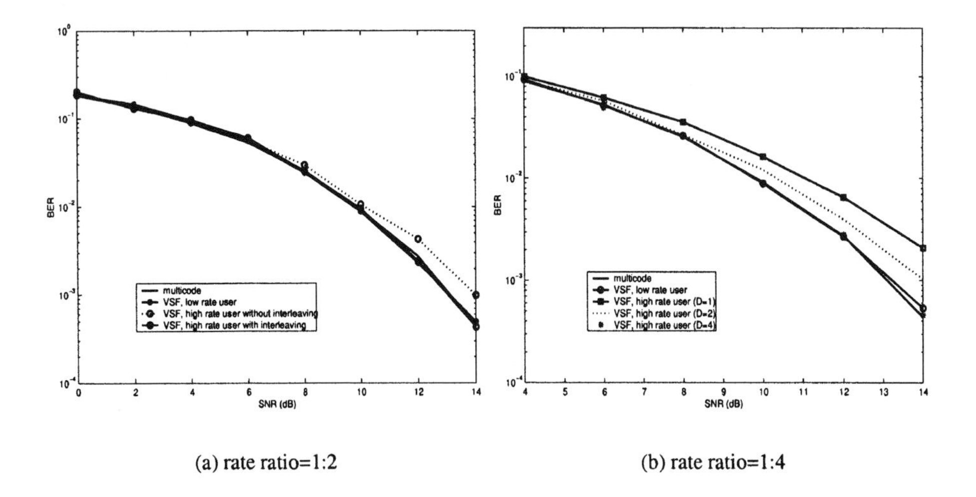

(a) rate ratio=1:2

(b) rate ratio=1:4

Figure 3. Performance of an MC-CDMA system with different multirate schemes and correlated Rayleigh fading channels ($N = 32$).

5. CONCLUSIONS

In this paper, we presented the unified system model for a multirate MC-CDMA system. The bit error rate performance of MC-CDMA systems with different multirate schemes was investigated. The impact of the subcarrier interleaving scheme was studied for VSF MC-CDMA. Results show that the two multirate systems can achieve similar performance.

ACKNOWLEDGEMENTS

This research was supported by the National Technology Agency of Finland, Nokia, the Finnish Defence Forces, Elektrobit and Instrumentointi.

AFFILIATIONS

Centre for Wireless Communications, P.O.Box 4500, FIN-90014, University of Oulu, Finland.

REFERENCES

[1] K. Fazel and S. Kaiser (Eds), *Multi-carrier Spread-Spectrum & Related Topics.* Boston: Kluwer Academic Publishers, 2002.

[2] S. Hara and R. Prasad, "Overview of multicarrier CDMA," *IEEE Commun. Mag.*, vol.35, pp. 126-133, Dec. 1997.

[3] S. Kaiser, "Multi-carrier CDMA Mobile Radio System-Analysis and Optimization of Detection, Decoding, and Channel Estimation," Ph.D. dissertation, Univ. of Munich, Germany, 1998.

[4] M. J. Juntti, "Performance of multiuser detection in multirate CDMA systems" *Wireless Personal Communications*, vol.11, pp. 293-311, Dec. 1999.

[5] M. Tan and Y. Bar-ness, "Performance comparison of the multi-code fixed spreading length (MFSL) scheme and the variable spreading length (VSL) scheme of multi-rate MC-CDMA" in *Proc. of IEEE ISSSTA'02*,Prague, Czech Republic, Sept. 2002, pp. 108-112.

[6] S. Hara and R. Prasad, "Design and performance of multicarrier CDMA systems in frequency-selective Rayleigh fading channels," *IEEE Trans. Veh. Technol.*, vol.48, pp. 1584-1595, Sept. 1999.

[7] W. C. Jakes, Ed., *Microwave Mobile Communication.* John Wiley&Sons, 1974.

[8] T. Kim et. al., "Performance of an MC-CDMA system with frequency offsets in correlated fading," *Prof. of IEEE ICC'2000*, pp. 1095-1099.

[9] N. C. Beaulieu, M. L. Merani, "Efficient simulation of correlated diversity channels," *Prof. of IEEE WCNC'2000*, pp. 207-210.

[10] L. L. Scharf, *Statistical Signal Processing: Detection, Estimation, and Time Series Analysis,* Addison-Wesley, Reading, MA, USA, 1991.

[11] H. V. Poor and S. Verdú, "Probability of error in MMSE multiuser detection," *IEEE Trans. Inform. Theory*, vol. 43, pp. 857-871, May 1997.

[12] M. Latva-aho, "Advanced receivers for wideband CDMA systems," Ph.D. dissertation, Univ. of Oulu, Finland, 1998.

[13] M. Chiani, M. Z. Win, A. Zanella, R. K. Mallik and J. H. Winters, "Bounds and approximations for optimum combining of signals in the presence of multiple cochannel interferers and thermal noise," *IEEE Trans. Commun.*, vol.51, pp. 296-307, Feb. 2003.

[14] J. G. Proakis, *Digital Communications.* 3rd ed. New York: McGraw-Hill, 1995.

[15] M. Schwartz, W. R. Bennett, and S. Stein, *Communications Systems and Techniques.* New York: McGraw-Hill, 1966.

E. AL-SUSA*, D.G.M CRUICKSHANK, S. MCLAUGHLIN & Y. LEE

PRIORITY SWAPPING SUBCARRIER-USER ALLOCATION TECHNIQUE FOR ADAPTIVE MULTICARRIER BASED SYSTEMS

Abstract: The aim of this paper is to discuss the use of adaptive subcarrier-user allocation in a multicellular environment. The paper also proposes a priority swapping based allocation algorithm that can utilise the diversity of the channel selectivity. The channel information required by the base-station for a successful detection of the received signals from the different users can be used by the proposed algorithm to enable the base-station to adaptively and speedily allocate the different subcarriers to its users such that the total throughput is maximised. This type of downlink transmission can best be likened to a frequency hopping (FH) system with the slight difference that the later uses a pseudorandom hopping pattern for each user while the former uses a dynamic-time-varying and channel-dependent orthogonal hopping pattern for each user. It will be shown that the algorithm proposed here has a very high convergence rate and achieves a diversity gain equivalent to that obtained using an optimum allocation algorithm based on the maximum likelihood criterion.

1. INTRODUCTION

It is expected that in the very near future various mobile cellular communication systems will be able to provide data rates of the orders of tens of mega bits per second (Mbps). Link enhancement techniques embodied in the form of high-speed downlink packet access (HSDPA) for extending the ability of the 3G universal mobile telecommunication systems (UMTS) is just one of the systems that will provide data rates up to 20Mbps [1][2]. Other future systems such as the mobile broadband services (MBS) are being designed to provide data rates well in excess of a 100Mbps for cellular mobile users in the foreseeable future [3][4][5]. The demand for such high data rates puts pressure on the system designers to produce better and more bandwidth efficient techniques with practical levels of complexity. It is now widely accepted that adaptive transmission techniques may well be a necessity to fulfil such high demands. Adaptive transmission in the form of a combination of modulation and coding has already been standardised for HSDPA systems [1]. Due to its resilience to multipath and inherent wideband nature, orthogonal frequency division multiplexing (OFDM) has been tipped as the most likely multiple access technique to be used for future 4G mobile communication systems [6]. This paper studies the use of adaptive subcarrier-user allocation and proposes an adaptive allocation algorithm based on priority swapping allocation (PSA) for a multicarrier multiuser system in the concept of multicellular environments. The study is based on a downlink time-division duplex (TDD) transmission. The hopping pattern for each user here is generated based on knowledge of the channels of the different users in the cell such that the total throughput is maximised.

K. Fazel and S. Kaiser (eds.), Multi-Carrier Spread-Spectrum, 65–72.

This paper is organised as follows. In the first section we described the simulation model for our multicellular system. Then a description of the PSA algorithm is provided. Finally, some simulation results assessing the convergence rate of the allocation algorithm and the throughput of different adaptation techniques are given before the paper concludes.

2. SYSTEM MODEL

A schematic diagram of the adaptive system implemented here is shown in Figure 1. The cellular structure modelled assumes packet-based transmission. Only one tier of interfering cells is considered which implies that seven base stations (BS) in total are simulated. A rap around technique is used to emulate a larger system and a hexagonal cell shape is implemented. We assume that all cells have the same statistical behaviour with equal number of users and similar user-distribution. The simulation is based on omni-directional antennas and a frequency reuse factor (FRF) of 1. In the case of adaptive transmission, we assume that each BS is aware of the average signal to noise plus interference ratio (SNIR) for each of its users on each of the individual subcarriers. In order to vary the data rate, different quadrature amplitude modulation (QAM) schemes are used. We also assume that the multipath channel is fixed during the transmission time of each frame, which is fixed at 1000 OFDM blocks, and that the allocation of the subcarriers to the users is updated per frame. To make the simulation feasible we fix the number of users to 60 and the number of subcarriers to 64. We assume that 60 of the subcarriers can be used and that each user only requires one subcarrier at a time. The users are always orthogonal in the frequency domain which implies that each subcarrier can only be used by one user at a time.

The received signal of each user is made up of 8 components with only one component being the useful part, 6 interference components from the surrounding cells and the additive noise component as given by equation (1).

$$R = X \cdot H + \sum_{c=1}^{6} I_c \cdot H_c^{'} + \sigma^2 \tag{1}$$

Where, X is a vector of the data symbols in the frequency domain, $H_c^{'}$ is the frequency response of the multipath channel, I_c is the interfering symbols from the cells in the first tier and σ^2 is the variance of the additive noise. The channel for each interfering signal goes through an independent multipath channel. This implies that a total of 420 (7 x 60) independent channels were simulated. We assume that each multipath channel has 6 independent rays and a maximum delay spread equal to 10% of the OFDM block duration, which is considered to be fully absorbed by the appended cyclic prefix. This corresponds to a loss of 10% of the total capacity as a result of the cyclic prefix. We also assume that the concerned cell is partitioned into six triangles with the cell facing each of the triangles being the dominant source of interference on the users within that triangle. Each cell is divided into five equal-area regions with the users in each region being uniformly distributed within their region. The desired mobiles are in the central cell. We assume that the BS in each cell is aware of the path-loss for each user and that the transmitted power is budgeted such that the average received power per user is equal regardless of its distance from the serving BS. This implies that users within the region close to the serving BS require less power, hence contribute less interference, and that users on the outer region require higher power, hence contribute and receive higher interference to/from neighbouring cells. Note that this arrangement does not increase the interference between users of the same cell as they are occupying orthogonal subcarriers. We assume that interfering cells transmit continuously and that the subcarrier-user matching in any cell appears random to other cells. We use rate ½ Turbo (2-parallel convolutional codes) forward error correction channel coding with MAP decoding. We use the Okumura-Hata propagation model for an urban macro cell with base station antenna height of 30m, mobile antenna of 1.5m and carrier frequency of 1950MHz and assume that the distance between two neighbouring base stations is 2Km.

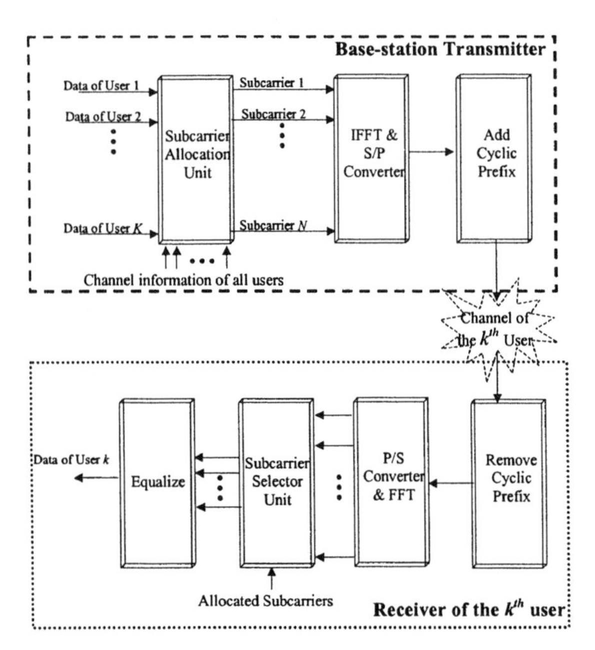

Figure 1: Schematic of the adaptive subcarrier-user transceiver

3. THE PRIORITY SWAPPING ALLOCATION (PSA) ALGORITHM

In order to obtain the maximum diversity gain from the channel selectivity, one can use an allocation algorithm based on the maximum likelihood criterion (MLA) in which all subcarrier-user combinations are tested in order to find the best allocation pattern. Such technique however entails high complexity, which is proportional to the factorial of the number of subcarriers and/or users and hence becomes impractical as this number increases beyond 10. Such complexity is further magnified when the allocation is required to be continuously updated in response to the time-varying channel conditions. The PSA algorithm proposed here can be viewed as a suboptimal version of the MLA algorithm designed with the aim to keep complexity at an acceptable level while maintaining a good diversity-gain performance. The idea behind the PSA algorithm is to focus the search for the best subcarrier-user allocation pattern (*ATP*) on the subcarrier-user combinations that match the users with their best corresponding subcarriers. Because several users may happen to have their best subcarriers at the same frequencies, the PSA algorithm performs the allocation based on a pre-generated set of K different priority levels (*PL*), corresponding to the K active users. The algorithm attempts to allocate the best subcarriers to their corresponding users taking into account the given priority level of each user. In the event when the same subcarriers happen to be the best for more than one user, these subcarriers are then allocated to the user with the highest given priority level. In order to obtain the best *ATP*, the PSA algorithm repeats the allocations based on several sets of priority levels. Similar to the case of the MLA algorithm, the total required transmission power, P_T, is used in the selection of the selection of the best *ATP* (which corresponds to the smallest required P_T). The operations involved in the PSA algorithm can be more clearly summarised into the following six steps:

Step 1: *for* $k = 1,2\cdots K, n = 1,2\cdots N$

Sort in a descending order the elements of each row of

$$A = \begin{Bmatrix} \alpha_{0,0} & \alpha_{0,1} & \cdots & \alpha_{0,N-1} \\ \alpha_{1,0} & \alpha_{1,1} & \cdots & \alpha_{0,N-1} \\ \ddots & \ddots & \ddots & \ddots \\ \alpha_{K-1,0} & \alpha_{K-1,1} & \cdots & \alpha_{K-1,N-1} \end{Bmatrix}$$

(Each column of A represents the attenuation on one subcarrier by the frequency response of the channel of one user, while each row represents the attenuation on the subcarriers of one user)

Step 2: Using an initial *PL* (e.g. user 0 with highest priority down to user K-1 with the lowest) and the sorted matrix A generate the corresponding *ATP*.
Step 3: Calculate the corresponding required transmit power, P
Step 4:Using a new set of *PL*, find a new allocation transmit pattern (*ATP'*) and calculate the new required transmit power (P')
Step 5:Compare P with P', if $P' < P$ replace *ATP* with *ATP'*, $P = P'$
Step 6:Repeat steps 4 and 5, each time using a new set of *PL*

3.1 PL Generation

In order to increase the speed of convergence of the PSA algorithm and maximise the probability of selecting the optimal *ATP*, it is essential to select the sets of priority levels wisely. Through trial and error, it was found that the best way to achieve this is to ensure that the generated *PL* sets allow the users to be equally allocated all the levels of priorities within any specified number of iterations. For example, starting from an initial set of a certain priority order, a minimum of K iterations is required if the K users are to be equally allocated all the different priority levels, which, in this case, results in each user being allocated once each of the priority levels.

In the simulations shown here, the sets of priority levels are generated in two steps. In the first instant a pseudorandom generator is used to produce an initial set of priorities. Then, $K-1$ new sets are produced by circularly shifting the priorities within this set so that each user is allocated all the different priorities, for this particular priority order, within the minimum of K iterations. Once the circular shift is complete, an independent set of priority levels is generated and then used in the initialisation of the circular shift register to produce more *PL* sets. To ensure that the users are statistically given equal priority, the sets should be generated in the form of independent groups of K sets, where the sets of each group are interdependent by the virtue of the circular shift. In the simulation section, it will be shown that the required number of *PL* sets to find the best *ATP* depends on the number of subcarriers/users under consideration.

4. SIMULATION RESULTS

The simulation results shown in this section have two purposes. The first one is to evaluate the performance of the PSA algorithm while the second is to evaluate the performance of an adaptive subcarrier-user allocation based OFDM system that uses the PSA algorithm. The results shown assume that a perfect estimate of the frequency response of the channel is available to the base station prior to the allocation.

Figure 2 *(a)* below shows the convergence rate of the PSA algorithm relative to the maximum likelihood allocation algorithm. The simulation test presented here is based on a system with $K = N = 8$ and the assumption that each user requires only one sub-carrier. The number of sub-carriers and users is kept relatively small for this test to enable us to perform the comparison with the MLA algorithm that requires a number of iterations equal to the factorial of the number of users. In this test, 8 uncorrelated multipath channels, corresponding to 8 users, where generated using a random generator. The PSA and MLA algorithms were then used to find the best subcarrier and user allocation pattern that results in the best diversity gain. It can be seen from this figure that after about 300 iterations the diversity gain achieved by the PSA algorithm is very closely matching that of the MLA algorithm, which requires 5040 iterations. In fact it can be seen from this figure that after only 100 iterations, the PSA algorithm reaches within less than 0.3dB of that achieved with the MLA algorithm. Figure 2 *(b)* compares the power gain achieved for a hundred uncorrelated channel sets using the PSA algorithm, while fixing the number of iterations to 300, with that using the full MLA algorithm. It can be seen from this figure that almost 90% of the times the PSA algorithms achieves the same diversity gain as the MLA algorithm.

As the complexity of the MLA algorithm is very high it is not possible to compare the convergence rate of the PSA algorithm with the MLA at $N > 8$.

Figure 3 shows the impact of increasing the number *PL* sets on the average power gain for a variety of scenarios. It is important to notice that the number of *PL* sets shown in this figure is normalised by the number of users/subcarriers. It is obvious from the figure that the achievable average power-gain saturates after between 2 to 10 (x *K*) *PL* sets, depending on whether there is many or few active users, respectively. It can also be seen that there is a consistent increase in the average power gain in response to increasing the number of users/subcarriers, which amounts to about 1.5dB when the number of users/subcarriers is varied from 8 to 512. This is because the diversity range of the system increases in response to increasing the number of users and subcarriers.

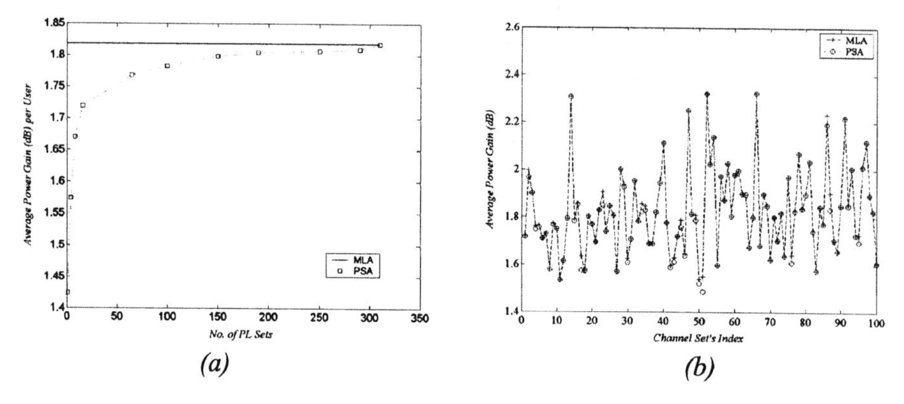

Figure 2: Convergence rate of the PSA algorithm relative to the MLA

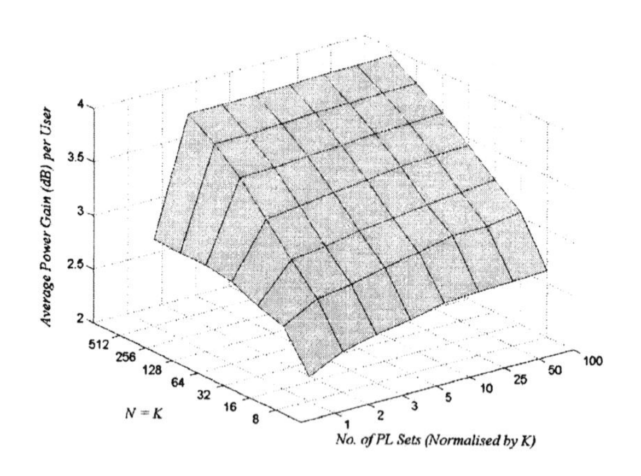

Figure 3: Average power gain versus number of PL for different N & K

In the second set of results shown below, we compare the performance of different adaptive schemes that use the PSA algorithm in a multiuser multicellular environment. The systems compared are the adaptive subcarrier, power and modulation (ASAP), adaptive subcarrier and modulation (ASAM), fixed subcarrier and adaptive modulation (FSAM) and random subcarrier allocation (RSA). The results also include a comparison to highlight the impact of using information only about the fading levels on each subcarrier with respect to each user and the impact of using full information about the average signal to noise plus interference ratio on each subcarrier with respect to each user, in the allocation process. The shadowing effect considered has a standard deviation of 8dB and a lognormal distribution. Both uniform and non-uniform user-distributions are considered in the simulation. In the case of the uniform distribution, each region of the cell under consideration is assumed to have an equal number of users while in the case of the non-uniform distribution 60% of the users is assumed to be concentrated in the outer two regions of the cell while only 20% of the users are in the most inner two regions.

Figure 4 and Figure 5 show a comparison between the aforementioned five schemes as a function of the cumulative distribution function (CDF) and data throughput. In Figure 4, the results are shown for the case when the power of the AWGN is significant relative to the power of the received signal and interference. In this case, we fixed the noise power such that the energy per bit to noise power only ratio (E_b/N_o) is 10dB. In the case of Figure 5, we remove the AWGN and show the results in terms of signal to interference ratio only. In both figures the results are shown for an average bit error rate (BER) of 10^{-4}.

In the case of Figure 4 it is clear that the best system to use varies depending on the user distribution and data throughput required. For example, in case of the uniform user-distribution *(a)*, the ASAP scheme is provides the best result up to a throughput of 8.4Mbps while in the case of *(b)* this is only true up a throughput of 4.2Mbps. This is believed to be due to the fact that with the ASAP scheme all the subcarriers have the same power level at the receiver and hence if the required SNIR is higher than the actual one, none of the subcarriers will be able to deliver that data rate. On the other hand, in the case of the ASAM scheme some subcarriers will have better SNIR than others and that is why their CDF does not fall as sharply.

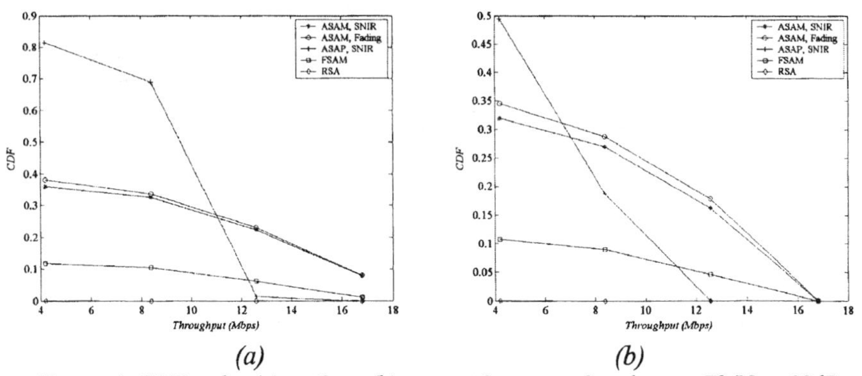

Figure 4: CDF under (a) uniform (b) non-uniform user distribution, Eb/No =10dB

In Figure 5, the comparison reveals that under good SNIR, the ASAP scheme provides the best performance, especially when the user-distribution is uniform.

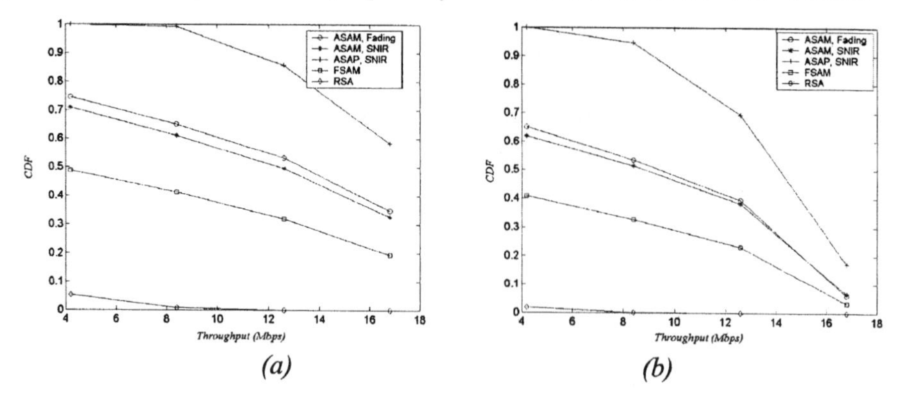

Figure 5: CDF under (a) uniform and (b) non-uniform user distribution, No AWGN

5. CONCLUSION

In this paper we have proposed an efficient subcarrier-user allocation technique and evaluated the performance gains achieved when using this algorithm in a multiuser multicellular environment.

6. 7. ACKNOLEDGEMENT

The work reported here has formed part of the Wireless Access area of the Core 2 Research Programme of the Virtual Centre of Excellence in Mobile & Personal Communications, Mobile VCE (www.mobilevce.com) whose funding support, including that of EPSRC, is gratefully acknowledged.

7. REFERENCES

[1] 3GPP TR 25.858 V1.0.4, January 2002.
[2] RAN WG1 #23, R1-02-0142. Lucent. MIMO system simulation methodology
[3] L. Fernandes (Feb 1995), Developing a System Concept and Technologies for Mobile Broadband Communications, IEEE Personal Communications Magazine, Vol. 2. No. 1, pp. 54- 59.
[4] M. Prögler and S. Svaet (eds. 1999), MBS Performance Evaluation, ACTS-SAMBA Deliverable A0204/TN/PK/DS/ P/014/b1, ACTS Central Office, Brussels, Belgium.
[5] Mobile Broadband Systems, Final Report, ERO July 1.997. www.ero.dk.
[6] H. Atarashi, S. Abeta and M. Sawahashi (2003), "Variable spreading factor orthogonal frequency code division multiple access (VSF-OFCDM) for broadband packet wireless access", IEICE Transactions on Communications, Vol. E86-B, 291-299.

8. AFFILIATIONS

Institute of Digital Communications, The University of Edinburgh, The King's Buildings, Mayfield Road, Edinburgh, EH9 3JL, Scotland, UK
* Department of E&EE, UMIST, PO BOX 88, Manchester, M60 1QD, UK, Email: alsusa@iee.org

N. BENVENUTO, P. BISAGLIA AND F. TOSATO

SUB-BAND LOADING FOR PRE-EQUALIZED UPLINK OFDM-CDMA SYSTEMS†

Abstract. In this paper a new pre-equalization technique is presented for uplink OFDM-CDMA systems. All users employ spreading both in frequency (on sub-bands of the available spectrum) and time; moreover, bit and power loading is used at the mobile terminals to accomplish throughput maximization. The proposed loading algorithm, to avoid interference among users, inherently inverts the channel amplitude on the sub-carriers of each sub-band, while the phase is adjusted separately. Numerical results are given to highlight the effectiveness of the proposed scheme with respect to existing solutions.

1. INTRODUCTION

Recently, for uplink OFDM-CDMA systems, as alternative to multi-user detection at the receiver, pre-equalization techniques at the transmitter have been proposed [1, 2], in conjunction with time division duplex (TDD). The idea behind pre-equalization is to vary the gain assigned to each sub-carrier on the basis of the channel state information (CSI). In particular, in [1] the simplest choice of pre-equalization, which consists in the inversion of the channel frequency response, is investigated, but the problem related to the constraint of the transmit power is not addressed. In [3], a normalization factor is introduced; this guarantees no multiple access interference (MAI), but causes a significant degradation in performance. Alternatively, in [2], a system is proposed which selects the pre-equalizer coefficients to minimize the signal-to-interference-plus-noise ratio (SINR) at the decision point. This technique limits the transmit power, but introduces MAI.

In this work a new system is presented where, for each user, the OFDM-CDMA sub-carriers are grouped into adjacent sub-bands of equal size [4] and spreading is done both in frequency and time. Moreover, absence of MAI is obtained through channel inversion at the transmitter, and the power constraint is met by bit and power loading [5], across the sub-bands. We note that to simplify the access method, avoiding MAI, the sub-band grouping is the same for all the users sharing the uplink channel.

The remainder of this paper is organized as follows. In Section 2 the structure of the uplink OFDM-CDMA considered is presented. The novel bit and power loading algorithm is described in Sections 3 and 4; numerical results are reported in Section 5 and conclusions are in Section 6.

†This work was supported in part by MIUR, under the FIRB project "Reconfigurable platforms for wideband wireless communications," prot. RBNE018RFY.

K. Fazel and S. Kaiser (eds.), Multi-Carrier Spread-Spectrum, 73–79.

2. THE OFDM-CDMA UPLINK SYSTEM

We consider an hybrid OFDM-CDMA uplink system where the mobile terminals (MTs) perform spreading both in frequency and time. Transmission is quasi-synchronous, i.e. the active users are synchronous within the margin of the cyclic prefix. Let N be the number of useful sub-carriers per OFDM symbol, and L_F and L_T be the frequency and time spreading factors, such that the overall spreading factor is $L = L_F L_T$. Moreover, we assume that $L_F = N/N_S$, so that the sub-carriers in each OFDM symbol can be grouped into N_S sub-bands, each composed of L_F sub-carriers.

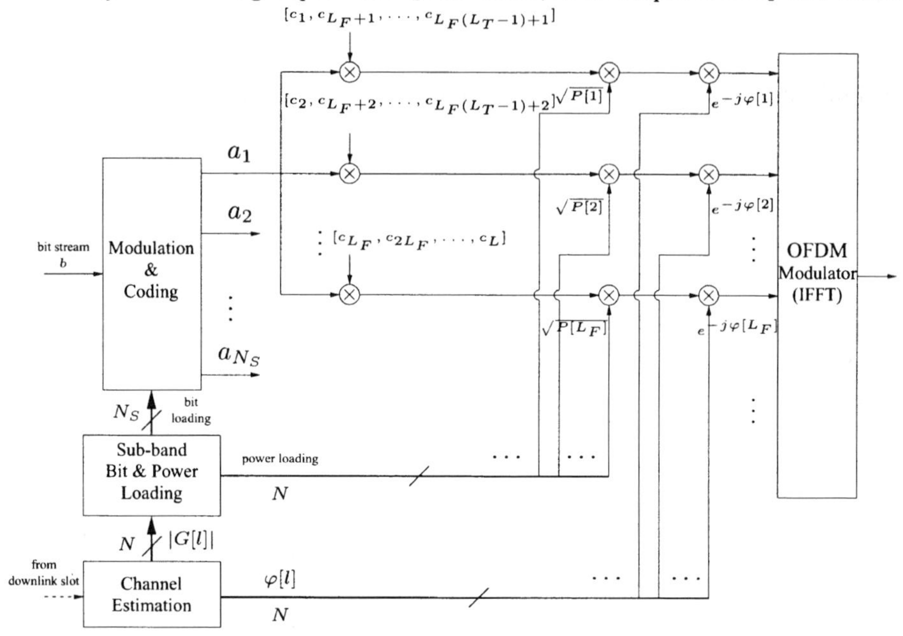

Figure 1. Block diagram of the transmitter at the MT, of a generic user. L_F and L_T are the frequency and time spreading factors. The overall spreading factor is $L = L_F L_T$. N is the number of useful sub-carriers of the OFDM modulator, N_S is the number of sub-bands ($N/N_S = L_F$). $|G[l]|$ and $\varphi[l]$, $l = 1, \ldots, N$, is the amplitude and phase of the estimated channel frequency response on the l-th sub-carrier.

2.1. Transmitter

In Fig. 1 the block diagram of a generic MT transmitter is depicted. The user's index has been omitted for simplicity. In the scheme, each user data stream is divided into N_S unequal-rate sub-streams, as the number of sub-bands. The symbol $a_s^{(u)}$ (belonging to a unitary power constellation) of user u, with $u = 1, \ldots, U$, and $s = 1, \ldots, N_S$, is spread by the spreading code $\mathbf{c}^{(u)} = [c_1^{(u)}, \ldots, c_L^{(u)}]$ and the resulting L chips are arranged into an $L_F \times L_T$ matrix. Orthogonal spreading codes are used. Each column of L_F chips is then transmitted by L_F sub-carriers (sub-band s) of an OFDM symbol. Hence, N chips are input to the OFDM modulator and L_T OFDM symbols are required to transmit N_S symbols.

To optimize resources, each MT adopts a bit and power loading algorithm across the sub-bands. The amplitude of the channel estimate is fed to the loading algorithm which calculates the bit distribution on the N_S sub-bands and the amplitude of the N sub-carrier pre-equalizer coefficients, $\sqrt{P[l]}, l = 1, \ldots, N$.

2.2. Channel

In the following, the user's index will be omitted for simplicity when referring to a generic user. Let $G[l] = |G[l]|e^{j\varphi[l]}, l = 1, \ldots, N$, be the channel frequency response of a generic user. The phase of the channel estimate is used for phase pre-equalization. Hence, in Fig. 1 the chip sequence is multiplied by $\sqrt{P[l]}e^{-j\varphi[l]}$. We define the received sub-carrier signal-to-noise ratio, normalized to a unitary transmitted power as

$$\widetilde{\Gamma}[l] = |G[l]|^2/\sigma_W^2 , \qquad (1)$$

where σ_W^2 is the sub-carrier noise power.

2.3. Receiver

In Fig. 2 the block diagram of the receiver at the BS for the reference user is sketched. After the OFDM demodulator and the despreading operation, the samples corresponding to the L chips are simply summed up. We note that the receiver has to wait for L_T OFDM symbols to combine all the chips, thus latency due to collecting the chips increases with the time spreading factor L_T.

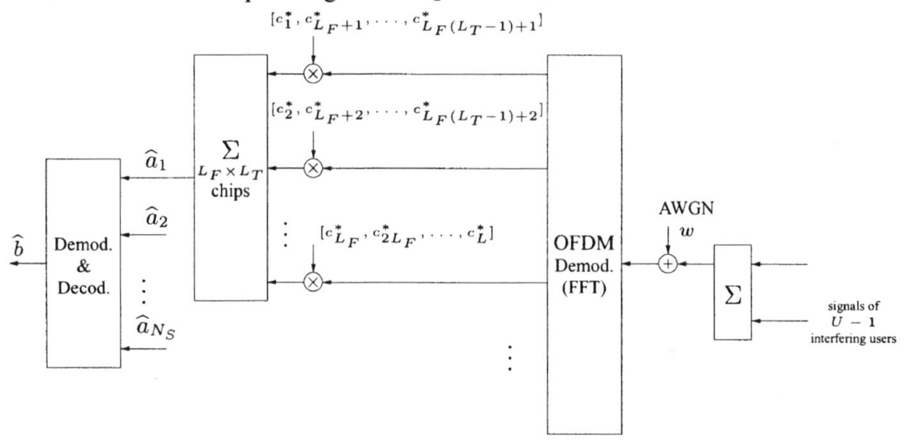

Figure 2. Block diagram of the receiver at the BS for the reference user of Fig. 1.

3. CHANNEL LOADING FOR OFDM

We now briefly recall the channel loading algorithm for OFDM [5, 6]. It assumes that a set of transmission modes[2] $\{\mathtt{M}_1, \ldots, \mathtt{M}_K\}$ can be loaded into the sub-carriers

[2]A combination of modulation format and code rate.

and $\{\Gamma_1, \ldots, \Gamma_K\}$ are the corresponding minimum signal-to-noise ratios per sub-carrier, required at the receiver decision point to guarantee a target bit error probability $\mathcal{P}_{e,ref}$[3]. Hence, from (1) the minimum transmit power required to use mode k in the sub-carrier l is given by

$$P_k[l] = \Gamma_k / \widetilde{\Gamma}[l] \quad , \quad k = 1, \ldots, K \quad , \quad l = 1, \ldots, N \, . \tag{2}$$

The channel loading algorithm [5], based on the normalized signal-to-noise ratios (1) and under the constraint that the overall transmit power is given, assigns the modes and the corresponding powers (2) to the various sub-carriers, with the objective of maximizing the number of information bits per OFDM symbol.

Unfortunately, this algorithm cannot be applied directly to our OFDM-CDMA system because it operates with single sub-carriers rather than sub-bands. Moreover, it does not consider the interference arising from multiple users.

4. CHANNEL LOADING FOR OFDM-CDMA

In the OFDM-CDMA system, for each of the N_S sub-bands, we have to select the mode which guarantees the bit error probability $\mathcal{P}_{e,ref}$, with a given overall transmit power constraint. Correspondingly, we have to calculate the power required on each sub-carrier, from which the pre-equalizer coefficients $\sqrt{P[l]}$ in Fig. 1 are derived.

We impose that no MAI is present at the BS after despreading of each user. This can be accomplished through pre-equalization, where each transmit sub-carrier is multiplied by a coefficient which performs channel inversion. However, while in OFDM, in the presence of deeply faded sub-carriers, channel inversion may cause the power level of the best sub-carriers to drop, in OFDM-CDMA, this is avoided if adjacent sub-carriers (having very similar attenuation) are grouped into sub-bands, and sub-carrier channel inversion is done after the loading algorithm has assigned the power level for each sub-band. For the following analysis we assume that the channel is time invariant during the transmission of L_T OFDM symbols.

For each sub-band, to meet the target bit error probability $\mathcal{P}_{e,ref}$, the signal-to-noise ratio Γ, at the detection point, must be greater or equal to a target Γ_k, $k \in \{1, \ldots, K\}$, associated to corresponding mode $\mathtt{M}_k$. Also, to guarantee absence of interference we impose a zero-forcing condition on the coefficients $\sqrt{P[l]}$, $l = 1, \ldots,$ L_F, of the various sub-carriers. Overall, for each sub-band, it must be

$$\begin{cases} \sqrt{P[l]} \cdot |G[l]| = R \, , & l = 1, \ldots, L_F \\ \Gamma = \Gamma_k \end{cases} , \tag{3}$$

where R is a parameter to be determined. We now relate the sub-channel signal-to-noise ratios to an equivalent sub-band signal-to noise ratio used in channel loading. In a synchronous system and with orthogonal spreading codes the received signal for the considered sub-band, after despreading, is given by:

[3] We note that these signal-to-noise ratios are evaluated for an AWGN channel, since an interference-free OFDM system can be regarded as a parallel of AWGN channels.

$$r = \sum_{i=1}^{L} a\sqrt{P[(i-1) \bmod L_F + 1]} \cdot |G[(i-1) \bmod L_F + 1]| c_i c_i^* + \sum_{i=1}^{L} c_i^* W_i$$
$$= a \cdot LR + \sum_{i=1}^{L} c_i^* W_i \,, \tag{4}$$

where W_i are the AWGN samples, with variance σ_W^2. The corresponding sub-band signal-to-noise ratio is given by $\Gamma = (LR^2)/\sigma_W^2$. On the other hand, introduced the sub-carrier signal-to-noise ratio, $\Gamma[l] = P_k[l]\widetilde{\Gamma}[l]$, where $\widetilde{\Gamma}[l]$ is defined in (1), from (3), it is $\Gamma[l] = \Gamma/L, l = 1, \ldots, L_F$.

For the generic mode k, set $\Gamma = \Gamma_k$, we derive the required sub-carrier power

$$P_k[l] = \frac{\Gamma[l]}{\widetilde{\Gamma}[l]} = \frac{1}{L} \cdot \frac{\Gamma_k}{\widetilde{\Gamma}[l]} \quad , \quad l = 1, \ldots, L_F \,. \tag{5}$$

Hence, the overall sub-band power, P_k, associated to mode k, is given by

$$P_k = \sum_{i=1}^{L} P_k[(i-1) \bmod L_F + 1] = \frac{1}{L_F} \sum_{l=1}^{L_F} \frac{\Gamma_k}{\widetilde{\Gamma}[l]} \,. \tag{6}$$

Introduced the normalized sub-band signal-to-noise ratio

$$\widetilde{\Gamma}_s = \left(\frac{1}{L_F} \sum_{l=1}^{L_F} \frac{1}{\widetilde{\Gamma}[l]} \right)^{-1} \quad , \quad s = 1, \ldots, N_S \,, \tag{7}$$

it is

$$P_k = \Gamma_k / \widetilde{\Gamma}_s \,. \tag{8}$$

Hence $\widetilde{\Gamma}_s$ plays the analogous role of $\widetilde{\Gamma}[l]$ in OFDM. Using (7) we can apply bit and power loading across the sub-bands under a total power constraint. Then, the pre-equalizer coefficients are determined by (5).

We note that, when spreading in time is present and the channel is time-varying, the MTs can exactly pre-equalize the channel for the first OFDM symbol only. However, the channel variations during the transmission of the following symbols will produce some MAI. Hence, in order to meet the bit error rate requirements, we increase the target signal-to-noise ratios at the receiver by a margin against interference and/or channel variations. That is to say, if Δ is the margin, with $\Delta \geq 0$, bit and power loading is done assuming that $\{\Gamma_1(1+\Delta), \ldots, \Gamma_K(1+\Delta)\}$ are the new minimum signal-to-noise ratios that meet the target bit error probability. Obviously, Δ is a system parameter and is a function of the Doppler frequency of the channel.

5. PERFORMANCE EVALUATION

Performance has been assessed by evaluating the achievable bit rate or throughput (i.e. the maximum number of bits that can be transmitted with a given bit error probability) in a Rayleigh fading channel. We have compared the proposed scheme with the SINR-based method described in [2], where, besides the channel frequency response and the noise power σ_W^2, also the number of active users is required. This technique, indeed, has been shown to achieve better performance than systems using either conventional pre-equalization techniques at the transmitter or multi-user detection at the receiver [2, 3]. Since in [2] only the case wherein the spreading factor equals the number of sub-carriers ($N = L$) is dealt with, here, the method has been extended to the case with $L < N$. Moreover, adaptation of transmission modes has been included.

5.1. Numerical Results

The HIPERLAN/2 (or 802.11a) framework has been used for the basic OFDM system parameters. In details, the system consists of 64 sub-carriers, of which only 48 carry useful information, and allows 7 different transmission modes. All the results are given in terms of maximum average system throughput versus the average signal-to-noise ratio (Γ_{avg}) of the OFDM system. The average throughput per user is calculated by averaging the allocated bits of the loading algorithm over many realizations of HIPERLAN/2 channel model A. Moreover, since loading is performed on an assumed SNR, and the channel may vary in the next OFDM symbols, bits are discarded if the corresponding SINR at the decision point gets lower than the target. Each channel is assumed time invariant over the duration of an OFDM symbol.

We start our simulation results by considering, in Figure 3, only spreading in frequency (i.e. $L_T = 1$), and for values of L_F equal to 2, 8, 16 and 48[4]. In particular, the latter corresponds to spreading a single symbol over all the sub-carriers. Results for the other spreading factors lie in between. For comparison, the case with $L_F = 1$, corresponding to a bare single-user OFDM system, is also reported; this can be seen as a bound for the system. Simulations are obtained with both our proposed algorithm (B&P pre-equalization), described in Section 4, and the modified SINR-based algorithm [2]. As expected, we note that the throughput decreases with the increase of the frequency spreading factor L_F. In any case, the B&P pre-equalization algorithm outperforms the modified SINR-based algorithm for all the possible spreading values, except for $L_F = 48$, where the proposed algorithm simply reduces to the PRE-ORC with threshold [3].

In Figure 4, instead, only spreading in time is performed (i.e. $L_F = 1$), with a user speed equal to 50 km/h. The same speed will be kept in all the remaining results. Clearly, if a time invariant channel was considered, no degradation in the throughput would be observed with an increase of L_T, as explained in Section 4. For each spreading factor L_T, the margin Δ (see Section 4) has been optimized, i.e. the value Δ that guarantees the maximum throughput, has been selected. From the results, we note that, by increasing the spreading factor up to $L_T = 12$, slight degradation in the maximum throughput is observed. For spreading factors higher than 12, however, a more significant reduction in throughput occurs, due to increased MAI caused by channel time variations. More importantly, a close observation of Figures 3 and 4

[4]The constraints on the length of the Walsh-Hadamard codes (which must be a power of two) and on the physical layer implementation of the MC-CDMA (involving 48 data sub-carriers) require the spreading factor L to be restricted to the values of 2, 4, 8, 12, 16, 24 and 48.

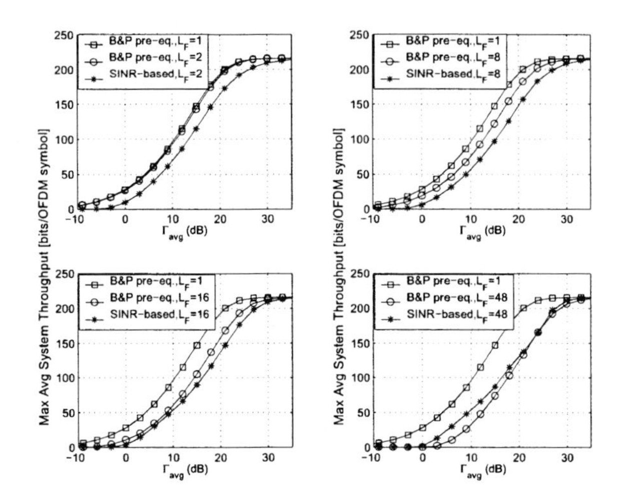

Figure 3. Maximum average system throughput versus average signal-to-noise ratio with $L_T = 1$ and values of L_F equal to 2, 8, 16 and 48. Comparison between the B&P pre-equalization algorithm and the modified SINR-based algorithm.

suggests that if an high number of users (L) has to be allocated, a compromise between spreading in frequency and spreading in time needs to be found.

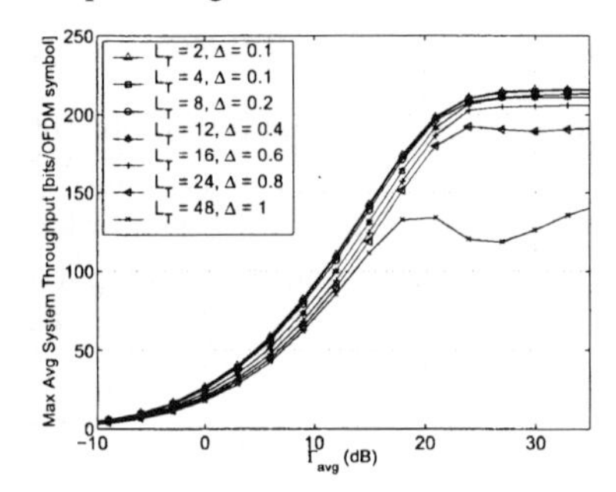

Figure 4. Maximum average system throughput versus average signal-to-noise ratio with $L_F = 1$ and values of L_T equal to 2, 4, 8, 12, 16, 24 and 48, using in each case the optimum value of the margin Δ.

In order to evaluate the potential of the proposed algorithm, throughput with spreading both in frequency and time, was evaluated (results not shown here). It is seen that, for a given L, hybrid combinations provide a maximum throughput higher than the throughput obtained with spreading in frequency only ($L_F \neq 1$).

The best hybrid combinations, namely $L_F = 4$ and $L_T = 4$ for the 16 user case, and $L_F = 8$ and $L_T = 6$ for the 48 user case, are reported in Figure 5 and compared with the SINR-based algorithm, already shown in Figure 3. For comparison, the single-user bound ($L_F = 1$ and $L_T = 1$) is also reported. At a $\Gamma_{avg} = 20$ dB our algorithm, with respect to the SINR-based algorithm, improves the maximum achievable throughput from 137 to 180 bits/OFDM symbols (an improvement of 31.4%), in the case of 16 users, and from 130 to 170 (30.8 %), in the case of 48 users. The loss with respect to the single-user bound is equal to 6.7% in the case of 16 users and to 12% in the case of 48 users.

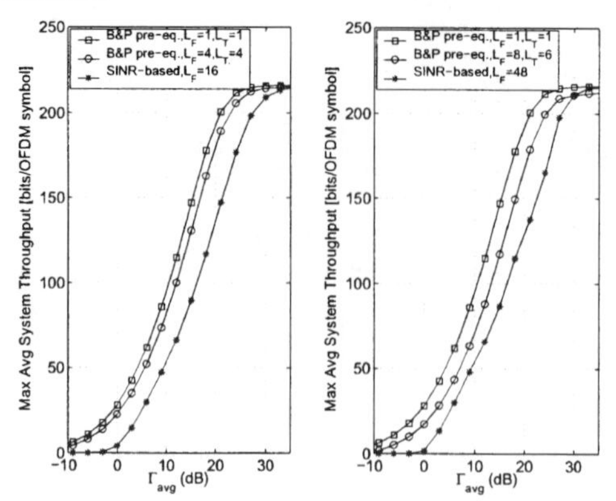

Figure 5. Maximum average system throughput versus average signal-to-noise ratio with spreading both in frequency and time. Results have been obtained in a full load system having 16 users, on the left, and 48 users, on the right. Comparison between the hybrid B&P pre-equalization algorithm and the modified SINR-based algorithm.

6. CONCLUSIONS

In this paper a new pre-equalization technique for uplink OFDM-CDMA systems has been presented. Results from computer simulations have shown that the proposed solution, in a full loaded system, outperforms other uplink schemes in terms of maximum achievable throughput and may yield a throughput very close to the OFDM single-user bound.

7. AFFILIATION

The authors are with the Dept. of Information Engineering, University of Padova, Via Gradenigo 6/B, 35131 Padova. Email: {nb, paola.bisaglia, filippo.tosato}@dei.unipd.it.

REFERENCES

[1] S. Nobilet and J.-F. Helard, "A pre-equalization technique for uplink MC-CDMA systems using TDD and FDD modes," *VTC Fall 2002*, pp. 346-350, Oct. 2002.

[2] D. Mottier and D. Castelain, "SINR-based channel pre-equalization for uplink multi-carrier CDMA systems," *PIMRC 2002*, pp. 1488-1492, Sept. 2002.

[3] P. Bisaglia, N. Benvenuto and S. Quitadamo, "Performance comparison of single-user pre-equalization techniques for uplink MC-CDMA systems," *to be presented at IEEE GLOBECOM'03*, Dec. 2003.

[4] T. Keller and L. Hanzo, "Sub-band adaptive pre-equalised OFDM transmission," *VTC Fall 1999*, pp. 334-338, Sept. 1999.

[5] B.S. Krongold, K. Ramchandran and D.L. Jones, "Computationally efficient optimal power allocation algorithms for multicarrier communication systems," *IEEE Trans. on Comm.*, vol. 48, pp. 23-27, Jan. 2000.

[6] N. Benvenuto and F. Tosato, "A loading algorithm for OFDM with modulation and coding adaptation and its application to selection of transmission modes," *Submitted to IEEE Trans. on Comm.*

S. KANAMORI, M. ITAMI, H. OHTA, AND K. ITOH

A STUDY ON SUBCARRIER POWER CONTROL OF OFDM TRANSMISSION DIVERSITY COMBINED WITH DATA SPREADING

Abstract. In OFDM transmission, transmission diversity and data symbol spreading are effective schemes to improve the bit error rate characteristics under frequency selective fading and more improvement can be achieved by combining them. Moreover, further improvement is expected by combining a subcarrier power control scheme with them, especially, under severe frequency selective fading environments. In this paper, a subcarrier power control of OFDM transmission diversity combined with a data symbol spreading scheme in the frequency domain is proposed and its characteristics are analyzed by computer simulations.

1 INTRODUCTION

OFDM (Orthogonal Frequency Division Multiplexing) is a digital modulation scheme that achieves very efficient frequency utilization and realizes high speed data transmission under limited band width. The OFDM signal consists of many carriers which are orthogonally arranged in the frequency axis and the data symbols are transmitted by modulating each carrier using usual digital modulation schemes such as QPSK, QAM and so on. The OFDM signal is suitable for limited band width transmission because the shape of its power spectrum is almost rectangular. Moreover, it is possible to reduce the affection of inter-symbol interference by inserting a guard interval without much degrading data transmission rate [1] [2] [3]. Therefore, it is widely used and considered for high speed data transmission such as digital television broadcasting, wireless LAN, power-line communication, the next generation mobile communication systems and so on. The primary factor of degradation in OFDM transmission is the influence of frequency selective fading. Under frequency selective fading, bit error rate of the carriers whose signal to noise ratio is made small by the fading much degrades. Especially, severe degradation of signal to noise ratio of specific carriers frequently occurs. This degrades the total performance of the system and an appropriate measure must be taken. In order to solve this problem, several methods are proposed such as use of error correcting codes, diversity reception, carrier transmission power control, adaptive loading schemes, data symbol distribution in frequency domain and so on. In this paper, a subcarrier power control of OFDM transmission diversity combined with a data

K. Fazel and S. Kaiser (eds.), Multi-Carrier Spread-Spectrum, 81–88.

symbol spreading scheme in the frequency domain is proposed. In proposed system, transmission power of each subcarrier in each transmission antenna is controlled so as to maximize the average signal to noise ratio of the data symbols after de-spreading the signals from DFT. As the result of computer simulations, it is confirmed that the proposed system can improve the bit error rate characteristics under multipath channel.

2 DATA SPREADING OFDM

In OFDM transmission, SNR(signal to noise ratio) of some carriers sometime becomes very low and under frequency selective fading caused by multipath and it is sometimes very difficult to retrieve correct symbols from such carriers. In the data spread OFDM system, it is possible to recover symbols by an inverse unitary even transform even if SNR of the specific carrier is very low, because each data symbol is spread over many carriers. This means that an inverse unitary transform averages influence of frequency selective fading to the specific carrier over the block of data symbols. Therefore, data spreading scheme improves influence by the carrier that SNR is very degraded. The concept of data spreading OFDM [4] is shown in Figure1 and Figure 2. If it

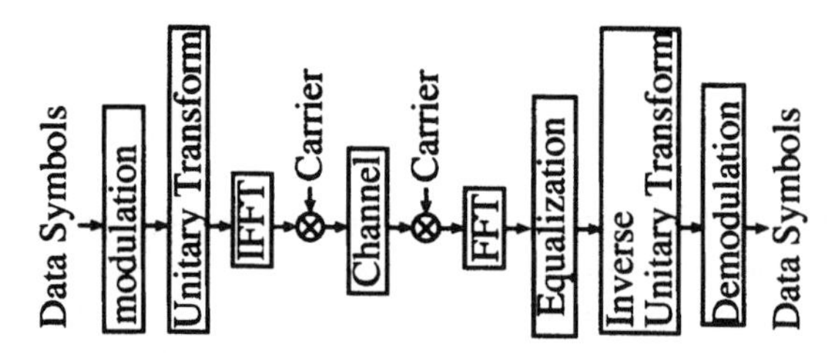

Figure 1: The concept of data spreading OFDM

is assumed that $d_n = a_n + jb_n$ $(n = 1, 2, \cdots, N-1)$ is the input data symbols modulated PSK or QAM, the block of new data symbols, D_n, is obtained by using a unitary transform as

$$D_n = \sum_{k=0}^{N-1} u_{n,k} d_k \qquad (1)$$

where N is the block size of the transform and the same value as the number of the carriers is used in this paper. $u_{n,k}$ is the coefficient that determines characteristic of a unitary transform and satisfies the following relation.

$$\sum_{n=0}^{N-1} u_{m,k} u_{n,k}^{*} = \sum_{n=0}^{N-1} u_{k,m} u_{k,n}^{*} = \begin{cases} 1 & (m = n) \\ 0 & (m \neq n) \end{cases} \qquad (2)$$

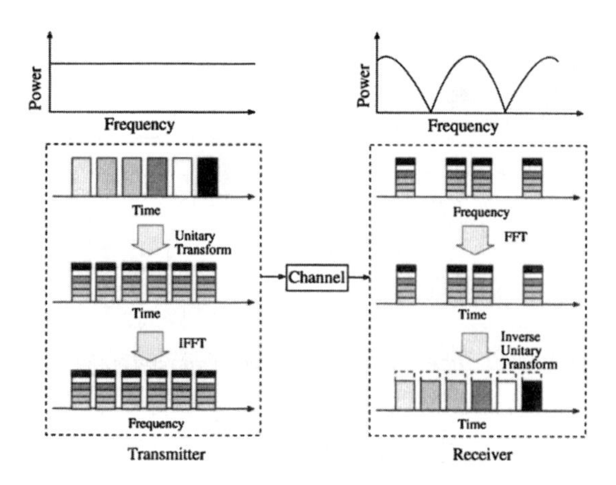

Figure 2: Data spreading OFDM

The spread data symbols, D_n, are transformed to the original data symbols, $\hat{d}_l$, by applying an inverse unitary transform as

$$\hat{d}_l = \sum_{n=0}^{N-1} u_{l,n}^* D_n \tag{3}$$

As shown in Figure 2, in data spreading OFDM, since each data symbol is spread over the whole carries, it is possible to recover data symbol even if signal to noise ratio of specific carrier is very degraded by frequency selective fading. The data spread OFDM scheme can use frequency diversity and it doesn't sacrifice data rate. Therefore, efficient data transmission is possible.

3 SYSTEM MODEL

In this paper, it is assumed that M-antenna transmission diversity is used and following $s_m(t)$ is transmitted from antenna #m. $m(1, 2 \cdots, M)$ is the number of transmission antenna. Figure 3 shows the concept of the proposed scheme.

$$s_m(t) = \mathrm{Re}\left[\sum_{l=0}^{N-1} w_{m,l} D_l e^{j2\pi(f_c + l f_0)t}\right] \tag{4}$$

Where N is the number of OFDM subcarriers and f_c is the lowest subcarrier frequency and f_0 is the subcarrier interval. In $s_m(t)$, the same block of data symbols, $D_0, D_1, \cdots, D_{N-1}$ are transmitted and transmission power of each subcarrier is controlled by multiplying a gain, $w_{m,l}$, to each data symbol as shown in eq.(4). Moreover, the block of data symbols, $D_0, D_1, \cdots, D_{N-1}$ is generated by converting the original block of data symbols, $d_0, d_1, \cdots, d_{N-1}$,

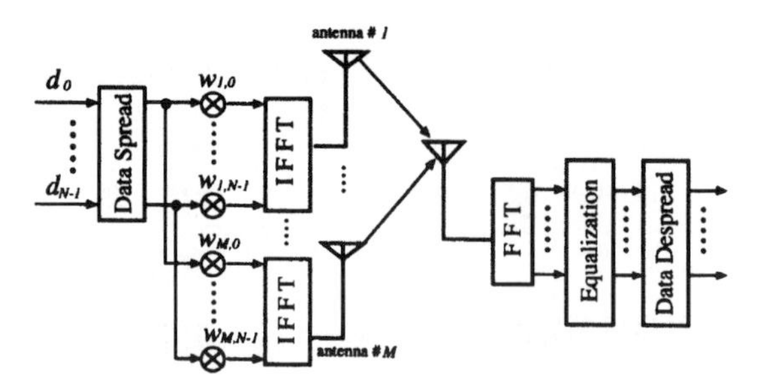

Figure 3: The concept of the proposed scheme

by using a orthogonal transform as shown in the following formula.

$$D_l = \sum_{n=0}^{N-1} u_{l,k} d_k \tag{5}$$

Where $u_{l,k}$ is the element of the orthogonal matrix and in this paper Walsh Hadamard Transform (WHT) is assumed. $d_0, d_1, \cdots, d_{N-1}$ are data symbols modulated by a specific modulation scheme such as PSK, QAM and so on. The transform in eq.(5) spreads original data symbols over the whole subcarriers and this is effective to improve characteristic of OFDM under frequency selective fading [4][5][6]. In the receiver, one antenna receiver is used and received signal is first transformed by DFT to generate each symbol transmitted on each carrier. The output of DFT correspond to each carrier, $\hat{D}_l$ is obtained by the following formula.

$$\hat{D}_l = \sum_{m=1}^{M} H_{m,l} w_{m,l} D_l + Z_l \tag{6}$$

Where $H_{m,l}$ is the transfer function corresponds to l'th carrier from antenna #m, and Z_l is additive white Gaussian noise and it is assumed that $E[Z_l] = 0$ and $E[Z_l Z_{l'}^*] = \delta_{l,l'} N_0$. Dividing $\hat{D}_l$ by $\sum_{m=1}^{M} H_{m,l} w_{m,l}$, the following symbol, X_l, is obtained.

$$X_l = D_l + \frac{Z_l}{\sum_{m=1}^{M} H_{m,l} w_{m,l}} \tag{7}$$

By applying inverse orthogonal transform in eq.(5), an estimate of the original data symbol, $\hat{d}_n$, is derived.

$$\hat{d}_n = \sum_{l=0}^{N-1} u_{l,n}^* X_l = d_n + \sum_{l=0}^{N-1} \frac{u_{l,n}^* Z_l}{\sum_{m=1}^{M} H_{m,l} w_{m,l}} \tag{8}$$

In the right hand of eq.(8), the first term is the desired data symbol and the second term is generated by influence of additive noise. Therefore, signal to noise ratio of d_n is obtained by the following formula.

$$SNR_n = \frac{E[|d_n|^2]}{\sum_{l=0}^{N-1} \frac{|u_{l,n}|^2 E[|Z_l|^2]}{\sum_{m=1}^{M} |H_{m,l} w_{m,l}|^2}} \tag{9}$$

eq.(9) can be simplified as shown in the following formula if WHT is used as the orthogonal transform.

$$SNR_n = \frac{N}{\sum_{l=0}^{N-1} \left| \frac{1}{\sum_{m=1}^{M} H_{m,l} w_{m,l}} \right|^2} \frac{P_s}{N_0} \tag{10}$$

Where $P_s = E[|d_n|^2]$. eq.(10) indicates that SNR_n is independent to the index n and all symbols have the same bit error rate. Therefore, it is possible to minimize total system bit error rate if SNR_n is maximized by controlling the transmission gains, $w_{m,l}$, appropriately. In the following discussion, optimal $w_{m,l}$ are obtained under the condition where channel transfer functions, $H_{m,l}$, are known in the transmitter. Moreover, the constraint of total transmission power is also assumed and the following two cases are examined.

1. Sum of total transmission power of each antenna is constant.(denoted as Constraint-1)

$$\sum_{l=0}^{N-1} \sum_{m=1}^{M} w_{m,l}^2 = MN \tag{11}$$

2. Total transmission power of each antenna is constant.(denoted as Constraint-2)

$$\sum_{l=0}^{N-1} w_{m,l}^2 = N \qquad (1 \le m \le M) \tag{12}$$

For Constraint-1 and Constraint-2 , $w_{m,l}$ that maximize SNR_n is obtained by using Lagrange multipliers. As the results, the following formulas are obtained. The detailed derivation of them are omitted here. (1) Constraint-1

$$w_{i,j} = \sqrt{\frac{MN|H_{i,j}|^2}{\sum_{l=0}^{N-1} \frac{\left(\sum_{m=1}^{M} |H_{m,j}|^2\right)^{3/2}}{\left(\sum_{m=1}^{M} |H_{m,l}|^2\right)^{1/2}}}} \tag{13}$$

(2) Constraint-2

$$w_{i,j} = \sqrt{\frac{N|H_{i,j}|^2}{\sum_{l=0}^{N-1} \frac{|H_{i,l}|^2 \left(\sum_{m=1}^{M} \sqrt{|H_{i,j}|^2 |H_{m,j}|^2}\right)^{3/2}}{\left(\sum_{m=1}^{M} \sqrt{|H_{i,l}|^2 |H_{m,l}|^2}\right)^{3/2}}}} \tag{14}$$

Where $w_{i,j}$ is the gain that is applied to j'th subcarrier of transmitter #i. By using these gains, optimal performance can be obtained under two constraints.

4 RESULT OF SIMULATIONS

In this section, the performance of the proposed system is analyzed by computer simulations. The parameters used in the simulations are shown in Table 4. In

Table 1: Simulation Parameters

Symbol Modulation	16QAM	
Number of subcarriers	1024	
Carrier Interval	4.0	[kHz]
Minimum Carrier Frequency	100.0	[MHz]
Length of Effective Symbol	250.0	[μs]
Length of Guard Interval	31.25	[μs]

this simulation, a simple 2-path frequency selective fading channels(denoted as Channel-1) and 3-path frequency selective fading channels(denoted as Channel-2) from each antenna to the receiver are assumed and their delay spreads are shown in Table 4. In this simulation, it is also assumed that channel transfer functions is perfectly known in the transmitter and receiver. The symbol modulation scheme used here are 16QAM. Bit error rate characteristics against

Table 2: Delay Time

	Path#1	Path#2	Path#3
antenna #1	0	5 [μs]	8.5 [μs]
antenna #2	0	7 [μs]	6.5 [μs]
antenna #3	0	6 [μs]	7.5 [μs]
antenna #4	0	8 [μs]	5.5 [μs]

carrier to noise ratio under static multipath channel in Table 2 is shown in Figure 4 and Figure 5. In Figure 6 and Figure 7, two, three and four transmission antennas are used respectively and 16QAM modulation is used. In the Figure 4 and Figure 5, 'CONV' denotes the case when simple 2-antenna transmission diversity is used. 'DS' denotes the case when data spreading is combined 2-antenna transmission diversity. 'PC+DS(1)' and 'PC+DS(2)' denote the case

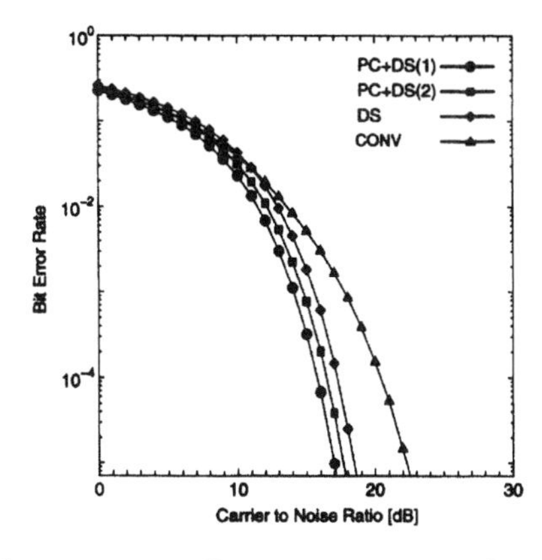

Figure 4: Bit error rate characteristic using 2-antenna transmission diversity(Channel-1)

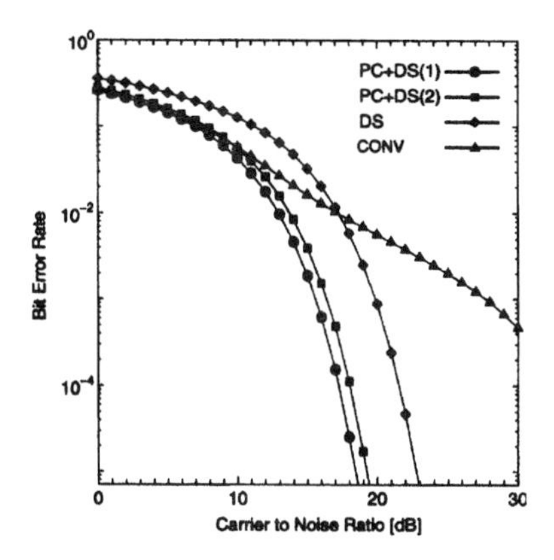

Figure 5: Bit error rate characteristic using 2-antenna transmission diversity(Channel-2)

when transmission power is determined by using Constraint-1 and Constraint-2 respectively. Figure 6 and Figure 7 shows the bit error rate characteristic of the proposed scheme. In this figure, 'TD2', 'TD3' and 'TD4' denote the case when simple 2, 3 and 4-antenna transmission diversity is used, respectively. As shown in the figures, performance of 'CONV' is worst under high carrier to noise ratio and performance of 'DS' is worst under low carrier to noise ratio because influence of noise is spread over the whole carriers by the de-spreading process. However, proposed scheme can improve performance over wide range of carrier to noise ratio. Furthermore, by increasing the number of antennas is more improved by diversity effect. The performance of the proposed scheme under the Constraint-1 has better performance than the one under the Constraint-2. This is because flexibility of transmission power allocation under Constraint-1 is larger than the one under Constraint-2.

5 Conclusion

In this paper, subcarrier power control of OFDM transmission diversity combined with data spreading scheme was proposed and good performance improvement was confirmed by computer simulations. In this analysis, it is assumed that perfect knowledge of channel transfer function at the transmitter is assumed. However, in the actual system, it is necessary to feedback information from the receiver to the transmitter. Therefore, in the further study, more actual system that transmits information bi-directionally is analyzed. Moreover, the system using multiple receiving antennas is analyzed. Moreover, use of more actual constraints of Transmission power such as spectral mask is also investigated.

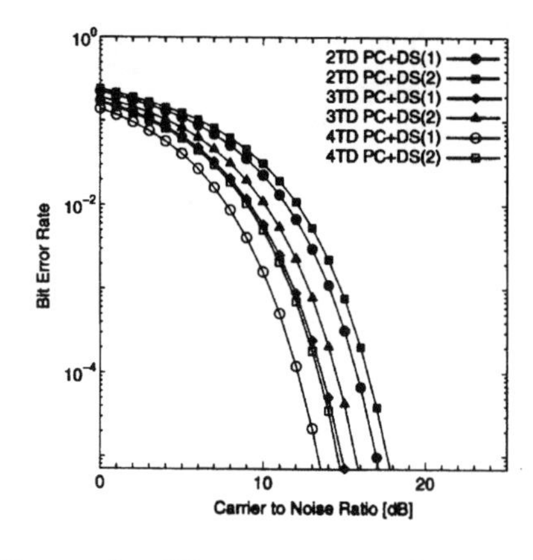

Figure 6: Bit error rate characteristic using multi antenna transmission diversity(Channel-1)

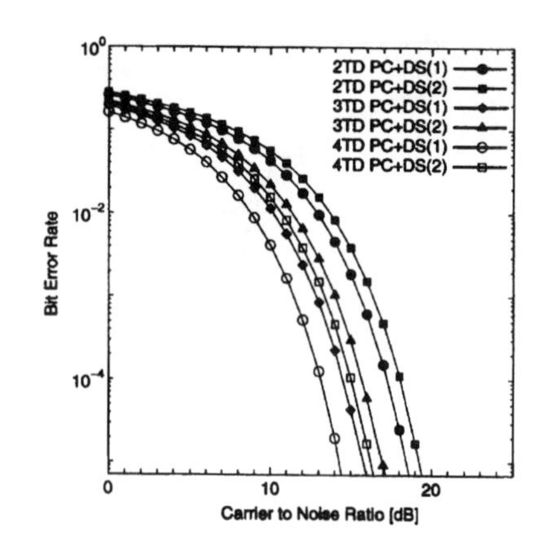

Figure 7: Bit error rate characteristic using multi antenna transmission diversity(Channel-2)

References

[1] R. W. Chang and R. A. Gabby: "A Theoretical Study of Performance of an Orthogonal Multiplexing Data Transmission Scheme", IEEE Trans. Comm., COM-16, pp.529-540, 1968

[2] S. B. Weinstein and P. W. Ebert: "Data Transmission by Frequency-Division Multiplexing using the Discrete Fourier Transform", IEEE Trans. Comm., COM-19, pp.628-634, 1971

[3] B.Hirosaki: "An Analysis of Automatic Equalizer for Orthogonally Multiplexed QAM Systems", IEEE Trans. Comm.,COM-28, pp.73-83, 1980

[4] M.Itami, T.Teramoto, K.Okada, H.Uesugi, S.Hatakeyama, and K.Itoh: "Improving the Error Characteristics of OFDM by Distributing Data Symbol in Frequency Domain", ITE, 51, 9,pp.1468-1475, 1997(Japanese)

[5] T.Nishiyama, S.Nomura, M.Itami, K.Itoh, and H.Aghvami: "A Study on Controlling Transmission Power of Carriers of OFDM Signal Combined with Data Symbol Spreading in Frequency Domain ", Proc. of ISPLC2002, pp.125-129, March 2002

[6] T.Nishiyama, T.Shirai, M.Itami, K.Itoh, and H.Aghvami: "A Study on Controlling Transmission Power of Carriers of OFDM Signal Combined with Data Symbol Spreading in Frequency Domain ", IEICE, VOL.E86-A, NO.8, August 2003

Department of Applied Electronics, Tokyo University of Science
2641 Yamazaki, Noda, Chiba 278-8510, Japan
Tel:+81-4-7124-1501 Ext. 4226, Fax:+81-4-7122-9195
E-Mail:{satoshi,itami,itoh}@itlb.te.noda.sut.ac.jp
Communications Research Laboratory
3-4 Hikarino-oka, Yokosuka, Kanagawa, 239-0847 Japan
E-Mail:ohta

ROBERT NOVAK AND WITOLD A. KRZYMIEŃ

PACKET RE-TRANSMISSION OPTIONS FOR THE SS-OFDM-F/TA SYSTEM

Abstract. This paper examines the benefits of adaptive and asynchronous re-transmission scheduling for the SS-OFDM-F/TA system in terms of data throughput and packet delay. The conditions under which asynchronous and adaptive re-transmission schemes ensure a performance gain are determined. The number of possible re-transmissions for a given format, the maximum allowed time between re-transmissions, and the packet selection criteria within a packet re-transmission scheme are examined.

1. INTRODUCTION

Wireless packet data access systems designed for delay tolerant best-effort services on the cellular downlink have received considerable attention for some time now. Single-carrier packet data systems for the evolution of cdma2000 and UMTS have been proposed and are in various stages of development [1][2]. These systems adaptively allocate radio resources in time, and take advantage of multi-user diversity to increase the average data throughput per sector.

Orthogonal frequency division multiplexing (OFDM) has long been considered in conjunction with link adaptation techniques such as adaptive bit loading and power allocation [3-5], and more recently with multi-user diversity [6-8]. An adaptive spread spectrum OFDM system using frequency and time allocation (SS-OFDM-F/TA) was recently proposed as a multi-carrier solution for future high bit rate packet data wireless access systems [9].

The SS-OFDM-F/TA system exploits multi-user diversity and exploits temporal variations of the transmission channel by transmitting to the mobile experiencing the best propagation conditions in a given sub-band and time slot, where a sub-band is a group of adjacent sub-carriers, over which fading is essentially flat. In addition, spectral nulls present in the transmission band are avoided by transmitting to a mobile with a relatively high signal-to-interference ratio (SIR) over a given frequency sub-band. In this manner, the system uses allocation of radio resources in time and frequency domains to increase the average cell throughput. Proportionally fair scheduling of packets is considered in this paper [10]. Data throughputs of 1.3 to 1.4 b/s/Hz/sector have been achieved using allocation of radio resources in two-dimensions in low-mobility environments with outdated channel information [11], and high-mobility environments with perfect channel prediction [12], using packet sizes identical to those of the IS-856 system [1].

Type II hybrid ARQ employing incremental redundancy and soft packet combining is an important feature of current best effort packet data systems. Transmission formats that allow for multi-slot packet transmission increase the

K. Fazel and S. Kaiser (eds.), Multi-Carrier Spread-Spectrum, 89–100.

number of effective transmission rates, and allow for better adaptation to the channel conditions through earlier termination of re-transmissions, at the cost of increased packet delay. In this paper, we examine synchronous, asynchronous and adaptive re-transmission schemes designed to increase throughput, while maintaining acceptable packet delays.

Section 2 of this paper describes the SS-OFDM-F/TA system structure. Specific configurations and algorithms are presented in Section 3. Section 4 describes the simulation structure, and Section 5 contains the results and discussion. The paper is concluded in Section 6.

2. SYSTEM DESCRIPTION

2.1. SS-OFDM

The SS-OFDM transmitter is shown in Figure 1. The transmission bandwidth occupied by N subcarriers is equally divided into M sub-bands, each containing $L=N/M$ subcarriers. Each of the M frequency sub-bands operates as an independent communications channel. Individual packets are assigned to each sub-band for transmission based on requested data rates from mobiles. QPSK, 8 PSK and 16 QAM modulation constellations and a turbo code with base rate 1/5 are considered as in [13]. The stream of encoded data symbols is serial to parallel converted to form L symbol streams. The L streams are code division multiplexed (CDM). Each stream is spread by an orthogonal Walsh code of length L concatenated with a base station sector specific pseudo-noise (PN) sequence. The L spread streams are synchronously summed to form a single chip stream. The chip stream is serial-to-parallel converted to form L parallel chip streams. The L chip streams are mapped onto adjacent subcarriers along with the chip streams from the other M-1 sub-bands to form the N subcarrier SS-OFDM symbol. Transmission power is divided uniformly across the subcarriers (and hence, sub-bands) in the system.

A guard interval containing a cyclic prefix of duration T_g is added to each SS-OFDM symbol to mitigate inter-symbol interference (ISI). The transmitted signal is disturbed by frequency selective fading and log-normal shadowing. Transmissions from interfering base stations further degrade the desired signal, introducing inter-cell interference (ICI) at the receiver.

The mobile user receives only signals on the sub-band(s) assigned to it. The receiver performs minimum mean squared error (MMSE) equalization to partially restore the orthogonality between packet data streams while minimizing ICI amplification. The packet is turbo decoded and a cyclic redundancy check (CRC) is performed to verify successful packet transmission. Either an acknowledgement (ACK) or negative acknowledgement (NAK) of successful packet reception is transmitted on the reverse link for use in the hybrid ARQ scheme.

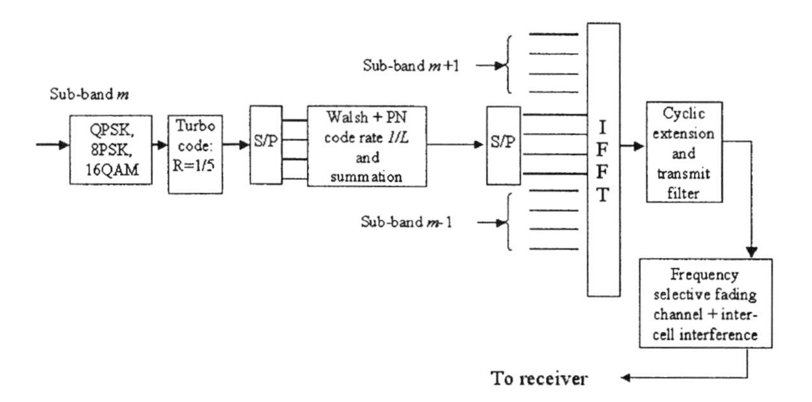

Figure 1. Transmitter structure

2.2. Adaptive Transmission and Single Sub-Band Allocation

Using pilot symbols embedded in each time-frequency frame, each mobile estimates the received signal to noise ratio (SIR) in each sub-band. These SIR estimates are transmitted to the base station by the mobile on the reverse link. For a given mobile, the SIR estimates are compared to a look-up table of possible transmission formats (Table 1). In this manner, a packet is adaptively modulated and coded according to the channel conditions of the sub-band. The delay between the channel estimation at the mobile and transmission at the base station is about 3 slots (one slot is approximately 1.34 ms long).

It should be noted that the requested data rate can be based on out-dated channel information (SIR at the time of request) or channel prediction may be used to predict the channel conditions 3 slots ahead. Adaptive margins are added to the SIR thresholds given in Table 1 to account for imperfect channel estimation and prediction.

Table 1. Packet format sets.

Packet format set A_1			*Packet format set B_2*			*Packet format set C_4*			*Packet format set D_8*		
Max. no. of slots	*Packet size (bits)*	*SIR at 1% PER (dB)*	*Max. no. of slots*	*Packet size (bits)*	*SIR at 1% PER (dB)*	*Max. no. of slots*	*Packet size (bits)*	*SIR at 1% PER (dB)*	*Max. no. of slots*	*Packet size (bits)*	*SIR at 1% PER (dB)*
8	408	-13.1	8	408	-13.1	8	408	-13.1	8	408	-13.1
4	408	-10.0	4	408	-10.0	4	408	-10.0	8	792	-10.5
2	408	-6.9	2	408	-6.9	4	792	-7.3	8	1560	-7.5
1	408	-3.7	2	792	-4.0	4	1560	-4.3	8	2328	-5.0
1	792	-0.7	2	1560	-1.0	4	3096	0.3	8	3864	-2.0
2	2328	1.7	2	2328	1.7	2	3096	3.7	4	3096	0.3
1	1560	5.0	2	3096	3.7	2	3864	5.9	2	3096	3.7
1	2328	8.3	2	3864	5.9	1	2328	8.3	2	3864	5.9
1	3096	10.5	1	2328	8.3	1	3096	10.5	1	2328	8.3
1	3864	13.7	1	3096	10.5	1	3864	13.7	1	3096	10.5
			1	3864	13.7				1	3864	13.7

Table 2. The modulation constellations used with different packet sizes.

Modulation constellation	*Packet sizes (bits)*
QPSK	408, 792, 1560
8 PSK	2328
16 QAM	3096, 3864

The allocation of transmit sub-bands to users in a given time slot is dependent upon the ratio of the data rate requested to the average data rate of packets delivered to a user over a suitable averaging interval (proportionally fair scheduling [10]). This ratio is calculated at the base station for every user. The mobile with the highest ratio will receive transmission in the sub-band(s) and time slot under consideration. The value of the average data rate delivered to each user is updated using a low pass filter with a time constant of N_{ave} = 1000 slots.

This process takes advantage of multi-user diversity, as channel conditions are practically independent for different mobiles. The evaluation and allocation process is completed for every sub-band. Allocation of sub-bands is done on a per packet basis, and hence, after a packet has been transmitted the sub-band is reallocated (except in the synchronous re-transmission case) according to the scheduling algorithm. Furthermore, a mobile may receive transmission on one or more sub-bands, whereas another mobile may not receive any transmission during a given time slot.

2.3. Allocation of Disjoint Sub-Bands

In highly frequency selective environments, a large number of sub-bands is required to fully take advantage of the diversity in the system. In addition, the time slots need to be sufficiently short to ensure the channel does not change significantly. These time-frequency constraints may lead to very small packet sizes, which are undesirable. Sufficiently large packet sizes can be used by allowing several disjoint sub-bands to be allocated together. The formation, as well as the allocation, of these groups of disjoint sub-bands is adaptive. This paper considers allocation of several disjoint sub-bands to a user in sets of M_s=M/M_g, where M_g is the number of sub-band groups. A packet transmission is spread over the M_s sub-bands allocated to it.

For each user, the M_s sub-bands with the highest SIRs are grouped together. The SIR over the group of sub-bands is compared to the data rate thresholds. The SIR of the group of sub-bands can be accurately determined from SIRs in each sub-band [11][12]. The user with the highest estimated to average data rate ratio will receive transmission as described for single sub-band allocation in Section 2.2. The process is repeated for the remaining M-M_s sub-bands until all sub-bands have been allocated.

3. RE-TRANSMISSION OPTIONS

3.1. Synchronous and Asynchronous Re-Transmission

Synchronous re-transmission was originally proposed for the IS-856 high bit rate packet data system [1]. After a packet transmission is scheduled to a user, re-transmissions occur every 4 slots regardless of the channel conditions until the packet is successfully delivered or the maximum allowed number of re-transmissions is reached. The 4-slot delay is necessary in order to receive positive or negative packet transmission acknowledgement from the receiver [1]. Provided the channel changes very slowly, using synchronous re-transmissions results in low packet delay, minimal signalling and high throughput.

At higher mobile speeds, synchronous re-transmissions do not take advantage of the multi-user diversity as the channel significantly changes between re-transmissions. Recently, an adaptive and asynchronous re-transmission method was proposed to ensure that the multi-user diversity is exploited by each re-transmission [14][15].

Asynchronous re-transmissions occur only when the user is selected by the scheduler (in the same manner as for new packet transmissions). In the SS-OFDM-F/TA system, the re-transmissions can also occur in any sub-band [11]. Due to the parallel structure of the system, there may be many partially transmitted packets to choose from for each user. In this paper, we consider the minimum mobile buffer memory size of $4M/M_g$ packets. The following methods are used to select a packet for re-transmission, or start a new packet transmission, once the scheduler has selected a user.

Minimum delay, re-transmit priority (MD-RP) - The packet that has gone without re-transmission the longest is selected for re-transmission.

Same format, no re-transmit priority (SF-NRP) – The packet of exactly the format appropriate for the channel condition is selected for re-transmission. If no such packet is awaiting re-transmission, a new packet transmission is started.

Minimum delay with supportable format, no re-transmit priority (MDSF-NRP) - The packet that has gone without re-transmission the longest, and is of a format that can be supported by the immediate channel conditions, is selected for re-transmission. If no such packet is awaiting re-transmission, a new packet transmission is started.

Minimum delay, re-transmit priority with parallel allocation (MD-RP-PA) – Let the number of sub-bands (or groups of sub-bands) assigned to the kth user in a given time slot be M_k. As with MD-RP, select the M_k packets that have been waiting the longest for transmission. Of these M_k packets, assign the one with the lowest transmission format to the sub-band (or group of sub-bands) with the lowest SIR. Continue this process for the next M_k–1 packets.

Minimum delay, with supportable format, no re-transmission priority with parallel allocation (MDSF-NRP-PA) – Find the M_k packets that meet the MDSF-NRP criteria. Of these M_k packets, assign the one with the lowest transmission format to the sub-band (or group of sub-bands) with the lowest SIR. Continue this process for the next M_k–1 packets. If less than M_k packets meet the criteria, start new packet transmissions.

A distinct alternative to these algorithms is to re-transmit a packet when the channel conditions are as good as, or better than when the packet was initially transmitted. To this end, we consider:

Minimum delay, with supportable format channel state information based (MDSF-CSIB) – For all users, find the packets that require re-transmission and of a format that can be supported by the immediate channel conditions. Of these packets, transmit the packet that has been waiting the longest. If no such packet exists, transmit a packet to the user selected by the proportionally fair scheduler.

3.2. Maximum Number of Re-Transmissions

A significant factor in the performance of asynchronous and synchronous re-transmission schemes is the number of allowable re-transmissions for a given packet. For example, if the SIR estimate is not reliable it may be advantageous to use a packet size of 792 bits over 8 slots, rather than a 408 bit packet over 4 slots. Four data packet format sets are used: A_1, B_2, C_4 and D_8, with the subscript of each set indicating the maximum number of allowed re-transmissions.

3.3. Maximum Re-Transmission Interval

Aside from maximizing the average throughput per sector, re-transmission schemes that minimize the packet delay and average packet delay per user are desirable. In this paper, we impose limits on the maximum delay between re-transmissions of a given packet to ensure a relatively small packet delay. If a packet has been awaiting re-transmission longer the maximum re-transmission interval, then re-transmission of the packet to the given user is immediately scheduled, regardless of channel conditions. Note that in the case of the allocation of disjoint sub-bands as described in Section 2.3, the sub-bands are re-allocated to offer the best possible channel conditions for the forced re-transmission.

3.4. Adaptive Re-Transmission

In general, this paper considers the use of all spreading sequences in a sub-band time slot for a single transmission or re-transmission. It is also possible to transmit two packets in the same sub-band time slot, to the same user, using code division. If the SIR is sufficiently large at a given time, a larger modulation constellation and only a subset of the spreading sequences can be used for the packet re-transmission (this concept was presented in [15] for a multi-service, single carrier system). The left over spreading sequences can then used for another (re)-transmission. This process requires feedback from the mobile of the extra energy required to successfully complete the transmission. Table 3 indicates the possible options.

Table 3. Possible formats of packet transmission to be completed using a larger constellation, and formats of the additional packet to be sent in the same time slot/sub-band

Original packet modulation	*Larger constellation applied*	*Packet sizes of the additional transmission*	*Constellation of the additional transmission*
QPSK	8 PSK	408, 792	8 PSK
	16 QAM	408, 792, 1560	QPSK or 16 QAM
8 PSK	16 QAM	408, 792	8 PSK

4. SIMULATION STRUCTURE AND PARAMETERS

The simulations are partitioned into link level and system level simulations. In the link level simulations, curves of the packet error rate (PER) in additive white Gaussian noise (AWGN) are generated for all the transmission formats given in Table 1, accounting also for early terminations. The PER curves are employed in the system level simulations to determine success or failure of packet transmissions. Employing the AWGN results in the system level simulations is reasonable, since the channel is practically time-invariant during each packet time slot.

The simulation is organized into 30-second runs. At the beginning of each run, users are dropped into an embedded sector. During the simulation run, shadowing and frequency selective Rayleigh fading vary accordingly. Path loss is assumed constant during each run. 100 sets of runs are completed for each case, with each run considering a different set of user positions. Users are distributed uniformly over the area of the sector.

Table 4. Simulation parameters.

	Indoor	*Pedestrian*
N (subcarriers)	256	512
M, M_s	8, 1	32, 8[1]
Bandwidth	10 MHz	5 MHz
Main lobe bandwidth	9.4787 MHz	4.7393 MHz
T_s (SS-OFDM symbol duration)	27.008 μs	108.032 μs
T_g (cyclic prefix)	1 μs	4 μs
Time slot length	1.344384 ms	1.344384 ms

The simulation parameters for the indoor and pedestrian environment are listed in Table 4. Path loss and fading models are taken from ITU recommendations for the Indoor Office B, and the Pedestrian B channels [16]. In this paper, we primarily consider mobiles in outdoor pedestrian environments, however these mobiles may be moving very slowly, or at low vehicular speeds (e.g. passengers in vehicles in an urban environment). We therefore consider the pedestrian B channel with maximum Doppler shifts from 5 Hz to 100 Hz, corresponding to mobile speeds of up to approximately 55 km/h at a carrier frequency of ~2 GHz. It can be noted that the

[1] Refers to M total sub-bands, allocated in sets of M_s=8 sub-bands.

ITU Pedestrian B channel is very similar to the ITU Vehicular A channel. The maximum Doppler shift in the Indoor channel is f_D =5 Hz. Autocorrelation of shadowing is exponential [16], with a decorrelation distance of 5 metres. The standard deviations of the log-normal shadowing processes are 12 dB and 10 dB in the Indoor and Pedestrian environments respectively [16]. Cross-correlation of the shadowing from sector antennas of the same cell is assumed equal to 1, and 0.5 between different base station sites. A 19 hexagonal cell cluster, with sectored cells (3 sectors per cell) is considered. A horizontal radiation pattern from a typical commercial antenna with a 65° 3 dB beamwidth is used.

Slow and fast sector selection is considered in this paper. In slow sector selection, the choice of a sector is based on values of path loss and shadowing at the beginning of each 30-second run. Fast sector selection (1 Hz) is based on a running average of path loss and shadowing values with a time constant of 1 second. It is assumed that the number of users communicating with each sector transmitter is constant during a given simulation run.

Each packet transmission over a sub-band, or group of sub-bands, during a given time slot consists of 1536 complex code chips, 1200 of which carry data, and 336 are reserved for pilot and MAC layer signalling (21.875%). Adaptive data rate selection margins are used to ensure a packet error rate of approximately 1% in each run, for each rate. Additional adaptive margins are employed in the adaptive re-transmission algorithm. As minimal reverse link signalling is desirable, SIR feedback with 3 dB and 6 dB quantization increments is also considered.

5. RESULTS

The re-transmission options described in Section 3 have been simulated extensively in the Pedestrian B channel. Asynchronous re-transmission of packets provides a higher average data throughput per sector than using synchronous re-transmission in most cases. Asynchronous re-transmissions are particularly useful in the cases of a maximum Doppler shift of 50 Hz and no channel prediction (7% throughput gain), and 100 Hz with channel prediction (10% throughput gain). A maximum of 3.9 Mb/s per sector was achieved at 100 Hz and with no prediction. Unless otherwise stated, asynchronous re-transmissions in this paper are assumed to be with MD-RP-PA.

All asynchronous re-transmission schemes examined in this paper provide relatively the same average throughput per sector. Nevertheless, the SF-NRP and MD-RP-PA algorithms perform the best. The algorithms employing parallel allocation (MD-RP-PA, MDSF-NRP-PA) perform slightly better than those without it (MD-RP, MDSF-NRP).

Packet delays are significantly different between different asynchronous re-transmission algorithms. The distribution of packet delays (which include head-of-queue delay and transmission delay) and the average packet delay per user reveal large delays associated with the SF-NRP algorithm, and small delays for the MD-RP-PA algorithm. Delays associated with the synchronous system are shorter than for all the asynchronous options, and hence, employing asynchronous re-transmissions trades-off more reliable re-transmissions for longer packet delays.

Imposing a maximum on re-transmission intervals reduces the delay associated with asynchronous re-transmission. Under a Doppler shift of 50 Hz, forcing a re-transmission every 40 slots reduces the packet delay per user by nearly half with a negligible loss in the average throughput per sector. In the case of perfect prediction and 100 Hz maximum Doppler shift, an *RTI* =10 slots results in a throughput loss of < 2%, with nearly the same delay as for the synchronous re-transmission scheme.

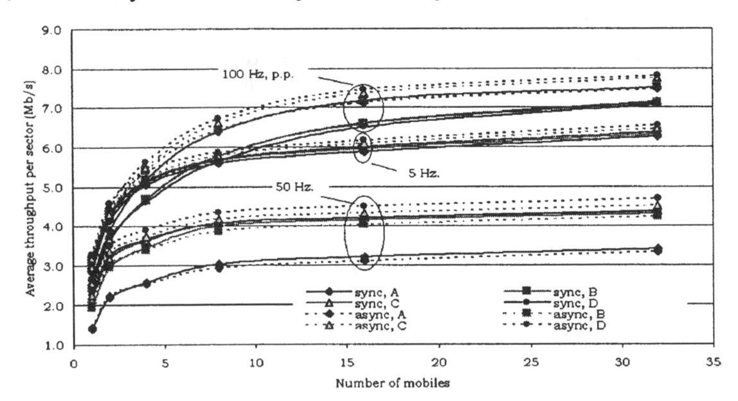

Figure 2. Average throughput per sector in the 5 MHz Pedestrian channel; asynchronous algorithm is MD-RP-PA.

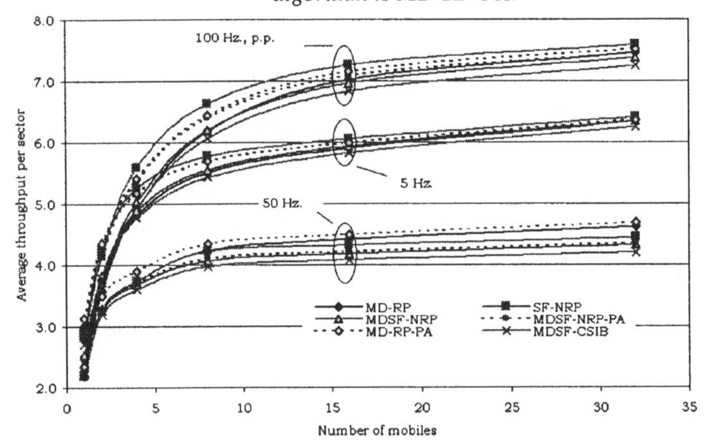

Figure 3. Throughput with different asynchronous algorithms; packet format set D_8.

Aside from providing a higher average throughput, asynchronous re-transmissions primarily benefit users with poorer channel conditions. For example, the average data throughput delivered to the 20th to 50th percentile users is increased by 17% over that for synchronous re-transmissions in a 100 Hz Doppler channel with prefect prediction.

SIR feedback quantization negatively affects the performance of the system, particularly when transmission formats with a low number of re-transmissions are used. However, asynchronous re-transmission with 3 dB quantization increments and packet format set D_8 results in the same throughput as synchronous re-transmission with perfect SIR estimates, and only a 2% loss in throughput compared to asynchronous re-transmission with perfect SIR estimates. Asynchronous re-transmission schemes perform better than synchronous for larger quantization intervals.

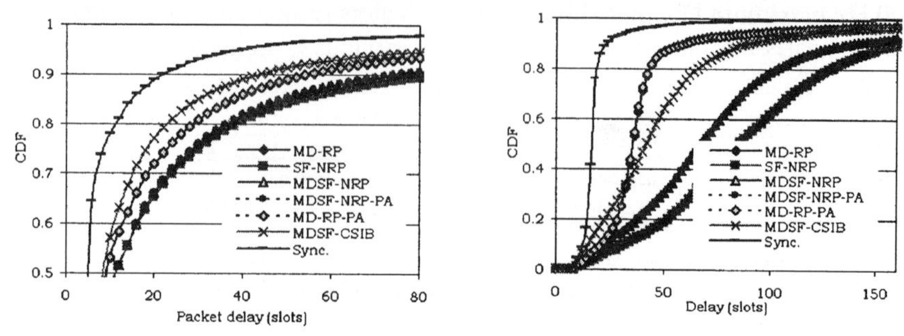

Figure 4. Packet delays (left) and average packet delays per user (right) in the Pedestrian channel; 50 Hz Doppler shift, packet format B_2, K=32.

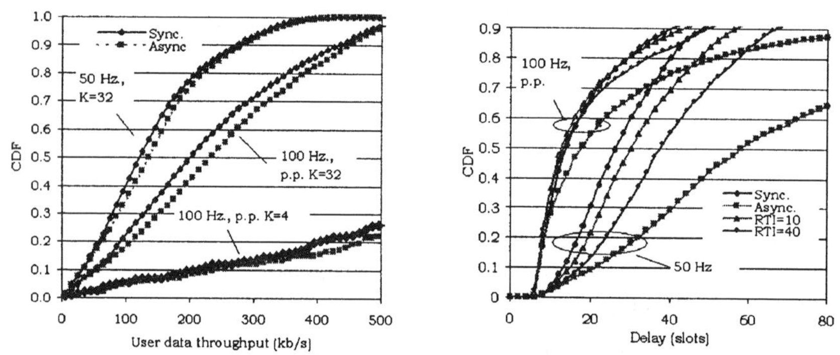

Figure 5. The distribution of average user throughputs with D_8.

Figure 6. Average delays per user with a maximum on re-transmission intervals (RTI); K=32, packet format D_8.

Using adaptive re-transmission in order to allow an additional packet transmission with code division provided only modest average throughput per sector gains (< 3%). This scheme increased throughput delivered to users with poorer channel conditions, as such users are more likely to select transmission formats with QSPK or 8 PSK constellations. For the case of K=32 users, a small increase in average throughput for these users was observed along with a decrease of up to 6% in the total average packet delay per user. The gain in the average throughput per user was more pronounced at K=4 users; for example, a nearly 20% (370 to 440 kb/s) gain was observed for the 20th percentile users with f_D =5 Hz.

In the indoor channel, the average data throughput per sector is lower with asynchronous re-transmissions. This is because users experiencing very poor channel conditions can be served with asynchronous re-transmissions, whereas with synchronous re-transmissions, these users are served less frequently, or not at all.

Serving such low throughput users increases fairness, but lowers the aggregate throughput per sector. Data throughput per sector with 3-dB SIR quantization is also shown.

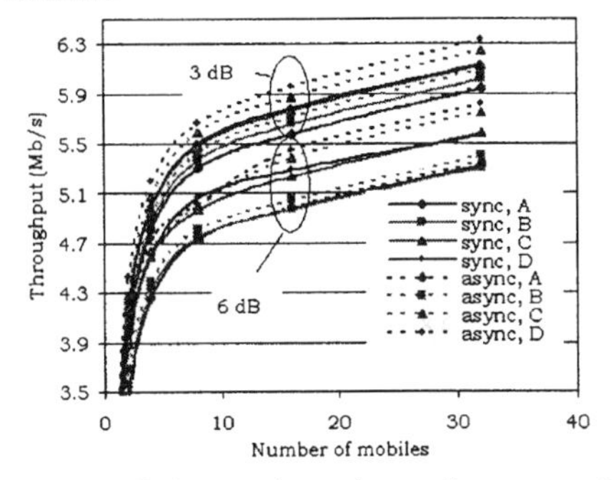

Figure 7. Sector throughput of systems with quantization; f_D =5 Hz.

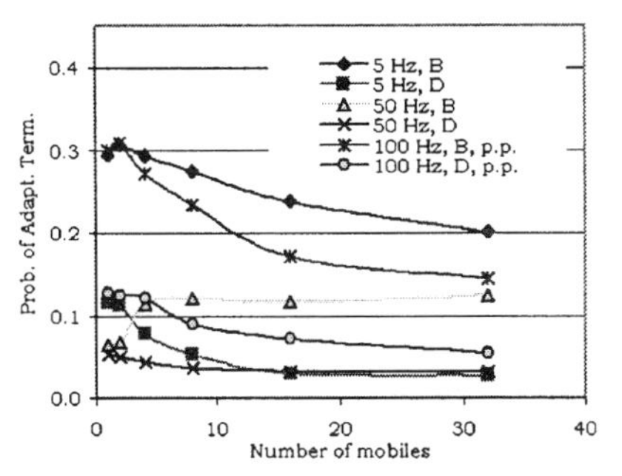

Figure 8. Probability that a packet will be terminated using a larger constellation size.

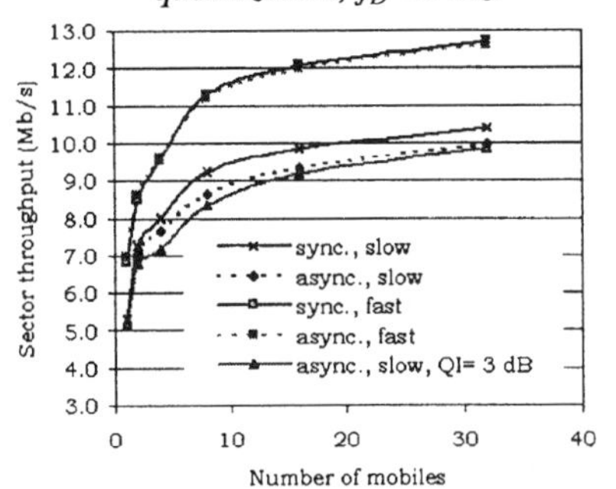

Figure 9. Throughput per sector in the Indoor B channel with fast and slow sector selection; packet format D_8.

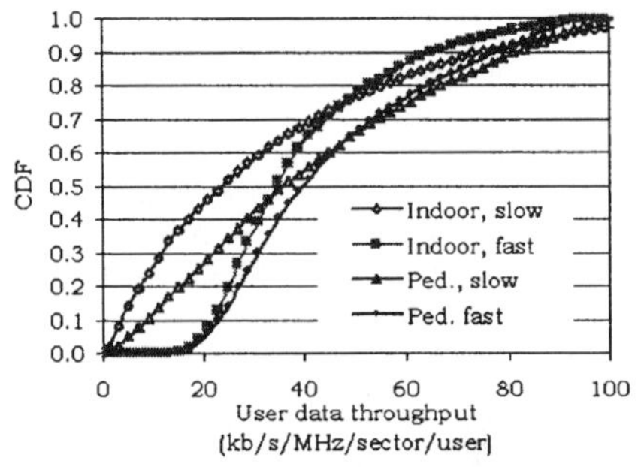

Figure 10. Distribution of throughputs per user with fast and slow sector selection; K=32 users, packet format D_8.

Fast sector selection dramatically increases throughput per sector, particularly for users experiencing poor channel conditions. For the pedestrian channel with fast sector selection, throughputs of 7.2 and 8.3 b/s/Hz/sector at 5 Hz and 100 Hz (with perfect prediction) maximum Doppler shifts, respectively, have been observed.

6. CONCLUSIONS

The paper has considered several asynchronous re-transmission algorithms, of which the MD-RP-PA scheme yields high throughput with relatively low delays. A maximum re-transmission time interval constraint can lower the delays associated with asynchronous re-transmission, without sacrificing throughput. An adaptive re-transmission scheme that uses additional information from the mobile was also examined. Only moderate gains have been observed with this scheme as asynchronous re-transmission and the large number of transmission formats allow for adequate adaptation to the channel.

It should be noted that data throughputs presented here could be greatly increased by using larger constellations. However, we have considered only constellations up to and including 16 QAM in order to avoid the need to use complex receiver structures.

7. ACKNOWLEDGMENTS

The authors gratefully acknowledge funding for this work provided by TRLabs, the Alberta Informatics Circle of Research Excellence (iCORE), and the Natural Sciences and Engineering Research Council (NSERC) of Canada.

Robert Novak and Witold A. Krzymień
University of Alberta / TRLabs
2nd Floor ECERF, 9107 – 116 Street
Edmonton, Alberta, Canada T6G 2V4
wak@ece.ualberta.ca

8. REFERENCES

[1] TIA/EIA/IS-856, "cdma2000 High Speed Packet Data Air Interface Specification", Telecommunications Industry Association, Arlington, VA, Nov. 2000.

[2] S. Parkvall et al, "The evolution of WCDMA towards higher speed downlink packet data access", in *Proc. VTC 2001 Spring*, Rhodes, Greece, May 2001.

[3] Kalet, "The multitone channel", *IEEE Trans. on Commun.*, vol. 37, no. 2, pp. 119-124, Feb. 1989.

[4] P. S. Chow, "A practical discrete multitone transceiver loading algorithm for data transmission over spectrally shaped channels", *IEEE Trans. on Commun.*, vol. 43, no. 2/3/4, pp. 773-775, Feb./Mar./Apr. 1995.

[5] T. Keller, L. Hanzo, "Adaptive multicarrier modulation: a convenient framework for frequency-time processing in wireless communications", *IEEE Proceedings*, May 2000, pp. 611-640.

[6] C. Y. Wong et al, "Multiuser OFDM with adaptive subcarrier, bit and power allocation", *IEEE JSAC*, vol. 17, no. 10, pp. 1747-1758, Oct. 1999.

[7] R. Grünheid, H. Rohling, "Adaptive modulation and multiple access for the OFDM transmission technique", *Wireless Personal Communications*, vol. 13, No. 1, pp. 5-13, May 2000.

[8] W. Rhee, J. M. Cioffi, "Increase in capacity of multiuser OFDM system using dynamic subchannel allocation", in *Proc.VTC'00 Spring*, Toyko, Japan. Sept. 2000.

[9] R. Novak, W. A. Krzymień, "A downlink SS-OFDM-F/TA packet data system employing multi-user diversity", in *Proc. of the 3rd Intl. Workshop on Multi-Carrier Spread-Spectrum (MCSS 2001)*, Oberpfaffenhofen, Germany, pp. 181-190, Sept. 2001.

[10] J. M. Holtzman, "CDMA forward link waterfilling power control", in *Proc. VTC2000 Spring*, Tokyo, Japan, May 2000.

[11] R. Novak, W. A. Krzymień, "An adaptive downlink spread spectrum OFDM packet data system with two-dimensional radio resource allocation: performance in low-mobility environments", in *Proc. WPMC'02*, Honolulu, HI, Oct. 2002.

[12] R. Novak, W. A. Krzymień, "SS-OFDM-F/TA system packet size and structure for high mobility cellular environments", in *Proc. VTC'03-Spring*, Jeju, Korea, April 2003.

[13] TIA/EIA/IS-2000, "Physical layer standard for cdma2000 spread spectrum systems; Release C", Telecommunications Industry Association, Arlington, VA, May 2002.

[14] A. Das, F. Khan, and S. Nanda, "A2IR: an asynchronous and adaptive hybrid ARQ scheme for 3G evolution", in *Proc. VTC'2001-Spring*, Rhodes, Greece, May 2001.

[15] A. Das, F. Khan, A. Sampath, and H.J. Su," Adaptive, asynchronous incremental redundancy with fixed transmission time intervals (TTI) for HDSPA", in *Proc. PIMRC'2002*, Lisbon, Portugal, Sept. 2002.

[16] ETSI, "Universal Mobile Telecommunications System (UMTS); Selection Procedures for the Choice of Radio Transmission Technologies of the UMTS", TR 101 112 v3.2.0, April 1998.

COMPARISON OF ITERATIVE DETECTION SCHEMES FOR MIMO SYSTEMS WITH SPREADING BASED ON OFDM

D. Y. YACOUB, M. A. DANGL, U. MARXMEIER, W. G. TEICH, J. LINDNER
University of Ulm, Dept. of Information Technology
Albert-Einstein-Allee 43, 89081 Ulm, Germany

Abstract. In this paper we consider a coded transmission over a time-invariant MIMO channel, where channel state information is present at the receiver but not at the transmitter. As basic modulation technique we apply OFDM. In addition to MC-CDM spreading techniques, we introduce a new class of spreading matrices that works both in antenna and frequency direction, referred to as Multi-Carrier Cyclic Antenna Frequency Spreading (MC-CAFS). In the case of OFDM crosstalk only occurs because of the use of multiple transmit antennas, whereas systems with spreading are additionally affected by interference among subcarriers. Therefore, we utilize advanced joint detection techniques with moderate computational complexity like iterative detection, demapping, and decoding to mitigate the interference in both cases.

1. INTRODUCTION

Multicarrier (MC) transmission schemes are considered to be promising candidates for the fourth generation of mobile communication systems. Among several MC variants, OFDM represents a well-known technique and is used in broadcast media like, e. g. , European terrestrial digital television (DVB-T) and digital audio broadcasting (DAB). However, it may happen that one or more subcarriers of an OFDM system are completely faded out due to the frequency-selective behavior of the channel. Spreading over the subcarriers (MC-CDM) is a technique to overcome this problem, see, e.g., [1] and [2]. In the following, we investigate a coded data transmission of a single user over a time-invariant MIMO (multiple input multiple output) channel with OFDM as basic modulation technique. We assume that channel state information is present at the receiver but not at the transmitter. Besides spreading in frequency direction (MC-CDM) as investigated in [3] we consider additional spreading in the spatial dimension [4] that represents a special case of space frequency coding [5]. Therefore, we propose a new class of spreading matrices operating on both the frequency and spatial dimension (Multi-Carrier Cyclic Antenna Frequency Spreading, MC-CAFS). Furthermore, we investigate the performance of soft interference cancellation schemes with moderate computational complexity for both systems with spreading (MC-CDM, MC-CAFS) and without spreading (OFDM). The applied detectors are known as Recurrent Neural Network (RNN) detector [6], [7], [8] and Soft Cholesky Block Decision Feedback Equalizer (SCE) [9]. In the coded case we apply detection techniques like iterative detection, demapping and decoding [10] to reduce interference. We demonstrate that MC-CAFS, combined with appropriate coding, improves the bit error rate (BER) performance for the applied detection techniques although the level of interference is higher.

K. Fazel and S. Kaiser (eds.), Multi-Carrier Spread-Spectrum, 101–112.

In this section, we introduce a MIMO-OFDM system model with n_T transmit and n_R receive antennas. We apply a discrete-time vector-valued transmission model that allows a compact notation of the system, see, e. g., [2]. Throughout this paper vectors are defined as column vectors.

First we start with a single input single output (SISO) system and then extend it to the MIMO case. In the SISO-OFDM system, the available bandwidth is divided into N subchannels (subcarriers) allowing N symbols to be simultaneously transmitted. The orthogonality between the subcarriers is obtained by adding a cyclic prefix to the transmitted vector. If the length of the cyclic prefix is greater than or equal to the length of the channel impulse response (CIR) each of the subchannels experiences flat fading even if the original channel is frequency-selective. In the following, we assume a time-invariant multipath channel using a T-spaced model of L taps with CIR $\underline{h}_{\text{ch}} = [h_0, \ldots, h_{L-1}]^T$. Furthermore, the length of the cyclic prefix shall be as large as the length of the CIR. The vector $\underline{x}[k]$ represents the OFDM symbol of size N at time k. Its elements are taken from the M-ary modulation alphabet $\mathcal{A}_x = \{a_1, \ldots, a_M\}$ that may be complex-valued. Modulation and demodulation can be efficiently implemented through the use of the inverse discrete Fourier transform (IDFT) and the discrete Fourier transform (DFT). The $N \times N$ diagonal matrix $\underline{H}$ contains the transfer function of the channel. $\underline{H}$ is given by:

$$\underline{H} = \underline{\text{diag}}(\text{DFT}(\underline{h})), \tag{1}$$

where $\underline{h} = [\underline{h}_{\text{ch}}^T, 0, \ldots, 0]^T$ is the zero padded CIR of length N and "$\underline{\text{diag}}$" stands for composing a diagonal matrix. At the receiver, perfect channel state information is assumed. Thus, the phase rotation caused by the channel is compensated by the matrix $\underline{H}^H$, where $(\cdot)^H$ denotes the conjugate complex transpose operation. Thus, the received OFDM symbol vector $\underline{\tilde{x}}[k]$ of size N at time k is obtained by:

$$\underline{\tilde{x}}[k] = \underline{R}_{\text{SO}}\, \underline{x}[k] + \underline{\tilde{n}}[k], \tag{2}$$

where $\underline{R}_{\text{SO}} = \underline{H}^H \underline{H}$ is the channel matrix for the SISO-OFDM system. The vector $\underline{\tilde{n}}[k]$ is a vector of colored noise samples with covariance matrix $2N_0\, \underline{R}_{\text{SO}}$. However, when using real-valued symbol alphabets only $N_0\, \underline{R}_{\text{SO}}$ has to be considered in the detector. The component variances of the complex AWGN process before the channel matched filter are $\sigma_n^2 = N_0$.

In the MIMO-OFDM case, N subcarriers are available to each transmit antenna and are the same for all antennas. Thus, the MIMO-OFDM system model can be thought of as four SISO-OFDM systems transmitting at the same time, while using the same frequencies, with however two important differences: transmit antenna cooperation and joint detection at the receiver. Accordingly, the $N \times (n_T N)$ MIMO-OFDM matrix for the ith receive antenna $\underline{H}_i$ can be defined as $\underline{H}_i = [\underline{H}_{i,1}, \ldots, \underline{H}_{i,n_T}]$. The submatrices $\underline{H}_{i,j}$ are $N \times N$ diagonal matrices and given by:

$$\underline{H}_{i,j} = \underline{\text{diag}}(\text{DFT}(\underline{h}_{ij})), \tag{3}$$

where $\underline{h}_{ij}$ is the zero padded CIR of length N between transmit antenna j ($j = 1, \ldots, n_T$) and receive antenna i ($i = 1, \ldots, n_R$). The channel matrix $\underline{R}_{\text{MO}}$ of the MIMO-OFDM system with n_R receive antennas has thus dimension $(N\, n_T) \times (N\, n_T)$ and is given by:

$$\underline{R}_{\text{MO}} = \sum_{i=1}^{n_R} \underline{R}_{\text{MO},i} = \sum_{i=1}^{n_R} \underline{H}_i^H\, \underline{H}_i. \tag{4}$$

Since $\underline{R}_{\text{MO}}$ is actually a sum of n_R channel matrices, each corresponding to one receive antenna, zero elements on the main diagonal become unlikely by increasing n_R. Spreading, whether in frequency and/or antenna direction, can be represented through multiplication by a real valued spreading matrix, $\underline{U}$, and the resulting channel matrix, R_{S}, is thus

$$\underline{R}_{\text{S}} = \underline{U}^H\, \underline{R}_{\text{MO}}\, \underline{U}. \tag{5}$$

In General, Space-Time Block Codes can also be presented by (5). However, in this case real valued notations are necessary [11]. Note that (2) can be also applied to the MIMO case by replacing $\underline{R}_{\text{SO}}$ with $\underline{R}_{\text{MO}}$ or $\underline{R}_{\text{S}}$, respectively and increasing the dimension of the vectors from N to $N \cdot n_T$.

Fig. 1 depicts the discrete-time vector-matrix model, when additional coding is included. For ease of notation the time index k is omitted. The source symbol vectors $\underline{q}$ are encoded with a terminated convolutional code or a turbo code, optionally punctured, and permuted by a random interleaver $\underline{\Pi}$. The code vectors $\underline{c}$ are mapped on to symbols from the M-ary symbol alphabet $\mathcal{A}_x$. We assume that the symbol vector $\underline{x}$ is subdivided into blocks of length $N \cdot n_T$, which are transmitted over the discrete-time MIMO channel matrix in parallel. $\underline{R}$ denotes either the channel matrix $\underline{R}_{\text{MO}}$ for MIMO-OFDM or $\underline{R}_{\text{S}}$ for MIMO systems with spreading as introduced above.

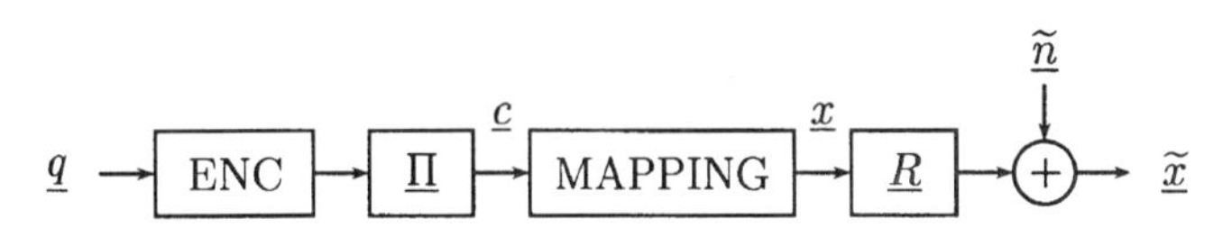

Figure 1. Matrix-vector model with coding.

3. SPREADING

In the uncoded SISO-OFDM transmission, the Rayleigh fading statistics of the subcarriers lead to poor bit error performance, which, however, can be improved through diversity transmission [12]. In this case, the energy of each transmit symbol is spread over several of the Rayleigh fading subcarriers. It was shown in [13] that, after equalization and due to diversity gain in the case of large spreading, the original channel is transformed into a set of parallel AWGN channels. Since, no channel knowledge was assumed at the transmitter, the elements of the spreading matrix were all chosen to be of the same amplitude. In addition, only orthogonal matrices were used for spreading. Similarly, spreading can also be applied to the MIMO-OFDM system. It is also assumed that no channel knowledge is available at the transmitter and accordingly only orthogonal spreading matrices are considered. There are, however, a few differences between the two OFDM systems. First, as described in Section 2, the MIMO-OFDM discrete-time channel matrix is non-diagonal. Second, it is unlikely for large n_R that one or more of the main diagonal elements suffers from total fading. Third, spreading in the SISO case is only done in frequency direction. Employing several transmit antennas provides a new dimension (spatial dimension) that can be exploited for spreading. In this section, several spreading matrices will be discussed.

3.1. Spreading Criteria

Before examining some of the spreading matrices for MIMO-OFDM, we first need to cover the criteria behind choosing them. For that, the following definitions are needed. In all what follows, we shall limit ourselves to orthogonal spreading matrices.

Definition 1: Mean Square Deviation of the diagonal elements, MSD
We define the MSD, α, as follows

$$\alpha = \frac{1}{n}\sum_{i=1}^{n}(r_{ii} - \bar{r})^2, \tag{6}$$

where $\bar{r}$ is the mean of the diagonal elements of any discrete time channel matrix, $\underline{R}_{MO}$ or $\underline{R}_S$, r_{ii} its diagonal elements and $n = N \cdot n_T$. The channel matrix was normalized such that $\bar{r} = 1$. In the ideal case, $\alpha = 0$, i. e. all diagonal elements are equal.

Definition 2: Average square value of the Off-Diagonal Elements, ODE
We define the ODE as

$$\beta = \frac{1}{n}\frac{1}{n-1}\sum_{i \neq j}|r_{ij}|^2, \tag{7}$$

where r_{ij} are the off-diagonal elements of $\underline{R}_{MO}$ or $\underline{R}_S$. The goal is to minimize the ODE (in the ideal case $\beta = 0$), which is directly related to the amount of interference. However, minimizing the ODE contradicts with minimizing the MSD. This is because the two are related via the Frobenius norm, $\| \underline{R} \|_F^2 = \sum_i \sum_j |r_{ij}|^2$, as follows:

$$\beta = -\frac{1}{n-1}\alpha + \frac{\| \underline{R} \|_F^2 - n}{n(n-1)}. \tag{8}$$

Clearly, (8) is a linear equation, whose slope is negative and inversely proportional to $n - 1$. Hence, the ODE decreases as the MSD increases. Nonetheless, the ODE becomes independent of MSD for large n and thus we can concentrate on minimizing the MSD as one of the criteria behind choosing a suitable spreading matrix.

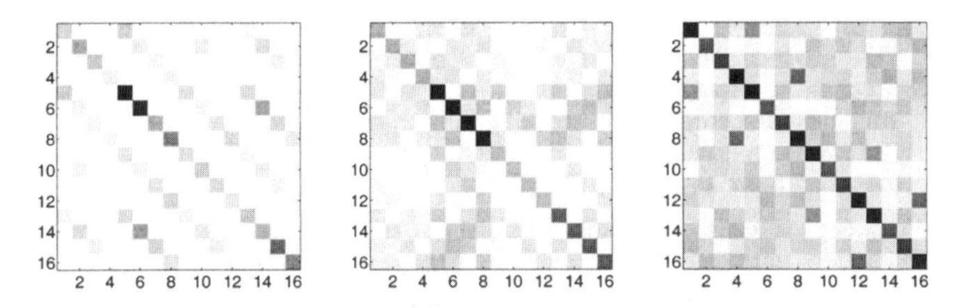

Figure 2. Channel matrices: from left to right OFDM ($\underline{R}_{\mathrm{MO}}$), MC-CDM ($\underline{R}_{\mathrm{MM}}$), MC-CAFS ($\underline{R}_{\mathrm{MCAF}}$). The matrices were normalized, such that the maximum amplitude is 1 (black boxes), white boxes denote zero amplitude.

In order to better describe the second criteria, it is advantageous to write the MIMO-OFDM channel matrix, $\underline{R}_{\mathrm{MO}}$, as a sum of two matrices, a diagonal matrix (D) and a non-diagonal matrix (ND),

$$\underline{R}_{\mathrm{MO}} = \underline{R}_{\mathrm{MO,D}} + \underline{R}_{\mathrm{MO,ND}} \,. \tag{9}$$

Accordingly, after spreading the resulting channel matrix, $\underline{R}_{\mathrm{S}}$, can also be expressed as a sum of 2 matrices, $\underline{R}_{\mathrm{S,D}}$ and $\underline{R}_{\mathrm{S,ND}}$:

$$\underline{R}_{\mathrm{S}} = \underline{U}^H \underline{R}_{\mathrm{MO,D}} \underline{U} + \underline{U}^H \underline{R}_{\mathrm{MO,ND}} \underline{U} = \underline{R}_{\mathrm{S,D}} + \underline{R}_{\mathrm{S,ND}}. \tag{10}$$

Note that neither of these two matrices is diagonal. If spreading is done over all possible dimensions, the diagonal elements of $\underline{R}_{\mathrm{S,D}}$ will all be equal. However, the diagonal elements of $\underline{R}_{\mathrm{S,ND}}$ will most likely contribute to those of the resulting $\underline{R}_{\mathrm{S}}$ matrix. The effect of the diagonal elements of $\underline{R}_{\mathrm{S,ND}}$ will depend on the magnitude and sign of the elements $\underline{R}_{\mathrm{MO,ND}}$. These are unpredictable to a transmitter that has no channel knowledge. Thus, and also based on the results obtained from (8), only spreading matrices that satisfy the following equation shall be considered:

$$\underline{\mathrm{diag}}(\underline{U}^H \underline{R}_{\mathrm{MO,ND}} \underline{U}) = \underline{\mathrm{diag}}(\underline{R}_{\mathrm{S,ND}}) = \underline{\mathrm{diag}}(\underline{0}). \tag{11}$$

That is, the diagonal elements of $\underline{R}_{\mathrm{S,ND}}$ have to vanish.

3.2. Spreading in Frequency Direction, MC-CDM

Similar to the SISO-MC-CDM case, symbols of the MIMO-OFDM system may also be spread over several or all of the available subcarriers. We consider a 4×4 ($n_T = n_R = 4$) MIMO system with $N = 4$ subcarriers per transmit antenna. The spreading is done at each transmit antenna and over all four subcarriers. That is, the first symbol of antenna 1 is spread over all 4 subcarriers of that antenna, the second symbol over the same subcarriers and so forth. The spreading matrix, $\underline{U}$, in this case is

$$\underline{U} = \begin{pmatrix} \underline{W} & \underline{0} & \underline{0} & \underline{0} \\ \underline{0} & \underline{W} & \underline{0} & \underline{0} \\ \underline{0} & \underline{0} & \underline{W} & \underline{0} \\ \underline{0} & \underline{0} & \underline{0} & \underline{W} \end{pmatrix}, \tag{12}$$

where $\underline{W}$ is the 4×4 Walsh Hadamard matrix, and $\underline{0}$ are zero matrices of the same size. Frequency spreading as described satisfies (11). Note that the aim of spreading in MIMO-MC-CDM systems, or in any MIMO system discussed here, is not to overcome zero diagonal elements, but to equally distribute the symbol energy over the subcarriers. In other words, to reduce the MSD. This advantage becomes obvious by considering the main diagonal elements of the resulting channel matrix. The main diagonal elements of $\underline{R}_{\mathrm{MO}}$ are, in general, not equal, whereas every four neighboring diagonal elements of the MIMO-MC-CDM matrix, $\underline{R}_{\mathrm{S}} = \underline{R}_{\mathrm{MM}}$, are identical. This is, however, only true if the spreading is done over all four subcarriers, (compare left and middle matrices of Fig. 2). The values of the main diagonal elements influence the matched filter bound (MFB, i. e., all interference is completely removed). If all diagonal elements were equal, the MFB BER curve would coincide with the AWGN curve. In addition, better BER performance can be expected if equalization adequately mitigates the interference. Here, spreading over all four subcarriers is applied. In a practical system with more subcarriers the size of the matrix, $\underline{W}$, should be adapted so that a trade-off between diversity and complexity is obtained.

3.3. Spreading in Antenna and Frequency Direction, MC-CAFS

Spreading in antenna direction alone is basically the same as frequency spreading, except that the spreading is done over the antennas (space) and not over frequency. This kind of spreading does not change the structure of the channel matrix, the diagonal and off-diagonal elements are just combinations of the original channel matrix elements, and on the average no BER performance improvement is expected (see Fig. 3 for the MFB, Antenna/OFDM). In addition, spreading in antenna direction alone does not satisfy (11). Based on the idea of spreading in antenna direction we introduce a new class of spreading matrices which employ a cyclic shift at each transmit antenna. By cyclic shift we mean that while a symbol is spread over all antennas, it is sent from each antenna on a different subcarrier, and thus the name Cyclic Antenna Frequency Spreading. This spreading matrix, $\underline{U}$, satisfies (11) and is given by:

$$\underline{U} = \begin{pmatrix} \underline{I} & \underline{I} & \underline{I} & \underline{I} \\ \underline{I}_{c1} & -\underline{I}_{c1} & \underline{I}_{c1} & -\underline{I}_{c1} \\ \underline{I}_{c2} & \underline{I}_{c2} & -\underline{I}_{c2} & -\underline{I}_{c2} \\ \underline{I}_{c3} & -\underline{I}_{c3} & -\underline{I}_{c3} & \underline{I}_{c3} \end{pmatrix}, \tag{13}$$

where $\underline{I}$ is a 4×4 identity matrix, and $\underline{I}_{c1}$, $\underline{I}_{c2}$ and $\underline{I}_{c3}$ are given by

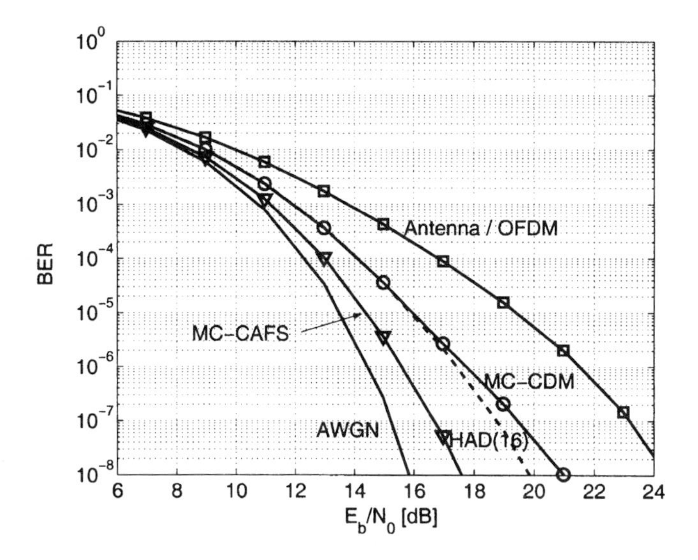

Figure 3. Average matched filter bound for a time variant channel for different spreading matrices.

$$\underline{I}_{c1} = \begin{pmatrix} 0 & 0 & 0 & 1 \\ 1 & 0 & 0 & 0 \\ 0 & 1 & 0 & 0 \\ 0 & 0 & 1 & 0 \end{pmatrix}, \quad \underline{I}_{c2} = \begin{pmatrix} 0 & 0 & 1 & 0 \\ 0 & 0 & 0 & 1 \\ 1 & 0 & 0 & 0 \\ 0 & 1 & 0 & 0 \end{pmatrix}, \quad \underline{I}_{c3} = \begin{pmatrix} 0 & 1 & 0 & 0 \\ 0 & 0 & 1 & 0 \\ 0 & 0 & 0 & 1 \\ 1 & 0 & 0 & 0 \end{pmatrix}.$$

As can be seen on the right of Fig. 2, this type of spreading introduces more interference, yet better energy distribution is achieved and the average MFB is more than 2 dB better than that of MC-CDM at high E_b/N_0, see Fig. 3. The new spreading matrix introduced here, requires that $N = n_T$ and $N \cdot n_T = 2^m$ with $m \in \{2, 3, \ldots\}$. The construction for more general parameters is part of ongoing work.

The above described spreading matrices, (12) and (13), are both based on the Hadamard matrix. None of them employs full diversity, that is, they do not exploit all of the available dimensions. The spreading matrices contain zero elements, and the non-zero elements all have an amplitude of 1. However, those matrices have the advantage that they do not reduce the transmission rate, i. e., transmission is at full rate in contrast to space frequency coding with full diversity proposed in [5]. The 16×16 Hadamard matrix would spread a symbol over all dimensions, but it does not satisfy (11). The average MFB for this matrix is shown in Fig. 3 for comparison.

4. ITERATIVE DETECTION TECHNIQUES

In the following, we consider two iterative soft interference cancellation schemes. The basic idea of both schemes, the RNN detector [7] and the SCE [9], is to estimate

Table 1. Constellation and Mapping of 8 *QAM.*

symbols	$1+j$	$-1+j$	$-1-j$	$1-j$	$1+\sqrt{3}$	$(1+\sqrt{3})j$	$-1-\sqrt{3}$	$(-1-\sqrt{3})j$
bits	100	111	001	010	000	110	101	011

the interference (represented by off-diagonal elements of the channel matrix $\underline{R}$) and subtract it during an iterative process. The difference between the RNN detector and the SCE lies in the soft decision function and an additional preprocessing step of the SCE. The soft decision function of the RNN detector processes a single symbol, whereas the SCE makes a decision based on the whole symbol vector. Furthermore, the SCE applies a whitening filter as preprocessing step in order to reduce the interference. Therefore, the SCE usually offers better performance. However, computational complexity of the SCE is higher compared to the RNN detector. We apply the RNN detector with serial update and the variance computation as defined in [7].

In the coded case, we apply iterative detection and decoding (turbo detection). Turbo detection was introduced for ISI channels in [14], where the optimum APP equalizer was used. Since its computational complexity would be much too large for the scenarios we consider here, we apply the RNN detector and SCE instead as described in [15], [10], and [16]. The decoder employs the BCJR algorithm. In the case of turbo codes we combine turbo detection with turbo decoding [17]. The performance of a turbo scheme may be further enhanced for higher modulation alphabets when additionally iterative demapping and decoding [18] for non-Gray mappings is included [10]. We use this technique for an 8 QAM alphabet. Table 1 depicts the symbol constellation and mapping for 8 QAM.

5. SIMULATION RESULTS & CONCLUSIONS

In the following we present BER simulation results for the classes of spreading matrices discussed in the previous section. We choose a 4×4 MIMO channel, where the channel impulse response from each transmit to each receive antenna is time-invariant and has length $L = 2$. The channel coefficients are complex-valued, randomly chosen, and kept constant during the whole simulation. A time-invariant scenario is more suitable to compare different detectors since there is no averaging effect. The underlying OFDM system has $N = 4$ subcarriers and thus the loss in SNR due to the cyclic prefix may be evaluated as $\gamma = 10\log_{10}(\frac{N+L-1}{N}) \approx 0.97\,\text{dB}$. Note that we considered this small OFDM system only, since it is sufficient to obtain essential properties in a MIMO environment and is computationally less demanding. However, in practice a much larger number of subcarriers would be used, which reduces the SNR loss γ. For the coded case, we consider non-systematic convolutional codes with memory 2 (rate $1/2$ and punctured to rate $3/4$) and memory 4 (rate $1/2$) as well as a rate $1/2$ punctured turbo code with constituent encoders of memory 4. The corresponding generator polynomials of the convolutional codes in octal notation are $[5,7]$ and $[23,35]$. The generator polynomials of the constituent recursive, systematic encoders of the turbo code are $[23,37]$ (see [19]), where the second polynomial denotes the feedback polynomial. The length of one codeword is adapted to the symbol alphabet resulting in $100 \cdot N \cdot n_T = 1600$ transmit symbols. We apply a random interleaver which has the length of one codeword. We use the SCE and RNN detector with 10 iterations as detection schemes for the uncoded case, whereas in the coded case 5 iterations (5 detection and 5 decoding steps) are employed. When applying the turbo code, for each decoding step 5 iterations between the constituent decoders are carried out, additionally. We assume Gray mapping for the 4 PSK alphabet.

Fig. 4 illustrates the performance of the SCE and RNN detector for the uncoded 4 PSK transmission. The SCE with MC-CDM performs best, whereas the new scheme MC-CAFS shows an error floor in the case of the RNN detector. The reason for this behavior is that the amount of interference is largest for MC-CAFS. However, MC-CAFS leads to lower BER curves than MC-CDM and OFDM when combined with a linear MMSE detector (results not shown here). Furthermore, the good MFB performance of MC-CAFS suggests that MC-CAFS has the potential to outperform the other schemes when combined with appropriate coding strategies to further mitigate interference.

Fig. 5 shows BER curves of the SCE for the memory 2 code with rate $1/2$ and $3/4$. For both code rates MC-CAFS outperforms OFDM and MC-CDM and approaches closely AWGN performance. As expected, the results indicate that the advantage of a spread transmission over an unspread one is larger with increasing code rate. Fig. 6 shows the case of the stronger code with memory 4. We observe again that MC-CAFS performs best beyond a certain E_b/N_0 threshold. However, when using an even stronger code like the turbo code, see Fig. 7, OFDM shows in the depicted SNR region better results than the other schemes. The reason for this behavior is that the turbo coded system is more sensitive to interference. In Fig. 8 the BER of the SCE and RNN detector for 8 QAM and the memory 2 code is depicted. We apply iterative detection, demapping, and decoding for both the SCE and the RNN detector. Beyond 8.5 dB and 10.2 dB MC-CAFS performs best in the case of the SCE and RNN detector, respectively, and comes close to the AWGN curve with 15 demapping and decoding steps.

As an example, in Fig. 9 an extrinsic information transfer chart (EXIT) chart [20] illustrates the convergence behavior of the RNN detector for the scenario with 8 QAM and the memory 2 code shown in Fig. 8. On the abscissa we have average mutual input information of the RNN detector ($I_{\mathrm{A},1}$) and average mutual output information of the decoder ($I_{\mathrm{E},2}$, transfer curve denoted as CC[5,7]). The ordinate shows average mutual output information of the RNN detector ($I_{\mathrm{E},1}$), and average mutual input information of the decoder ($I_{\mathrm{A},2}$), respectively. Note that the RNN transfer curves can only be obtained approximately [10], but are sufficient for an analysis of the convergence behavior. The RNN MC-CAFS curve starts at lowest mutual output information due to the largest interference. Therefore, if little input information (a priori information) is available (low SNR regime), MC-CAFS performs worse than the other schemes as shown in Fig. 8. Similar curves and thus similar conclusions could be obtained for the 4 PSK scenarios and the SCE. Furthermore, a larger memory/higher rate of the channel code or stronger interference reduces the gap between detector and decoder curve, leading to a possible early stop of convergence of the iterative process. Therefore, in the case of the turbo code OFDM showed the turbo cliff at lower E_b/N_0 than the other schemes (Fig. 7). However, for high a priori information the MC-CAFS curve lies above the MC-CDM and OFDM curves (Fig. 9), leading to the better performance of MC-CAFS for large E_b/N_0 (Fig. 8). Additionally, the 8 QAM demapper curve for the AWGN channel is shown in Fig. 9 (dashed line), which is closely approached for perfect a priori information by the MC-CAFS curve.

6. REFERENCES

[1] N. Yee, J. P. Linnartz, and G. Fettweis, "Multicarrier CDMA in indoor wireless radio networks," in *Proc. PIMRC*, 1993, pp. 975–979.

[2] J. Lindner, "MC-CDMA in the context of general multiuser/multisubchannel transmission methods," *European Trans. Telecomm.*, vol. 10, no. 5, pp. 351–367, July/Aug. 1999.

[3] M. A. Dangl, D. Yacoub, U. Marxmeier, W. G. Teich, and J. Lindner, "Performance of joint detection techniques for coded MIMO-OFDM and MIMO-MC-CDM," in *Proc. COST 273 Workshop on Broadband Wireless Local Access*, Paris/France, May 2003, pp. 17/1–17/6.

[4] J. Lindner and C. Pietsch, "The spatial dimension in the case of MC-CDMA," *European Trans. Telecomm.*, no. 5, pp. 431–438, Sept./Oct. 2002.

[5] H. Bölcskei and A. J. Paulraj, "Space-frequency codes for broadband fading channels," in *Proc. IEEE Int. Symposium on Information Theory*, Washington/US, June 2001, p. 219.

[6] W. G. Teich and M. Seidl, "Code division multiple access communications: Multiuser detection based on a recurrent neural network structure," in *Proc. IEEE ISSSTA*, vol. 3, Mainz/Germany, Sept. 1996, pp. 979–984.

[7] C. Sgraja, A. Engelhart, W. G. Teich, and J. Lindner, "Multiuser/multisubchannel detection based on recurrent neural network structures for linear modulation schemes with general complex-valued symbol alphabet," in *Proc. COST 262 Workshop on Multiuser Detection in Spread Spectrum Communications*, Reisensburg/Germany, Jan. 2001, pp. 45–52.

[8] A. Engelhart, W. Teich, and J. Lindner, "A survey of multiuser/multisubchannel detection schemes based on recurrent nerual networks," *Wirel. Commun. Mob. Comput.*, vol. 2, pp. 269–284, 2002.

[9] J. Egle, C. Sgraja, and J. Lindner, "Iterative soft cholesky block decision feedback equalizer – a promising approach to combat interference," in *Proc. IEEE VTC*, vol. 3, Rhodes/Greece, May 2001, pp. 1604–1608.

[10] M. A. Dangl, W. G. Teich, J. Lindner, and J. Egle, "Joint iterative equalization, demapping, and decoding with a soft interference canceler," in *Proc. 7th Int. Symposium on Communication Theory and Applications*, Ambleside/UK, July 2003, pp. 36–41.

[11] C. Pietsch, S. Sand, W. G. Teich, and J. Lindner, "Modeling and performance evaluation of multiuser MIMO systems using real-valued matrices," *IEEE Journal on Selected Areas in Communications*, June 2003, to appear.

[12] A. Bury, J. Egle, and J. Lindner, "Diversity comparison of spreading transforms for multicarrier spread spectrum transmission," *IEEE Trans. Commun.*, vol. 51, no. 5, pp. 774–781, May 2003.

[13] M. Reinhardt and J. Lindner, "Transformation of a Rayleigh fading channel into a set of parallel AWGN channels and its advantage for coded transmission," *Electronics Letters*, vol. 31, no. 25, pp. 2154–2155, Dec. 1995.

[14] C. Douillard, M. Jezequel, C. Berrou, A. Picart, P. Didier, and A. Glavieux, "Iterative correction of intersymbol interference: Turbo-equalization," *European Trans. Telecomm.*, vol. 6, no. 5, pp. 507–511, Sept./Oct. 1995.

[15] C. Sgraja, A. Engelhart, W. G. Teich, and J. Lindner, "Combined equalization and decoding for BFDM packet transmission schemes," in *Proc. 1st Int. OFDM-Workshop*, Hamburg/Germany, Sept. 1999, pp. 19–1ff.

[16] J. Egle and J. Lindner, "Iterative joint equalization and decoding based on soft cholesky equalization for general complex valued modulation symbols," in *Proc. 6th Int. Symposium on DSP for Communication Systems*, Sydney-Manly/Australia, Jan. 2002, pp. 163–170.

[17] D. Raphaeli and Y. Zarai, "Combined turbo equalization and turbo decoding," *IEEE Communications Letters*, vol. 2, pp. 107–109, Apr. 1998.

[18] S. ten Brink, J. Speidel, and R. Yan, "Iterative demapping and decoding for multilevel modulation," in *Proc. IEEE GLOBECOM*, vol. 1, Sydney/Australia, Nov. 1998, pp. 579–584.

[19] S. Benedetto, R. Garello, and G. Montorsi, "A search for good convolutional codes to be used in the construction of turbo codes," *IEEE Trans. Commun.*, vol. 46, no. 9, pp. 1101–1105, Sept. 1998.

[20] S. ten Brink, "Designing iterative decoding schemes with the extrinsic information transfer chart," *AEÜ Int. Jour. Electron. Commun.*, vol. 54, no. 6, pp. 389–398, Nov. 2000.

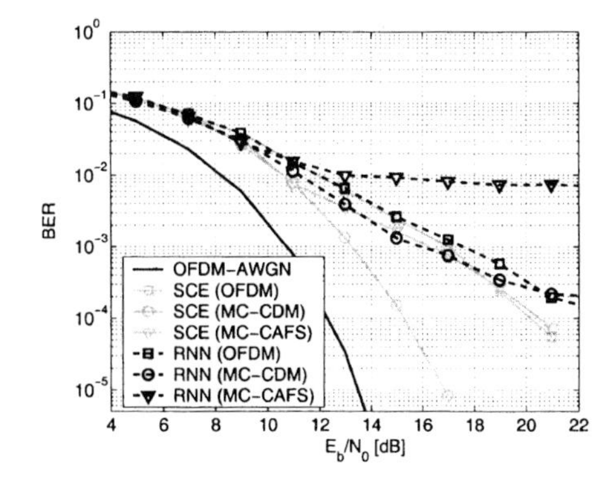

Figure 4. BER of SCE (solid lines) and RNN detector (dashed lines) for uncoded 4 PSK.

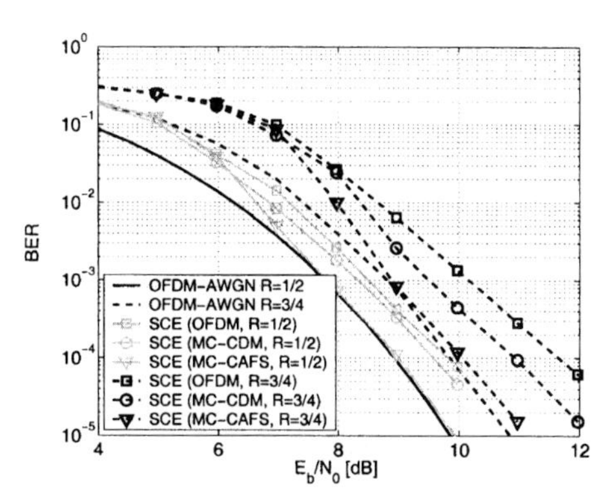

Figure 5. BER of SCE for 4 PSK, convolutional code with memory 2, rate 1/2 (solid lines) and rate 3/4 (dashed lines).

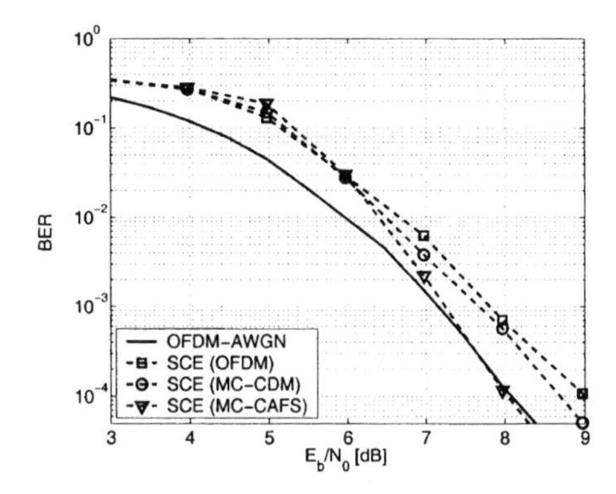

Figure 6. BER of SCE for 4 PSK, convolutional code with memory 4, rate 1/2.

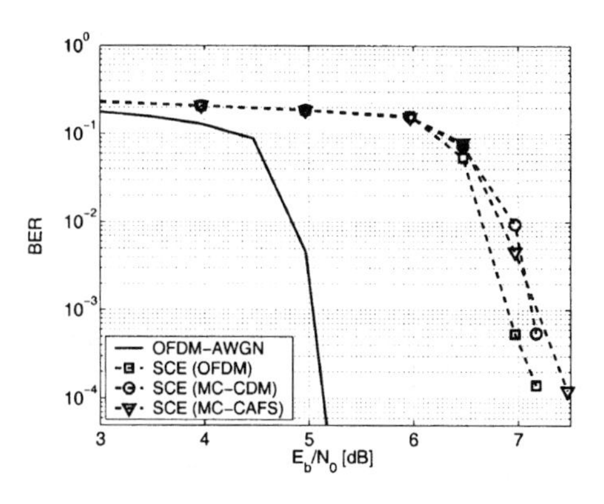

Figure 7. BER of SCE for 4 PSK, turbo code, rate 1/2.

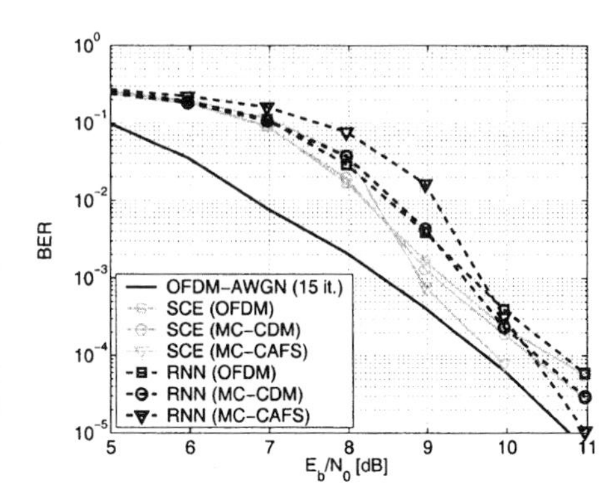

Figure 8. BER of SCE (solid lines) and RNN detector (dashed lines) for 8 QAM, convolutional code with memory 2, rate 1/2.

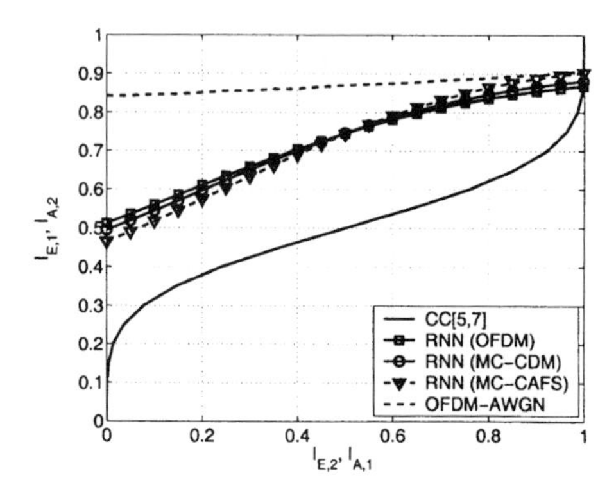

Figure 9. EXIT curves for 8 QAM (E_b/N_0 = 11 dB).

Section II

CODING, MODULATION AND SPREADING

D. A. GUIMARÃES AND J. PORTUGHEIS

TURBO PRODUCT CODES FOR AN ORTHOGONAL MULTICARRIER DS-CDMA SYSTEM

Abstract. In this article it is suggested a class of product codes and its application to a multicarrier DS-CDMA system. The iterative (turbo) decoding of this class is based on a very simple minimum distance decoding of the component codes. Performance simulation results for various coded systems are compared with capacity calculations and reveal a good trade-off between complexity and performance.

1. INTRODUCTION

It is somewhat a consensus that multicarrier systems, especially those combined with the code-division multiple access technique, are potential candidates to be used in fourth-generation wireless communication systems. Therefore, it is of interest to develop supporting techniques for them that will actually turn this potential into reality. Among these techniques, efficient and simple channel coding/decoding strategies represent one challenge.

In this context, this article describes a class of low rate multidimensional product codes and its iterative (turbo) decoding applied to the orthogonal multicarrier DS-CDMA system suggested in [1] for multi-path Rayleigh fading channels.

The article is organized as follows: in Section 2 the multicarrier system is described and in Section 3 some results concerning the channel capacity analysis for this system are presented. Section 4 describes the proposed class of product codes whereas Section 5 presents the iterative (turbo) decoding process. In Section 6, the performance of the turbo product codes when applied to the multicarrier CDMA system is investigated. Finally, Section 7 is devoted to the concluding remarks.

2. DESCRIPTION OF THE MC-DS-CDMA SYSTEM

In the system suggested in [1], transmitted data bits are serial-to-parallel converted to M parallel branches. On each branch, each bit is repeated S times and the replicas feed different block interleavers. Then, these identical bits are direct sequence spread spectrum BPSK modulated and transmitted using orthogonal carries. Hence, there are a total number of MS carriers and time-frequency diversity is achieved. At the receiver, the matched filter outputs of the S identical-bit carriers are combined prior to decoding. The system of [1] can be viewed as a combination or generalization of the *copy-type* and *S/P-type* configurations described in [2]. The main attributes of this system are: 1) the possibility of overcoming the performance of the conventional single-carrier CDMA system, and 2) the reduction of complexity through the use of one matched filter per carrier, instead of a RAKE receiver, situation that is achieved if the number of carriers satisfies [1]

K. Fazel and S. Kaiser (eds.), Multi-Carrier Spread-Spectrum, 115–122.

$$MS \geq 2L_1 - 2 \tag{1}$$

where L_1 is the number of resolvable propagation paths for a single-carrier CDMA system with the same total bandwidth as that of the MC-DS-CDMA system. In [1] it is further assumed a 50% of spectrum overlap of adjacent modulated and orthogonal carriers.

3. CHANNEL CAPACITY ANALYSIS

In this article it is assumed that the receiver has perfect knowledge of the channel state information and that there is no transmit power adaptation scheme. It is further assumed that the compatibility constraint of [5] is satisfied, that is, the channel gains are independent and identically distributed (i.i.d.) random variables, and the input distribution that maximizes mutual information is the same, regardless of the channel state.

Let $g[i]$ represent the channel state information at the discrete-time instant i, and assume that it is possible to generate by computer a sufficient large number X of values for g, based on a known probability distribution. Then, the Shannon capacity of the fading channel with side information at the *receiver only* given in [5], can be estimated as follows

$$C = \frac{1}{X}\sum_{i=1}^{X} B\log_2\left(1+\gamma g^2[i]\right) \tag{2}$$

where B is the channel bandwidth and γ is the average received signal-to-noise ratio (SNR). The value of X is the one enough for convergence in (2).

For BPSK signalling on a fading channel, the capacity can be estimated through

$$C_{\text{BPSK}} = \frac{1}{X}\sum_{i=1}^{X}\int_{-\infty}^{\infty} p(y\,|\,\psi[i])\log_2\frac{p(y\,|\,\psi[i])}{p(y)}dy \tag{3}$$

where

$$\psi[i] = g[i]\sqrt{E_s} \tag{4}$$

$$p(y\,|\,\psi[i]) = \frac{1}{\sqrt{2\pi\sigma^2}}\exp\left[\frac{-(y-\psi[i])^2}{2\sigma^2}\right] \quad \text{and} \tag{5}$$

$$p(y) = \frac{\exp\left[\frac{-(y-\psi[i])^2}{2\sigma^2}\right]+\exp\left[\frac{-(y+\psi[i])^2}{2\sigma^2}\right]}{\sqrt{8\pi\sigma^2}} \tag{6}$$

and where E_s is the BPSK symbol energy and σ^2 accounts for the variances of the interference plus noise. For an AWGN channel $g[i] = 1$. The results obtained through (2) and (3) demonstrate perfect agreement [6] with those obtained through their analytical counterparts, showing the applicability of the method for both unconstrained and constrained input alphabets.

The reverse link of the MC-DS-CDMA system for a user of reference can be interpreted, in one hand, as a set of M parallel channels with BPSK signalling. These channels are defined from each of the M serial-to-parallel converter outputs at the transmitter to each of the M combiner's outputs at the receiver. On the other hand, this link can be interpreted as a set of MS parallel channels defined from each of the MS modulator inputs at the transmitter to each of the MS matched filters outputs at the receiver. Then the channel capacity for the system can be estimated as the sum of M or MS individual capacities [6], depending on the case under consideration. This sum is possible if it is presumed independence among the M or MS channels, a reasonable assumption when the bandwidth occupied by each modulated carrier is smaller than the coherence bandwidth of the channel.

In this article we consider three situations for the MC-DS-CDMA system configuration: Equal Gain Combining (EGC), Maximal Ratio Combining (MRC) and no combining. The first two situations are interpreted as M parallel channels. The last one is interpreted as MS parallel channels. It is worth noting that, even when EGC combining is considered for capacity estimation, the receiver has knowledge of the channel state information.

If the sum of the interference at the receiver input is modelled as Gaussian, the capacity for each of the M or MS channels of the MC-DS-CDMA system can be approximately estimated using (3). However, the value of $\sqrt{E_S}$ in this expression should be substituted by [1] $T\sqrt{P/2}$, where P is the average transmitted power per carrier and T is the BPSK symbol duration. Furthermore, Table I shows the values for $g[i]$ and for the variances of the interference plus noise, σ^2, to be operated in (3), according to each case taken into consideration here. The values of the average signal-to-noise ratio, γ, in each situation are also given in Table I. In this table, the value of $\beta[i]_\nu$, $i = 1, 2, \ldots X$, corresponds to the i-th value of the computer generated Rayleigh random variable, i.i.d. for all i and ν, and J_ν accounts for the interference variances at the output of each matched filter, at the receiver side. In fact, the values of J_ν are different for different combiner's outputs [1], but if this difference is not taken into account, the channel capacity results are not significantly affected [6].

Figure 1 shows capacity results, in terms of spectral efficiency versus the minimum average SNR per information bit for error-free transmission on the Rayleigh channel, for $MS = 6$ and variable M and S. For M channels with diversity, the results shown are for MRC combining. The systems with EGC combining have average capacities identical to those with MRC combining, although, for a given SNR, the necessary transmitted power per carrier with EGC is grater than that necessary for MRC combining. The length N_1 of the spreading code for a single-carrier CDMA system taken for reference was made equal to 60, and the number of resolvable paths L_1 for this single-carrier case was made equal to 4. In this case (1) is satisfied and the number L of resolvable paths per carrier reduces to 1. The multi-path intensity profile was considered uniformly distributed and the number of active users, K, was made equal to 10 (the capacity reduces, as the number of users increases, as expected).

Table I. Values operated in (3), (4), (5) *and* (6).

	γ	$g[i]$	σ^2
M channels EGC combiner	$\frac{PT}{N_0 S}\mathrm{E}\left[\left(\sum_{v=1}^{S}\beta_v\right)^2\right]$	$\sum_{v=1}^{S}\beta[i]_v$	$\sum_{v=1}^{S}J_v + \frac{N_0 TS}{4}$
M channels MRC combiner	$\frac{PT}{N_0}\mathrm{E}\left(\sum_{v=1}^{S}\beta_v^2\right)$	$\sum_{v=1}^{S}\beta[i]_v^2$	$g[i]\left(\frac{1}{S}\sum_{v=1}^{S}J_v + \frac{N_0 T}{4}\right)$
MS channels	$\frac{PT}{N_0}\mathrm{E}(\beta^2)$	$\beta[i]$	$\frac{1}{S}\sum_{v=1}^{S}J_v + \frac{N_0 T}{4}$

where [1]: $J_v = \frac{PT^2}{6N^3}r + \frac{1}{M}\sum_{p=1}^{M}\frac{PT^2}{2\pi^2 N^3}\mu Q_{p,v}$ and where: $N = \frac{2M}{MS+1}N_1$,

$r = 2(K-1)N^2$, $\mu = (K-1)N^2$ and $Q_{p,v} = \sum_{\substack{m=1 \\ \neq p+(v-1)M}}^{MS}\frac{1}{[m-p-(v-1)M]^2}$

$N_0/2$ is the received noise p.s.d. and $\mathrm{E}(x)$ is the expectation operator.

It can be observed from Figure 1 that it is more advantageous to explore the maximum order of diversity (M = 1 and S = 6) instead of exploring the maximum parallelism of the data stream and that, as the order of diversity, S, increases, the channel capacity of the MC-DS-CDMA channel approximates the AWGN one. The observation of MS channels, without diversity, significantly reduces the capacity, especially for high information rates.

The results shown in Figure 1 indicate that it should be preferable to use a low-rate outer code concatenated with the inner repetition of the MC-DS-CDMA system in order to aim the best performance, since the capacity is increased with increasing S and it is changed less than 1 dB for code rates below 0.2. At the receiver, the outputs of the combiners can be viewed as soft inputs for the outer decoder. Furthermore, the length of the spreading code per carrier, N, can be adjusted [6] to compensate for the reduced coded symbols duration $R_c T$ due to coding of rate R_c, keeping unchanged the transmission bandwidth and the information data rate relative to the uncoded system. In this case, however, the channel capacity is reduced, since the total variance of interference in the decision variable is increased.

4. DESCRIPTION OF THE CLASS OF PRODUCT CODES

Let $\boldsymbol{c}_1$ be a codeword of the binary repetition code $\boldsymbol{C}_1 = (n/2, 1, n/2)$ and $\boldsymbol{c}_2$ be a codeword of the binary single parity-check code $\boldsymbol{C}_2 = (n/2, n/2-1, 2)$. A codeword $\boldsymbol{c}$ of the non-systematic code $\boldsymbol{C} = (n, k, d_{\min}) = (n, n/2, 4)$ can be expressed as [7]

$$c = [01]c_1 \oplus [11]c_2 \tag{7}$$

where the sum $\oplus$ is over GF(2) and the product $[01]c_1$ is calculated by substituting a 0 in c_1 by 00 and a 1 by 01. The same is done for $[11]c_2$, where now a 1 in c_2 becomes 11. By using the same non-systematic code $\boldsymbol{C}$ as the component code in each of the D dimensions, a product code of length n^D, rate $(½)^D$ and minimum distance 4^D is obtained.

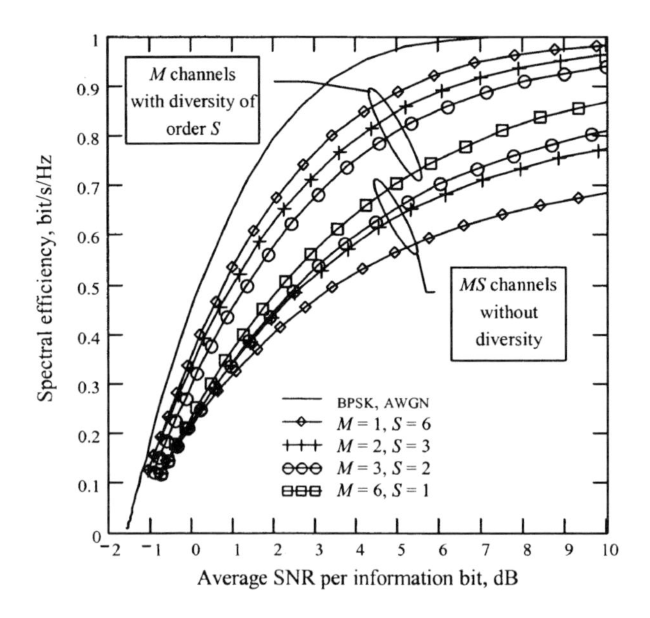

Figure 1. Spectral efficiency for the MC-DS-CDMA system, MS = 6, variable M and S.

A D-dimensional product code can be interpreted as a serial concatenation of D codes separated by D–1 row-column type block interleavers in which the number of columns is $N_c = n$ and the number of rows of the interleaver between the code d and d+1 is $N_r = n^{d-1}k^{D-d}$ [6]. If this rule is accomplished, it is possible to verify [6] that all the n^{D-1} n-element vectors oriented in the "direction" of each dimension of the D-dimensional hypercube of n^D coded bits are codewords of $\boldsymbol{C}$. The key feature of this component code $\boldsymbol{C}$ is that it is possible to derive for it a very simple minimum distance decoding algorithm: set $c_1 = \mathbf{0}$ (the all-zero codeword) and apply Wagner decoding [3] for a single parity-check code of length $n/2$ over the binary alphabet {00, 11}. The decision is $\hat{c}$. Then set $c_1 = \mathbf{1}$ (the all-one codeword) and again apply Wagner decoding for a parity-check code over the alphabet {01, 10}. The decision is $\hat{c}$'. Compare the Euclidean distances from $\hat{c}$ and $\hat{c}$' to the received codeword $\boldsymbol{r}$ and choose as the final decision the shortest one.

5. TURBO DECODING ALGORITHM

In the initialisation phase of the iterative decoding algorithm, the channel likelihood ratios for all received noisy symbols at the discrete-time instant i are estimated using $\Lambda(c \mid r, g[i]) = g[i]r$, where $g[i]$ is the fading amplitude estimated according to Table I, c is the transmitted codeword symbol and r is the received channel value. For notational simplicity, hereafter the index i in $g[i]$ will be dropped.

The turbo-decoding algorithm is essentially the Pyndiah's one [4]. However, instead of using Chase algorithm to decode the codewords in each dimension, the Wagner algorithm is applied here. Figure 2 shows a block diagram representing operations for the j-th decoding step of the Pyndiah's SISO algorithm, where the maximum value of j, say j_{max}, is the total number of iterations multiplied by D. The vector $\boldsymbol{R}$ represents all n^D received noisy symbols and $\overrightarrow{\boldsymbol{gR}}$ represents the symbol-by-symbol multiplication by the respective fading amplitudes. The expression "decoding in one dimension" in this figure means decoding n^{D-2} $n \times n$ arrays of the soft input $\boldsymbol{E}(j)$ in the "direction" of one dimension. Decoding an array consists of decoding n rows (or n columns). Hence, "decoding in one dimension" implies applying the SISO decoding algorithm n^{D-1} times.

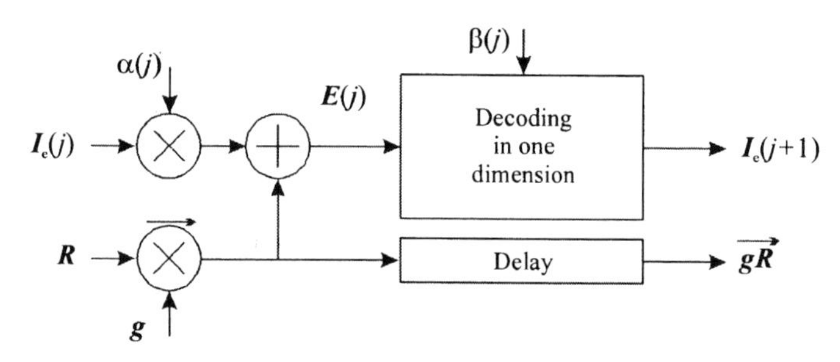

Figure 2. Turbo decoding structure for the j-th decoding step.

The variations of the parameter $\beta(j)$ were chosen to follow a linear rule, and the variations of the parameter $\alpha(j)$ used to weight the extrinsic information, $\boldsymbol{I}_e(j)$, were chosen to follow a logarithmic rule [7]. The reliabilities of decisions, r_d, were always obtained through $r_d = \beta \hat{c}_d$, where $\hat{c}_d$ represents a symbol of the final decision $\hat{\boldsymbol{c}}_d$.

6. SIMULATION RESULTS

Figure 3 presents some simulation results for the uncoded and coded MC-DS-CDMA system for $MS = 6$, variable M and S, $N_1 = 60$, $L_1 = 4$, $K = 10$, uniform multi-path power delay profile and uncorrelated channel gains in time and frequency. The number of iterations in the turbo decoding process was made equal to 10. The performance of the uncoded single-carrier system with a four-tap RAKE receiver is also presented. It can be seen that for $M = 1$, $S = 6$ and $M = 2$, $S = 3$ the performance of the multicarrier system overcomes the performance of the single-carrier one. It also can be seen that the performance of the uncoded system with MRC combining overcomes the performance with the EGC combining rule. This is shown for $M = 1$ and $S = 6$. However, the use of MRC combining did not bring performance gains in the case of the coded system, as compared to the use of EGC combining. It can be further observed in Figure 3 that, for bit error rates below 2×10^{-4}, infinite coding gains can be obtained for all cases considered in this figure. The best performance result is 5.8 dB away from capacity (approximately –0.8 dB for code rate 1/8, $M = 1$ and $S = 6$, according to Figure 1), for a bit error rate of 10^{-5}.

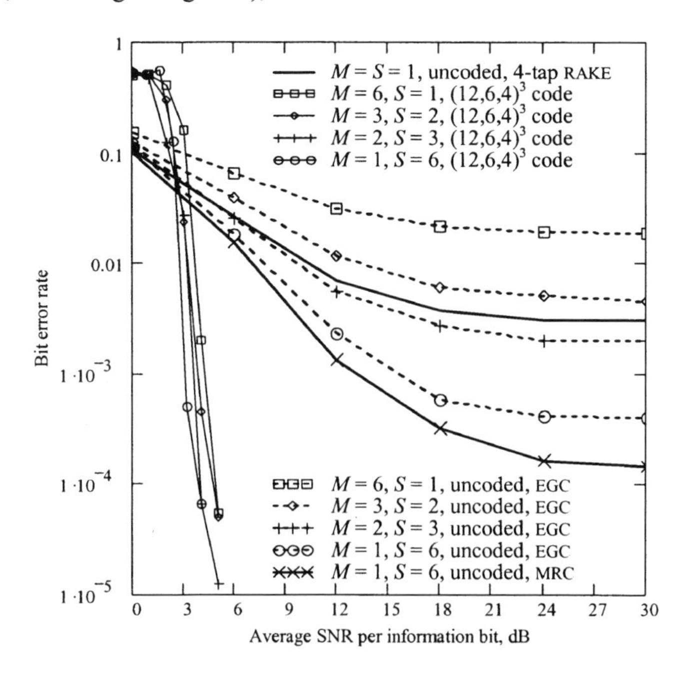

Figure 3. Uncoded and coded MC-DS-CDMA system, MS = 6 and variable M and S.

For the coded system, it is possible to modify the length of the spreading code for each carrier in such way that the information rate and the total occupied bandwidth are kept unchanged, as compared to the uncoded system. This situation was investigated in [6] and demonstrated a decrease in performance less than 0.5 dB for all the cases considered in Figure 3. Although this is an attractive solution, the channel capacity is reduced, since the variance of the total interference is increased. However, in this case, the gap between performance and capacity decreases.

Another situation investigated in [6] was the modification of the channel parameters in such a way that the performance of the uncoded system becomes better than those showed in Figure 3. It was verified that, in this situation, the performance of the coded system is roughly the same as in the cases reported in Figure 3. This verification indicates that greater coding gains can be achieved if the channel becomes worse.

7. CONCLUDING REMARKS

This work described a class of low rate multidimensional product codes and its iterative (turbo) decoding applied to the orthogonal multicarrier DS-CDMA system suggested in [1] for multi-path fading channels. The key feature of this class is that the component code can be decoded through a very simple minimum-distance algorithm based on applying the Wagner rule [3]. The turbo decoding algorithm used a simplified form of Pyndiah's SISO decoding algorithm [4]. Some performance results for this class of codes on the MC-DS-CDMA system of [1] were reported here, unveiling a good trade-off between performance and complexity. It was verified that the best choice for the system parameters corresponds to the use of the maximum allowable frequency diversity, like the one attained by the *copy-type* configuration [2], instead of the maximum parallelism of the data stream, as is the case for the *S/P-type* configuration (OFDM-CDMA) [2]. It was also pointed out that, for the coded system, a very low decrease in performance is observed if the length of the spreading code is changed in order to keep unchanged the information rate and the occupied bandwidth, as compared to the uncoded system.

8. REFERENCES

[1] E. Sourour and M. Nakagawa, "Performance of Orthogonal Multicarrier CDMA in a Multipath Fading Channel", *IEEE Transactions on Communications*, vol. 44, no. 3, pp. 356-367, Mar. 1996.

[2] S. Kaiser, *Multi-Carrier CDMA Mobile Radio Systems - Analysis and Optimization of Detection, Decoding and Channel Estimation*, Ph.D. Thesis: VDI Verlag GmbH. Düsseldorf, 1998.

[3] R. A. Silverman and M. Balser, "Coding for Constant Data-Rate Systems", *IRE Trans. Inform. Theory*, PGIT-4, pp. 50-63, 1954.

[4] R. M. Pyndiah, "Near-Optimum Decoding of Product Codes: Block Turbo Codes", *IEEE Trans. Commun.*, vol. 46, no. 8, pp. 1003-1010, Aug. 1998.

[5] A. Goldsmith & P. P. Varaiya, "Capacity of Fading Channels with Channel Side Information", *IEEE Trans. Inf. Theory*, vol. 43, no. 6, pp. 1986-1992, Nov. 1997.

[6] D. A. Guimarães, "A Class of Product Codes and its Turbo Decoding Applied to a Multicarrier CDMA System", *Ph.D. Thesis*, State University of Campinas - Unicamp, SP, Brazil, June 2003 (*in Portuguese*).

[7] D. A. Guimarães and J. Portugheis, "A Class of Product Codes and Its Iterative (Turbo) Decoding" in: *Proceedings of the 3rd International Symposium on Turbo Codes & Related Topics*, pp. 431-434, Brest, France, September 1-5, 2003.

9. AFFILIATIONS

D. A. Guimarães: DTE, Inatel, Av. João de Camargo, 510, CEP 37540-000, S. R. Sapucaí, MG, Brazil. E-mail: dayan@inatel.br.
J. Portugheis: DECOM, FEEC, Unicamp, CP 6101, CEP 13083-970, Campinas, SP, Brazil. E-mail: jaime@decom.fee.unicamp.br.

M. FUJII, M. ITAMI AND K. ITOH

PERFORMANCE EVALUATION OF DIVERSITY GAIN AND CODING GAIN IN CODED ORTHOGONAL MULTI-CARRIER MODULATION SYSTEMS

Abstract. This paper presents a performance evaluation of diversity and coding gains in coded orthogonal multi-carrier modulation systems. Assuming maximum likelihood decoding, we analyze contribution of channel encoding to pairwise error rate in terms of the Hamming distance for coded-OFDM systems and coded-OFDM-CDM systems. Moreover, the average bit error rate is upper-bounded using the pairwise error rate and is compared with the results obtained by computer simulations.

1 INTRODUCTION

Coded-OFDM systems in time and frequency selective fading channels, in sharp contrast to uncoded-OFDM systems, turn out to enjoy equivalent diversity effect due to the redundancy of the channel coding in addition to the coding gain. The performance, however, seems sensitive to the fading situation. On the other hand, OFDM-Code Division Multiplexing (CDM) (or MC-CDMA) systems in the down-link are known to achieve the full diversity gain owing to the scheme in which each data symbol is spread over the channel bandwidth by modulating OFDM carrier signals according to the signature sequence in spite of Self-Interference (SI) (or multiple access interference in CDMA) from each of the data symbols due to multiplexing. The same thing may be said of coded-OFDM-CDM systems. Performance gap between coded-OFDM and coded-OFDM-CDM systems is considered to result from trade-off between the diversity gain, the coding gain and the effect of SI. Recently this fact has come to be known by computer simulations[1]. In this paper, we introduce Pairwise Error Rate (PER) using Maximum Likelihood (ML) decoding algorithm in order to theoretically evaluate the gap in the optimal reception. For linear block code (including equivalent codes for turbo code), it is known that bit error probability can be upper-bounded by PER and distance spectra which can be characterized by the number of codewords as a function of Hamming weight of the codewords[2][3]. In general, the distance spectra can be determined by the scheme of channel coding. On the other hand, it is important for evaluating the PER to consider the effect of the channel characteristics and the interleaving scheme as well as the Hamming distance. From the viewpoint of the PER,

K. Fazel and S. Kaiser (eds.), Multi-Carrier Spread-Spectrum, 123–132.

in this paper, we explain a relationship between the diversity gain and the coding gain for the coded orthogonal MC systems in frequency selective fading channel.

2 SYSTEM MODEL

We assume that the orthogonal MC modulation system investigated in this paper is able to make use of different frequency sinusoidal carriers whose total number is equal to N. Each of the carriers is spaced by f_0 from adjacent carriers where f_0 is equal to the inverse of the effective OFDM symbol period T_s. Figure 1 shows the assumed transmitter model for coded orthogonal MC

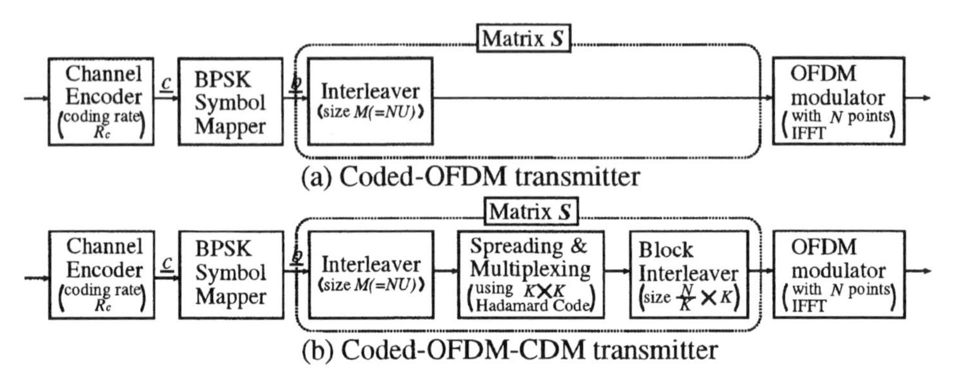

Figure 1: Transmitter Model

modulation system.

Let us consider a linear code with code rate R_c. A binary codeword vector with length M is denoted by $\underline{c} = [c_0, \cdots, c_{M-1}]^T, c_m \in \{0,1\}$. BPSK modulation is used to map each bit of the codeword into a modulated baseband equivalent signal from the $\{\pm 1\}$ set. The modulated vector $\underline{b}$ is given by $\underline{b} = 2\underline{c} - \underline{1}$ where $\underline{1}$ is the vector each entry of which is equal to one. In order to transmit the modulated vector, when $M > N$, $U = M/N$ OFDM modulated symbol periods are required. Provided the length of the guard intervals be sufficient to avoid the effect of the inter-symbol-interference in multi-path channels, the fast Fourier transform operation in the effective symbol periods T_s at the receiver gives the signal vector $\underline{Y}$ as follows.

$$\underline{Y} = \boldsymbol{AHS}\underline{b} + \underline{\eta}.$$

Each component is given by

$$\begin{aligned}
\underline{Y} &= [\underline{y}_0^T, \cdots, \underline{y}_{U-1}^T]^T, & \underline{y}_u &= [y_{uN}, \cdots, y_{uN+N-1}]^T \\
\boldsymbol{H} &= \mathrm{diag}\,[\boldsymbol{h}_0, \cdots, \boldsymbol{h}_{U-1}], & \boldsymbol{h}_u &= \mathrm{diag}\,[h_{uN}, \cdots, h_{uN+N-1}] \\
\underline{\eta} &= [\underline{\eta}_0^T, \cdots, \underline{\eta}_{U-1}^T]^T, & \underline{\eta}_u &= [\eta_{uN}, \cdots, \eta_{uN+N-1}]^T
\end{aligned}$$

y_{uN+n}, h_{uN+n} and η_{uN+n} are, respectively, the received signal, the channel frequency response and the additive noise component, on the n-th frequency carrier at the u-th OFDM symbol period. We assume that the effective transmission bandwidth f_0 occupied by one carrier frequency channel is smaller than the coherence bandwidth of the mobile radio channel. Therefore, fading in the bandwidth of one carrier frequency channel is flat. We also assume fading due to uncorrelated scatters with any delay-power profile. A is the signal amplitude. $\boldsymbol{S}$ is a $M \times M$ matrix. The received signal energy per bit denoted by E_b is defined as

$$E_b = \frac{E\{\|A\boldsymbol{H}\boldsymbol{S}(2\underline{c} - \underline{1})\|^2\}T_s}{R_c M} = \frac{A^2 T_s}{R_c}$$

where $E\{\boldsymbol{H}^H \boldsymbol{H}\} = \boldsymbol{I}$ and $\boldsymbol{S}^H \boldsymbol{S} = \boldsymbol{I}$, respectively.

3 PERFORMANCE EVALUATION OF DIVERSITY GAIN AND CODING GAIN

When ideal channel state information is available at the receiver, the PER that the decoder prefer codeword $\underline{e}$ to the transmitted codeword $\underline{c}$ is given by

$$P(\underline{c} \to \underline{e}) = Q\left(\sqrt{2R_c \frac{E_b}{N_0} \|\boldsymbol{H}\boldsymbol{S}(\underline{c} - \underline{e})\|^2}\right)$$

where we use $E\{\underline{\eta\eta}^H\} = (N_0/T_s)\boldsymbol{I}$[4]. The Euclidean distance in this formula can be expressed as

$$\|\boldsymbol{H}\boldsymbol{S}(\underline{c} - \underline{e})\|^2 = \sum_{q=0}^{M-1} \left| h_q \sum_{m=0}^{M-1} s_{q,m}(c_m - e_m) \right|^2 \tag{1}$$

where $s_{q,m}$ denotes the (q, m) entry of $\boldsymbol{S}$. For the Rayleigh fading channels, when we introduce the random variable $X_q = \sqrt{2R_c \frac{E_b}{N_0}} h_q \sum_{m=0}^{M-1} s_{q,m}(c_m - e_m)$ and the vector $\underline{X} = [X_0, \cdots, X_{M-1}]^T$, the average PER can be determined by averaging over the correlated complex zero-mean Gaussian random variable h_q and is given by

$$\begin{aligned} \bar{P}(\underline{c} \to \underline{e}) &= E\{Q(\|\underline{X}\|)\} \\ &= \frac{1}{2} \sum_{i=0}^{\mathrm{r}\{\boldsymbol{R}_{xx}\}-1} \prod_{\substack{j=0 \\ j \neq i}}^{\mathrm{r}\{\boldsymbol{R}_{xx}\}-1} \frac{\lambda_i}{\lambda_j - \lambda_i} \left(1 - \sqrt{\frac{\lambda_i}{2 + \lambda_i}}\right) \end{aligned} \tag{2}$$

and $\lambda_i, (i = 0, \cdots, \mathrm{r}\{\boldsymbol{R}_{xx}\} - 1)$ are the distinct eigenvalues of the covariance matrix $\boldsymbol{R}_{xx} = E\{\underline{X}\underline{X}^H\}$ and $\mathrm{r}\{\boldsymbol{R}_{xx}\}$ is the rank of $\boldsymbol{R}_{xx}$[5]. The rank of $\boldsymbol{R}_{xx}$ substantially determines the diversity gain.

3.1 In case of coded-OFDM systems

Let the matrix $\boldsymbol{S}$ shown in figure 1 (a) denote a symbol interleaver in coded-OFDM system. We assume that the interleaving process rearranges the input sequence into a one-to-one deterministic format based on the mapping function $f_1(\cdot)$ in order to reduce the effect of burst error events. If we define the mapping rule as

$$s_{q,m} = \begin{cases} 1 \;, & (q = f_1(m)) \\ 0 \;, & (q \neq f_1(m)) \end{cases}, \tag{3}$$

from equation (1), we obtain

$$\|\boldsymbol{HS}(\underline{c} - \underline{e})\|^2 = \sum_{q=0}^{M-1} \left| h_q \left(c_{f_1^{-1}(q)} - e_{f_1^{-1}(q)} \right) \right|^2 = \sum_{m=0}^{M-1} |h_{f_1(m)}|^2 (c_m - e_m)^2 . \tag{4}$$

Letting d denote the Hamming weight of $\underline{e} - \underline{c}$, equation (4) can be simplified as

$$\|\boldsymbol{HS}(\underline{c} - \underline{e})\|^2 = \sum_{i=0}^{d-1} |h_{f_1(m_i)}|^2 \tag{5}$$

where m_i, $(i = 0, \cdots, d-1)$ denotes index m corresponding to $c_m \neq e_m$ contributing to the weight. Using $X_i = \sqrt{2R_c \frac{E_b}{N_0}} h_{f_1(m_i)}$ in place of the original X_q, the average PER is given by equation (2). The rank of $\boldsymbol{R}_{xx}$ is limited by the Hamming distance d. As the correlation of the channel frequency characteristics corresponding to the indices $\{f_1(m_i)\}$ gets smaller, more of the received signal survive (i.e. $|h_{f_1(m_i)}|^2 >$ noise level) with higher probability. In such cases, the Hamming distance contributes not only the coding gain but also the diversity gain. Figure 2 (a) shows an example of Hamming distance distribution with $d = 7$ on the frequency and time unit signal space. When fading of the channel exhibits time-invariance and frequency selectiveness, the channel encoding, in addition to the coding gain, provides frequency diversity effect with the rank 6 of $\boldsymbol{R}_{xx}$ less than $d = 7$ because 2 of the weight contributing signal units share the same frequency carrier.

In addition to the Hamming distance between two codewords, therefore, it is important to consider the positions in which they differ. We introduce the notion of uniform interleaver to evaluate the average PER. The average PER which depends on d can be determined by averaging over possible interleaver patterns with equal event probability, and is given by

$$\bar{P}_d = E\{\bar{P}(\underline{c} \to \underline{e})\}.$$

Moreover, the overall Bit Error Rate (BER) can be upper-bounded by

$$P_b \leq \sum_{d \geq d_{\min}} B_d \bar{P}_d \tag{6}$$

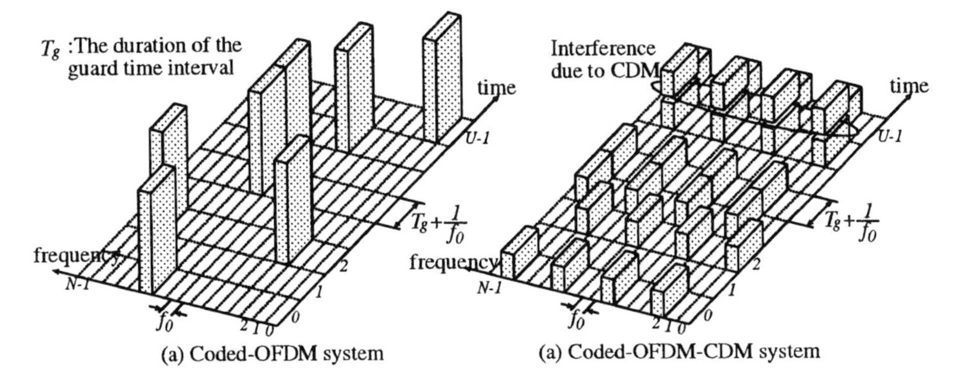

Figure 2: An example of Hamming distance distribution on frequency and time domains ($d = 7$)

where B_d is the average number of bit errors caused by codewords with Hamming distance equal to d and the set of pairs of $\{d, B_d\}$ is called the distance spectrum.

3.2 In case of coded-OFDM-CDM systems

In case of coded-OFDM-CDM systems, let the matrix $\boldsymbol{S}$ shown in figure 1(b) denote the spreading and multiplexing the bits of a codeword in addition to a symbol interleaver. In comparison with coded-OFDM systems, coded-OFDM-CDM systems employ spreading in order to achieve the diversity gain directly. In this paper, we make use of the Walsh-Hadamard matrix given by

$$\boldsymbol{W}_K = \begin{bmatrix} \boldsymbol{W}_{K/2} & \boldsymbol{W}_{K/2} \\ \boldsymbol{W}_{K/2} & -\boldsymbol{W}_{K/2} \end{bmatrix}, \forall K = 2^k, k \geq 1, \boldsymbol{W}_1 = [1]$$

where K is the length of the signature sequences and is smaller than N. Moreover, we define an integer number L which is equal to UN/K. For coded-OFDM-CDM system described in this paper, we redefine $\boldsymbol{S}$ shown in figure 1(b) as

$$\boldsymbol{S} = \frac{1}{\sqrt{K}} \begin{bmatrix} \boldsymbol{P}_N & & 0 \\ & \boldsymbol{P}_N & \\ 0 & & \ddots & \\ & & & \boldsymbol{P}_N \end{bmatrix} \begin{bmatrix} \boldsymbol{W}_K & & 0 \\ & \boldsymbol{W}_K & \\ 0 & & \ddots & \\ & & & \boldsymbol{W}_K \end{bmatrix} \boldsymbol{Q}$$

where $\boldsymbol{P}_N$ denotes a $N \times N$ block interleaver matrix in order to assign signal sequences with length K to frequency carriers spaced by $(N/K)f_0$ for getting frequency effective diversity gain and its (i,j) entry $p_{i,j}$ is defined as

$$p_{i,j} = \begin{cases} 1 \ , \ (i = f_2(j) = (N/K)(j \bmod K) + \lfloor j/K \rfloor) \\ 0 \ , \ (i \neq f_2(j)) \end{cases} .$$

When we assume that $\boldsymbol{Q}$ is the symbol interleaver with mapping function $f_1(\cdot)$ in equation (3), we obtain

$$\begin{aligned} \|\boldsymbol{H}\boldsymbol{S}(\underline{c} - \underline{e})\|^2 &= \frac{1}{K} \sum_{l=0}^{L-1} \sum_{k=0}^{K-1} \left| h_{f_2(lK+k)} \right|^2 \left| \sum_{i=0}^{K-1} w_{k,i} \left(c_{f_1^{-1}(lK+i)} - e_{f_1^{-1}(lK+i)} \right) \right|^2 \\ &\leq \frac{1}{K} \sum_{i=0}^{d-1} \sum_{k=0}^{K-1} \left| h_{f_2(l_i K+k)} \right|^2 . \end{aligned}$$

Compared with equation (5) in coded-OFDM system, it is possible for the coded-OFDM-CDM system to achieve $\mathrm{r}\{\boldsymbol{R}_{xx}\} \leq dK$. Figure 2 (b) shows an example of Hamming distance distribution with $d = 7$ in case of coded-OFDM-CDM system. In this case, it is possible to get greater diversity gain than coded-OFDM system because of extensive distribution of the signal units contributing to the codeword distance. According to the scheme, however, it may happen that two or more distance-contributing signal units duplicate on every frequency-time unit of a CDM set. Such a phenomenon leads to the decreased Euclidean distance and rank of $\boldsymbol{R}_{xx}$. Consequently, it causes the deteriorated diversity and coding gains.

4 NUMERICAL EXAMPLES

In this paper, we assume that the delay-power profile $\rho(\tau)$ with exponential decay, expressed in the normalized form as

$$\rho(\tau) = \frac{1}{\sigma_d} \exp\left(-\frac{\tau}{\sigma_d} \right)$$

where σ_d denotes the mean delay spread. Assuming uncorrelated scatters with $1/(Nf_0)$ delay intervals, the correlation value between channel frequency responses is calculated as

$$E\{h_{n_1} h_{n_2}^*\} = \int \rho(\tau) e^{-j2\pi(n_1 - n_2) f_0 \tau} d\tau = \frac{1}{1 + j\frac{n_1 - n_2}{N}\mathrm{SCR}}$$

where we define the coherence bandwidth B_c as $B_c \equiv 1/(2\pi\sigma_d)$, and the Signal bandwidth to Coherence bandwidth Ratio (SCR) is defined as

$$\mathrm{SCR} \equiv \frac{Nf_0}{B_c} = 2\pi\sigma_d N f_0. \tag{7}$$

In analyzing frequency diversity effect for MC modulation, we use SCR as the parameter representing the effective number of uncorrelated scatters. As SCR gets larger, the channel fades selectively on frequency domain and the correlation between different channel frequency responses has smaller value.

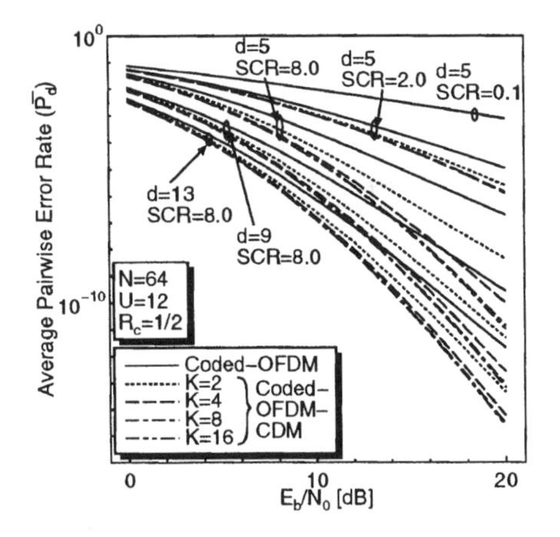

Figure 3: PER performance versus E_b/N_0

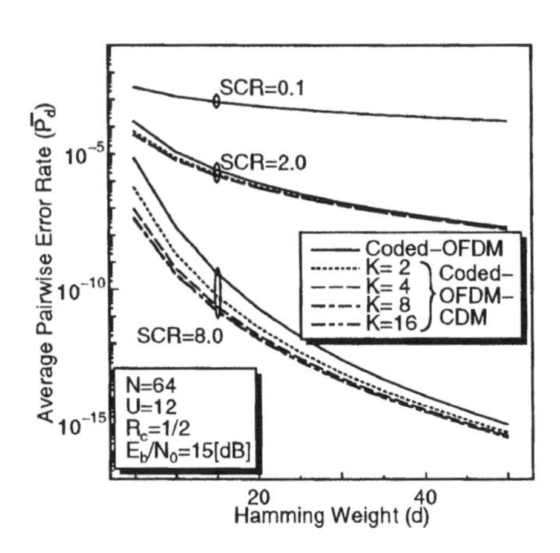

Figure 4: PER performance versus Hamming distance d

In figure 3, average PER performances for coded-OFDM system and coded-OFDM-CDM systems with varying K are shown as a function of E_b/N_0 for $N = 64$, $U = 12$ and $R_c = 1/2$. In case $d = 5$, for example, as SCR gets larger, it is possible to improve the performances owing to the frequency diversity effect for all the systems. In the larger SCR, the performance of coded-OFDM-CDM system with the larger K well outperforms that of code-OFDM system because an extension of redundancy by spreading codeword helps to get more frequency diversity effect. We cannot, however, expect any more improvement in performance, if we provide larger K than SCR. For smaller value of SCR, the degree of improvement with larger K becomes smaller since the distance d and K help each other.

Figure 4 shows average PER performances versus Hamming distance d under the same conditions as figure 3 except for the fixed $E_b/N_0 = 15$[dB]. The diversity gain for all the systems grows with the increasing d especially for the larger SCR. At the larger E_b/N_0 for which $\bar{P}_d$ with smaller d influences the whole performance, it is expected that coded-OFDM-CDM system well outperforms coded-OFDM system because duplication of distance-contributing signal units is less probable.

Next, in order to show a reasonableness of upper bounded BER in equation (6), we introduce an approximated BER $\tilde{P}_b$ by taking into account erroneous

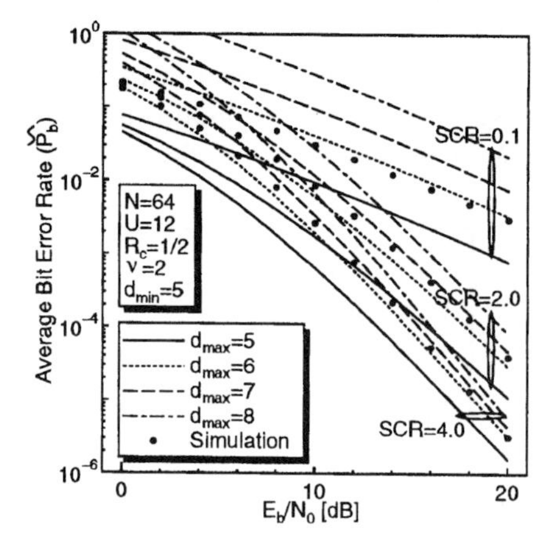

Figure 5: BER performance versus E_b/N_0 for coded-OFDM

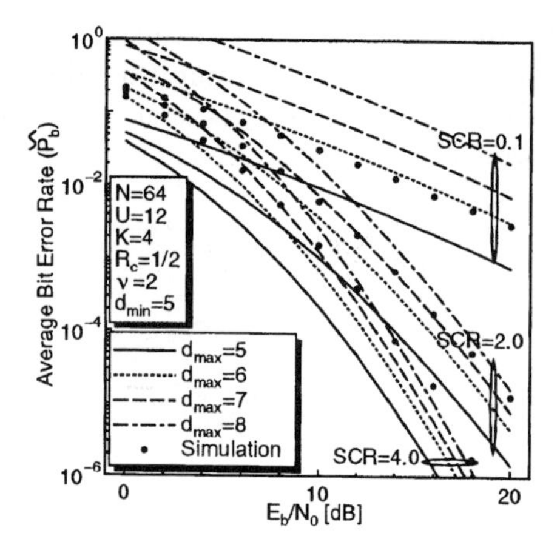

Figure 6: BER performance versus E_b/N_0 for coded-OFDM-CDM $K = 4$

codewords with a range of distance from $d_{\min}$ to $d_{\max}$.

$$\tilde{P}_b \leq \sum_{d=d_{\min}}^{d_{\max}} B_d \bar{P}_d. \tag{8}$$

Let us consider a simple rate 1/2 non-systematic convolutional code generated by two generator sequences (111, 101) with memory order $\nu = 2$ and minimum distance $d_{\min} = 5$. Figure 5 and 6, respectively, show the average BER performances for the coded-OFDM system and the code-OFDM-CDM system in case $K = 4$ under the same conditions as in figure 3. In figure 7, upper bounds of BER by $d_{\max} = 8$ are compared between coded-OFDM and coded-OFDM-CDM systems. In case of the higher SCR at the higher E_b/N_0, the BER's are dominated by $B_d\bar{P}_d$ of low distance. Especially, under these conditions, coded-OFDM-CDM systems are observed to well outperform code-OFDM systems, verifying the expectation.

5 CONCLUSIONS

This paper provided a performance evaluation of diversity and coding gains in coded orthogonal multi-carrier modulation systems. We analyzed the contribution of channel encoding to Pairwise Error Rate (PER) in terms of Hamming distance for coded-OFDM systems and coded-OFDM-CDM systems. Moreover, the average bit error rate upper bounds were evaluated using the PER and were compared with the results obtained by computer simulations. Especially, under there condition of the higher degree of frequency-selective fading and the higher E_b/N_0, coded-OFDM-CDM systems were observed to well outperform coded-OFDM systems. The results of the present study gave the ground

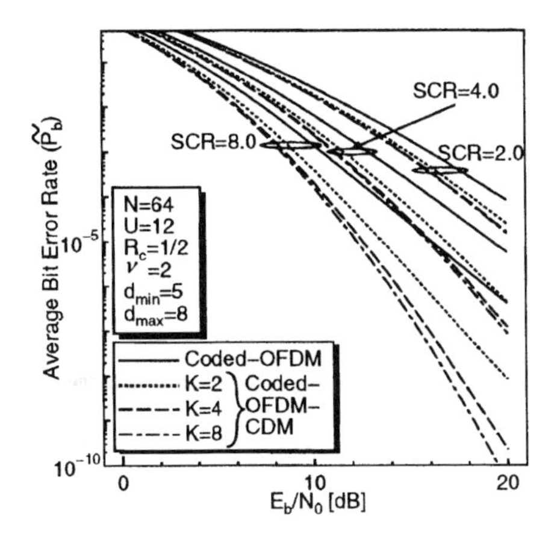

Figure 7: BER upper-bounds; comparison of coded-OFDMs with and without CDM

for investigation of coded-OFDM-CDM systems to cope with broadband channel impairment. They also gave the upper limits of performance such systems could aim at. The system with maximum likelihood detector, however, requires large computational complexity. We are studying new encoding and alternative decoding schemes for coded-OFDM-CDM systems.

References

[1] S. Kaiser, "OFDM Code-Division Multiplexing in Fading Channels", *IEEE Trans., Commun.*, vol. 50, no. 8, pp. 1266-1273, Aug. , 2002.

[2] S. Benedetto and G. Montorsi, "Unveiling Turbo Codes: Some Results on Parallel Concatenated Coding Schemes", *IEEE Trans. Inform. Theory*, vol. 42, no. 2, pp. 409-428, March, 1996.

[3] B. Vucetic and J. Yuan, *Turbo Codes Principles and Applications*, Kluwer Academic Publishers, 2000.

[4] V. Tarokh, N. Seshadri and A. R. Calderbank, "Space-Time Codes for High Data Rate Wireless Communication: Performance Criterion and Code Construction", *IEEE Trans. Inform. Theory*, vol. 44, no. 2, pp. 744-765, March, 1998.

[5] S. Verdú *Multiuser Detection*, Cambridge University Press, 1998.

Department of Applied Electronics, Tokyo University of Science
2641 Yamazaki, Noda, Chiba 278-8510, Japan
Tel:+81-4-7124-1501 Ext. 4226, Fax:+81-4-7122-9195
E-Mail:{fujii,itami,itoh}@itlb.te.noda.sut.ac.jp

Adaptive Coding in MC-CDMA/FDMA Systems with Adaptive Sub-Band Allocation

P. Trifonov, E. Costa and A. Filippi
Siemens AG, ICM N PG SP RC FR, D-81739- Munich

Abstract. The MC-CDMA/FDMA scheme is a candidate for the air-interface of beyond 3G mobile communications. An efficient adaptive sub-band allocation (ASBA) approach has been recently shown to provide a considerable gain in the uncoded system performance. In this paper, adaptive coding is proposed for application in conjunction with ASBA. This is proved to yield a significant performance improvement, especially if a user-service prioritisation is considered in the ASBA.

Keywords: OFDM, MC-CDMA, Adaptive Frequency Allocation, Adaptive Coding.

1. Introduction

The OFDM-based frequency and code division multiple access concepts OFDMA and MC-CDMA and the derived hybrid solutions are regarded as leading candidate for beyond 3G mobile communications, especially for the synchronous downlink [1]-[2].This certainly owns to the high robustness of OFDM to the radio channel time-dispersion, but also to the flexibility and adaptability offered by such schemes in the assignment of the frequency resources.

In this work, we consider the hybrid MC-CDMA/FDMA scheme [3]. The transmission bandwidth is sub-divided into a number of sub-bands, each allocated to a group of users (FDMA), which transmit in a MC-CDMA fashion. An efficient adaptive frequency mapping, referred to as adaptive sub-band allocation (ASBA) has been proposed in [4] for the downlink of MC-CDMA/FDMA. Under the assumption that an estimate of the channel experienced by each user over the whole bandwidth is available at the base station (e.g. from the uplink received signal in a time division duplex system), an optimisation algorithm produces a combination of user-grouping and sub-carrier grouping that maximises the overall link capacity. The ASBA has been shown to provide a significant gain in the uncoded system performance as compared to the usual fixed frequency mapping, based on the interleaving of the sub-carriers assigned to different user-groups [4].

Starting point of this work has been the observation that the ASBA provides indeed a significant gain also in a coded system. Hence, our goal is that of finding a proper channel coding scheme for MC-CDMA/FDMA systems with ASBA. It has to be observed that, since

K. Fazel and S. Kaiser (eds.), Multi-Carrier Spread-Spectrum, 133–140.

the considered ASBA optimises the overall link capacity, it is likely to produce a combination of user-grouping and sub-carrier grouping for which the signal-to-noise-plus-interference ratio (SNIR) experienced by the different users differ significantly. As a consequence, if a too low coding rate is chosen according to the performance of the users with the lowest SNIR, a waste in bandwidth efficiency may occur. We propose to use adaptive coding in conjunction with ASBA. By adaptive coding we mean here the adjustment of the coding rate for each user according to the actual SNIR seen by that user after the ASBA.

Moreover, we note that, within a cell, some services may need to be guaranteed a certain throughput, while others may have very low throughput demands. We consider here a slight modification of the ASBA algorithm proposed in [4] to cope with a prioritisation among classes of user-services on the basis of their throughput requirements. Different but fixed coding rates can be used for users with different priority. We will show, however, that adaptive coding enables a significant performance gain over a much larger signal-to-noise ratio (SNR) range.

In Section 2, we briefly review the principle of ASBA. In Section 3, we explain how adaptive coding can be applied in conjunction with ASBA. In Section 4, simulation results on the performance of MC-CDMA/FDMA with joint adaptive coding and ASBA are reported and discussed. Finally, conclusions are drawn in Section 5.

2. Adaptive Sub-Band Allocation in MC-CDMA/FDMA

Let $B = \{B_1, B_2, \ldots, B_Q\}$ be a partition of the whole set of N sub-carriers within the transmission bandwidth. The q−th sub-band B_q, $q = 1, 2, \ldots, Q$, consists of N_{sb} not necessarily adjacent sub-carriers. Let then $U = \{U_1, U_2, \ldots, U_Q\}$ be a partition of the set of all K active users. The q−th user-group U_q consists of up to K_{MC} users, which spread their data symbols in frequency direction over the same sub-band, only separated by orthogonal codes of length K_{MC}. Without loss of generality, the user-group U_q is assigned the sub-band B_q.

The normalised capacity of user k, $k = 1, 2, \ldots, K$, over sub-carrier n, $n = 1, 2, \ldots, N$, can be expressed as

$$C_{k,n} = \log_2\left(1 + \frac{|H_{k,n}|^2}{\sigma_\eta^2}\right) \quad \text{bit/s/Hz}, \tag{1}$$

where $H_{k,n}$ is the channel transfer factor experienced by user k over sub-carrier n and σ_η^2 is the variance of an additive white Gaussian noise (AWGN) including both the AWG channel noise and the multi-user interference (MUI) [4]. Under the hypothesis of transmit signal power equal to 1, we have SNIR $= \frac{|H_{k,n}|^2}{\sigma_\eta^2}$.

Thus, the capacity of user k over the sub-band B_q, hereafter referred to as user-capacity per sub-band, is given by

$$C_{k,B_q} = \sum_{n \in B_q} C_{k,n}, \tag{2}$$

from which the capacity of the user-group U_q over the sub-band B_q can be derived as $C_{U_q,B_q} = \sum_{k \in U_q} \sum_{n \in B_q} C_{k,n}$. The overall link capacity reads then as $C_{TOT} = \sum_{q=1}^{Q} C_{U_q,B_q}$. The optimisation addressed by the ASBA consists in selecting the pair of partitions B and U which maximises C_{TOT} for a given channel estimate. For details on the optimisation algorithm the reader is referred to [4].

In order to let the ASBA take a given prioritisation among users into account, the user-capacity per sub-carrier $C_{k,n}$ can be multiplied by a proper weighting factor $F > 1$. Let us assume that the set of K active users is sub-divided into P priority classes, in such a way that the users in class $P-1$ have the highest priority and the users in class 0 have the lowest priority. Then, in order to guarantee that highly-prioritised users are allocated the sub-bands where they experience the highest SNIR, we assign to their user-capacity higher weights than for other, lower priority, users. That is, the ASBA optimisation algorithm is fed with the modified user-capacity per sub-carrier $C'_{k,n} = F^{i_k} C_{k,n}$, where i_k is the priority of user k and F is the chosen weighting factor.

3. Adaptive Coding

By adaptive coding we mean the adjustment of the coding rate for each user according to its SNIR, while keeping the codeword length and the decoding parameters fixed. Since we assume to apply adaptive coding jointly with ASBA, the single user-capacity per sub-band (cf. (2)) provided by the ASBA can be used as an indicator of the SNIR. According to its value, the coding rate of the single user is selected in order to optimise the system performance with respect to a given criterion. A block diagram of the transmission scheme with adaptive coding in conjunction with ASBA is depicted in Fig. 1 for the general case in which the code is given by the concatenation of an outer and an inner code.

We observe that the BER or the Frame Error Rate (FER) are not appropriate performance measures. Indeed, the probability of decoding error can be minimized by selecting the lowest possible coding rate.

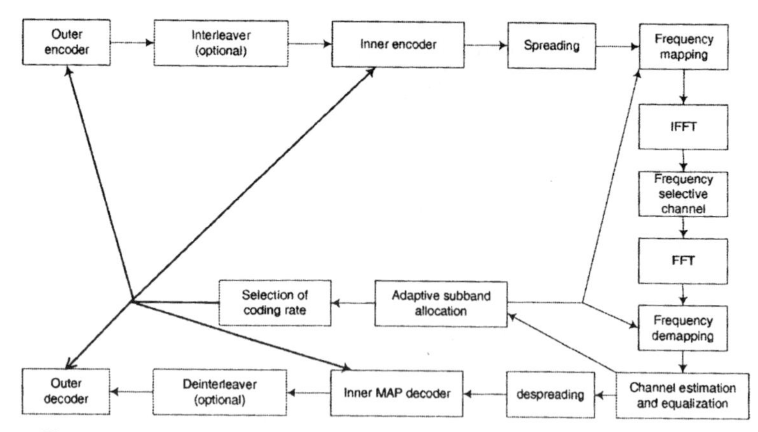

Figure 1. Transmission scheme with adaptive coding and ASBA

This would introduce, however, very high redundancy. If, on the one hand, it is desirable to minimize the FER, on the other hand, as much as possible data should be transmitted within each data packet, i.e. codeword. Therefore, by means of adaptive coding we aim to maximise the user throughput defined as the average number of received data symbols per codeword. For a code of dimension s and codeword length n over the Galois Field GF(2^m), the throughput can be expressed as

$$S(s,n,m,C) = s \cdot (1 - P_e(s,n,m,C)), \tag{3}$$

where s denotes also the number of received data symbols in case of successful decoding and $P_e(s,n,m,C)$ is the probability of incorrect decoding for a given user-capacity per sub-band C (cf. (2)).

Given a fixed codeword length n, in order to maximise the average user throughput an optimisation process is carried out that results in a list of intervals of user-capacity per sub-band, $[T_i;T_{i+1})\,, i = 1..L$, with each interval associated to a code dimension s_i. That is, a code with parameters (n, s_i) is used whenever the user capacity per sub-band yielded by the ASBA is $C \in [T_i, T_{i+1})$. The values T_i are the switching thresholds for which the system throughput is maximised for the considered operational environment. This includes the channel propagation conditions, e.g. the noise level and the fading characteristics, and the ASBA settings, e.g. the initial assumptions and the number of iterations of the optimisation algorithm.

Since both the propagation conditions and the ASBA setting represent random variables, the resulting user-capacity per sub-band C is also a random variable with some probability density function $p(C)$. As a consequence, by defining the set $\{\mathbf{T}\}$ of all possible lists of switching thresholds $\mathbf{T} = (T_1, \ldots, T_L)$, with $T_i < T_{i+1}$, the optimisation problem can be stated as finding

$$\bar{\mathbf{T}} = \arg\max_{\mathbf{T}} \int_0^\infty \tilde{S}(n,m,C,\mathbf{T})p(C)dC, \tag{4}$$

where $\tilde{S}(n,m,C,\mathbf{T})$ is the average user throughput obtained by the adaptive system for given switching thresholds $\mathbf{T}$ and user-capacity per sub-band C, that is

$$\tilde{S}(n,m,C,\mathbf{T}) = S(s_i,n,m,C),\ i : C \in [T_i; T_{i+1})\,. \tag{5}$$

Since it is difficult to find an analytical expression for this function (cf. (3)), we have approximated it by means of simulations. More specifically, the optimisation has been performed as follows:

- Simulations have been run for different values of the code dimension s and for different values of SNR $= E_s/N_0$, where E_s is the average received energy per symbol and N_0 is the one-sided power spectral density of the channel noise.

- For each value of s, the obtained values of user-capacity per sub-band and the corresponding throughput values have been recorded.

- For each considered list of capacity intervals $\mathbf{T} = (T_1, \ldots, T_L)$, the code dimension s_i yielding the highest throughput has been determined for the single capacity intervals to approximate the function (5) as shown in Fig 2.

- The last step has been repeated for many different lists of capacity intervals and the maximum of (4) has been determined.

Two adaptive coding schemes have been considered. The first is constructed through the concatenation of a Reed-Solomon (RS) code and a convolutional code (CC) and it is referred to as ARSCC in the sequel. The second is given by an adaptive turbo coding (ATC). We note that the rate of the ARSCC scheme can be changed either by changing the dimension of the RS code or by changing the rate of the CC. The latter task is usually accomplished by puncturing and/or by changing the number of generator polynomials. However, to achieve a sufficient number of different coding rates, it might be necessary to use very long puncturing patterns and/or very high number of generator polynomials, which may represent design and implementation issues. In this work, we restrict ourselves to changing the rate of the RS code. Since the turbo codes are given by the parallel concatenation of convolutional codes [5], their rate can be changed as for the CC.

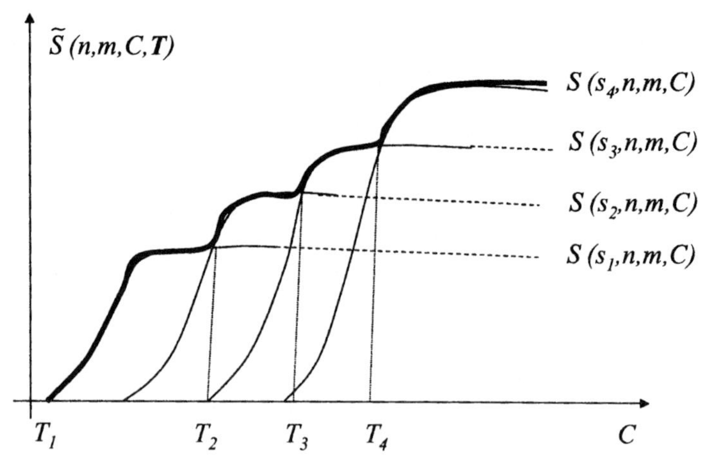

Figure 2. Approximation of the average user-throughput as a function of the user-capacity per sub-band and of the set of switching thresholds.

4. Simulation Results

In the simulations, a bandwidth $B = 20$ MHz and a carrier frequency $f_c = 5.5$ GHz are assumed. The channel is chosen to be a Rayleigh fading channel with exponential power delay profile and maximum delay spread $\tau_{\max} = 5\,\mu$s. $N = 512$ sub-carriers and $Q = 8$ sub-bands with up to $K_{\mathrm{MC}} = 8$ users each are considered as in [4]. Moreover, 16−QAM bit mapping is assumed. For a fair comparison of the different coding schemes, the codeword length is fixed to 512 bytes. The considered Reed-Solomon codes are RS$(255, s, 256 - s)$ over GF(2^8) [5]. The concatenated convolutional code is the rate 1/2 recursive systematic code based on the generator polynomials 0133 and 0171 (octal). The turbo code has the encoder structure specified in [6]. Performance results are reported in terms of average user throughput versus SNR.

Fig. 3.a illustrates the results for a MC-CDMA/FDMA system with ASBA when only fixed convolutional coding with different coding rates is adopted. It can be seen that the system without coding can achieve the maximum throughput of 512 bytes per packet for very high SNR. However, for SNR below 15 dB the system fails to transmit any data. On the other hand, the system with coding is able to transmit data at low SNR, but it is efficient only in a narrow SNR range.

The beneficial effect of adaptive coding can be observed in Fig. 3.b, which reports the results achieved with the two considered adaptive coding methods. For the ARSCC, the comparison of the results obtained with and without ASBA proves that ASBA provides a significant gain also in the presence of channel coding. Moreover, the ARSCC yields a considerable gain as compared to a fixed rate 1/4 code given

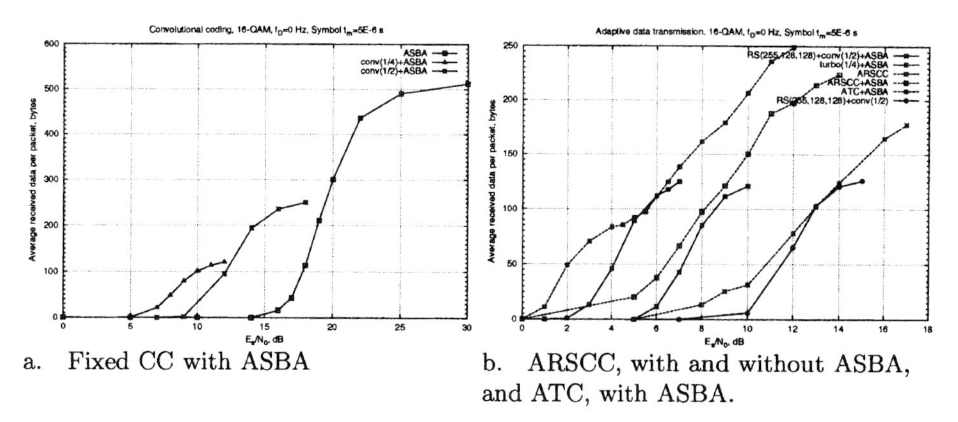

a. Fixed CC with ASBA

b. ARSCC, with and without ASBA, and ATC, with ASBA.

Figure 3. Performance of the unprioritized system

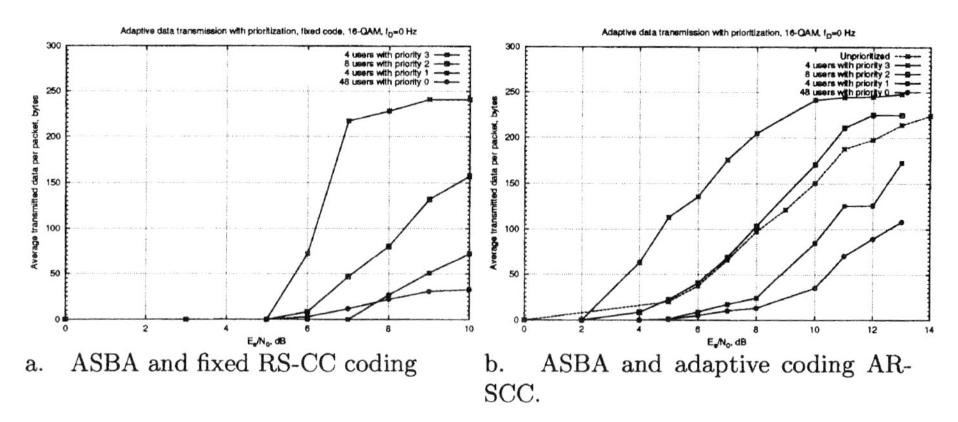

a. ASBA and fixed RS-CC coding

b. ASBA and adaptive coding ARSCC.

Figure 4. Performance of different priority users

by the concatenation of the RS(255, 128, 128) code with the rate 1/2 CC. The ATC gives even a higher gain. However, a lack of flexibility in the selection of the turbo coding parameters leads to a non-smooth behaviour near 4 dB. In this SNR region, in fact, the sub-carrier assignment yielded by the ASBA is good enough to achieve almost error-free transmission using a rate 1/6 turbo code, but still too bad to use a rate 1/4 turbo code. Hence, the lower coding rate has to be chosen, so limiting the throughput.

The results obtained with ASBA in the presence of a user prioritisation are reported in Fig. 4.a and Fig. 4.b, for fixed and adaptive coding, respectively. $P = 4$ priority classes are assumed over $K = 64$ users, of which 4 with priority 3, 8 with priority 2, 4 with priority 1 and 48 with priority 0.

In case of fixed coding, the concatenated RS-CC scheme has been chosen, and fixed, but different, coding rates have been selected for users with different priority. More specifically, on the basis of the average user throughput at $E_s/N_0 = 10$ dB, the dimension s of the RS code equals 241, 170, 85 and 35 for users with priority 3, 2, 1, 0, respectively. From Fig. 4.a it can be inferred that fixed average throughput values are achieved depending on the user priority at high SNR. With fixed coding it is not possible neither to improve the throughput in case of very good channels (higher E_s/N_0), e.g. when the ASBA provides a favourable allocation, nor to achieve satisfactory throughputs in bad channels (lower E_s/N_0). By using adaptive coding, in contrast, a gain in throughput is observed over a large SNR region in Fig. 4.b. In particular, for lower priority users a significant throughput improvement is achieved at high SNR. Moreover, adaptive coding enables a full exploitation of the prioritisation, i.e. a significant difference in the throughput of users belonging to different priority groups can be noticed.

5. Conclusions

An adaptive coding approach has been proposed and investigated for application in MC-CDMA/FDMA systems jointly with the adaptive frequency mapping known as ASBA. Through adaptive coding, the coding rate of the single users is changed according to the SNIR provided by the ASBA, in such a way that the average user throughput is maximised. Simulation results have shown that adaptive coding yields in general a significant improvement of the throughput at higher SNRs, while enabling satisfactory throughput at low SNRs. Moreover, the application of adaptive coding results to be particularly advantageous when a user prioritisation is considered in conjunction with ASBA.

References

[1] H. Rohling, T. May, K. Brueninghaus and R. Gruenheid, "Broad-Band OFDM Radio Transmission for Multimedia Applications", in *IEEE Proceedings*, vol.87, Oct. 1999.

[2] K. Fazel and S. Kaiser, *Multi-Carrier Spread-Spectrum and Related Topics*, Kluwer Academic Publishers, Boston, 2002.

[3] S. Kaiser, *Multi-carrier CDMA Mobile Radio Systems- Analysis and Optimisation of Detection, Decoding, and Channel Estimation*, Number 531 in Fortschrittberichte VDI, Reihe 10. VDI- Verlag, Dusseldorf, 1998.

[4] E. Costa, H. Haas, E. Schulz and A. Filippi, "Capacity Optimisation in MC-CDMA Systems", *ETT Eur. Trans. on Telecomm.*, vol. 13, Sept./Oct. 2002.

[5] M. Bossert, *Channel Coding for Telecommunications*, Wiley and Sons, 1999.

[6] C. Berrou, A. Glavieux and P. Thitimajshima, "Near Shannon-limit Error-Correcting Coding and Decoding: Turbo Codes", in *Proc. of International Communication Conference ICC 1993*, pp. 1064-1070.

HONGNIAN XING & MARKKU RENFORS

A STUDY OF MULTICARRIER CDMA SYSTEMS WITH DIFFERENTIAL MODULATION

Abstract. In this paper, we investigate the performance of downlink multicarrier CDMA systems with differential modulation methods in multipath fading environments by analytical and simulation tools. Compared with basic coherent detection methods, the numerical results show that differential modulation provides robust performance and can actually result in competitive performance in cases where the channel estimation error in coherent detection is significant. Especially, with fast fading channels, differential modulation and detection could be viable alternative to coherent modulation, resulting in reduced implementation complexity and reduced pilot overhead.

1. INTRODUCTION

The idea of multicarrier CDMA appeared in 1993 [1][2]. Since its introduction, there have been a large number of papers on different variants of MC-CDMA systems [3][4][5]. Among them, the basic idea of MC-CDMA in which spreading is performed in frequency domain only is the most attractive one [1]. In this paper, we study the detection methods for such MC-CDMA systems.

Optimum linear solution to the multi-user MC-CDMA detection problem has been presented in the literature (see, e.g., [12]), but it is worth to consider also simplified approaches. Since the basic MC-CDMA system is a parallel transmission system that contains the signal from a user in all subcarriers, the study of detection methods could be based on the usual channel equalization methods for subcarriers and diversity combining strategies. Considering the channel equalizer at each subcarrier, the normal zero-forcing (ZF) and minimum mean-square error (MMSE) criterions could be applied for coherent detection. Since the channel can be thought as a flat fading one at each subcarrier, the MMSE equalizer usually gives a better performance than the ZF equalizer [6].

The combining strategy is another important issue since diversity combining is naturally obtained in MC-CDMA systems. Several combining methods, such as maximum ratio combining (MRC), equal gain combining (EGC), and controlled equalization (CE) have been investigated [7]. The performance depends critically on the code orthogonality after equalization. So the EGC has better performance than MRC. CE enhances the EGC by discarding some deeply faded subcarriers and has some additional gain since the code orthogonality can be restored better. It can be noted that in case of a fully loaded system, the optimum MMSE approach is equivalent to per-carrier MMSE equalization and EGC [12].

In most analysis concerning the channel equalization methods, the channel information is assumed to be known perfectly. In practice, however, the channel information is estimated from the training symbols (or pilots). The estimation in noisy environment will usually give an error floor, due to the errors in the channel

K. Fazel and S. Kaiser (eds.), Multi-Carrier Spread-Spectrum, 141–150.

estimation. In fading channels, the loss due to the estimation error will become increasingly significant as the speed of fading increases. Such a loss could be partly compensated by applying more training symbols (or pilots). However, the system efficiency is reduced with increased pilot density. Interpolation helps to estimate the fading effect and reduce the error floor (or pilot density). However, the estimation error is still significant in fast fading cases.

One of the possible solutions is applying differential modulation and detection schemes at each subcarrier. Since differential detection gets the signal information from the difference between two consecutive symbols, the effect of channel fading could be very well eliminated.

The performance of the differential modulation is 3 dB worse than that of coherent modulation in an AWGN channel [6]. However, the relative performance of differential modulation will be improved if the channel estimation errors dominate in coherent detection. One particular reason for applying differential modulation at each subcarrier is that the intersymbol interference (ISI) can be cancelled at each subcarrier by using guard intervals. The key point is that the multi-access interference (MAI) due to ISI can be eliminated so that the differential detection is possible at each subcarrier in the downlink multi-user case.

Another advantage of using differential modulation is the possibility of avoiding channel estimation and equalization at the receiver. From the implementation point of view, this will reduce the system complexity significantly at baseband.

There are two basic ways of combining the idea of differential modulation and the MC-CDMA scheme: applying the differential modulation to user symbol sequences or to the subcarrier signals after combining. It is clear that the first form of differential modulation is applicable (at least in principle) to both uplink and downlink, whereas the latter form can be used only in downlink.

In the first approach, the differential modulation is directly applied to each user symbol sequence of a MC-CDMA system [8]. At the receiver side, differential symbol detection is then done at the point after de-spreading and combining. However, the performance of such a system is only acceptable for mildly frequency selective channels. There are two possible ways to improve the performance. One is to divide the total frequency band into several groups so that fading within each group is relatively flat. The non-coherent combining and demodulation is applied for an individual group. The combining in this case is constructive, but the overall diversity gain is lost to some extent. A special case (with spreading factor 1) of such a method is the differential OFDMA system [9]. Another way is to apply a simple equalization scheme, such as EGC so that the phases can be aligned for constructive combining. Compared to the normal coherent EGC system, differential detection with EGC could be more robust to channel estimation errors and fast fading.

In the second approach, non-coherent modulation/demodulation is used at subcarrier level. Here, the idea of differential modulation should be modified since the amplitude of the combined multi-user symbol of a subcarrier signal is irregular. So only the phase of the combined symbol would be modulated differentially. At the receiver side, the differential demodulation of phase is applied at each subcarrier before de-spreading and combining. Such a system has been investigated for binary differential modulation in AWGN channels [10]. In this paper, we study such a system with differential QPSK modulation, in downlink, using the independently

Rayleigh fading channel model, by both analytical and simulation methods. The performance of such a system is also compared to several coherent detection methods, in order to show the possible benefits obtained by differential modulation.

The paper is organized as follows: The differential MC-CDMA system model is presented in Section 2 and the performance of such a system is analysed in Section 3 in case of independently fading subcarriers. Section 4 gives performance results for the studied differential MC-CDMA system by simulations. Results for coherent MC-CDMA systems are also given for comparison. The conclusions are given in Section 5.

2. THE SYSTEM DESCRIPTION

The simplified baseband block diagram is shown in Figure 1. A data symbol of the desired user is first repeated for all (used) subcarriers. Then the spreading operation is done in frequency, i.e., the data symbol is multiplied by a chip of the spreading code for the desired user at each subcarrier. After combining with the data symbols (which are also multiplied by their corresponding chips) from other users, the combined symbol is differentially modulated. Specifically, the phase of the transmitted symbol is the phase of the previous transmitted symbol plus the phase of the new symbol after combining, whereas the amplitude of the transmitted symbol is the same as the amplitude of the current symbol from combining. After the multi-carrier (IDFT) modulation, the signal is converted into serial format. At each conversion, a symbol (so called OFDM symbol) is generated. The cyclic extension of the symbol is used as the guard interval.

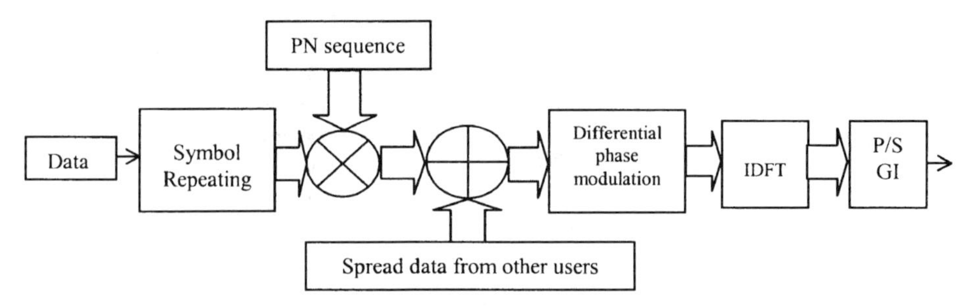

Figure 1. Differentially phase modulated MC-CDMA transmitter.

The sampled version of the j th transmitted data symbol is given as

$$x_j(n) = \text{IDFT}(\tilde{S}_j(k)), \quad \text{for } n = 0, \ldots, N-1 \tag{1}$$

where $S_j(k)$ and $\tilde{S}_j(k)$ are the kth subcarrier samples before and after differential phase modulation, respectively,

$$\begin{aligned} S_j(k) &= \sum_{i=0}^{U-1} X_{i,j} P_i(k) \\ \tilde{S}_j(k) &= S_j(k)\tilde{S}_{j-1}(k) \big/ \left|\tilde{S}_{j-1}(k)\right| \end{aligned} \tag{2}$$

Here i is the index of user, $P_i(k)$ is the user specific code for the i th user, and $X_{i,j}$ is the j th symbol of i th user.

The amplitude of the combined (multi-user) symbol at each subcarrier is irregular. It is possible that such a combination will generate a symbol, which has the amplitude equal to zero (or close to zero). Then the reference phase for the next symbol would be poor, even if the corresponding subcarrier is not deeply faded [10].

There are various ways to avoid such a situation. For example, the amplitude levels for different codes/users could be slightly tuned, if the zero amplitude is detected at the transmitted. Or one could use different but known phase rotations for different users. In this paper, we just re-scale the amplitude of the one user by multiplying with a real constant (of value 1.2) if zero amplitude is found at any subcarrier.

The channel is described as a multipath fading channel with the discrete impulse response

$$h(n) = \sum_{m=0}^{M-1} a_{m,n} \delta(m-n) . \tag{3}$$

In this case, the channel has a finite impulse response of length M, and $a_{m,n}$ is the attenuation factor for path m at time n. For slowly fading channels, $a_{m,n}$ can be viewed as a constant during several (or at least one) OFDM symbols, in which case,

$$a_{m,n} = a_{m,0} = a_{m,1} = \ldots = a_m . \tag{4}$$

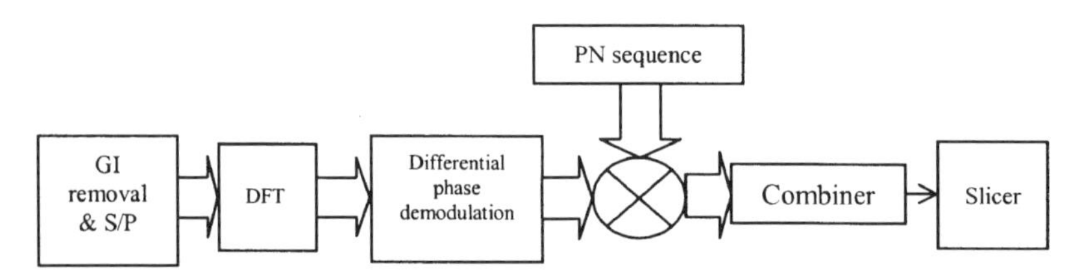

Figure 2. Differentially phase modulated MC-CDMA receiver.

The simplified block diagram of the receiver side is shown in Figure 2. Differential phase demodulation is performed at each subcarrier after the DFT. After the despreading operation (with local sequence $P_i^*(k)$), the data symbols from all subcarriers are summed together and then fed to the slicer. This acts just as a correlator, not only for separating different users, but also for collecting the symbol energy from all subcarriers.

3. PERFORMANCE ANALYSIS

Since the differential phase demodulation at each subcarrier provides the corrected phase of the combined symbol at the subcarrier, the following despreading and combining will combine the signal from the desired user constructively. The amplitude of the combined signal is not affected by the phase demodulation. In this case, the overall system is similar to the EGC MC-CDMA system [7]. Therefore, the system analysis can be based on the analysis of the EGC system. The only difference is that in case of coherent EGC, the phase is estimated and corrected, whereas in the differential system, the phase difference between consecutive subcarrier samples is used.

The received signal of the 0th user ($i = 0$) can be given as

$$r_{0,j}(n) = (x_{0,j}(n) + \sum_{i=1}^{U-1} x_{i,j}(n)) * h(n) + w(n) \tag{5}$$

where $w(n)$ is the sampled version of AWGN noise and $*$ is the convolution operator. The recovered subcarrier signals are then

$$X'_{0,j} = \sum_{k=0}^{N-1} (P_0^*(k)\mathrm{DFT}(r_{0,j}(n)) \tag{6}$$

If (4) holds, then (6) can be rewritten as

$$\begin{aligned} X'_{0,j} &= X_{0,j} \sum_{k=0}^{N-1} (P_0(k)P_0^*(k) \sum_{m=0}^{M-1} a_m e^{\frac{-j2\pi mk}{N}}) \\ &+ \sum_{i=1}^{U-1} (X_{i,j} \sum_{k=0}^{N-1} (P_i(k)P_0^*(k) \sum_{m=0}^{M-1} a_m e^{\frac{-j2\pi mk}{N}})) \\ &+ \sum_{k=0}^{N-1} W(k) \end{aligned} \tag{7}$$

The first term in the pervious equation is the distorted desired signal. The second term is the multi-access interference (MAI), and the third term is AWGN noise with variance $\sigma_n^{\ 2}$. The amplitude of the combined attenuation term $\sum_{m=0}^{M-1} a_m e^{\frac{-j2\pi mk}{N}}$ is defined as ρ_k. The local-mean power at the k th subcarrier is then defined to be

$$\bar{p}_k = \frac{1}{2} E((\rho_k)^2) \tag{8}$$

The total local-mean power for the i th user is then $\bar{p} = N\bar{p}_k$. Since the codes are orthogonal, half of the chip products should be positive and the other half should be negative. Applying the Central Limit Theorem (CLT) for both the positive and negative parts, the MAI term can be approximated by a zero-mean Gaussian random variable with a variance of [7]

$$\sigma_{MAI}^{\ 2} = 2(U-1)(1-\gamma)\bar{p} \tag{9}$$

where

$$\gamma = \frac{\pi}{4}(\frac{e^{-K}}{K+1})\left[(1+K)I_0(\frac{K}{2}) + K \times I_1(\frac{K}{2})\right]^2 \tag{10}$$

and $I_0(\ )$ and $I_1(\ )$ represents the zeroth and first order modified Bessel functions, repectively. This model applies in the case where the subchannels are independently fading with identical Rician statistics and Rician parameter K.

Then, with QPSK modulation, the probability of making a decision error can be written as

$$\Pr(error \mid \bar{p}, K, \{\rho_k\}_{k=0}^{N-1}) = \frac{1}{2}\text{erfc}\left(\sqrt{\frac{\frac{1}{2}(\sum_{k=0}^{N-1}\rho_k)^2}{2(U-1)(1-\gamma)\bar{p}+\sigma_n^{\ 2}}}\right) \tag{11}$$

For an outdoor multipath fading channel, the amplitude of the combined signal at each subcarrier can be assumed to be Rayleigh distributed (the Ricean factor $K=0$). In [1], it was shown that using the CLT to approximate the sum of iid Rayleigh random variables leads to an adequate approximation. The sum can be approximated as a zero-mean Gaussian distribution. Then the average BER can be obtained by averaging Eq. (11) over the Gaussian distribution, leading to

$$\Pr(error \mid \bar{p}) = \frac{1}{2}\text{erfc}\left(\sqrt{\frac{\pi}{4}\frac{N\bar{p}}{(2(U-1)(1-\frac{\pi}{4})\bar{p})+\sigma_n^{\ 2}}}\right) \tag{12}$$

For the differentially phase modulated MC-CDMA system, the MAI is similar to that in the coherent EGC case. The main difference is the channel noise. Basically, compared to the coherent EGC case, the channel noise is doubled in the differentially phase modulated case, so the average BER can be approximated as

$$\Pr(error \mid \bar{p}) = \frac{1}{2}\text{erfc}\left(\sqrt{\frac{\pi}{4}\frac{N\bar{p}}{(2(U-1)(1-\frac{\pi}{4})\bar{p})+2\sigma_n^{\ 2}+\sigma_{ext}^{\ 2}}}\right) \tag{13}$$

The extra noise term $\sigma_{ext}^{\ 2}$ is induced by the possible poor phase reference, due to the channel fading and irregular amplitude signals. It can be seen from (13) that the average BER is a function of spreading factor N, number of active users U, and the channel noise variance.

4. THE PERFORMANCE ANALYSIS BY SIMULATIONS

The system performance has also been investigated by simulations in the QPSK case. Two reference systems, a coherent EGC MC-CDMA system and a per-carrier MMSE equalized MC-CDMA system have been included in the comparisons [7] [11]. In simulations, the subcarriers are assumed to be independently Rayleigh fading. In practice, there are correlations between nearby subcarriers. However, such an independent fading model is valid if a suitable frequency interleaving scheme is used. Also based on such an assumption, the theoretical analysis can be simplified to the form presented in Section 3.

Figure 3 shows the performance of the differentially phase modulated MC-CDMA system in case of different spreading factors in the fully loaded case (i.e., the number of users equals the spreading factor). The theoretical curve corresponding to (13) matches closely with the simulation results with the highest spreading factor.

As discussed previously, there is a loss due to the doubled channel noise (at subcarrier level) in differential modulation, compared to the ideal coherent detection. Although the zero amplitude case for the combined symbol (at subcarrier level) can be avoided by different methods at the transmitter (the performance doesn't seem to depend critically on the method used), the joint effect of irregular transmitted amplitudes of signals and irregular (overall) channel frequency response will possibly induce more errors (due to the poor reference phase) than the coherent EGC method. It can also be seen from the simulations that in the fully loaded case, the performance of the system is rather independent of the spreading factor. The difference between coherent EGC and differential system is consistently about 5 dB.

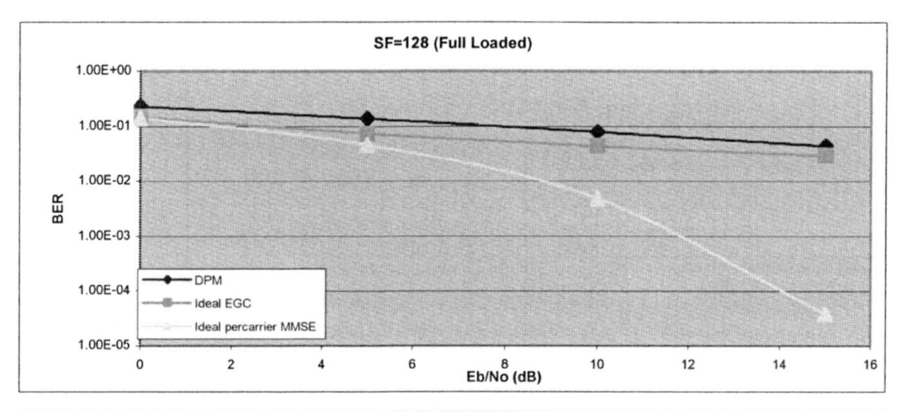

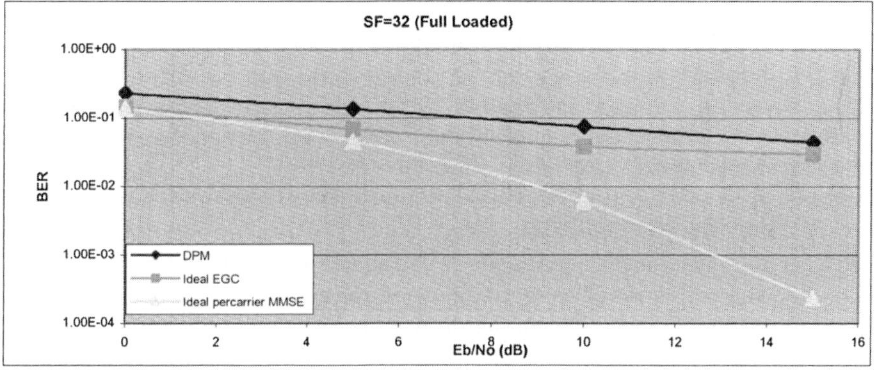

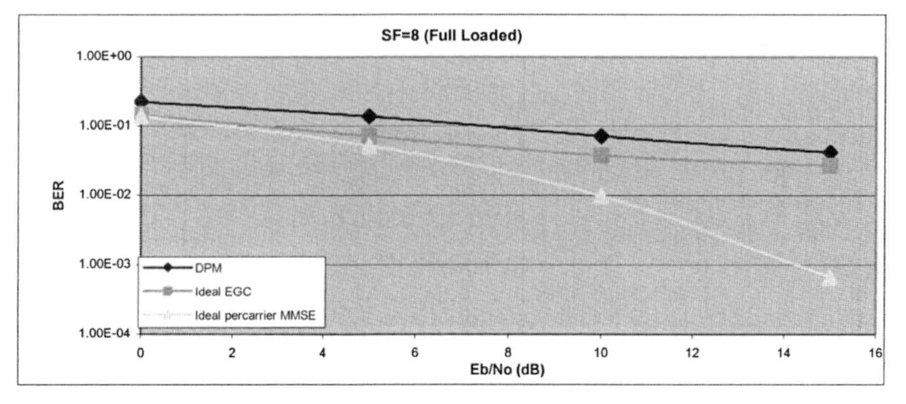

Figure 3. The performance of the differentially phase modulated (DPM) MC-CDMA system compared with coherent detection methods (fully loaded case).

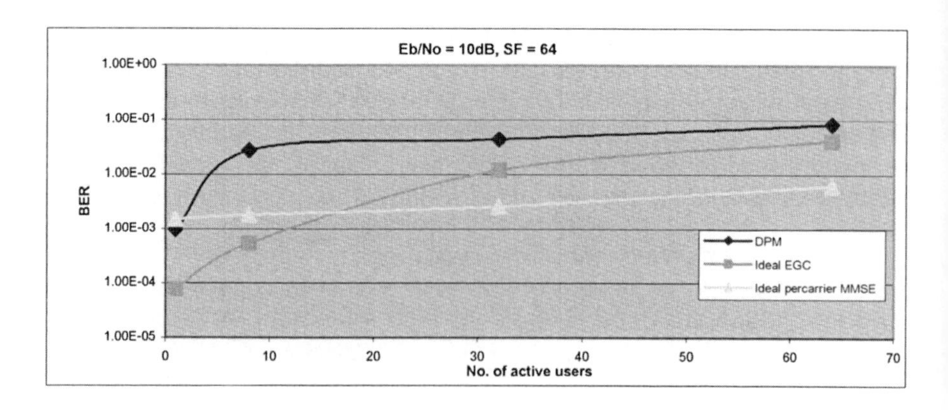

Figure 4. System performance as a function of number of active users (SF=64, E_b/N_o=10 dB).

Figure 4 shows the performance of the simulated systems as a function of the load with E_b/N_o=10 dB. As expected, the differential phase modulation system gives worse results compared to the ideal coherent methods. However, the BER performance can still be expected to be sufficient with the kind of error control coding techniques commonly used in mobile communication systems.

Due to the extra noise induced by applying the differential modulation at each subcarrier, the performance of such a system is always worse compared to the ideal coherent detection methods. In practice, however, the estimation of the phase and amplitude of the subchannel response is not ideal. As discussed in the introduction part, the channel noise and fading effect in time domain will induce estimation errors in coherent detection. Figure 5 shows the effect of inaccurate channel estimation with coherent detection methods.

Inaccuracies in channel estimation will induce an additional noise source in coherent detectors. In case of using interpolated pilots for subchannel estimation, the estimation quality is reduced when the distance between the pilot and the estimated symbol is increased. Then the total SNR is reduced. Such an effect could be simulated by applying an additional noise component. Figure 5 shows the performance comparison in two cases using different spread factors and E_b/N_o=10 dB. It is shown that if the ratio of signal power to channel estimation error variance is lower than about 9 dB, the differential phase modulation method will start to have better performance than the coherent EGC method.

Basically, on this axis a value of 10 dB (equal to the E_b/N_o value) would correspond to the case where the channel estimate is based on a single pilot at the same level as the data subcarrier samples. This is the case of minimum pilot density from the channel estimation point of view, where a single pilot is included in the coherence region in the time-frequency domain. Increasing the pilot density would improve the channel estimate (using a proper estimation filter) and the coherent system performance. The differential detection method seems to give a performance close to the performance that can be achieved in coherent EGC system with minimum pilot density (and without any boosting of the pilot symbols). However, in the differential system there are no overheads in the transmission capacity because of pilots.

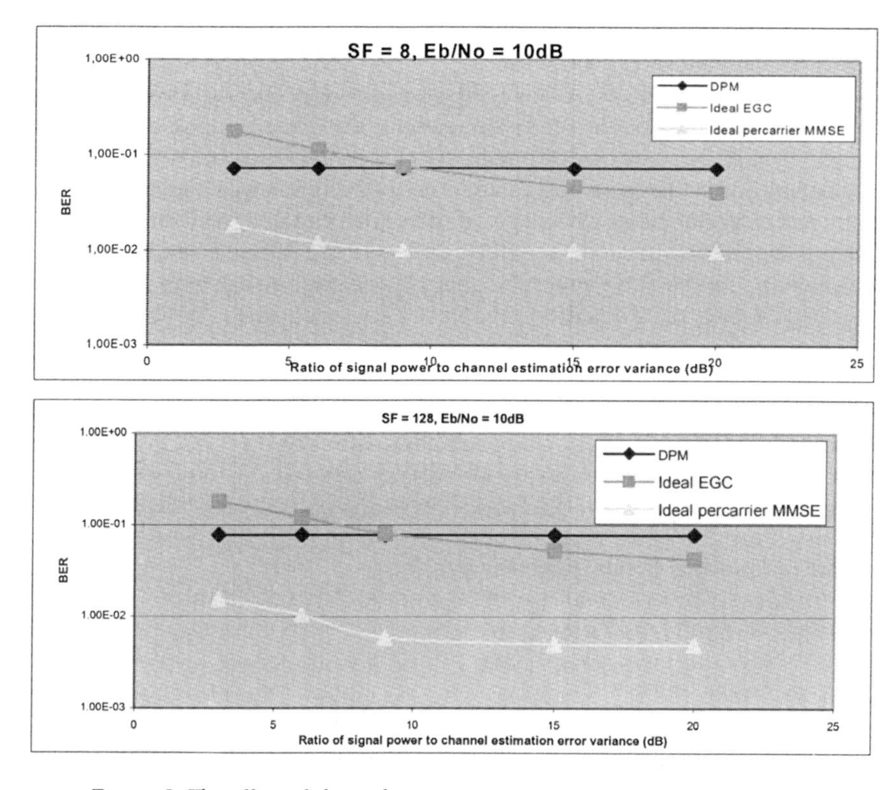

Figure 5. The effect of channel estimation errors in coherent detectors.

5. CONCLUSIONS

In this paper, a differential phase modulated downlink MC-CDMA system has been studied in case of independent Rayleigh fading environment. The different user symbols are combined at each subcarrier and the phase of the combined signal is differentially modulated. At the receiver side, the demodulation is applied at each subcarrier, so that the right phase of the combined symbol is provided. A simple approach to avoid possible zero amplitude subcarrier due to the multi-user symbol combining is proposed and the scheme is proven to have good performance through simulations.

In [10], a binary DPSK MC-CDMA system was investigated in AWGN channel. In our paper, the analysis of the differential QPSK MC-CDMA system is given by both analytical and simulation methods in multipath fading environment. The similarity of the proposed scheme to the coherent EGC MC-CDMA system was utilized in the analysis, so the overall analytical study is simplified. Compared to the ideal coherent EGC MC-CDMA system, additional errors are caused by the doubled channel noise and joint effect of irregular amplitudes and frequency selective fading. The simulation shows that the performance gap between the differential phase modulated MC-CDMA system and the coherent EGC MC-CDMA system with ideal channel estimation is about 5 dB.

In practice, however, the channel noise and fading effect will induce errors to the channel estimates. It was shown by simulations, that there is a possible performance gain obtained by applying the differentially phase modulated MC-CDMA method, if the channel estimation error in the coherent method (such as the EGC method) is significant. So the main advantage of the differentially modulated MC-CDMA system is that it provides robust performance with fast fading channels with no pilot overhead. In this case, in order to improve the performance, or to increase the system efficiency, the studied differentially phase modulated MC-CDMA could be proposed.

In future work, the exact relations between the fading effect (Doppler effect) and extra noise level (SNR degradation due to channel estimation errors) will be studied further, so that proper conditions for applying differential MC-CDMA systems can be stated more accurately.

Hongnian Xing & Markku Renfors
Institute of Communications Engineering,
Tampere University of Technology,
P.O. Box 553, Fin-33101 Tampere
Finland

REFERENCES

[1] N. Yee, J.-P Linnartz, and G. Fettweis, "Multicarrier CDMA for Indoor Wireless Radio Networks," *Proc. of PIMRC'93*, pp. 109-113, September, 1993, Yokohama, Japan.

[2] K. Fazel and L. Papke, "On the Performance of Conveolutionally-Coded CDMA/OFDM for Mobile Communication Systems," *Proc. of PIMRC'93*, pp. 468-472, September, 1993, Yokohama, Japan.

[3] V. DaSilva and E.S. Sousa, "Performance of Orthogonal CDMA Codes for Quasi-Synchronous Communications System," *Proc. of ICUPC'93*, pp. 995-999, October, 1993, Ottawa, Canada.

[4] L. Vandendorpe, "Multitone Direct Sequence CDMA System in an Indoor Wireless Environment," *Proc. of First IEEE Benelux Symp. on Comm. & Vehic. Techn.*, pp. 4.1.1-4.1.8, October, 1993, Delft, Netherlands.

[5] S. Kaiser, and K. Fazel, "A Spread-Spectrum Multi-Carrier Multiple Access System for Mobile Communications," *Multi-Carrier Spread Spectrum*, K. Fazel and G.P. Fettweis (Eds.), pp. 49-56, 1997, Kluer Academic Publishers, The Netherlands.

[6] John G. Proakis, *Digital Communications*, McGraw-Hill, 1989.

[7] Nathan Yee and J. P. Linnartz, "Controlled Equalization of Multi-Carrier CDMA in an Indoor Rician Fading Channel," *Proc. of VTC'94*, pp. 1665-1669, 1994.

[8] Hongnian Xing, Petri Jarske, and Markku Renfors, "The Performance Analysis of a Multi-Carrier CDMA System Using DQPSK Modulation for Frequency Selective Fading Channels," pp. 245 – 248, Proceedings of SPAWC'97, Paris, France, April, 1997

[9] Gunnar Wetzker, Martin Dukek, Harald Ernst, and Friedrich Jondral, "Multi-Carrier Modulation Schemes for Frequency-Selective Fading Channels," *Proc. of ICUPC'98*, pp. 939-943, 1998, Florence, Italy.

[10] Andrew C. McCormick, Peter M. Grant, and Gordon J. R. Povey, "A Differential Phase-Shift Keying Multicarrier Code Division Multiple Access System with an Equal Gain Combining Receiver," *IEEE Trans. on Vehicular Tech.*, pp. 1907-1917, vol. 49, no. 5, September, 2000.

[11] Z. Li, and M. Latva-aho, "Performance Comparison of Frequency Domain Equalizers for MC-CDMA Systems," *Proc. IEEE Conf. Mobile Wireless Networks*, Recife, Brazil, August, 2001.

[12] Tobias Hidalgo Stitz, Mikko Valkama, Jukka Rinne, and Markku Renfors, "Performance of MC-CDMA vs. OFDM in Rayleigh Fading Channels," in these proceedings.

M. SAITO, T. HARA, T. GIMA, M. OKADA, and H. YAMAMOTO

SPREADING SEQUENCES FOR MC-CDMA SYSTEMS WITH NONLINEAR AMPLIFIER

1. INTRODUCTION

Multi-carrier CDMA (MC-CDMA) systems have been studied enormously in recent years aiming to the next generation mobile communications [1]. In MC-CDMA systems, the spreading sequences are multiplied by the data in frequency domain instead of spreading in time domain as DS-CDMA systems.

One of the serious problems of MC-CDMA systems is the large amount of amplitude fluctuation caused by multi-carrier modulation. It reduces the transmission capacity and power efficiency due to the nonlinearity of the power amplifier. Power consumption of mobile terminals and the size of base stations should be reduced as small as possible in order to increase the battery life of the terminals and minimize the costs of building networks, respectively.

The spreading sequence have a crucial role to solve the problems, because the spreading sequences characterize the PAPR (Peak-to-Average Power Ration) of transmitted signals [2], [3]. Of course, other important role of spreading sequences is their ability of identification, that is, the auto- and cross-correlation properties [2], [3].

In this paper, we evaluate the PAPR properties of MC-CDMA signal with various binary spreading sequences for both up-link and down-link. M-sequences and their derivations are included as the spreading sequences which have not been reported ever.

We also show that the degradation properties due to the power loss and nonlinearity of nonlinear amplifier. It is concluded that the amplifier is feasibly used in the nonlinear region by adequately selecting the spreading sequence for both up- and down-link.

2. SYSTEM MODEL

Figure 1 shows the system model of MC-CDMA system employed in this study. We consider the transmitters of both mobile terminal and base station in MC-CDMA systems, that is, up-link and down-link. The number of active user K is 1 for the case of up-link, and M down-link, where M is the maximum number of orthogonal spreading sequences.

The input data are copied by the copier to N_c parallel streams. Where the number of copied data is the same as the number of subcarriers and the length of spreading sequences. At each branch, the data is multiplied by the corresponding element of

K. Fazel and S. Kaiser (eds.), Multi-Carrier Spread-Spectrum, 151–158.

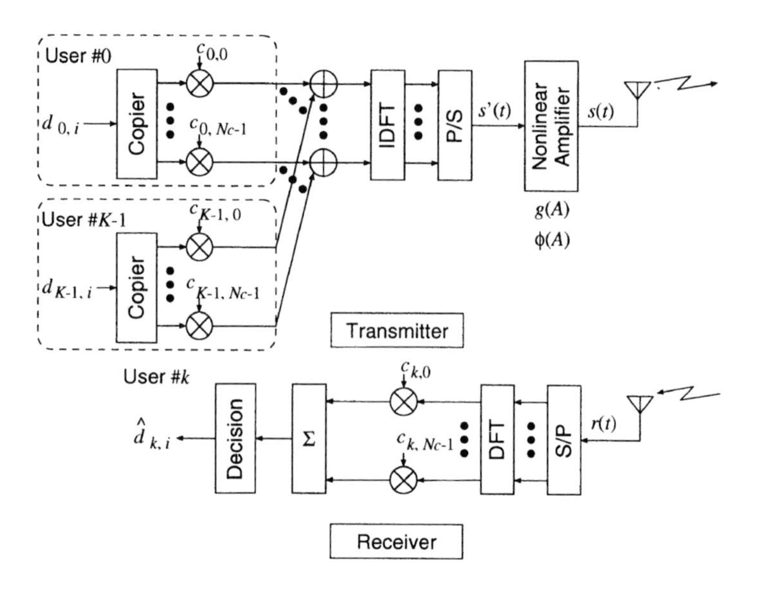

Figure 1: *System Model*

the spreading sequence. The spreading sequence assigned to k-th user is shown as $c_k = (c_{k,0}, c_{k,1}, \dots, c_{k,L-1})$, where the length of the sequence is L. The spread data are provided into the N_{DFT}-point IDFT (Inverse Discrete Fourier Transform).

The IDFT output signal of user k of up-link is given as follows.

$$s'_k(t) = \sum_{i=-\infty}^{\infty} \sum_{m=0}^{N_c-1} d_{k,i} c_{k,m} \exp(j2\pi mt/T_s), \tag{1}$$

where T_s is the symbol duration, and $d_{k,i}$ is the i-th data of the user, and we assume the sequence length is the same as the number of subcarrier, that is, $N_c = L$.

On the other hand, the IDFT output signal of down-link is shown as follows.

$$s'(t) = \sum_{i=-\infty}^{\infty} \sum_{m=0}^{N_c-1} \sum_{k=0}^{K-1} d_{k,i} c_{k,m} \exp(j2\pi mt/T_s), \tag{2}$$

where we assume K active users.

The IDFT output signal (2) is provided to the SSPA whose AM/AM and AM/PM conversions are shown as follows (See also Fig. 2).

$$g(A) = \frac{A}{\left\{1 + (A/A_0)^{2p}\right\}^{1/2p}} \tag{3}$$

$$\phi(A) = \begin{cases} 0 & (A < 0.25) \\ 0.5259 \cdot (A - 0.25) & (A \geq 0.25) \end{cases}, \tag{4}$$

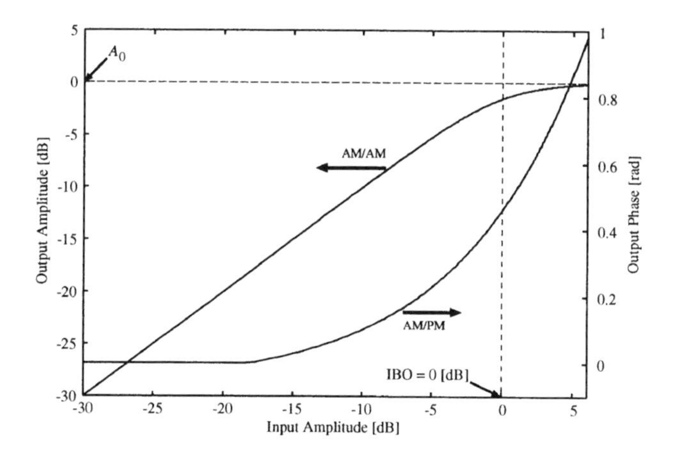

Figure 2: *AM/AM and AM/PM characteristics of SSPA.* $p = 2,\ A_0 = 1$.

where A_0 is the saturation amplitude, and p is the smoothness factor of the amplifier. In this study, we set as $p = 2$. The AM/PM conversion of SSPA is neglected in some previous papers [4]–[6]. The measured data introduced in Ref. [8], however, looks significant changing of the output phase. The AM/PM conversion shown in (4) is the approximation of the data of the SSPA operated at 2 GHz [8].

The amplified signal can be obtained by the previous functions (3) and (4), and shown as follows.

$$s(t) = g(|s'(t)|) \exp j(\theta(t) + \phi(|s'(t)|)), \tag{5}$$

where $\theta(t)$ is the angle of the input signal $s'(t)$.

The operation point of power amplifiers is determined by the IBO (Input Back Off) or OBO (Output Back Off). We define that IBO and OBO as [4],

$$\mathrm{IBO} = \frac{\mathrm{E}\left\{|s'_i(t)|^2\right\}}{A_0^2}, \quad \mathrm{OBO} = \frac{A_0^2}{\mathrm{E}\left\{|s_i(t)|^2\right\}}. \tag{6}$$

When we take back-off, the output level is decreased by the amount of corresponding OBO. Therefore, the reduction shall be adjusted so that the output power of the base station is kept constant to evaluate the BER at equal basis for any value of IBO.

3. PAPR PROPERTIES FOR VARIOUS SPREADING SEQUENCES

The PAPR properties of MC-CDMA up-link signal is characterized by the spreading sequence. Therefore, selecting the sequence of small PAPR is suitable for reducing the power consumption of mobile terminals.

On the other hand, that of down-link signal is indicated by the set of spreading sequences as well as carried data. The number of combinations of transmitted data is so enormous that it is impossible to analyze the whole combinations for more than several tens of sequence length and the number of active users. Statistically investigating the PAPR property is crucial for MC-CDMA down-link.

The PAPR of MC-CDMA signal $s(t)$ is defined as follows

$$\mathrm{PAPR} = \frac{\max |s(t)|^2}{\frac{1}{T_s}\int_0^{T_s} |s(t)|^2}. \tag{7}$$

In the following subsections, we will evaluate the PAPR properties for various kinds of binary spreading sequences. It is difficult to calculate the PAPR of (true) continuous signals, we set the number of DFT points as 4096. It is mentioned that the oversampling factor is required to be as 4 or 8 times as the number of subcarriers, therefore, the DFT points might be sufficient for evaluating the PAPR.

3.1. Spreading Sequences

The sets of binary sequences listed in Table 1 are treated in the PAPR analysis. The sequence length L is set at 31 for non-orthogonal sequences and 32 for orthogonal sequences. The number of available sequences M is also denoted in the table.

When the sequence length is odd number, the first chip (element) is carried by direct current component, and the half of the others are assigned to the subcarriers of positive frequencies and the rest of the chips to those of negative frequencies. On the other hand, when the length is even number, the former chips are assigned to the positive, and the latter to the negative, that is, dc component is not used.

Some sequences in the table have been already evaluated in the previous studies [2], [3]. What is different from those is M-sequences, CS (cyclically-shifted) M-sequences, and orthogonal M-sequences.

The M-sequences are well-known as their nearly ideal randomness. There are 6 primitive polynomials of degree 5, that is, the sequence length is 31. These are characterized by the connections of shift-registers which generate the M-sequences, as $45, 51, 57, 67, 73$ and 75 in octal form.

The CS M-sequences are defined as L kinds of cyclically-shifted version of one M-sequence. The sequences have the ideal cross-correlation properties due to the ideal auto-correlation properties of M-sequences. A CS M-sequence and its cyclically-shifted one is the same sequence. This property is so harmful in the conventional DS-CDMA systems, because the ability of user-identification by the sequence might be very weak. On the other hand, spreading sequences can be used in frequency region in MC-CDMA systems. Same weakness would occur in frequency region, however, such a large Doppler shift hardly occurs if several or several tens of giga hertz of carrier waves are used.

Adding element $+1$ after the last chip of each CS M-sequence and all $+1$ sequence of $L = 32$, we can obtain a set of orthogonal M-sequences. The all one sequence are orthogonal to the other sequences, but naturally, it has the worst PAPR property.

There exists 24 kinds of preferred pairs for $L = 32$, 24 kinds of Gold sequences can be generated. By inserting one chip ($+1$ is assumed here to balance the number

Table 1: *The list of spreading sequences treated in this study*

Sequence	Length L	#Sequences M
M-seq.	31	6
Cyclically-shifted M-seq.	31	31
Orthogonal M-seq.	32	32
Gold seq.	31	33
Orthogonal Gold seq.	32	32
Golay complementary seq.	32	32
Hadamard seq.	32	32

of $+1$ and -1) into the specific 32 Gold sequences, we can obtain orthogonal Gold sequences of length $L = 32$. For a Gold sequence of length $L = 31$, there are 31 places where the additional chip can be inserted. Therefore, we can obtain 31 different kinds of orthogonal Gold sequences. For the length of 32, we have evaluated the comprehensive PAPR of Gold sequences and orthogonal Gold sequence.

Golay complementary sequences are known as the sequences which can reduce the PAPR of multi-carrier signals [3]. Hadamard sequences are defined as the row vector of an Hadamard matrix. Golay complementary sequences and Hadamard sequences are sets of orthogonal sequences.

3.2. PAPR Properties of MC-CDMA transmitted signal

The PAPR values of 6 different M-sequences are $3.55, 3.68, 3.98, 5.54, 5.60, 5.85$ dB. Even though the PAPR values of some sequences are small, it is impractical to use the M-sequences as it is, because the number of sequences is small. It is just a reference that the PAPR of MC-CDMA signal with M-sequences derived by L-point IDFT is almost 0 dB. However, the number of DFT point increases, the PAPR also increases to the values described above.

The left of Fig. 3 shows the PAPR versus the sequence number of CS M-sequences and Gold sequences. The base M-sequence of CS M-sequences is generated by the primitive polynomial of 57 in octal form. The Gold sequences are generated by the primitive polynomials 45 and 51. As mentioned above, there exists many kinds of CS M-sequences Gold sequences, and orthogonal Gold sequences. The illustrated PAPR values of those sequences in Fig. 3 are the sequences which have the minimum average PAPR.

CS M-sequences tend to have lower PAPR than Gold sequences. The values are distributed between 3 dB and 6 dB, on the other hand, those of Gold sequences between 4 dB and 8 dB. For cross correlation properties, CS M-sequences and Gold sequences have almost the same correlation value under the down-link and the up-link with synchronous transmission among different users. Therefore, CS M-sequences are the better spreading sequences of binary non-orthogonal sequences on the PAPR.

The right of Fig. 3 shows the PAPR versus the sequence number of Hadamard sequences, Golay complementary sequences, orthogonal M-sequences, and orthogo-

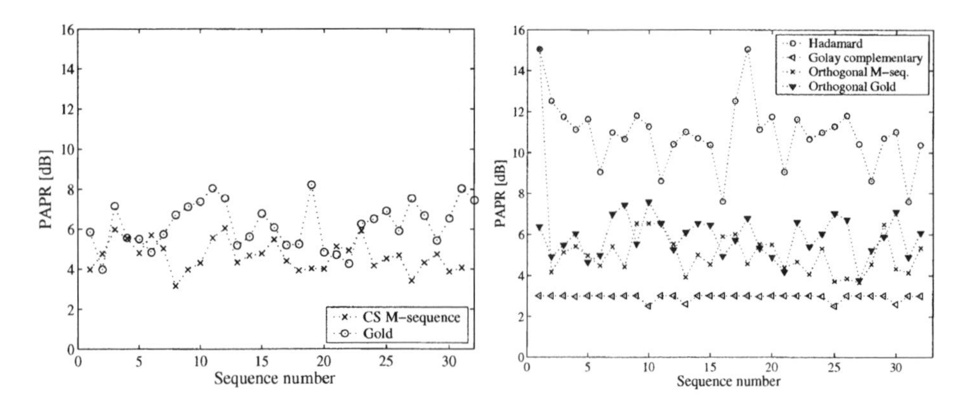

Figure 3: *PAPR properties of non-orthogonal (left) and orthogonal (right) spreading sequences*

nal Gold sequences. We have selected one of the orthogonal Gold and M-sequences based on the minimum mean PAPR among the sequence set. Among the orthogonal sequences, Golay complementary sequences have the best PAPR values which is about 3 dB, and Hadamard sequences the worst ones.

Figure 4 shows the cumulative distribution functions of PAPR values of Hadamard sequences, Golay complementary sequences, and orthogonal Gold sequences. The sequence length and the number of active users are set as $L = K = 32$. QPSK modulation is adopted as a modulation scheme. It is nearly impossible to analyze the PAPR for the whole combination of input data of all users, we have taken 2^{15} sample signals. From the figure, it can be observed that down-link MC-CDMA signals with Hadamard sequences take about 1 dB lower PAPR than the other sequences.

4. NUMERICAL RESULTS

The transmission performances of up-link are evaluated by both auto-correlation as the level of desired signal component and the cross-correlation from 31 interference signals as multiple access interference. The Golay complementary sequences are used for this evaluation whose results are shown in the left of Fig. 5. In Fig. 5, the equivalent SNR degradation due to the decrease of OBO and that due to the interferences are illustrated for various IBO values. The total degradation, which is the sum of degradation described above, is also shown in the figure. From the figure, it can be observed that the total degradation is not sensitive to IBO near the saturating region. When the amplifier is operated at the region, we can save the power by several dB comparing with the case that operating the amplifier at linear region. The amount of saving the power depends on the PAPR of the spreading sequences. The effect of interference caused by asynchronous transmission and multi-path fading is ignored in this simulation, considering these effects are required for precise analysis.

The right of Fig. 5 shows the equivalent SNR degradation of MC-CDMA system down-link due to the power loss and nonlinearity of amplifier at BER $= 10^{-3}$. The

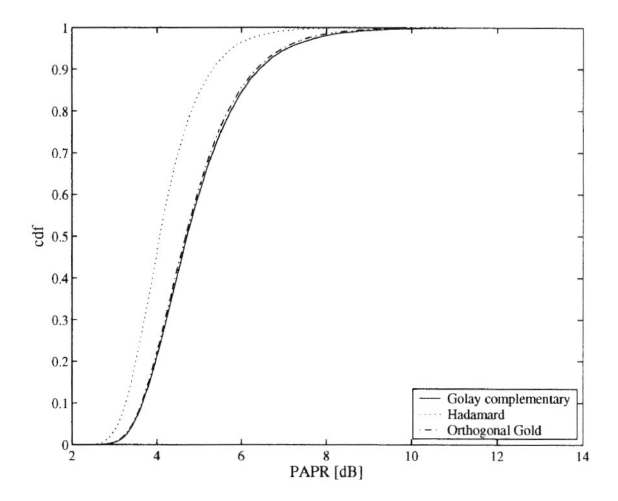

Figure 4: *Cumulative distribution function of PAPR of MC-CDMA signals with Golay complementary sequence (solid line), Hadamard sequence (dotted line), and Orthogonal Gold sequence (dashed line).* $N_c = 32, K = 32$

figure illustrates the power loss due to OBO as a degradation factor and BER degradation at BER $= 10^{-3}$ due to nonlinearity as a function of IBO. The total degradation is also evaluated as a summation of these two. Hadamard sequences are used as spreading sequences because of the good PAPR performance.

The total degradation is minimum at IBO $=$ 0 to 3 dB and insensitive to IBO at the region. We have chosen IBO $=$ 3 dB as a reasonable value of IBO considering those degradations. Comparing to use of SSPA within linear region (the maximum PAPR by Hadamard sequences is about 10 dB as shown in Fig. 4), we can save the power by about 7 dB, that is, the required size of the amplifier is reduced to $1/5$. This should be crucial considering that the power amplifier drastically increases its costs as the power becomes large.

5. CONCLUSIONS

In this study, we have evaluated the transmission performances of multi-carrier CDMA systems with nonlinear amplifier aiming at decreasing the size and power consumption at amplifier used in the mobile and base stations. This paper shows that the MC-CDMA signals can be transmitted at the saturation point of the nonlinear amplifier without severe degradation of performance. By selecting the spreading sequences which have small PAPR, we can save the power by several dB in up- and down-link.

We also have studied the PAPR of MC-CDMA systems with various binary spreading sequences for both non-orthogonal and orthogonal sequences. Totally, Golay complementary sequences have the best PAPR properties. On the other hand, cyclically-shifted M-sequences have good PAPR properties in non-orthogonal sequences.

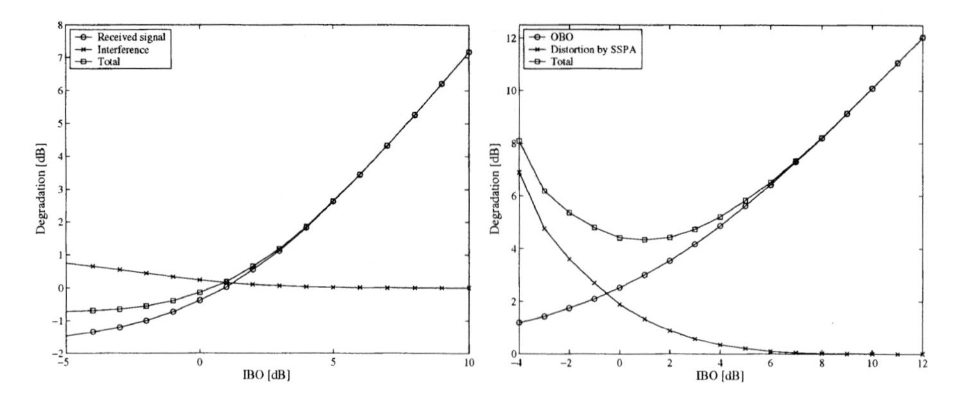

Figure 5: *Degradation of BER at* $\mathrm{BER} = 10^{-3}$ *in MC-CDMA up-link (left figure) and down-link (right figure), where a Golay complementary sequence and Hadamard sequences are used,* $N_c = 32$ *and* $K = 1$, *and* $K = 32$, *respectively.*

As a future work, we would like to investigate the effect of filtering, which is employed to decrease the out-of-band emission, of MC-CDMA systems with nonlinear amplifier in the fading channel.

Lastly it shall be noticed here that a part of this study has been conducted under the COE (Center Of Excellence) program sponsored by the Ministry of Education, Culture, Sports, Science and Technology of Japan.

References

[1] Hara, S., and Prasad, R. (1997), Overview of multicarrier CDMA, *IEEE Commun. Mag. 35*, 126–133.

[2] Popović, B.M. (1999), Spreading sequences for multicarrier CDMA systems, *IEEE Trans. Commun. 47*, 918–926.

[3] Nobilet, S., Helard, J.F., and Mottier, D. (2002), Spreading sequences for uplink and downlink MC-CDMA systems: PAPR and MAI minimization, *Euro. Trans. Telecommun. 47*, 465–474.

[4] Fazel, K. and Kaiser, S. (1998), Analysis of non-linear distortions on MC-CDMA, *Proceedings of ICC'98*, 1028–1034.

[5] Costa, E., and Pupolin, S. (2002), M-QAM-OFDM system performance in the presence of a nonlinear amplifier and phase noise, *IEEE Trans. Commun. 50*, 462–472.

[6] van Nee, R., and Prasad, R. (2000), OFDM for wireless multimedia communications, Artech House Publishers, 2000.

[7] Saito, M., Hara, T., Gima, T., Okada,, M., and Yamamoto, H. (2003), Selection of spreading sequence for multi-carrier CDMA systems, *in Proc. The First Joint Workshop on Mobile Multimedia Communicaions*, 46–50.

[8] Mizuta, S., Kawaguchi, H., and Akaiwa, Y. (2003), Performance of an adaptive predistorted power amplifier with peak limited baseband code-division multiplexed signal, *IEICE Trans. Commun. JB86–B*, 129–136 (in Japanese).

Nara Institute of Science and Technology (NAIST), Japan.

ANALYSIS OF LINEAR RECEIVERS FOR MC-CDMA WITH DIGITAL PROLATE FUNCTIONS

I. Raos, S. Zazo, A. Del Cacho

Department of Signals, Systems and Radiocommunications
Technical University of Madrid, Spain
E-Mail: ivana@gaps.ssr.upm.es

Keywords: spreading codes, MC-CDMA, Digital Prolate Functions

Abstract. Research presented in this contribution deals with efficiency improvement in downlink MC-CDMA system by applying slightly modified Digital Prolate Functions (DPF) as orthogonal spreading codes. Modified DPF applied as orthogonal codes in time-continuos MC-CDMA result in well-concentrated users' signals in interval smaller than symbol's. [1] The benefit of this signals' property is clear; time dispersive channels will not affect contiguous symbols seriously if no time redundancy (e.g. Cyclic Prefix) is added, that is, intersymbol interference (ISI) will be minimized. [2][3] The price that is to be paid for this efficiency improvement is somewhat increased receivers' complexity. In this contribution, analysis of different receivers for downlink will be performed, applying Filter Bank (FB) as digital front-end and linear detectors: decorrelator and MMSE detector. Eventually, it turns out that the performance, using a simple decorrelator, is satisfactory enough, making this type of systems feasible to implement.

1. INTRODUCTION

Future mobile multimedia systems must be designed for high data rates, especially in downlink where major traffic flow is expected. Multicarrier modulations are capable of coping with frequency selective channels and this put MC-CDMA systems as strong candidates for future communication mobile systems. In order to cope with mobile channels time spread, some kind of time redundancy is usually added to useful information part for ISI elimination and conversion of channel into multiplicative model in frequency domain. However, this leads to information rate loss, although its maximization is the primary goal in downlink communication channel. Therefore, we focused our research on spreading codes that could avoid the need for time redundancy.

Digital Prolate Functions (DPF) are orthogonal functions concentrated in time domain. By applying them as spreading codes in MC-CDMA systems the goal of time redundancy avoidance is accomplished, as codes can be designed with

K. Fazel and S. Kaiser (eds.), Multi-Carrier Spread-Spectrum, 159–166.

"incorporated" silent period inside the symbol period. Analysis of some linear receivers for this MC-CDMA system is developed in this contribution.

2. DIGITAL PROLATE FUNCIONS

Digital Prolate Functions are obtained as solution to the problem of time-domain concentration of trigonometric polynomial (frequency domain presentation). The total number of obtained orthogonal functions from this minimization (concentration) problem equals the number of frequencies used for the polynomial.

$$y(t) = \sum_{n=0}^{K-1} Y_n e^{jn\omega_0 t} \tag{1}$$

$$\alpha = \frac{1}{E_y} \int_0^{\tau} |y(t)|^2 dt = \frac{1}{E_y} \int_0^{T} |v(t) y(t)|^2 dt \tag{2}$$

Single user MC-CDMA signal is written in the same trigonometric polynomial format, meaning that these functions can be applied as users spreading codes. If signal duration is concentrated, channel time spread will not produce Inter Symbol Interference.

In "continuos world" concentration of all users' signals is not possible; concentration ratios (CR) will gradually decay from *1* (completely concentrated signal) to *0* (signal completely outside the smaller interval). Bearing in mind that full load is hardly achievable in real system, this restriction does not impose serious drawback. If system is designed to satisfy the relation between channel's time dispersion and symbol's interval of *20%*, "sufficient" concentration of approximately *80%* of users' signals is achieved.

3. LINEAR RECEIVERS

In a MC-CDMA system, difference between standard MC-CDMA system and proposed one, besides the different spreading code election, **lays** in no redundancy addition in transmitter in the latter case. At receiver, digital front-end is Filter Bank (FB) consisting of N exponentials followed by N integrators (N is the number of subcarriers). Discrete signal processing part is characteristic for each FB integration period (it can be chosen to be equal to symbol's interval or to the restricted interval). In [2] some receivers depending on integration period were proposed. In this contribution, integration period equals symbol interval while the smaller one was investigated in [3].

Matrix form of received signal at FB end is:

$$\mathbf{r}[n] = \mathbf{HYb}[n] + \mathbf{AY}(\mathbf{b}[n-1] - \mathbf{b}[n]) + \mathbf{n}[n] \tag{3}$$

where $\mathbf{H}$ is diagonal channel's matrix recollecting its frequency response per subcarrier, columns of $\mathbf{Y}$ are spreading codes of active users, $\mathbf{b}[n]$ is vector with users' information-bearing symbols, and $\mathbf{A}$ is hermitic Toeplitz matrix whose entries depend on channel's partial frequency response $H(n\omega_0, a, b)$:

$$\mathbf{A}(m,n) = \int_0^{T_c} e^{j(m-n)\omega_0 t} H(n\omega_0, t, T_c) dt \tag{4}$$

$$H(n\omega_0, a, b) = \int_a^b h(\tau) e^{-jn\omega_0 \tau} d\tau \tag{5}$$

with $h(\tau)$ being channel's impulse response.

The signal at the output of FB has two components. The first part is the signal transmitted through channel's steady-state (SS); it depends only on $\mathbf{H}$. The second one corresponds to the transient regimes (TR), the ones causing intra and inter symbol interference (iSI and ISI).

Two types of linear detectors applied after FB will be analyzed, decorrelator and Minimum Mean Square Error (MMSE).

3.1. ZF/Decorrelator

Decorrelator or Zero-Forcing (ZF) equalizer restores users' orthogonality destroyed by frequency selective channel fading. With signal model given by Eq. (3), ZF equalizer can be designed to equalize just SS channel's influence, or both SS and TR parts; both approaches are analyzed.

3.1.1. ZF1

This receiver is similar to standard MC receivers, equalizing just SS channel's part, $\mathbf{G} = \mathbf{H}^{-1}$, and is welcomed because of its computational simplicity. However, there exists a drawback, in case of no concentrated signals, since it leads to an error floor due to non-equalized transient period. For signals that are well concentrated, Eq. (4) fulfills:

$$\mathbf{AY} = \mathbf{0} \tag{6}$$

and no iSI nor ISI appear.

3.1.2. ZF2

Equalization of both SS and TR leads to the equalizer's form $\mathbf{G} = (\mathbf{H} - \mathbf{A})^{-1}$. This receiver has a drawback that estimation and inversion of non-diagonal channel dependent matrix requires more complex receiver.

3.2. Linear MMSE

MMSE detector is linear detector that jointly minimizes effects of MAI, ISI, iSI and noise. Depending on error that is to be minimized and information available at receiver, 2 types will be analyzed. If only the desired user signal's mean error is to be minimized, it is denominated MMSE Single User Detector (SUD). If mean error considering all active users is to be minimized, we are obtaining MMSE Multi User Detector (MUD).

3.2.1. MMSE1

MMSE equalizer when ISI, iSI and all active users are taken into account is denoted MMSE Multi User Detector (MUD). This solution requires knowledge of all users' waveforms and increases complexity of receivers. Equalizer matrix $\mathbf{G}$ has the following form:

$$\mathbf{G} = \left(\mathbf{H}^H - \mathbf{A}^H\right)\left(\mathbf{Y}_{HA} + \sigma^2 \mathbf{I}\right)^{-1} \tag{7}$$

$$\mathbf{Y}_{HA} = (\mathbf{H} - \mathbf{A})\mathbf{Y}\mathbf{Y}^H\left(\mathbf{H}^H - \mathbf{A}^H\right) + \mathbf{A}\mathbf{Y}(\mathbf{A}\mathbf{Y})^H \tag{8}$$

3.2.2. MMSE2

MMSE SUD takes into account only ISI and iSI originated from desired user, and knows only his spreading code:

$$\mathbf{G} = \left(\mathbf{H}^H - \mathbf{A}^H\right)\left(\mathbf{y}_{HA} + \sigma^2 \mathbf{I}\right)^{-1} \tag{9}$$

where $\mathbf{y}_{HA}$ is defined in Eq. (6) by replacing $\mathbf{Y}$ with desired user code $\mathbf{y}$.

4. ANALYSIS OF DPF AND RECEIVERS

An example of CRs of modified DPF and Hadamard codes, when *N=16*, symbols interval is *10ms*, channel's time dispersion is *2ms* (concentration of users' symbols in *8ms* interval, at the beginning of symbol's interval), are shown in Fig. *1*.

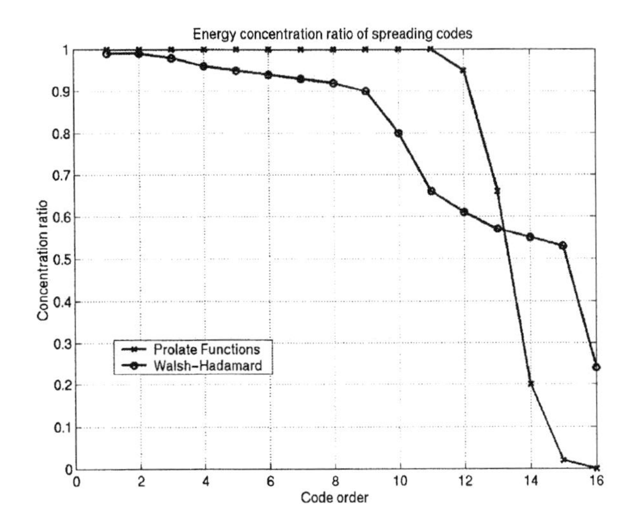

Fig 1. Concentration ratios of modified DPF and Hadamard spreading codes

Comparison of detectors is performed in terms of SINR at slicer and is shown in Fig. (2) and (3). Scenario for all simulations is 2-rays Rayleigh fading channel, with power profile *[0,-3]dB*. Results are averaged over each user and over active users. At the beginning of transmission, the most convenient spreading code (maximally concentrated among available ones) is assigned to the user.

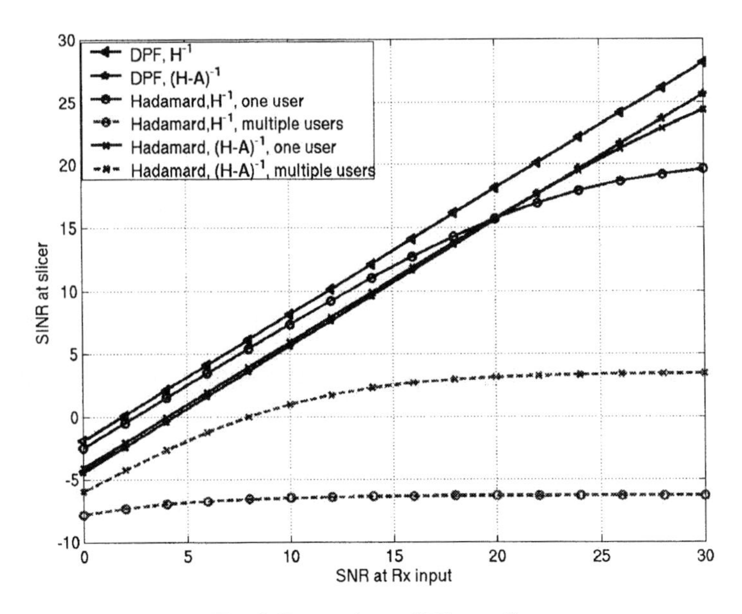

Fig. 2 Comparison of ZF equalizers

Systems with DPF codes keep the same behavior no matter if there is *1* active user or all concentrated, thanks to codes' concentration property $\mathbf{AY}_{act} = \mathbf{0}$. In this case, use of non-diagonal matrix $\mathbf{A}$ only enhances noise, while standard ZF equalizer has better performances. Both of those receivers have linear dependence of SNR at receiver input, as interference does not limit their performance.
In the case of Hadamard codes, even in a single user case ISI and iSI worsen performances when their power is comparable to noise power. Comparing different ZF equalizers, it can be concluded that the one with inversion of diagonal channel matrix behaves poorly when power of iSI and ISI is dominant over white noise power, and ZF equalizer $(\mathbf{H}-\mathbf{A})^{-1}$ has superior behavior for greater SNR ratios.

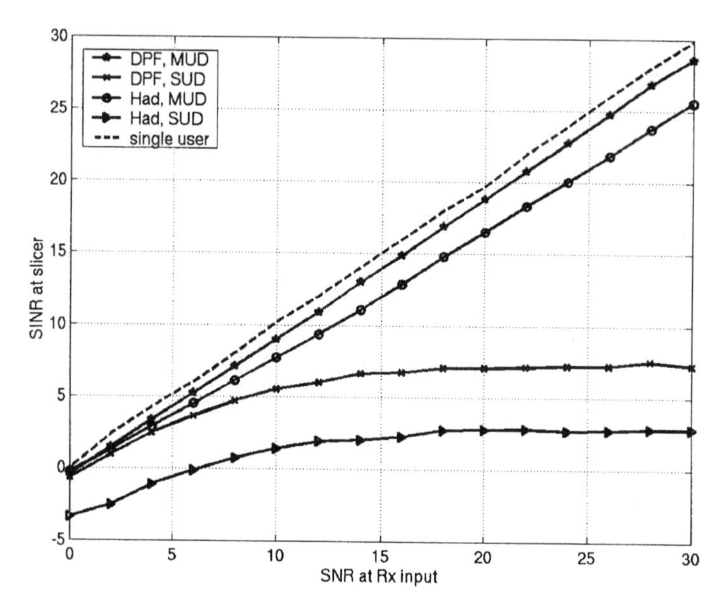

Fig. 3 Comparison of MMSE equalizers

Simplification of MMSE detectors, as obviating use of non-diagonal matrix $\mathbf{A}$, results in same performances for DPF system and worse for Hadamard. SUD cannot combat MAI, while MUD does not lose much with increment of users' number. Knowledge of active users' codes is required, as otherwise, interference saturates the system.
It should be noted that DPF-system with ZF has superior behavior compared to Hadamard and MMSE MUD Detector. This means that ZF detector, that does not need knowledge of active users' codes, can be applied, resulting in computationally simple and adequate receiver for downlink. If information about other users' codes is available in receiver, system with MMSE MUD with DPF shows 3dB's superiority with respect to Hadamard codes and same equalizer.

5. CONCLUSIONS

Modified DPF as spreading codes were proposed for downlink MC-CDMA channel. These codes enable system functioning without necessity for time redundancy, therefore increasing useful data rate. However, full load system is not possible to reach.

This system can be compared with standard MC-CDMA system with Hadamard codes. If the same full load is considered, useful data rate will be smaller in standard MC-CDMA system. If ZF equalization is applied, users remain orthogonal; however, if channel is not perfectly known, this equalization type is not adequate. Therefore, other equalizers, like MMSE, should be used. In this case, users' codes are not orthogonal and interference limits systems performance in terms of number of active users.

Therefore, in realistic case, with channel estimation errors, the adequate receiver would use MMSE equalization, and MC-CDMA with Hadamard and modified DPF codes which are comparable in performances, except that higher information data rates are possible in system with DPF due to time redundancy avoidance.

6. ACKNOWLEDGMENT

The European IST-2001-32620 project (www.ist-matrice.org) and TIC200-1395-C02-02 and 07T/0032/2000 supported the work presented in this contribution.

7. REFERENCES

[1] Slepian, D., Pollak, H. O. Prolate Spheroidal Wave Functions, Fourier Analysis and Uncertainty (I) In *The Bell System Technical Journal* Vol 40, N0 1, Jan 1961, 43-61

[2] Raos, I., Zazo, S., Bader, F. Prolate Spheroidal Functions: A General Framework for MC-CDMA Waveforms In Proc. of PIMRC 2002, Vol 5, pp. 2342-2346

[3] Raos, I., Zazo, S. Advanced Receivers for MC-CDMA with modified Digital Prolate Functions In Proc. of ICASSP 2003

ANA GARCÍA-ARMADA, J. RAMÓN DE TORRE, VÍCTOR P. GIL JIMÉNEZ, M. JULIA FERNÁNDEZ-GETINO GARCÍA

EVALUATION OF DIFFERENT SPREADING SEQUENCES FOR MC-CDMA IN WLAN ENVIRONMENTS

Abstract. In this paper we analyse the performance of several types of spreading sequences for MC-CDMA in terms of Peak-to-Average power Ratio (PAR) and Bit Error Rate (BER). The feasibility of MC-CDMA for WLAN systems is explored by considering its performance in Hiperlan channels. It can be concluded that choosing the adequate spreading sequence and signal parameters, multiple users can be accommodated in Hiperlan channels maintaining the high bit rates that characterize WLAN scenarios.

1. INTRODUCTION

DS-CDMA is a well-known technique that allows multiplexing different users that simultaneously transmit at the same time in the same frequency band with a potential increase in efficiency in comparison to TDMA or FDMA schemes. However, the loss of orthogonality caused by time-dispersive channels gives rise to multiple access interference that severely degrades the performance. Since OFDM is well suited for multipath environments, a combination of DS-CDMA and OFDM can be the solution to the above mentioned problem.

Several combinations of DS-CDMA and OFDM have been suggested in the literature: MT-CDMA (Vandendorpe, 1993), MC-DS-CDMA (Sousa, 1993), MC-CDMA (Yee, 1993). These techniques have been compared by Hara (1997) for different numbers of users sharing the downlink channel. MC-CDMA appears to be the best scheme in terms of Bit Error Rate (BER) performance when the number of users is allowed to increase, provided that the proper equalization is used. This fact motivated us to the evaluation of the feasibility of using such an scheme for Wireless LAN environments.

Wireless LANs are experiencing an increase in popularity as the only wireless systems that provide data rates up to 54 Mbps (IEEE 802.11a, Hiperlan 2). The transmission techniques that are able to provide such rates are seen as possible candidates for an extension of wireless systems towards 4G.

The use of MC-CDMA for WLAN environments is challenging in the sense that higher data rates than actual ones should be achieved while maintaining the goal of low-cost and low-power devices. This performance improvement goes through the adequate selection of spreading sequences in order to improve one of the worst characteristics of OFDM signals, namely, their high Peak-to-Average power Ratio

K. Fazel and S. Kaiser (eds.), Multi-Carrier Spread-Spectrum, 167–174.

(PAR) (Van Nee, 2000), while achieving the best possible Bit Error Rate (BER) performance in multipath environments.

In this paper we first introduce the MC-CDMA signal model and show how to select the best values of the signal parameters in frequency-selective channels such as WLAN environments. After that, in section 3 the performance in terms of PAR is analysed while section 4 deals with BER performance in multipath environments considering the example of one of the channel models developed for Hiperlan 2. The paper finishes with some concluding remarks.

2. MC-CDMA SIGNAL MODEL

In MC-CDMA each user's signal is spread in the frequency domain with a different sequence and transmitted simultaneously with other users' signals. If we denote K_{MC} the processing gain, each base-band information symbol will be multiplied by K_{MC} chips and each of these chips will be transmitted in one of $N_{MC}=K_{MC}$ orthogonal sub-carriers. The separation of these sub-carriers will be equal to $1/T_s$ being T_s the information symbol duration.

If we denote $a^u[k]$ u-th user's k-th information symbol and c_l^u l-th chip of u-th user's spreading sequence, the MC-CDMA signal for u-th user is:

$$s^u{}_{MC}(t) = \sum_{k=-\infty}^{\infty} \sum_{l=0}^{N_{MC}-1} a^u[k] c_l{}^u p(t-kT_s) e^{j2\pi f_l t} \tag{1}$$

Here $p(t)$ is a rectangular pulse and $f_l = \dfrac{l}{T_s}$.

In pure OFDM the parallel transmission of frequency-multiplexed information symbols causes the transmitted symbols to be longer than the multipath channel duration, thus combating the effects of inter-symbol interference. The spacing between orthogonal sub-carriers is such that sub-channels can be considered to be locally flat. However, in MC-CDMA the effect of parallel transmission is compensated by spreading so that the separation between sub-carriers is K_{MC} times bigger and therefore, in general, fading cannot be considered locally flat inside sub-channels. Moreover, high speed WLANs need high bandwidth allocations and the problem worsens.

This problem can be solved by slightly modifying the signal structure. If more sub-carriers are sent in parallel so that several information symbols (up to P) are multiplexed in frequency ($N_{MC}=P{\cdot}K_{MC}$), the transmitted symbol will be P times longer and the spacing between sub-carriers P times smaller, so that an adequate choice of P will solve multipath problem. The new MC-CDMA signal for u-th user is:

$$s^u{}_{MC}(t) = \sum_{k=-\infty}^{\infty} \sum_{p=0}^{P-1} \sum_{l=0}^{K_{MC}-1} a_p{}^u[k] c_l{}^u p(t-kT_{MC}) e^{j2\pi f_{p,l} t} \tag{2}$$

where

$$f_{p,l} = \frac{p \cdot K_{MC} + l}{P \cdot T_s} \tag{3}$$

Figure 1 shows the structure of this modified transmitter.

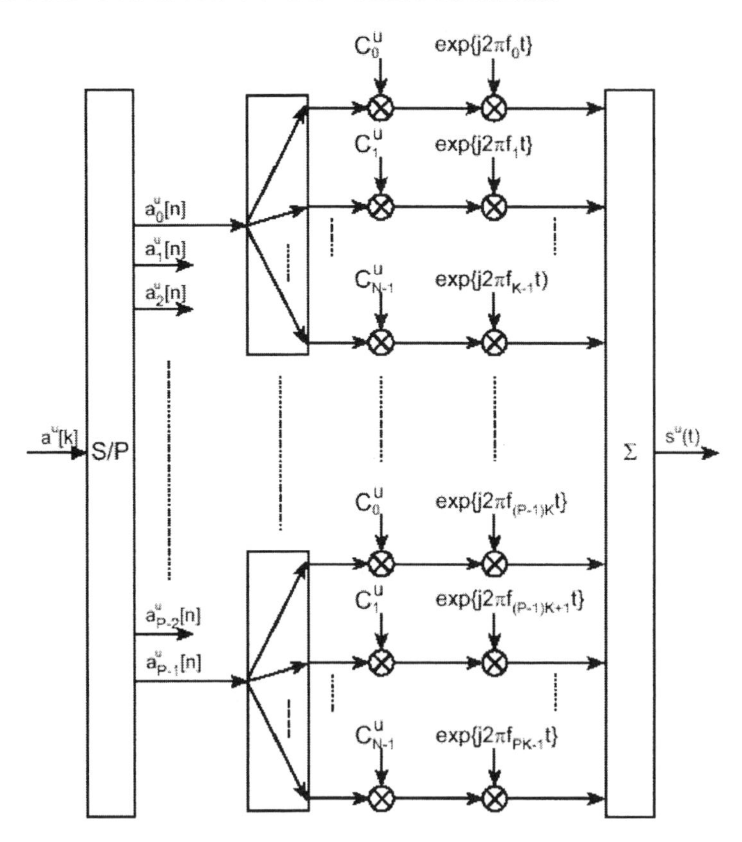

Figure 1. MC-CDMA transmitter with P branches

Finally, in order to ensure that frequency diversity is maximized, a scrambling can be performed so that the chips corresponding to the same information symbol are not sent in contiguous sub-carriers but separated *P* apart.

The choice of the value of *P* depends on the characteristics of the channel that is expected. In order to guarantee the orthogonality of the sub-carriers in the presence of a multipath channel, a cyclic prefix must be added to the MC-CDMA signal. If we denote by T_{CP} the cyclic prefix duration, we can define the ratio between the MC-CDMA symbol duration and the cyclic prefix length as:

$$T_{CP} = \frac{T_{MC}}{C} \tag{4}$$

Since the cyclic prefix introduces a loss in transmission efficiency, we would like the value of C to be as high as possible. A typical value is 4 or 8.

This value depends on the duration of the multipath channel response. If we denote this channel length by τ, the cyclic prefix must satisfy:

$$T_{CP} > \tau \Rightarrow C < \frac{T_{MC}}{\tau} \tag{5}$$

Substituting the value of T_{MC}, this relationship is satisfied if:

$$C < \frac{T_s}{\tau} P \tag{6}$$

Clearly, the inefficiency due to the cyclic prefix insertion is decreased by a factor of P if we use the parallel transmission scheme. If we take an example of a typical WLAN scenario with τ=200 ns and we wish to transmit at an information rate of $20 \cdot 10^6$ symbols per second, keeping the value of C=8, this can be fulfilled with a value of P equal to 32, while we should use a cyclic prefix four-times the length of the MC-CDMA symbol in order to cope with multipath without parallel transmission.

Using this value of P=32 and setting K_{MC}=32, that is, 32 devices can simultaneously share the MC-CDMA WLAN signal, the number of sub-carriers is N_{MC}=1024.

Once that we can cope with multipath, the spreading sequence c_l^u must be chosen in order to have the best PAR and BER properties in the application scenario. In this paper we analyse the following sequences: Walsh-Hadamard and polyphase Walsh-Hadamard (Seberry, 1992), orthogonal Golay (Popovic, 1999) and Zadoff-Chu sequences (Frank, 1962 and Chu, 1972).

3. PAR PERFORMANCE

In WLAN we have to face with two situations: one given user is transmitting towards the access point (uplink) or the access point is transmitting towards the different users (downlink). In terms of PAR, the best situation is the uplink, since the transmitter is only amplifying one signal spread with the user's sequence. However, in the downlink all user's signals are transmitted simultaneously from the access point, and therefore PAR is potentially higher.

Figure 2 shows PAR in K_{MC} = 32 uplink without multipath. Figure 3 shows PAR in K_{MC} = 32 downlink without multipath.

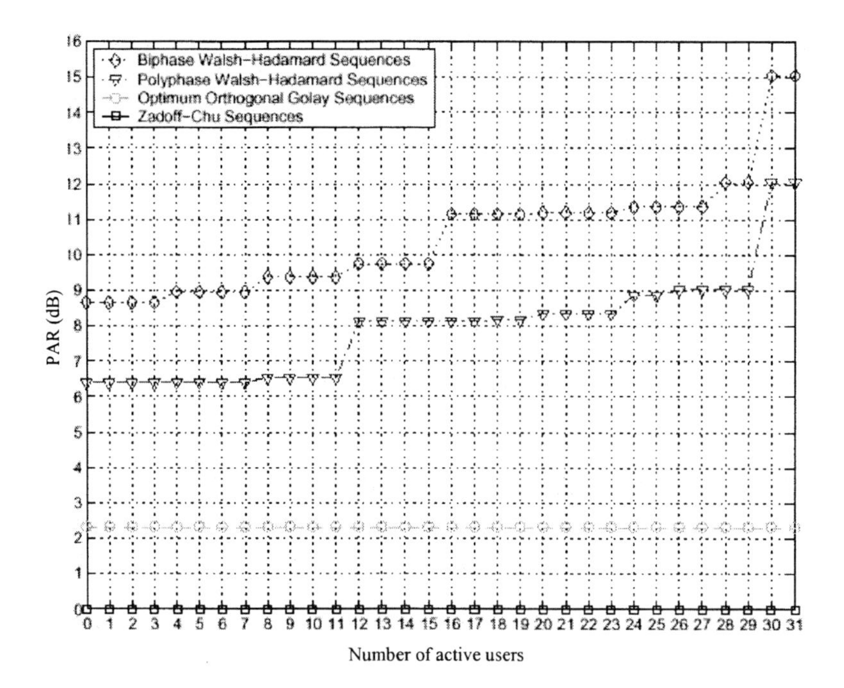

Figure 2. PAR in UL with 32-length sequences

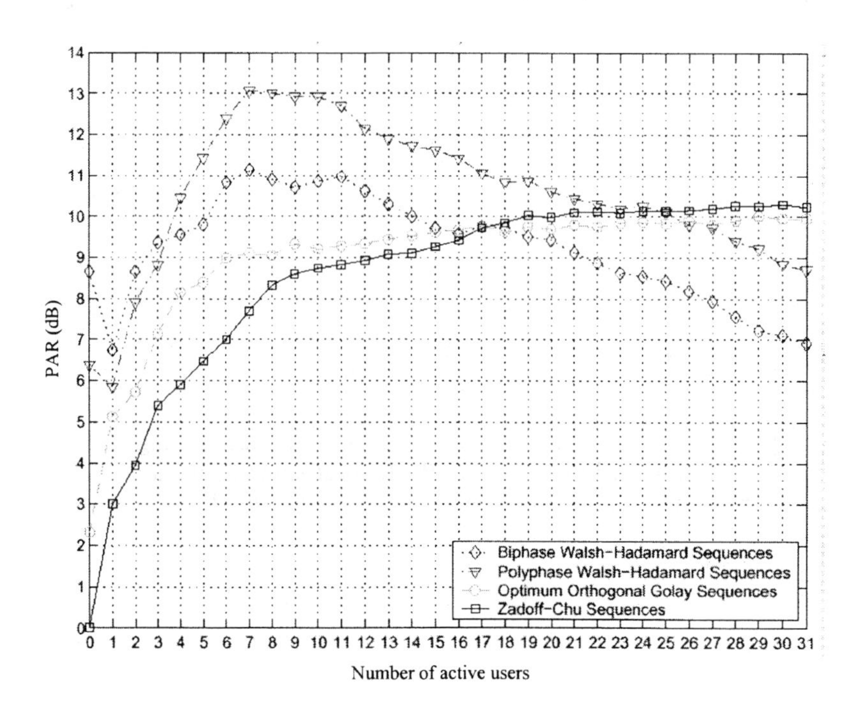

Figure 3. PAR in DL with 32-length sequences

Although Zadoff-Chu and orthogonal Golay sequences perform very well in uplink, keeping PAR constant and low independently of the number of active users, their behaviour in downlink is not so good, since PAR increases very fast up to medium load and it continues increasing, more slowly though, up to a fully loaded situation. Walsh-Hadamard sequences show a bad performance in uplink with PAR increasing with the number of users. However they show a slightly better behaviour in downlink and they perform better than Zadoff-Chu and Golay when the load of the system is above medium,

Therefore, according to their performance in terms of PAR, if the number of simultaneous users can be expected to be low, below medium load during most of the time, Zadoff-Chu sequences are the best choice. If the system is expected to be fully loaded most of the time, polyphase Walsh-Hadamard sequences should be chosen, but it must be taken into account that their average performance is poorer.

4. BER PERFORMANCE

In order to compare the performance of the different sequences in terms of BER, Figure 4 shows simulation results in the downlink of BPSK-modulated MC-CDMA with $K_{MC} = 32$, P=32, with severe multipath ($\tau=T_s$) and $E_b/N_0 = 6$ dB. Equal Gain Combining (EGC) has been used for compensation of the channel effects.

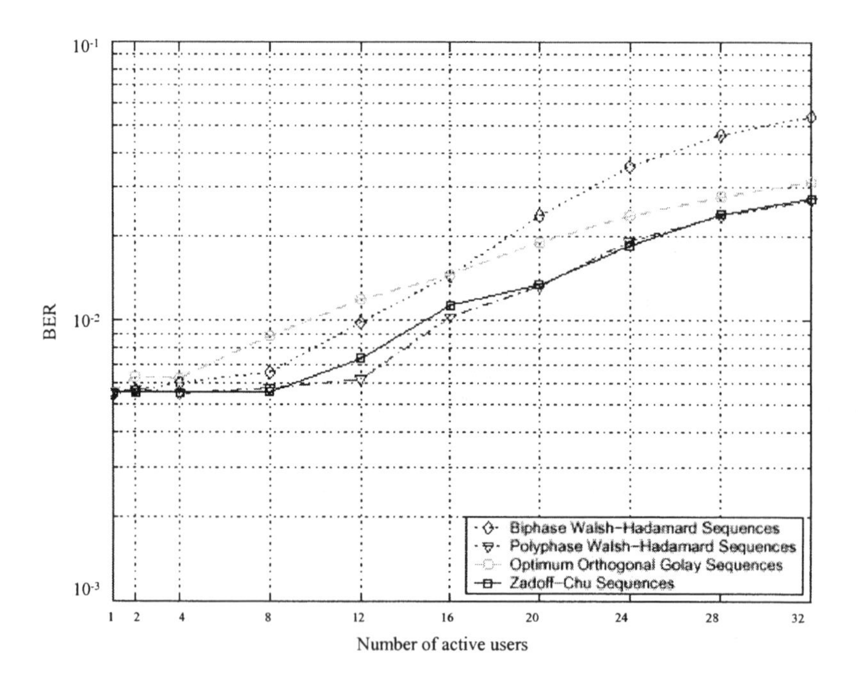

Figure 4. BER in multipath DL with 32-length sequences

Zadof-Chu and polyphase Walsh-Hadamard show the best performance. Golay sequences are the worst, more noticeably when the number of active users is high. This fact confirms the choice of Zadof-Chu or polyphase Walsh-Hadamard made in the previous section.

It is also interesting to consider the performance of MC-CDMA in real channels. In order to analyse the feasibility of applying MC-CDMA to WLAN, an Hiperlan 2 scenario has been chosen. Figure 5 shows BER performance of BPSK-modulated MC-CDMA with Zadoff-Chu sequences in Hiperlan channel A (Medbo, 1998) downlink. Since the length of channel model A is 390 ns, in order to keep C=8 we must use P=64. Again, we consider 32 maximum simultaneous users. Minimum Mean Square Error Combining (MMSEC) optimised per sub-carrier has been used for compensation of the channel effects, because it has been found to perform better than EGC (this can be seen comparing 6 dB-curve of Figure 5 with Figure 4). Although it is more complex to implement, it will give us an idea of the best achievable performance.

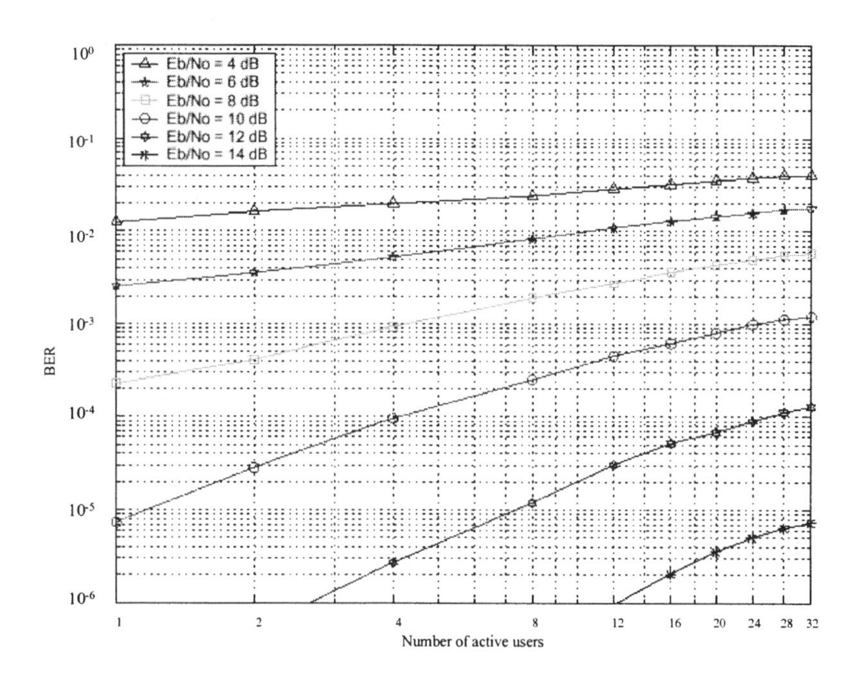

Figure 5. Performance in Hiperlan channel A

We can see in this Figure that the increase of the number of users causes the multiple access interference to increase and the BER performance is degraded. The effect is more important when signal-to-noise ratio is high because noise effect dominates in the error probability for low SNR values. However, in spite of this degradation, if E_b/N_0 values of 12-14 dB are used, 32 simultaneous users can share the channel with BER maintained below 10^{-4}.

5. CONCLUSIONS

In this paper we have considered the use of MC-CDMA in WLAN systems looking for transmission techniques that are able to extend the capabilities of wireless systems towards 4G. The MC-CDMA signal model that is able to cope with frequency selectivity of multipath channels has been presented and the way to choose the values of signal parameters has been shown.

Several spreading sequences have been analysed looking at their performance in terms of Peak-to-Average power-Ratio (PAR) and Bit Error Rate (BER).

It can be concluded that choosing the adequate spreading sequence and signal parameters, multiple users can be accommodated in Hiperlan channels maintaining the high bit rates that characterize WLAN scenarios.

6. REFERENCES

Chu, D.C. (1972). Polyphase Codes with Good Periodic Correlation Properties. *IEEE Transactions on Information Theory, 18*, 531-532.

Frank, R.L., & Zadoff, S.A. (1962). Phase Shift Pulse Codes with Good Periodic Correlation Properties. *IRE Transactions on Information Theory, 8*, 381-382.

Hara, S., & Prasad R. (1997). Overview of Multicarrier CDMA. *IEEE Communications Magazine, 35* (12), 126-133.

Medbo, J., & Schramm, P. (1998). *Channel Models for HIPERLAN 2.* ETSI/BRAN document no. 3ERI085B.

Popovic, B.M. (1999). Spreading Sequences for Multicarrier CDMA Systems. *IEEE Transactions on Communications, 47*(6), 918-926.

Seberry, J., & Yamada, M. (1992). *Hadamard Matrices, Sequences and Block Designs*. New York, USA: Wiley.

Sousa, E.S., & DaSilva, V.M. (1993). Performance of orthogonal CDMA codes for quasi synchronous communication systems. *Proceedings of IEEE ICUPC, Ottawa, Canada*, 995-999.

Vandendorpe L. (1993). Multitone direct sequence CDMA system in an indoor wireless environment. *Proceedings of IEEE First Symposium of Communications and Vehicular Technology in the Benelux, Delft, The Netherlands*, 4.1.1-4.1.8.

Van Nee, R., & Prasad, R. (2000). *OFDM wireless multimedia communications*. Boston, USA: Artech House.

Yee, N., Linnartz, L.P., & Fettweis, G. (1993). Multi-carrier CDMA in Indoor Wireless Radio Networks. *Proceedings of IEEE PIMRC, Yokohama, Japan*, 109-113.

7. AFFILIATIONS

Universidad Carlos III de Madrid, Dept. of Signal Theory and Communications

Butarque 15, Leganés 28911, Spain, Tel. +34 91 6249172, Fax: +34 91 6248749

email: {agarcia, jra, vjimenez, mjulia}@tsc.uc3m.es

Section III

SYNCHRONIZATION AND CHANNEL ESTIMATION

HEIDI STEENDAM AND MARC MOENECLAEY

UPLINK AND DOWNLINK MC-DS-CDMA SYNCHRONIZATION SENSITIVITY

Abstract. In this paper, we consider the effect of synchronization errors on the performance of the multicarrier direct-sequence code-division multiple access (MC-DS-CDMA) system and compare the results for downlink and uplink transmission. To evaluate the effect of small synchronization errors on the BER performance of the MC-DS-CDMA system, we derive simple analytical expressions for the BER degradation that are based upon truncated Taylor series expansions. We point out that a constant carrier phase offset or a constant timing offset do not give rise to performance degradation, for neither uplink nor downlink MC-DS-CDMA. The MC-DS-CDMA system is strongly degraded in the presence of a carrier frequency offset or a clock frequency offset. This degradation is proportional to the squares of the frequency offset and the number of carriers. Further, the degradation in the uplink is a factor N_s^2 (N_s is the spreading factor) larger than in the downlink, because the former suffers from a higher level of multi-user interference. The degradation caused by carrier phase jitter or timing jitter is the same in the uplink and the downlink, when the spectrum of the jitter is the same for all users. Further, the degradation is independent of the spectral contents of the jitter, the spreading factor and the number of carriers, but only depends on the jitter variance.

1. THE MC-DS-CDMA SYSTEM

The MC-DS-CDMA system is a combination of the MC transmission technique and the CDMA multiple access (CDMA) technique. In MC-DS-CDMA, the serial-to-parallel converted data stream is multiplied with the spreading sequence and then the chips belonging to the same symbol modulate the same carrier: the spreading is done in the time domain. MC-DS-CDMA has been proposed for mobile communications [1]-[3].

Without loss of generality, we use the terminology for the uplink. In MC-DS-CDMA, the symbol sequence to be transmitted at rate R_s is split into N_c lower rate symbol sequences $\{a_{i,k,\ell}\}$, where $a_{i,k,\ell}$ denotes the ith data symbol transmitted by user ℓ on the kth carrier of the multicarrier system; k belongs to a set I_c of N_c carrier indices. The symbol $a_{i,k,\ell}$ is multiplied with a spreading sequence $\{c_{i,n,\ell}|n=0,\ldots,N_s-1\}$ with spreading factor N_s. Each user is assigned a unique orthogonal spreading sequence. The resulting N_s components of the spread data symbol $a_{i,k,\ell}$, i.e. $\{a_{i,k,\ell}c_{i,n,\ell}| n=0,\ldots,N_s-1\}$ are then serially transmitted on the kth carrier of the multicarrier system. Hence, the spreading is accomplished in the time domain. To modulate the spread data symbols on the orthogonal carriers, an N_F-point inverse fast Fourier transform (inverse FFT) is used. To avoid that the multipath channel causes interference between the data symbols at the receiver, each FFT block is cyclically extended with a prefix of N_p samples. The resulting sequence $\{s_{i,n,m,\ell}\}$ is fed to a square root raised cosine filter $P(f)$ with roll off α and unit-energy impulse response

K. Fazel and S. Kaiser (eds.), Multi-Carrier Spread-Spectrum, 177–185.

$p(t)$ at a nominal rate $1/T=(N_F+N_p)N_sR_s/N_c$, resulting in the signal $s_\ell(t)$. We assume the carriers inside the roll off area are not modulated. Hence, of the N_F available carriers, only $N_c \leq N_F$ carriers are actually used. Without loss of generality, we focus on the detection of the data symbols transmitted by the reference user ($\ell=0$).

The transmitted signal $s_\ell(t)$ reaches the receiver through a slowly multipath fading channel. Assuming the path gains are constant over the duration N_c/R_s of N_s FFT blocks, the corresponding channel transfer function experienced by the ith symbol from user ℓ can be denoted by $H_{ch\ell}(f,i)$. Restricting our attention to wide-sense stationary uncorrelated scattering (WSSUS), the second-order moment $E[|H_{ch\ell}(f,i)|^2]$ is independent of both f and i. The output of the channel is disturbed by additive white Gaussian noise (AWGN) $w(t)$ with uncorrelated real and imaginary parts, each having a power spectral density of $N_0/2$. Further, the signal of user ℓ is affected by a carrier phase error $\phi_\ell(t)$. The sum of the different user signals is applied to the receiver filter, which is matched to the transmit filter, and sampled at nominal rate $1/T$. The contribution of user ℓ is disturbed by a timing offset $\varepsilon_{i,n,m,\ell}T$. In the uplink, where the contribution of each user is generated with a different transmit clock, upconverted by a different carrier oscillator and transmitted over a different multipath channel, the carrier phase error $\phi_\ell(t)$ and timing offset $\varepsilon_{i,n,m,\ell}T$ generally depend on the user index ℓ. In the downlink, on the other hand, the base station synchronizes the different user signals, and upconverts the sum of the different user signals with the same carrier oscillator. Further, as the different user signals reach the receiver of the reference user through the same multipath channel, the carrier phase error $\phi(t)$ and timing offset $\varepsilon_{i,n,m}T$ are the same for all users.

In the following, we assume that the transmitter (uplink) or receiver (downlink) of each user adapts its transmit clock phase such that N_F samples can be found outside the cyclic prefix that are free from interference from neighbouring FFT blocks. The resulting N_F samples are kept for further processing. As the removal of the cyclic prefix eliminates the interference between neighbouring blocks, the data symbols $a_{i,k,\ell}$ transmitted during symbol interval i are not affected by intersymbol interference from other symbol intervals. Hence, we omit the symbol index i in the sequel.

The N_F selected samples are applied to an N_F-point FFT, followed by one-tap equalizers $g_{n,k}$ that scale and rotate the FFT outputs. We denote by $g_{n,k}$ the coefficient of the equalizer, operating on the kth FFT output during the nth FFT block. We consider the case of the maximum ratio combiner (MRC). Each equalizer output is multiplied with the corresponding chip of the reference user's spreading sequence, and the N_s consecutive values are summed to yield the samples z_k at the input of the decision device. Based on the sample z_k, a decision is made about the data symbol $a_{k,0}$.

To measure the performance, we use the signal to interference and noise ratio (SINR), defined by $SINR_k(\phi,\varepsilon)=\beta P_{U,k}/(P_{N,k}+\beta P_{I,k}))$, where $\beta=N_F/(N_F+N_p)$ and $P_{U,k}$, $P_{I,k}$ and $P_{N,k}$ are the powers of the average useful component, the interference and the noise, respectively. Note that in general these powers depend on the carrier index k. In the absence of synchronization errors, the SINR reduces to

$SINR_k(0)=\beta|H_{k,0}|^2E_{s,k,0}/N_0$, where $E_{s,k,\ell}$ is the symbol energy transmitted on carrier k by user ℓ, $H_{k,\ell}=H_\ell(\text{mod}(k;N_F)/(N_FT))/T$, mod$(x;N_F)$ is the modulo-N_F reduction of x, yielding a result in the interval $[-N_F/2,N_F/2]$, and $H_\ell(f)=|P(f)|^2H_{ch,\ell}(f)$. The quantity $SINR_k$ still depends on the particular realization of the transfer functions $H_{k,\ell}$ ($k\in I_c$, $\ell=0,\ldots,N_u$-1) and the spreading sequences during the considered sequence of N_s FFT blocks. Hence, a more convenient performance indicator is $\overline{SINR_k}$, which is obtained by replacing $P_{X,k}$ ($X=U$, I, N) by their averages $\overline{P}_{X,k}$ over the fading characteristics and over all possible assignments of spreading sequences to the users. Because of the WSSUS assumption, $E[|H_{k,\ell}|^2]$ does not depend on the carrier index. We assume perfect power control: $E_s=E_{s,k,\ell}E[|H_{k,\ell}|^2]$. In this case, $\overline{SINR_k}(0)$ reduces to $\beta E_s/N_0$, which is independent of the carrier index k. The degradation (in dB) caused by the synchronization errors is defined by $Deg_k=(\overline{SINR}(0)/\overline{SINR_k}(\phi,\varepsilon))$.

2. CARRIER PHASE ERRORS

In this section, we investigate the sensitivity of MC-DS-CDMA to carrier phase errors in the absence of timing errors ($\varepsilon_{n,m,\ell}=0$).

2.1 *Carrier Frequency Offset*

In the case of small carrier frequency offsets ΔF_ℓ ($\ell=0,\ldots,N_u$-1), the carrier phase error linearly increases in time [4]: $\phi_\ell(t)=\phi_\ell(0)+2\pi\Delta F_\ell t$. For small carrier frequency offsets ($|N_F\Delta F_\ell T|<<1$), the useful power and noise power can be approximated by $\overline{P}_{U,k}=E_s$ and $\overline{P}_{N,k}=N_0$. The contribution of user ℓ to the interference power is proportional to R_ℓ, which is the correlation between the sequences $\{\tilde{c}_{n,\ell}\}$ and $\{\tilde{c}_{n,0}\}$, where $\tilde{c}_{n,\ell}=c_{n,\ell}\exp(j(2\pi n(N_F+N_p)\Delta F_\ell T+\phi_\ell(0)))$. In the uplink, for $\ell\neq0$, the chips $c_{n,\ell}$ of user ℓ and the chips $c_{n,0}$ of the reference user are rotated over different angles, so that the orthogonality between the different user signals is destroyed: $R_\ell\neq0$ for $\ell\neq0$. Hence, the carrier frequency offsets give rise to intercarrier and multi-user interference. In the downlink, however, all chips are rotated over the *same* angle, so that the orthogonality between the different user signals is maintained: $R_\ell=0$. Hence, in the downlink, multi-user interference is absent, and the carrier frequency offset only introduces intercarrier interference. In the following, we approximate the interference power by a truncated Taylor series (keeping up to quadratic terms) around $\Delta F_\ell=0$.

We assume that the carrier frequency offsets are within the interval $[-F_{max},F_{max}]$, where F_{max} is smaller than the carrier spacing: $N_FF_{max}T<<1$. In the downlink, where multi-user interference is absent, the total interference power yields

$$\overline{P}_{I,k} \approx E_s(\pi N_F \Delta F_0 T)^2/3 \tag{2}$$

Note that the degradation becomes independent of the carrier index k. The maximum degradation occurs for $|\Delta F_0|=F_{max}$.

In uplink MC-DS-CDMA, multi-user interference is present. For $N_s(N_F+N_p)F_{max}T \ll 1$, the orthogonality between the rotated chip sequences $\{\tilde{c}_{n,\ell}\}$, $\ell=0,\ldots,N_u-1$ is only slightly affected. Assuming the carrier frequency offsets ΔF_ℓ of the interfering users are uniformly distributed in the interval $[-F_{max},F_{max}]$, one obtains

$$\overline{P}_{I,k} \approx E_s \frac{N_u-1}{N_s-1} B\left((\Delta F_0 T)^2 + \frac{1}{3}(F_{max}T)^2\right) \tag{3}$$

where N_u is the number of active users and $B=(\pi N_s(N_F+N_p))^2/3$. When $N_s(N_F+N_p)F_{max}T \ll 1$ is no longer valid, the orthogonality between the rotated chip sequences $\{\tilde{c}_{n,\ell}\}$, $\ell=0,\ldots,N_u-1$ is strongly affected. In this case, the interference power can be approximated by

$$\overline{P}_{I,k} \approx E_s(N_u-1)/(N_s-1) \tag{4}$$

From (4) and (5) it follows that the degradation in the uplink is independent of the carrier index k. From (10), we observe that $|\Delta F_0|=F_{max}$ yields the maximum degradation. For $N_s(N_F+N_p)F_{max}T \ll 1$, the maximum degradation is proportional to $(N_s(N_F+N_p)F_{max}T)^2$, whereas for $N_s(N_F+N_p)F_{max}T>1$ the maximum degradation becomes essentially independent of N_s, N_F and $F_{max}T$. When $N_s(N_F+N_p)F_{max}T \ll 1$, the maximum degradation in the uplink is a factor N_s^2 larger than in the downlink. This can also be observed in figure 1, where the maximum degradation, obtained with the approximations (2) and the minimum of (3) and (4), is shown along with the actual degradation, for the maximum load ($N_u=N_s$). As we observe, the approximation is close to the actual degradation. Hence, the truncated Taylor series expansions can be used to compute the actual degradation caused by carrier frequency offsets in uplink and downlink MC-DS-CDMA.

When $\Delta F_\ell=0$, $\ell=0,\ldots,N_u-1$, the carrier phase error reduces to a constant phase offset $\phi_\ell(t)=\phi_\ell(0)$. The only effect of a constant phase offset is a rotation over an angle $\phi_\ell(0)$ of the contribution of user ℓ at the FFT outputs. As this rotation can be compensated for by the equalizer, a constant phase does not introduce a performance degradation.

2.2 Carrier Phase Jitter

To get rid of the strong degradation caused by carrier frequency offsets, a phase-locked loop (PLL) can be used in the upconversion and downconversion of the signals. The residual carrier phase error $\phi_\ell(t)$ can be modelled as a zero-mean stationary process with jitter spectrum $S_{\phi,\ell}(f)$ and jitter variance $\sigma^2_{\phi,\ell}$ [5]. For small jitter variances, i.e. $\sigma^2_{\phi,\ell} \ll 1$, the phase rotation $\exp\{j\phi_\ell(t)\}$ of the data symbols of user ℓ at the FFT outputs can be approximated by a truncated Taylor series: $\exp\{j\phi_\ell(t)\} \approx 1 + j\phi_\ell(t)$.

When all jitter processes in the uplink have the same jitter spectrum $S_{\phi,\ell}(f) = S_\phi(f)$, thus jitter variances $\sigma^2_{\phi,\ell} = \sigma^2_\phi$, and the load is maximum ($N_u = N_s$) [5], the degradation in the uplink and the downlink is the same, and given by

$$Deg \approx 10\log\left(1 + SINR(0)\sigma_\phi^2\right) \tag{5}$$

This degradation is independent of the carrier index k, the number of carriers, the spreading factor and the spectral contents of the jitter, but only depends on the jitter variance.

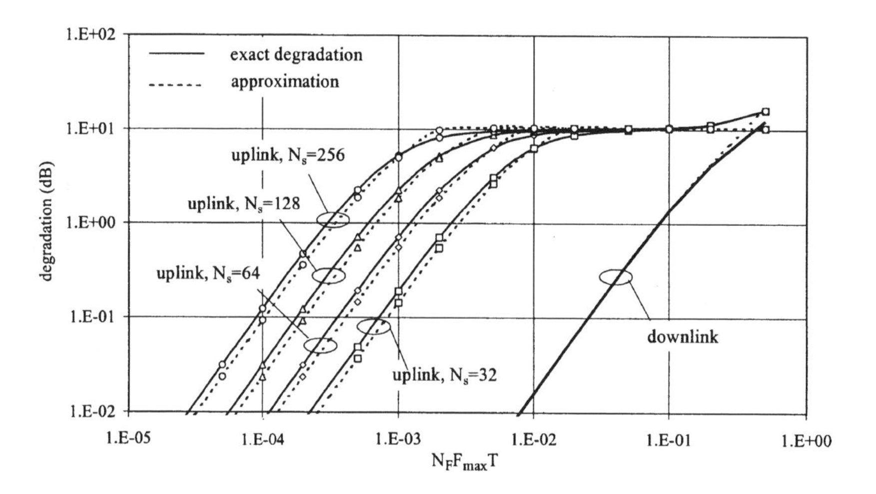

Figure 1. Carrier frequency offset, N_u=N_s, N_p=0, SINR(0)=10 dB

3. TIMING ERRORS

In this section, we investigate the sensitivity of MC-DS-CDMA to carrier phase errors in the absence of carrier phase errors ($\phi_\ell(t)$=0).

3.1 Clock Frequency Offset

Assuming the transmitter (uplink) and receiver (downlink) of each user has a free-running clock with a relative clock frequency offset $\Delta T_\ell/T$ as compared to the frequency $1/T$ of the base station clock, the timing deviation linearly increases in time: $\varepsilon_{n,m,\ell}=\varepsilon_{n,\ell}+m\Delta T_\ell/T$, m=0,...,$N_F$-1, where $\varepsilon_{n,\ell}$ is the timing deviation of the first of the N_F samples of the considered FFT block that are processed at the receiver [6]. For small clock frequency offsets ($|N_F\Delta T_\ell/T|<<1$), the useful power and noise power can be approximated by $\overline{P}_{U,k}=E_s$ and $\overline{P}_{N,k}=N_0$. The contribution of the data symbol $a_{k',\ell}$ to the interference power on the kth carrier is proportional to $R_\ell(k,k')$, which is the correlation between the sequences $\{\tilde{c}_{n,k',\ell}\}$ and $\{\tilde{c}_{n,k,0}\}$. In the sequence $\tilde{c}_{n,k,\ell}$, the chips $c_{n,\ell}$ are rotated over an angle $\exp(j2\pi\,\varepsilon_{n,\ell}\mathrm{mod}(k;N_F)\Delta T_\ell/T)$. As the chip sequences belonging to different users or transmitted on different carriers are rotated over different angles, the orthogonality between the signals transmitted by different users or on different carriers is lost: the clock frequency offset introduces intercarrier and multi-user interference, for both uplink and downlink MC-DS-CDMA. Note however that in the downlink, $R_\ell(k,k)=\delta_\ell$: symbols transmitted on carrier k to non-reference users do not give rise to interference at the kth output at the receiver of the reference user. As for given k the largest interference in the uplink comes from the multi-user interference from the contributions transmitted on carrier k, it follows that the multi-user interference in the uplink is substantially larger than in the downlink. In the following, we approximate the interference power by a truncated Taylor series (keeping up to the quadratic terms) around $\Delta T_\ell/T=0$.

We assume that the clock frequency offsets $\Delta T_\ell/T$, $\ell=0,\ldots,N_F-1$ are restricted to the interval $[-(\Delta T/T)_{max},(\Delta T/T)_{max}]$, where $|N_F(\Delta T/T)_{max}|<<1$ to keep the degradation sufficiently small. In downlink MC-DS-CDMA, the interference power can be approximated by

$$\overline{P}_{I,k} \approx \frac{E_s}{3}\left(\pi\mathrm{mod}(k;N_F)\frac{\Delta T_0}{T}\right)^2 \tag{6}$$

The degradation depends on the carrier index k, and becomes maximum for carriers at the edge of the roll off area. Further, $|\Delta T_0/T|=(\Delta T/T)_{max}$ maximizes the degradation.

In the uplink, we can distinguish two cases. First, when $N_s(N_F+N_p)(\Delta T/T)_{max}<<1$, the chip sequences $\tilde{c}_{n,k,\ell}$ are rotated over nearly the same angle, so that the orthogonality between the different carriers and the different users is only slightly affected. Assuming that the clock frequency offsets $\Delta T_\ell/T$ of the interfering users are uniformly distributed in the interval $[-(\Delta T/T)_{max},(\Delta T/T)_{max}]$, the interference power is approximated by

$$\overline{P}_{I,k} \approx E_s \frac{N_u - 1}{N_s - 1} B(\mathrm{mod}(k; N_F))^2 \left(\left(\frac{\Delta T_0}{T} \right)^2 + \frac{1}{3} \left(\frac{\Delta T}{T} \right)^2_{2\,max} \right) \tag{7}$$

where $B=(\pi N_s)^2/3$. The degradation depends on the carrier index k and becomes maximum for carriers at the edge of the roll off area. Further, the degradation is maximum when $|\Delta T_0/T|= (\Delta T/T)_{max}$. From (6) and (7), it follows that the maximum degradation in the uplink is a factor N_s^2 larger than in the downlink, when $N_s(N_F+N_p)(\Delta T/T)_{max}<<1$. When $N_s(N_F+N_p)(\Delta T/T)_{max}<<1$ is no longer valid, the orthogonality between the signals from the different users and on different carriers is strongly affected. In this case, one obtains

$$\overline{P}_{I,k} \approx E_s (N_u - 1)/(N_s - 1) \tag{8}$$

which is essentially independent of N_s, N_F and $(\Delta T/T)_{max}$. In figure 2, the maximum degradation obtained with the approximations (6) and the minimum of (7) and (8) is shown, along with the actual degradation. It follows that (6) and the minimum of (7) and (8) yield accurate approximations for the actual degradations.

When $\Delta T_\ell/T=0$, the timing error reduces to a constant timing offset $\varepsilon_{n,m,\ell}=\varepsilon_{0,\ell}$. As only carriers outside the roll off area are used, the only effect of the constant timing offset is a rotation over an angle $2\pi\ \varepsilon_{0,\ell}\mathrm{mod}(k;N_F)/N_F$ of the contribution of user ℓ at the kth FFT output. As the equalizer can compensate for this rotation, a constant timing offset does not yield performance degradation.

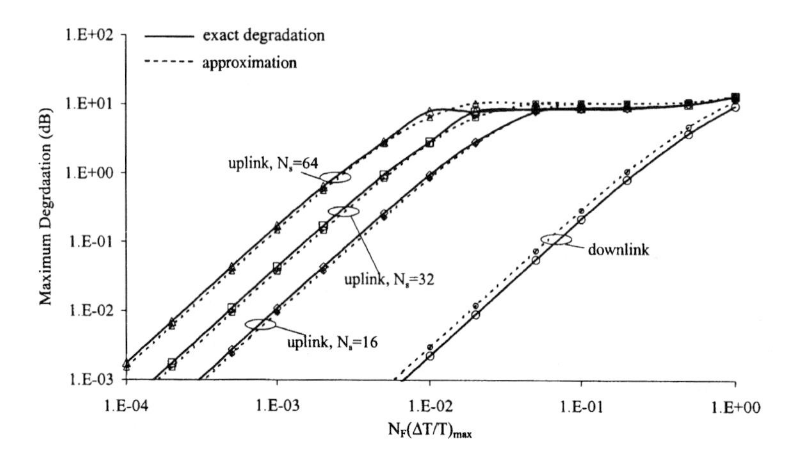

Figure 2. Clock frequency offset, $N_u=N_s$, $N_p=5$, $SINR(0)=10$ dB

3.2 *Timing Jitter*

When the strong degradation caused by clock frequency offsets cannot be tolerated, the transmitter (uplink) and receiver (downlink) clock phase can be adjusted using a PLL. The timing jitter $\varepsilon_{i,n,m,\ell}T$ introduced by this adaptation process can be modelled as a zero-mean stationary process with jitter spectrum $S_{\varepsilon,\ell}(f)$ and jitter variance $\sigma^2_{\varepsilon,\ell}$ [7]. For small jitter variances ($\sigma^2_{\varepsilon,\ell} << 1$), the phase rotation $\exp\{j\text{mod}(k;N_F)/N_F\, \varepsilon_{i,n,m,\ell}T\}$ at the FFT outputs can be approximated by a truncated Taylor series: $\exp\{j\text{mod}(k;N_F)/N_F\, \varepsilon_{i,n,m,\ell}T\} \approx 1 + j\text{mod}(k;N_F)/N_F\, \varepsilon_{i,n,m,\ell}T$.

When the load is maximum ($N_u = N_s$) and the jitter spectrum is the same for all users: $S_{\varepsilon,\ell}(f) = S_\varepsilon(f)$, (hence $\sigma^2_{\varepsilon,\ell} = \sigma^2_\varepsilon$), the degradation in the uplink is the same as in the downlink. This degradation depends on the carrier index k. The average degradation, which is obtained by replacing in the SINR the powers of the useful component, the interference and the noise by their arithmetical average over all carriers is given by:

$$Deg_{av} \approx 10\log\left(1 + SINR(0)\frac{\pi^2}{3}\sigma_\varepsilon^2\right) \tag{9}$$

which is independent of the number of carriers, the spreading factor and the spectral contents of the jitter, but only depends on the jitter variance.

4. CONCLUSIONS

To evaluate the effect of small synchronization errors on the BER performance of the MC-DS-CDMA system, we derive simple analytical expressions for the BER degradation that are based upon truncated Taylor series expansions. Computer simulations indicate that the degradation obtained from the Taylor series expansion yields a good approximation of the actual degradation. For both the uplink and the downlink, we compare the degradation caused by different types of synchronization errors. Assuming the load is maximum ($N_u = N_s$), and noting that the number of carriers N_c is proportional to the FFT length N_F, the results can be summarized as follows:

a) Constant phase offsets or constant timing offsets do not give rise to performance degradation, for neither uplink nor downlink MC-DS-CDMA, because these offsets can be compensated for at the FFT outputs.
b) The MC-DS-CDMA system is strongly degraded in the presence of carrier frequency offsets. This degradation is proportional to $(N_c\Delta FT)^2$, and the degradation in the uplink is N_s^2 times higher than in the downlink, as in the downlink the amount of multi-user interference is much higher than in the downlink.
c) For both the uplink and the downlink, the degradation caused by clock frequency offsets strongly increases with $(N_c\Delta T/T)^2$. The degradation in the uplink is a

factor N_s^2 higher than in the downlink, as the uplink is affected by a larger amount of multi-user interference.

d) When the spectrum of the carrier phase jitter or timing jitter is the same for all users, the degradation caused by carrier phase jitter or timing jitter is the same in the uplink and the downlink. The corresponding degradation is independent of the spectral contents of the jitter, the spreading factor and the number of carriers, but only depends on the jitter variance.

5. ACKNOWLEDGMENT

This work has been supported by the Interuniversity Attraction Poles Program - Belgian State - Federal Office for Scientific, Technical and Cultural Affairs.

6. REFERENCES

[1] Santella, G. (1995). Bit error rate performances of M-QAM orthogonal multicarrier modulation in presence of time-selective multipath fading. Paper presented at *IEEE International Conference on Communications*, Seattle, WA, 1638-1688.

[2] Kondo, S., & Milstein, L.B. (1996). Performance of multicarrier CDMA systems. *IEEE Transactions on Communications*, 44, (2), 238-246.

[3] DaSilva, V.M., & Sousa, E.S. (1993). Performance of orthogonal CDMA sequences for quasi-synchronous communication systems. Paper presented at *IEEE ICUPC*, Ottawa Canada, 995-999.

[4] Steendam, H., & Moeneclaey, M. (2001). The effect of carrier frequency offsets on downlink and uplink MC-DS-CDMA. *IEEE Journal on Selected Areas in Communications*, 19, (12), 2528-2536.

[5] Steendam, H., & Moeneclaey, M. (2001). The effect of carrier phase jitter on MC-DS-CDMA. Paper presented at *IEEE International Conference on Communications*, Helsinki, Finland, 1881-1884.

[6] Steendam, H., & Moeneclaey, M. (2001). Downlink and uplink MC-DS-CDMA sensitivity to timing jitter. Paper presented at *IEEE Symposium on Communications and Vehicular Technology in the BENELUX*, Delft, The Netherlands, 24-29.

[7] Steendam, H., & Moeneclaey, M. (2002). Uplink and downlink MC-DS-CDMA sensitivity to static clock frequency offsets, Paper presented at *IASTED International Conference on Communications Systems and Networks*, Malaga, Spain, 254-259.

7. AFFILIATIONS

H. Steendam and M. Moeneclaey are with the Department of Telecommunications and Information Processing, Ghent University, B-9000 Gent, Belgium (e-mail: Heidi.Steendam@telin.ugent.be, Marc.Moeneclaey@telin.ugent.be)

Y. ZHANG, R. HOSHYAR AND R. TAFAZOLLI

STUDY OF SYMBOL SYNCHRONIZATION IN MC-CDMA SYSTEMS

Abstract: In this paper, the requirement of symbol synchronization in MC-CDMA systems is investigated. Two data aided synchronization schemes are proposed which can be exploited both in downlink and uplink. In *Scenario I,* we spread the pilot data on some selected subcarriers and utilize the cyclic nature of pilot data in time domain for downlink synchronization and the correlation characteristic of spreading codes for uplink synchronization. In *Scenario II,* we allocate a longer PN code to each user which obtains a better cross correlation in time domain and apply a similar algorithm to *Scenario I* uplink case. Their performance are evaluated and compared in multipath fading channels.

1. INTRODUCTION

MC-CDMA is one of the multiple access methods of multicarrier systems, which combines the advantages of OFDM and CDMA. Accurate synchronization is necessary for MC-CDMA systems to keep the orthogonality of the subcarriers, which is essential for reliable transmission. In downlink connection, since all users experience the same fading and arrive at the receiver simultaneously, synchronization techniques used in single user OFDM systems [1]-[4] can be exploited without much modification. However, in uplink case, different users pass through different channels and reach the base station asynchronously. It is necessary to correct time and frequency offset separately for each user.

In OFDM systems, usually some periodicity of transmitted signals are utilized for symbol synchronization, such as the cyclic prefix of symbols [1]-[2] and the cyclic nature of training symbols [3]-[4], However, these methods are difficult to estimate different delays for each user in uplink case. A natural idea is to exploit the correlation property of spreading code for synchronization which is used in conventional CDMA systems. This method has been investigated in [5]-[7]. In this paper, we try to combine the advantages of these two methods, and give two different schemes nominated by *Scenario 1* and *Scenario 2* respectively.

2. SYSTEM MODEL

The transmitter model of MC-CDMA system we consider is depicted in Figure 1. Modulated data symbols are serial/parallel converted first, then each output of the S/P converter is multiplied by a higher rate chip sequence. After frequency domain interleaving, OFDM modulation is performed over all the subcarriers by IFFT block. Finally, cyclic prefix is inserted and time domain samples are P/S converted. Here we use Walsh-Hadamard sequences for spreading code.

K. Fazel and S. Kaiser (eds.), Multi-Carrier Spread-Spectrum, 186–194.

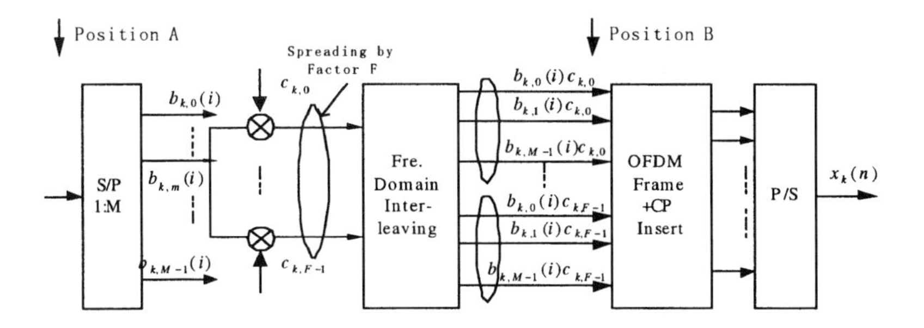

Figure 1 Transmitter model of MC-CDMA system

The time domain samples of the transmitted MC-CDMA baseband signal by user k can be expressed as

$$x_k(n) = \sum_{m=0}^{M-1} b_{k,m} \sum_{f=0}^{F-1} c_{k,f} \cdot e^{j\frac{2\pi}{N}(M\cdot f+m)\cdot n}, \quad -N_G \le n \le N-1 \tag{1}$$

where $b_{k,m}$ is the modulated data of the m*th* subcarrier, $c_{k,f}$ is the f*th* chip of user i's spreading code, F is the length of spreading code, N is the number of subcarriers, and N_G is the number of guard samples.

The multipath fading channel can be modelled as

$$h(n) = \sum_{p=0}^{P-1} h_p \cdot \delta(n - n_p) \tag{2}$$

where h_p is the complex gain of each path, and n_p is the corresponding path delay. Here we assume that the path delays are integer samples. In multicarrier systems when the channel spread delay is less than the duration of the cyclic prefix, the channel fading in each subcarrier can be modelled as flat fading, which means the channel effect on one subcarrier can be characterized by an amplitude scaling and a phase rotation.

Therefore, the received signal passing through the downlink channel can be written as

$$r(n) = \sum_{k=0}^{K-1}\sum_{m=0}^{M-1} b_{m,k} \sum_{f=0}^{F-1} c_{f,k} \cdot H_{Mf+m} e^{j\frac{2\pi}{N}(M\cdot f+m)\cdot n} + W(n), \quad -N_G \le n \le N-1 \tag{3}$$

where H_{Mf+m} is the channel effect and $W(n)$ is the sample of complex AWGN noise. In uplink, since users experiences different channel fadings and delays, the received signal of K active users is written as

$$r(n)=\sum_{k=0}^{K-1}\sum_{m=0}^{M-1}b_{m,k}\sum_{f=0}^{F-1}c_{f,k}\cdot H_{Mf+m,k}e^{j\frac{2\pi}{N}(M\cdot f+m)\cdot(n-\tau_k)}+W(n),\ -N_G\le n\le N-1 \quad (4)$$

where τ_k is the relative delay of k*th* user.

3. SYMBOL SYNCHRONIZATION REQUIREMENT

Accurate demodulation and detection of the multicarrier signal requires subcarrier orthogonality. Variations of sample clocks, uncorrected symbol timing and carrier frequency offset may distort the orthogonality of the subcarriers and cause intersymbol interference (ISI) and intercarrier interference (ICI). Hence, the multicarrier systems are much more sensitive to synchronization errors than single carrier systems. In this paper we focus on the effect of symbol timing offset and algorithms to counteract it, and assume other factors ideal.

At the receiver, the OFDM demodulator processes a block of N samples in the FFT at one time. The task of the symbol synchronization is to accurately estimate the start point of this block. The cyclic prefix provides some tolerance for the symbol timing offset in broadcast multicarrier systems. Let the sample indexes of a MC-CDMA symbol be {-Ng, ..., 0, 1, ..., N-1}, the timing offset be ε, and the maximum channel delay be τ. Consider a system Ng > τ, if $\varepsilon \in$ {-Ng+τ, ..., -1, 0}, the orthogonality is preserved and the offset appears as a linear phase rotation over subcarrier outputs. This phase rotation is identified by the channel estimator, which views it as a channel-induced phase shift. The channel estimator works as fine synchronizer in this condition. If ε exceeds the above range, orthogonality is distorted, and ISI and ICI are induced. Therefore, the performance of symbol synchronization algorithm is evaluated by whether the detected symbol start point is beyond the linear phase distortion area or not.

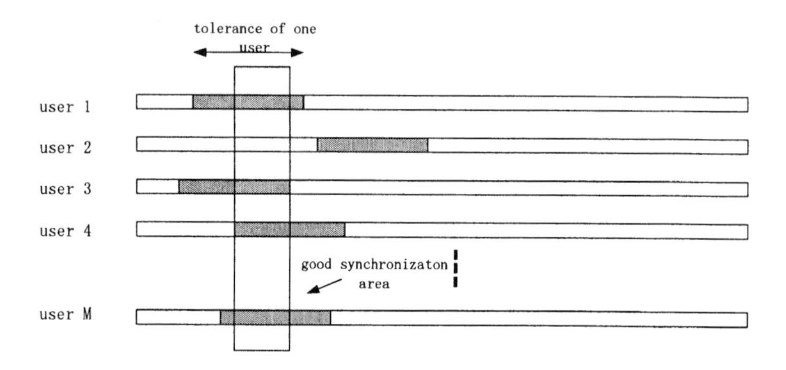

Figure 2 Synchronization requirement in uplink

In downlink MC-CDMA, since all users pass through the same channel, the timing offset effect is similar to the broadcast case. However in uplink, the ability of the cyclic prefix to compensate the timing error is largely reduced because signals from different users are not time-aligned. This effect is illustrated in Figure 2. The shaded area is the timing error tolerance provided by cyclic prefix of each user. The total tolerance of timing error is reduced comparing with the downlink case, which require the smaller variance of the timing estimator. We also notice that even at the good synchronization point, some large delay users such as user 2 causes the ISI and ICI and introduce more multiuser interference than other users. This effect can not be easily compensated by synchronization algorithms, and some feedback strategy is needed. It also indicates that synchronization in uplink is more challenging than downlink.

4. SYNCHRONIZATION SCHEMES

The synchronization methods we propose in this paper are data aided ones. Although non-data aided synchronization methods for multicarrier systems have also been investigated widely in literature, data aided ones are more suitable for fast and reliable synchronization and facing the great challenges in MC-CDMA uplink case. We assume a frame structure at the start of which some training symbols are inserted for synchronization as well as channel estimation. The length of frame is shorter than the coherence time of the channel. Hence in one frame time the channel delay keeps invariant.

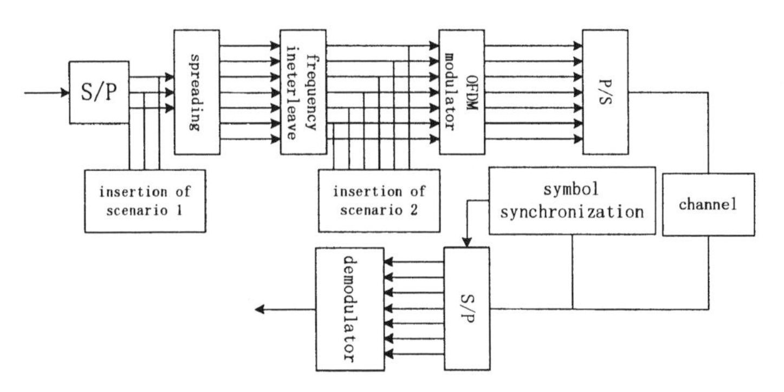

Figure 3 Block diagram of training symbol insertion

4.1 Scenario 1

In *Scenario 1*, the training symbol P is inserted before spreading, as shown in Figure 3. P can be expressed as $P = [\, P_1, P_2, \dots P_M \,]$, where there is

$$P_m = \begin{cases} A \cdot (1+i), m = Qj \\ 0, m = Qj + q \end{cases} \tag{5}$$

Here A, Q, j, q are integers, and $1 \le q \le Q-1$, $0 \le j \le M/Q$. After spreading, frequency domain interleaving and IFFT, the corresponding time-domain samples appear Q identical parts. Samples in each part can be expressed as

$$X_k = IFFT_{N/Q}(C_k \otimes \{P_{Qj}\}) \tag{6}$$

where C_k is the spreading code vector of the kth user. $\otimes$ denotes the Kronecker product, and $X_k = [x_{1,k}, x_{2,k}, \ldots x_{N/Q,k}]$.

In downlink, we search these Q identical parts of training symbols by autocorrelation algorithm in [3]. The start of the symbol is the maximum point of the metric given by

$$MD(d) = \frac{|P(d)|^2}{(G(d))^2} \tag{7}$$

where d is the sample index corresponding to the first sample in the observation window and

$$P(d) = \sum_{m=0}^{N-L-1} [r^*(d+m) \cdot r\ (d+m+L)],\ G(d) = \sum_{m=0}^{N-L-1} |r(d+m+L)|^2 \tag{8}$$

Here L is the delay step of autocorrelation. L can be chosen as $L = k \cdot N/Q$, k = 1, 2, …, Q/2. Advanced correlation metric which obverses a sharper peak value can be further studied based on the flexibility that training symbols provide.

In uplink, we notice that in the time domain samples of the training symbols, the information of spreading code is contained, to see (6). This information can be exploited for uplink synchronization for the following reasons: Firstly, each mobile station of MC-CDMA system has its own spreading code, which both the base station and mobile station know. Besides, according to the signal processing theory, the correlation properties of codes in frequency domain is related in big part with those in time domain.

We introduce a N/Q length user specific sequence $T_k = [X_k]^*$ to match the time domain training symbols of user k. $[\ ]^*$ denotes conjugate here. T_k is produced at receivers by the same process as X_k is generated. The estimation is defined as

$$\hat{\theta}_k = \arg\max_d |R_d \cdot T_k^T| \tag{9}$$

where $[]^T$ denotes vector transpose and R_d is a N/Q length vector of the received samples in (4). $R_d = [r(0+d), r(1+d),..., r(N/Q-1+d)]$. Correlation is applied Q times in one symbol time to get an average performance. We choose Q in the range of 2—6.

4.2 Scenario 2

In *Scenario 1* uplink case, correlation characteristic of spreading code in time domain is utilized. The time domain samples (IFFT of spreading code) have a perfect autocorrelation, while not very good cross correlation because of repetition of the same code in frequency domain [5]. Cross correlation characteristic is critical since we need distinguish different code for different users. Thus, instead of the insertion of *Scenario 1*, a longer PN code transmitted over all of the subcarriers is used and inserted at the position shown in Figure 3. The time domain cross-correlation property of the long PN code is better than the spreading code, which is shown in Figure 4. To use the long PN code to search for the start point of MC-CDMA symbols, a similar algorithm to (9) is used to search for the start point of MC-CDMA symbols. And different codes are allocated to different users in uplink and all users share a common code in downlink.

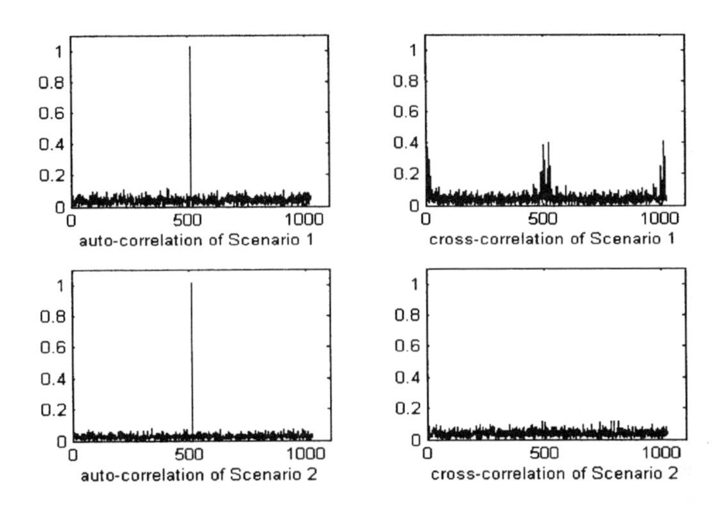

Figure 4. Time domain correlation characteristics of two scenarios

5. SIMULATION RESULTS

In simulation, we generate MC-CDMA signals using 1024 point IFFT and spreading sequences of length 32. The length of the cyclic prefix is 216 samples. The channel is modelled as a fading channel with 17 paths, the maximum channel delay of which is 170ns. In our case, cyclic prefix length is larger than the maximum delay spread and there is about 40 samples redundancy. Frame Error Rate (FER) is used as a performance measurement of the proposed method, which is defined as estimated timing error differs from real one by more than 5 sample times.

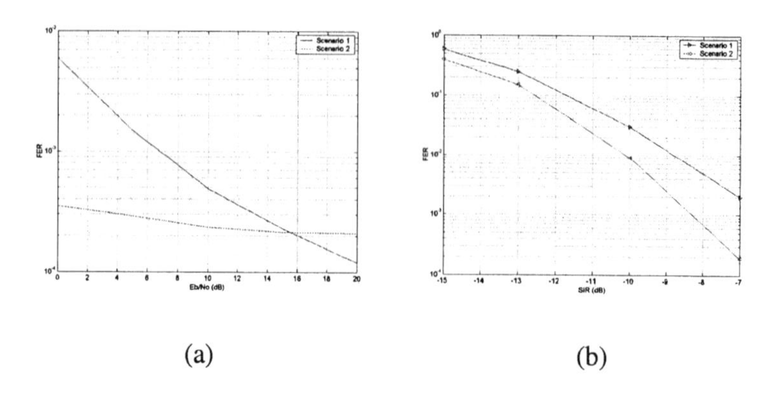

(a) (b)

Figure 5. Frame error performance of two scenarios: (a) downlink; (b) uplink

In downlink, 6 multiple users are assumed and FER performance with varying Eb/No is depicted in Figure 5 (a). In uplink, the channel noise is neglected since it is by far smaller than the multiuser interference. Thus we give a FER performance versus signal to interference ratio (Figure 5 (b)). *Scenario 2* achieves a superior one for the reason of the better cross-correlation characteristic of the training data.

6. CONCLUSIONS

In this paper, the requirement of symbol synchronization is studied and two synchronization schemes are proposed which can be applied both in uplink and downlink. Simulation shows that they achieve satisfactory performances in multipath fading channel. Frequency synchronization and channel estimation based on our schemes will be further studied.

Y. Zhang, R. Hoshyar and R. Tafazolli (Centre for Communication Systems Research, University of Surrey, Guildford GU2 7XH, United Kingdom)

REFERENCES

[1] J. J. van de Book, M. Sandell and P. O. Borjesson, "ML estimation of Time and Frequency offset in OFDM systems." IEEE Trans. on Signal Processing, pp. 1800-1805, July 1997.

[2] S. Barbarossa, M. Pompili and B. Giannakis, "Channel-Independent Synchronization of Orthogonal Frequency Division Multiple Access Systems," IEEE J. Select. Areas Commun., vol. 20, pp. 474-486, February 2002

[3] T. M Schimdl and Donald Cox, "Robust Frequency and Timing Synchronization for OFDM', IEEE Trans. Commun., vol. 45, pp 1613-1621, December 1997.
[4] T. Keller, L. Piazzo, P. Mandarini and L. Hanzo, "Orthogonal Frequency Division Multiplex Synchronization Techniques for Frequency-Selective Fading Channels", IEEE Journal on Selected Areas in Communications, vol. 19, pp.999-1008, June 2001
[5] S. H. Kim, K. Ha and C. W. Lee, "A Frame Synchronization Scheme for Uplink MC-CDMA", in *Proc. IEEE VTC*, pp. 2188-2192, 1999
[6] Sun, Y. Jiao, C. Honghong and Z. Zhou, "A Synchronization Scheme with less complexity on MC-CDMA Systems", In *Proc. IEEE VTC 2001 Fall*, vol: 1, pp. 57-61, 2001
[7] Ying Jiao, Chang Hong Hong ' An Low-Complex and Faster Synchronization Method for MC-CDMA Systems', *In Proc. IEEE VTC 2002,* pp. 1482-1486, 2002

HAKAN A. ÇIRPAN[1], ERDAL PANAYIRCI[2], HAKAN DOĞAN[3]

ITERATIVE CHANNEL ESTIMATION APPROACH FOR SPACE-FREQUENCY CODED OFDM SYSTEMS WITH TRANSMITTER DIVERSITY [(*)]

[1,3] *Department of Electrical-Electronics Engineering, Istanbul University Avcılar 34850, Istanbul, Turkey*

[2] *Department of Electronics Engineering, IŞIK University Maslak 80670 Istanbul, Turkey*

Abstract- Focusing on space-frequency transmit diversity OFDM transmission through frequency selective channels, this paper proposes a computationally efficient, maximum a posteriori(MAP) channel estimation algorithm. The algorithm requires a convenient representation of the discrete multipath fading channel based on the Karhunen-Loeve orthogonal expansion and estimates the complex channel parameters of each subcarriers iteratively using the Expectation Maximization(EM) method. In order to explore the performance, the closed-form expression for the average symbol error rate (SER) probability is derived for the zero-forcing equalizer. Furthermore, the MAP channel estimator's performance is studied based on the evaluation of the modified Cramer-Rao bound. Simulation results confirm our theoretical analysis, and illustrate that the proposed algorithm is capable of tracking fast fading and improving overall performance.

1. INTRODUCTION

Traditional wireless technologies are not very well suited to meet the demanding requirements of providing very high data rates with the ubiquity and mobility. Given the scarcity and exorbitant cost of radio spectrum, such data rates dictate the need for extremely high spectral efficient coding and modulation schemes. The combined application of space-time coding (STC) and OFDM modulation appears to be capable of enabling the types of capacities and data rates needed for broadband wireless services [1], [2].

The use of OFDM in transmitter diversity systems also offers the possibility of coding in a form of space-frequency OFDM (SF-OFDM) [3]. In [2], it was shown that the SF-OFDM system has the same performance as a previously reported ST-OFDM scheme in slow fading environments but shows better performance in the more difficult fast fading environments. This paper therefore focuses on iterative channel estimation approach for SF-OFDM systems. In this paper, a computationally efficient, maximum a posteriori (MAP) channel estimation

[(*)] This work was supported by Research Fund of the University of Istanbul. Project number: (UDP-186/06082003) and (T-131/11112002)

K. Fazel and S. Kaiser (eds.), Multi-Carrier Spread-Spectrum, 195–202.

algorithm is proposed for orthogonal frequency division multiplexing (OFDM) systems with transmitter diversity using space-frequency block coding. In the development of the MAP channel estimation algorithm, the channel taps are assumed to be random processes. Moreover, orthogonal series representation based on the Karhunen-Loeve expansion of a random process is applied which makes the expansion coefficient r.v.'s uncorrelated. Thus, the algorithm estimates the uncorrelated complex expansion coefficients iteratively using the Expectation Maximization(EM) method.

2. ALAMOUTI'S TRANSMIT DIVERSITY SCHEME FOR OFDM SYSTEMS

In this paper, we consider a transmitter diversity scheme in conjunction with OFDM modulation. It employs the Alamouti STBC system with 2 transmit antennas and 1 receive antenna, where utilizing N_c subcarriers per antenna transmissions. The fading channel between the μ th transmit antenna and the receive antenna is assumed to be frequency selective but time-flat and is described by the discrete-time baseband equivalent impulse response $\mathbf{h}_\mu = [h_{\mu,0}(n),\cdots,h_{\mu,L}(n)]$ standing for the channel order.

Let $\mathbf{A}_{k,\mu}(n)$ be the data symbol transmitted on the k th subcarrier frequency (frequency bin) from the μ th transmit antenna during the n th OFDM symbol interval. As defined, the symbols $\{\mathbf{A}_{k,\mu}(n),\ \mu=1,2\ ,\ k=0,1,\cdots,N_c\}$ are transmitted in parallel on N_c subcarriers by 2 transmit antennas. The generation of $\mathbf{A}_{k,\mu}(n)$ from the information symbols lead to corresponding transmit diversity OFDM scheme. In our system, the generation of $\mathbf{A}_{k,\mu}(n)$ is performed via space-frequency coding, which was first suggested in [3].

We consider a strategy which basically consists of coding across OFDM tones and is therefore called space-frequency coding. Since an OFDM communication system can be considered as a block transmission system, the serial input data symbols is converted into a data vector $\mathbf{A}(n)=\left[A_0(n),A_1(n),\cdots,A_{N_c-1}(n)\right]^T$. The space-frequency encoder then codes data symbol vector into two vectors $\mathbf{A}_1(n)$ and $\mathbf{A}_2(n)$ as $\mathbf{A}_1(n)=\left[A_0(n),-A_1^*(n),\cdots,A_{N_c-2}(n),-A_{N_c-1}^*(n)\right]^T$ $\mathbf{A}_2(n)=\left[A_1(n),A_0^*(n),\cdots,A_{N_c-1}(n),A_{N_c-2}^*(n)\right]^T$ respectively. In space-frequency Alamouti scheme, $\mathbf{A}_1(n)$ and $\mathbf{A}_2(n)$ are transmitted through the first and second antenna element respectively during the block instant n.

The operations of the space-frequency encoder can best be described in terms of even and odd polyphase component vectors. If the received signal sequence is parsed in even and odd blocks of N_c tones, $\mathbf{R}_e=\left[R_0(n),R_2(n),\cdots,R_{N_c-2}(n)\right]^T$ and

$\mathbf{R}_o = \left[R_1(n), R_3(n), \cdots, R_{N_c-1}(n)\right]^T$, the received signal can be expressed in vector form as

$$\mathbf{R}_e(n) = A_e(n)\mathbf{H}_{1,e} + A_o(n)\mathbf{H}_{2,e} + \mathbf{W}_e(n)$$
$$\mathbf{R}_o(n) = -A_o^*(n)\mathbf{H}_{1,o} + A_e^*(n)\mathbf{H}_{2,o} + \mathbf{W}_o(n) \tag{1}$$

or more in succinct form

$$\mathbf{R}(n) = \mathbf{A}(n)\mathbf{H}(n) + \mathbf{W}(n) \tag{2}$$

where $A_e(n)$ and $A_o(n)$ are an $N_c/2 \times N_c/2$ diagonal matrices with $diagA_e(n) = \mathbf{A}_e$ and $diagA_o(n) = \mathbf{A}_o$ respectively. $\mathbf{H}_{\mu,e}(n) = \left[H_{\mu,0}(n), H_{\mu,2}(n), \cdots, H_{\mu,N_c-2}(n)\right]^T$ $\mathbf{H}_{\mu,o}(n) = \left[H_{\mu,1}(n), H_{\mu,3}(n), \cdots, H_{\mu,N_c-1}(n)\right]^T$ be $N_c/2$ length vectors denoting the even and odd component vectors of the channel attenuations between the μ th transmitter and the receiver. Finally, $\mathbf{W}_e(n)$ and $\mathbf{W}_o(n)$ are an $N_c/2 \times 1$ zero-mean, i.i.d. Gaussian vectors that model additive noise in the N_c tones.

Equation (1) shows that the information symbols $A_e(n)$ and $A_o(n)$ are transmitted twice in two consecutive adjacent subchannel groups through two different channels. In order to estimate the channels and decode A with the embedded diversity gain through the repeated transmission, for each n, we can write the following from (1) according to assumption $\boldsymbol{H}_{1,e} \approx \boldsymbol{H}_{1,o}$, *and* $\boldsymbol{H}_{1,e} \approx \boldsymbol{H}_{1,o}$

$$\begin{bmatrix} \mathbf{R}_e(n) \\ \mathbf{R}_o(n) \end{bmatrix} = \begin{bmatrix} A_e(n) & A_o(n) \\ -A_o^*(n) & A_e^*(n) \end{bmatrix} \begin{bmatrix} \mathbf{H}_{1,e} \\ \mathbf{H}_{2,e} \end{bmatrix} + \begin{bmatrix} \mathbf{W}_e(n) \\ \mathbf{W}_o(n) \end{bmatrix} \tag{3}$$

Based on the model (3), our objective in this paper is to develop a channel estimation algorithm according to the MAP criterion. A different approach is adapted here to explicitly model the channel parameters by the Karhunen-Loeve (KL) series representation since, KL expansion allows one to tackle the estimation of correlated parameters as a parameter estimation problem of the uncorrelated coefficients.

4. EM-BASED MAP CHANNEL ESTIMATION

In the MAP estimation approach we choose $\hat{\mathbf{G}}$ to maximize the posterior probability density function (PDF) or

$$\hat{\mathbf{G}} = \arg\max_{\mathbf{G}} \left[In\, p(\mathbf{R}/\mathbf{G}) + In\, p(\mathbf{G}) \right] \tag{4}$$

where $\boldsymbol{G} = \left[\boldsymbol{G}_{1,e}^{T}, \boldsymbol{G}_{1,e}^{T} \right]^{T}$.

Given the transmitted signals $\mathbf{A}$ as coded according to space-frequency transmit diversity scheme and the discrete channel orthonormal series expansion representation coefficients $\boldsymbol{G}$ and taking into account the independence of the noise components, the conditional PDF of the received signal $\boldsymbol{R}$ can be expressed as,

$$p(\mathbf{R}/\mathbf{A},\mathbf{G}) \approx \exp\left[-(\mathbf{R} - \mathbf{A}\tilde{\psi}\mathbf{G})^{H} \boldsymbol{\Sigma}^{-1} (\mathbf{R} - \mathbf{A}\tilde{\psi}\mathbf{G}) \right] \tag{5}$$

where $\tilde{\boldsymbol{\Sigma}} = diag(\,\boldsymbol{\Sigma}\,\boldsymbol{\Sigma})$ and $\boldsymbol{\Sigma}$ is an $N \times N$ diagonal matrix with $\boldsymbol{\Sigma}[\mathbf{k},\mathbf{k}] = \sigma^{2}$, $for\ k = 0,1,\cdots,N-1$ and $\tilde{\psi} = diag(\boldsymbol{\psi},\boldsymbol{\psi})$

Obtaining MAP estimate of $\boldsymbol{G}$ from (5) is a complicated optimization problem and does not yield to a closed form solution. Solutions of such problems usually requires numerical methods, we therefore use the iterative EM algorithm. This algorithm inductively reestimate $\boldsymbol{G}$ so that a monotonic increase in the *a posteriori* conditional pdf in (4) is guaranteed. The monotonic increase is realized via the maximization of the auxiliary function

$$Q(\mathbf{G}/\mathbf{G}^{(i)}) = \sum_{A} p(\mathbf{R},\mathbf{G},\mathbf{A}) \log p(\mathbf{R},\mathbf{A},\mathbf{G}^{(i)}) \tag{6}$$

where $\boldsymbol{G}^{(i)}$ is the estimation of $\boldsymbol{G}$ at the i th iteration.

Note that $p(\mathbf{R},\mathbf{G},\mathbf{A}) \approx p(\mathbf{R}/\mathbf{A},\mathbf{G})p(\mathbf{G})$ since the data symbols $\boldsymbol{A} = \left\{ A_{k,\mu}(n) \right\}$ are assumed to be independent of each other and identically distributed and the fact that $\boldsymbol{A}$ is independent of $\boldsymbol{G}$. Therefore, (6) can be evaluated by means of the expressions (4) and (5).

Given the received signal $\boldsymbol{R}$ the EM algorithm starts with an initial value $\mathbf{G}^{(0)}$ of the unknown channel parameters $\boldsymbol{G}$. The $(i+1)$ th estimate of $\boldsymbol{G}$ is obtained by the $\mathbf{G}^{(i+1)} = \arg\max_{G} Q(\mathbf{G}/\mathbf{G}^{(i)})$ maximization step described by . Taking the pilot symbols into account, after long algebraic manipulations, the expression of the reestimate $\boldsymbol{G}_{\mu,e}^{(i+1)}(n)$ $(\mu = 1,2)$ can be obtained as follows:

$$\mathbf{G}_{1,e}^{(i+1)} = (\mathbf{I} + \Sigma\boldsymbol{\Lambda}^{-1})^{-1} \psi^{H} \left[\mathbf{V}_{1}^{(i)} \mathbf{R}_{e}(n) - \mathbf{V}_{2}^{\dagger(i)} \mathbf{R}_{o}(n) \right]$$

$$\mathbf{G}_{2,e}^{(i+1)} = (\mathbf{I} + \Sigma\boldsymbol{\Lambda}^{-1})^{-1} \psi^{H} \left[\mathbf{V}_{2}^{(i)} \mathbf{R}_{e}(n) + \mathbf{V}_{1}^{\dagger(i)} \mathbf{R}_{o}(n) \right] \tag{7}$$

where $(\mathbf{I}+\Sigma\mathbf{\Lambda}^{-1})^{-1} = diag(\left[\ (1+\sigma^2/\lambda_{\mu,0})^{-1},\cdots,(1+\sigma^2/\lambda_{\mu,N_c-2})^{-1}\ \right])$

$\mathbf{V}_l^{(i)} = diag\left[v_\mu^{(i)}(0), v_\mu^{(i)}(2),\cdots,v_\mu^{(i)}(N_{c-2})\right]$ and $\upsilon_\mu^{(i)}(k)$, is given as

$$v_1^{(i)}(k) = \begin{cases} A_{k,1}^*(n); & k \in S_{PS} \\ \Gamma_1^{(i)}(k); & k \in S_{PS}^c \end{cases} \quad v_2^{(i)}(k) = \begin{cases} A_{k,2}^*(n); & k \in S_{PS} \\ \Gamma_2^{(i)}(k); & k \in S_{PS}^c \end{cases}$$

where denotes the set of pilot symbols indices. Here, for $k \in S_{PS}^c, \mathbf{\Gamma}_\mu^i(k)$, represents the *a posteriori* probabilities of the data symbols at the i th iteration step and is defined by $\mathbf{\Gamma}_\mu^i(k) = \sum_{a_1,a_2 \in S_k} a_\mu^* P(\mathbf{A}_{k,1}(n) = a_1, \mathbf{A}_{k,1}(n) = a_1 / \mathbf{R}, \mathbf{G}^{(i)})$ and S_k denotes alphabet set taken by the k th OFDM symbol.

5. SYMBOL-ERROR RATE

In order to decode $\mathbf{A}_{k,e}(n)$ and $\mathbf{A}_{k,o}(n)$ with the embedded diversity gain through the repeated transmission, let us use (1) to write $\tilde{\boldsymbol{R}}_k(n)=\left[R_{k,e}(n), R_{k,o}^*(n)\right]^T$ as

$$\tilde{\boldsymbol{R}}_k(n) = \tilde{\boldsymbol{H}}_k \tilde{\boldsymbol{A}}_k(n) + \tilde{\boldsymbol{W}}_k(n) \tag{8}$$

where $\tilde{\boldsymbol{A}}_k(n) = \left[A_{k,e}(n), A_{k,o}(n)\right]^T$, $\tilde{\boldsymbol{W}}_k(n) = \left[W_{k,e}(n), W_{k,o}^*(n)\right]^T$

$$\tilde{\boldsymbol{H}}_k = \begin{bmatrix} \boldsymbol{H}_{1,k,e} & \boldsymbol{H}_{2,k,e} \\ \boldsymbol{H}_{2,k,o}^* & -\boldsymbol{H}_{1,k,o}^* \end{bmatrix} \tag{9}$$

where $\boldsymbol{H}_{\mu,k,e}$ and $\boldsymbol{H}_{\mu,k,o}$ are k th element of $\boldsymbol{H}_{\mu,e}$ and $\boldsymbol{H}_{\mu,o}$ respectively.

Depending on complexity versus performance tradeoffs, any linear equalizer can be applied to retrieve $\tilde{\boldsymbol{A}}_k(n)$ from (8). For example we may choose the ZF equalizer which is given by $\mathbf{Z}_k(n) = \tilde{\mathbf{H}}_k^H \tilde{\mathbf{R}}_k(n)$. When we adopt zero-forcing equalizer, theoretical SER evaluation is possible for a given constellation assuming $\tilde{\boldsymbol{W}}_k(n)$ AWGN with variance σ^2. Starting with (8), the decision vector $\mathbf{Z}_k(n)$ for $\tilde{\boldsymbol{A}}_k(n)$ is given by $\mathbf{Z}_k(n) = \tilde{\mathbf{H}}_k^H \tilde{\mathbf{R}}_k(n) = \tilde{\mathbf{H}}_k^H \tilde{\mathbf{H}}_k \tilde{\mathbf{A}}_k(n) + \boldsymbol{\eta}_k(n)$ where $\boldsymbol{\eta}_k(n) = \tilde{\mathbf{H}}_k^H \tilde{\mathbf{W}}_k(n)$.

To proceed further with our performance analysis, we assume QPSK modulated transmissions and use average SER defined as $\overline{P}_e = \frac{1}{2N_c}\sum_{k=0}^{N_{c-1}}\sum_{\mu=1}^{2}\overline{P}_{\mu,k}$ where $\overline{P}_{1,k}$ and $\overline{P}_{1,k}$ denote the SER's for the sequences $\mathbf{A}_{k,e}(n)$ and $\mathbf{A}_{k,o}(n)$ respectively.

SER is proportional to the square root of the instantaneous SNR, the corresponding SNR at the ZF output is thus

$$SNR = \frac{E\left\{\left|\tilde{\mathbf{H}}_k^H \tilde{\mathbf{H}}_k \tilde{\mathbf{A}}_k(n)\right|^2\right\}}{E\left\{\left|\tilde{\mathbf{H}}_k^H \tilde{\mathbf{W}}_k(n)\right|^2\right\}} = \frac{E_s \left|\tilde{\mathbf{H}}_k^H \tilde{\mathbf{H}}_k\right|^2}{\mathbf{C}_\eta} \tag{10}$$

where $\mathbf{C}_\eta = E\{\eta_k(n)\eta_k^H(n)\}$ and E_s denote the symbol energy of $\mathbf{A}_{k,\mu}(n)$.

With a scaled unitary matrix $\tilde{\boldsymbol{H}}_k$ and approximately constant complex channel gains assumptions, we can simplify $\tilde{\mathbf{H}}_k^H \tilde{\mathbf{H}}_k$ as $\tilde{\boldsymbol{H}}_k^H \tilde{\boldsymbol{H}}_k = \left\{\left|\boldsymbol{H}_{1,k,e}\right|^2 + \left|\boldsymbol{H}_{2,k,e}\right|^2\right\} \boldsymbol{I}_{2\times 2}$ and where $\boldsymbol{I}_{2\times 2}$ is the 2×2 identity matrix. Furthermore, the covariance matrix $\mathbf{C}_\eta$ can be computed as $\mathbf{C}_\eta = \left\{\left|\boldsymbol{H}_{1,k,e}\right|^2 + \left|\boldsymbol{H}_{2,k,e}\right|^2\right\} \sigma^2 \boldsymbol{I}_{2\times 2}$.

Based on (10), the corresponding closed form SER expressions for QPSK can be written as

$$\overline{P}_e = \frac{1}{N_c} \sum_{k=0}^{N_c - 1} Q\left(\sqrt{\frac{\left|\mathbf{H}_{1,k,e}\right|^2 + \left|\mathbf{H}_{2,k,e}\right|^2 E_s}{\sigma^2}}\right) \tag{11}$$

where $Q(.)$ denotes Q function.

6. MODIFIED-CRAMER-RAO BOUND

Let $L(\mathbf{R};\mathbf{G}) = In\, p(\mathbf{R}/\mathbf{G}) + In\, p(\mathbf{G})$ then the FIM is

$$J(\mathbf{G}) = E\left[\left(\frac{\partial L(\mathbf{R};\mathbf{G})}{\partial \mathbf{G}}\right)\left(\frac{\partial L(\mathbf{R};\mathbf{G})}{\partial \mathbf{G}}\right)^T\right]. \tag{12}$$

Assuming $\mathbf{A}_{k,\mu}(n)$'s are adopting finite complex values, $p(\mathbf{R}/\mathbf{G})$ can be obtained after averaging $p(\mathbf{R}/\mathbf{A},\mathbf{G})$ over all A's: $p(\mathbf{R}/\mathbf{G}) = E_A\{p(\mathbf{R}/\mathbf{A},\mathbf{G})\}$ which is computationally intensive. However, approximate $p(\mathbf{R}/\mathbf{A},\mathbf{G})$ can still be obtained from $p(\mathbf{R}/\mathbf{A},\mathbf{G})$.Since the logarithmic function is concave, we have by Jensen's inequality $In\, p(\boldsymbol{R/G}) \le E_A In\{p(\boldsymbol{R}/\boldsymbol{A},\boldsymbol{G})\}$. Therefore, we get a valid CRB from $E_A\{In(\mathbf{R}/\mathbf{A},\mathbf{G})\} + Inp(\mathbf{G})$, which may not be tight but is much easier to compute.

Let us then introduce the approximate MAP cost function

$$L \approx -\frac{1}{\partial^2} E_A \left\{ (R - A\tilde{\psi}G)^\dagger (R - A\psi G) \right\} - G^\dagger \tilde{\Lambda}^{-1} G \tag{13}$$

We now start constructing the FIM by calculating the derivative of (13) with respect to $\mathbf{G}' = \left[\mathbf{G}_{re}^T, \mathbf{G}_{im}^T \right]^T$ where $\mathbf{G}_{re}^T = \mathrm{Re}\left\{ \mathbf{G}^T \right\}$, $\mathbf{G}_{im}^T = \mathrm{Im}\left\{ \mathbf{G}^T \right\}$. Taking the partial derivatives of (13), we have

$$\frac{\partial \mathbf{L}}{\partial \mathbf{G}_{re}} = \mathrm{Re}\left\{ \frac{2}{\partial^2} \tilde{\psi}^{\mathrm{H}} \mathbf{A}^{\mathrm{H}} \mathbf{W} - 2\tilde{\Lambda}^{-1}\mathbf{G} \right\}, \quad \frac{\partial \mathbf{L}}{\partial \mathbf{G}_{im}} = \mathrm{Re}\left\{ \frac{2}{\partial^2} \tilde{\psi}^{\mathrm{H}} \mathbf{A}^{\mathrm{H}} \mathbf{W} - 2\tilde{\Lambda}^{-1}\mathbf{G} \right\} \tag{14}$$

Using the results proven in [7] and partial derivatives above, the information matrix can be obtained as

$$J(G) = \begin{bmatrix} 2\left\{ \frac{2}{\sigma^2}\mathbf{I} + \tilde{\Lambda}^{-1} \right\} & 0 \\ 0 & 2\left\{ \frac{2}{\sigma^2}\mathbf{I} + \tilde{\Lambda}^{-1} \right\} \end{bmatrix} \tag{15}$$

since the Alamouti's scheme imposes an orthogonal structure on the transmitted symbols, i.e., $E_A \left\{ \mathbf{A}^{\mathrm{H}} \mathbf{A} \right\} = 2\mathbf{I}$. This property is quite useful in that becomes diagonal matrix which is easily inverted.

5.SIMULATION

In this section, we presented simulation results for the performance analysis of the proposed iterative channel estimation algorithm. We choose the symbol error rate (SER) as well as mean square error (MSE) as our figure of merit to investigate the performance. The scenario for our SF-OFDM simulation study consists of a wireless QPSK OFDM system. The system has a 2MHz bandwidth and is divided into 256 tones with a total period (Ts) of 136 μ s, of which 8 μ s constitute the cyclic prefix (L=4). The uncoded data rate is 3.76 Mbit/s. We assume that the rms width is $\tau_{rms} = 4$ for the power-delay profile. We compare the MSE performance of the SF-OFDM channel estimator with ST-OFDM for different Doppler frequencies f_d =50,100,200 Hz . In addition to MSE results, we also provide modified Cramer-Rao Bound (CRB). In the full paper, we will give SER performance comparison of ST-OFDM and SF-OFDM systems and test our proposed approach for different scenarios.

Fig.1(a) presents SER performance of SF-OFDM systems according to both theoretical and experimental results. Moreover, the average MSE performance of the iterative channel estimation algorithm as a function of the average SNR is presented in (b) for different Doppler frequencies. From our simulations, it is observed that SF-OFDM shows better performance than ST-OFDM scheme in the more

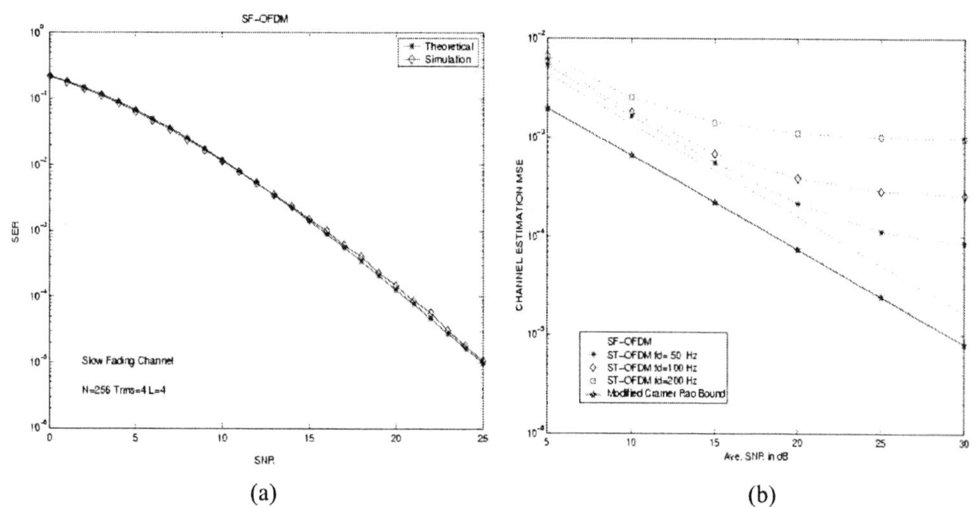

(a) (b)

Fig. 1. (a) SER performance of SF-OFDM systems as a function of average SNR (b) MSE performance of the iterative channel estimation algorithm as a function of average SNR

7. REFERENCES

[1] Y. (G.) Li, L. J. Cimini, N. Seshadri and S. Ariyavistakul, ``Channel estimation for OFDM systems with transmitter diversity in mobile wireless channels", *IEEE Journal Selected Areas on Commun*, vol. 17, no.3, pp. 461-471, July 1999.

[2] K.F. Lee and D.B. Douglas,``A Space-Frequency Transmitter Diversity Technique for OFDM Systems", *in Proc. IEEE Globecom*, San Francisco, CA, November 2000, pp. 1473-1477.

[3] S. M. Alamouti, ``A simple transmitter diversity scheme for wireless communications", *IEEE J. Selected Areas Commun.* , vol. 16, pp. 1451-1458, October 1998.

[4] V. Tarokh, N. Seshadri, and A. R. Calderbank ``Space-time codes for high data rate wireless communications: Performance analysis and code construction,", *IEEE Trans.Inform. Theory*, vol. 44, pp. 744-765, March 1998.

[5] V. Tarokh, Hamid Jafarkhani, and A. R. Calderbank, ``Space-time block Codes from orthogonal designs", *IEEE Trans. Inform. Theory*, vol. 45, no.5 pp. 1456-1467, July 1999.

[6] K. Yip and T. Ng, "Karhunen-Loeve Expansion of the WSSUS Channel Output and its Application to Efficient Simulation," *IEEE Journal on Selected Areas in Communications*, vol. 15, no 4, pp.640-646, May 1997.

[7] Stoica, P.; Nehorai, A.: MUSIC, Maximum Likelihood, and Cramer-Rao Bound. *IEEE Trans. on Acoustics, Speech, and Signal Processing,* May 1989, vol.37, no.5, 720-741.

[8] A.P. Dempster, N.M. Laird and D. B. Rubin, "Maximum Likelihood from Incomplete data via the EM algorithm. *Roy. Statist. Soc.*, vol.39, pp. 1-17, 1977

[9] E. Panayirci and H. A. Cirpan, "Maximum A Posteriori Multipath Fading Channel Estimation for OFDM Systems," *European Trans. on Telecommunications (ETT):* Special Issue on Multi Carrier Spread Spectrum & Related, Vol. 13, No. 5, pp. 487-494 September/October 2002.

COMPARISON OF PILOT MULTIPLEXING SCHEMES FOR ML CHANNEL ESTIMATION IN CODED OFDM-CDMA

Martin Feuersänger, Florian Hasenknopf, Volker Kühn and Karl-Dirk Kammeyer

Abstract As bandwidth is becoming precious, spectral efficiency is nowadays one of the key parameters of mobile radio communications. To reach high spectral efficiency in Code Division Multiple Access systems, combating multi-user interference is inevitable. Especially in an uplink scenario this can be done by applying multi-user detection. This paper compares two multiplexing schemes for pilot data in a system with combined maximum likelihood channel estimation (MLCE) and successive interference cancellation (SIC).

It is shown that even though preceding pilots deliver a better initial channel estimate a scheme with IQ-mapped pilots leads to better performance when using a combined MLCE-SIC receiver.

1. Introduciton

Today Code Division Multiple Access (CDMA) is widely used as an efficient scheme to acquire multiple access to mobile radio channels. In combination with OFDM (Orthogonal Frequency Division Multiplex) a Multi-Carrier-CDMA realization can be achieved where chips are only affected by flat fading [1, 2] thus leading to efficient transmitter and receiver structures.

Interference caused by access of multiple users (MUI) is therefore the prime reason for a limited system capacity in an uplink scenario. Multi-user detection (MUD) [3, 4], especially when integrated in iterative structures [5, 6, 7, 8, 9], largely improves spectral efficiency of MUI-degraded systems.

In this paper coarse synchronization is assumed thus leading to a quasi-synchronous system where remaining asynchronism is compensated by the inherent cyclic prefix. Nonlinear successive interference cancellation (SIC) is applied to combat multi-user interference. A maximum likelihood channel estimation (MLCE) and SIC are combined as described in [10] yielding iteratively improved channel and data estimates. Transmission is organized in frames where each frame again is split in a number of fading blocks where the number depends on the

K. Fazel and S. Kaiser (eds.), Multi-Carrier Spread-Spectrum, 203–210.

channel characteristics.
We compare two approaches of training data structures for channel estimation. In a first setup we apply pilot symbols preceding data symbols of each fading block thus generating a time multiplexed system of pilots and data (TM-system). In a second setup, the pilot data is mapped onto the imaginary part of the QPSK symbols which is based on training structure setup of UMTS [11] - we will refer to it as IQ-multiplexed system (IQ-system)
The paper is organized as follows: Section 2 describes the OFDM-CDMA system. Next, section 3 explains the combination of channel estimation with SIC and their adaptation to the two different training data realizations. In section 4 we present simulation results of these two realizations and analyze their performance. Section 5 summarizes the results of this paper.

2. OFDM-CDMA System

Figure 1 depicts the considered OFDM-CDMA transmitter. Each user u is transmitting L_d information bits $\mathbf{d}_u$ that are encoded by identical convolutional codes of rate $R_c = 1/2$ and constraint length $L_c = 3$. These coded bits $\mathbf{b}_u$ are interleaved by user-specific interleavers $\Pi_t^{(u)}$ with a length of $L_b = 2L_d$.
While $\mathbf{b}_u$ is mapped onto QPSK symbols in the TM case, the first $L_{p,\text{IQ}}$ QPSK symbols of the IQ-system consist of $L_{p,\text{IQ}}$ bits out of $\mathbf{b}_u$ determining the real part and $L_{p,\text{IQ}}$ pilot bits out of $\mathbf{p}_u$ for the imaginary part. The remaining $L_b - L_p$ bits in $\mathbf{b}_u$ are mapped to QPSK symbols . This leads to $L_{p,\text{IQ}} = 2L_{p,\text{TM}}$ OFDM symbols in the IQ-system carrying pilot data.
The complex QPSK symbols are spread separately in real and imaginary part by a factor of $N_s = 32$ using pseudo random long codes $\mathbf{c}_u$. The load of the system is defined as $\beta = U/N_s$ where U is the number of users in the system $(1 \leq u \leq U)$.
The spread sequence $\mathbf{x}_u$ is then transformed into the time domain. The number

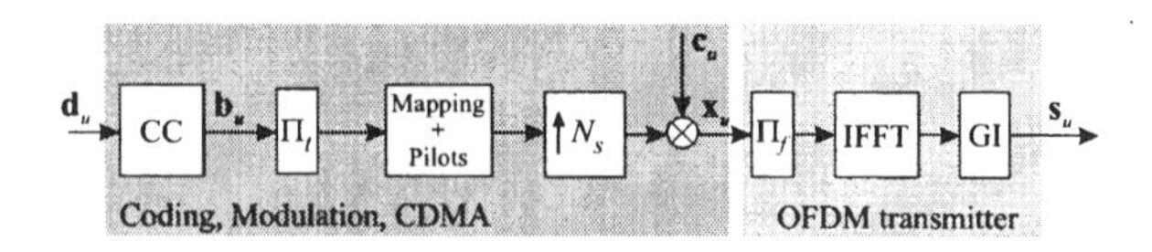

Figure 1. Block diagram of OFDM-CDMA transmitter

of carriers N_c is equal to $N_c = nN_s = 64$. After interleaving over N_c chips in frequency direction an inverse Fourier transform is applied. Each OFDM symbol is preheaded by a cyclic prefix with the duration T_g. The resulting signal $\mathbf{s}_u$ is then transmitted over the channel. After inversing the transform at the receiver this implies that each chip is only affected by multiplicative flat fading.

2.1 Channel Model

As we are investigating an uplink scenario all users are transmitted over U individual 4-path Rayleigh block fading channels. The impulse response of the channel remains constant over a fading block consisting of L_f OFDM symbols. Impulse responses of successive fading blocks are statistically independent. At the receiver, the cyclic prefix is removed. One FFT window is adequate for transforming all users back into the frequency domain when assuming that the guard interval compensates residual delays which are left over after coarse synchronization. This leads to the received vector $\mathbf{r}$.

2.2 Simulation Parameters and Pilot Setup

We assume a signal bandwidth of $B = 5$ MHz, i.e. OFDM symbols are $T_s = 6.4\ \mu$s in duration. The chosen block length is $L_f = 20$ OFDM symbols. The proportions of pilots to data are based on slot format #3 of the uplink FDD mode defined in [11], i.e. a frame incorporates 35% training data. Hence, the number of QPSK training symbols for the TM-system equals $L_{p,\text{TM}} = 7$. In the IQ-system 70% of the QPSK symbols carry training information in the imaginary part. The maximum number of iterations N_{it} for the SIC is 16.

3. Combined Channel Estimation and Multi-user Detection

Figure 2 shows a schematic of the combination of channel estimation and successive interference cancellation. After initial ML channel estimation over all U channel impulse responses the first SIC iteration is executed. After FEC decoding of all user signals the reconstructed data is fed back to the channel estimation where it is used as additional pseudo-training data for an improved channel estimation that can now exploit the whole length of the fading block. Based on these estimates the next iteration of the SIC is executed. This process is repeated until no further improvement is achieved or the maximum number of iterations N_{it} is reached.

3.1 Initial Channel Estimation

The initial estimate of the U user-specific channels differs for the TM- and the IQ-system. Training data is obtained from different positions in the QPSK symbol-stream. The training data of the IQ-system is additionaly disturbed by the data component in the real part of the QPSK symbols.

Let $\mathbf{P}$ be a $L_{p,\text{TM}} \times U$ resp. $L_{p,\text{IQ}} \times U$ matrix containing as diagonal submatrices those OFDM symbols carrying training information. $\mathbf{T}_{\text{DFT}}\mathbf{h}$ represents the channel transfer function in its time representation taking into account a L-tap

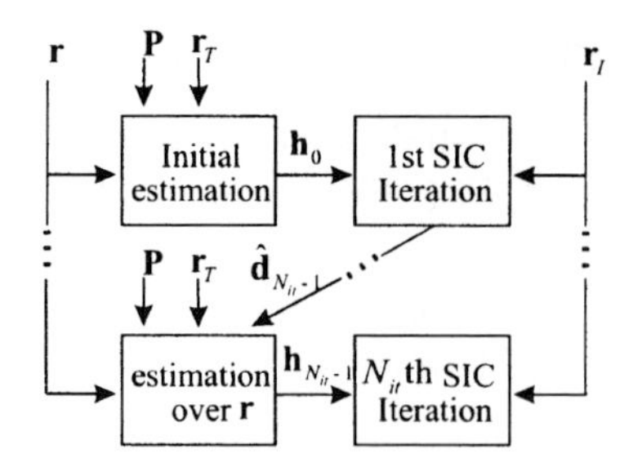

Figure 2. Combined ML Channel Estimation and Successive Interference Cancellation

impulse response. For the TM-system we receive during the training period

$$\mathbf{r}_{T,\mathrm{TM}} = \mathbf{P}\mathbf{T}_{\mathrm{DFT}}\mathbf{h} + \mathbf{n}_T \tag{1}$$

If we define $\tilde{\mathbf{P}} := \mathbf{P}\mathbf{T}_{\mathrm{DFT}}$, the maximum likelihood (ML) estimate of the channel impulse response for the TM-system is

$$\hat{\mathbf{h}}_{\mathrm{MLCE,TM}} = (\tilde{\mathbf{P}}^H\tilde{\mathbf{P}})^{-1}\tilde{\mathbf{P}}^H\mathbf{r}_T = \mathbf{h} + (\tilde{\mathbf{P}}^H\tilde{\mathbf{P}})^{-1}\tilde{\mathbf{P}}^H\mathbf{n}_T \tag{2}$$

showing that the estimate of h is disturbed by a modified AWGN term $\tilde{\mathbf{n}}_{T,\mathrm{TM}} = (\tilde{\mathbf{P}}^H\tilde{\mathbf{P}})^{-1}\tilde{\mathbf{P}}^H\mathbf{n}_T$. For the IQ-system we receive during the training period

$$\mathbf{r}_{T,\mathrm{IQ}} = (\mathbf{X} + j\mathbf{P})\mathbf{T}_{\mathrm{DFT}}\mathbf{h} + \mathbf{n}_T, \tag{3}$$

where $\mathbf{X}$ represents the unknown data in the real part of the QPSK symbols. The ML channel estimate for the IQ-system can now be derived as

$$\hat{\mathbf{h}}_{\mathrm{MLCE,IQ}} = -j(\tilde{\mathbf{P}}^H\tilde{\mathbf{P}})^{-1}\tilde{\mathbf{P}}^H\mathbf{r}_T. \tag{4}$$

The IQ-system contains additional interference caused by unknown data in the real part of the QPSK symbols containing also the training data. The initial estimate of the IQ-system is of the form

$$\hat{\mathbf{h}}_{\mathrm{MLCE,IQ}} = \mathbf{h} - j(\tilde{\mathbf{P}}^H\tilde{\mathbf{P}})^{-1}\tilde{\mathbf{P}}^H\tilde{\mathbf{X}} - j(\tilde{\mathbf{P}}^H\tilde{\mathbf{P}})^{-1}\tilde{\mathbf{P}}^H\mathbf{n}_T. \tag{5}$$

Just like $\tilde{\mathbf{P}}$ we have defined $\tilde{\mathbf{X}} := \mathbf{X}\mathbf{T}_{\mathrm{DFT}}$. Besides the modified AWGN contribution $\tilde{\mathbf{n}}_{T,\mathrm{IQ}} = -j(\tilde{\mathbf{P}}^H\tilde{\mathbf{P}})^{-1}\tilde{\mathbf{P}}^H\mathbf{n}_T$ we get an aditional interference term $\tilde{\mathbf{n}}_{X,\mathrm{IQ}} = -j(\tilde{\mathbf{P}}^H\tilde{\mathbf{P}})^{-1}\tilde{\mathbf{P}}^H\tilde{\mathbf{X}}$. Since $\tilde{\mathbf{P}}$ for the IQ-system is twice as long for the TM-system but each element contains only half power since only the imaginary part is considered we can state that $\tilde{\mathbf{n}}_{T,\mathrm{IQ}}$ and $\tilde{\mathbf{n}}_{T,\mathrm{TM}}$ have equal power. This leaves the IQ-system with the additional interference term $\tilde{\mathbf{n}}_{X,\mathrm{IQ}}$ thus suffering from greater degradation than the TM-system. In figure 3 this is shown by comparing the mean squared error (MSE) of the initial channel

estimation for both systems at a SNR of $E_b/N_0 = 10$ dB. The higher the load β the larger the influence of $\tilde{\mathbf{n}}_{X,\mathrm{IQ}}$ gets.
Even though the IQ-system provides a worse initial estimation this estimation is still good enough to ensure convergence of the SIC iterations.

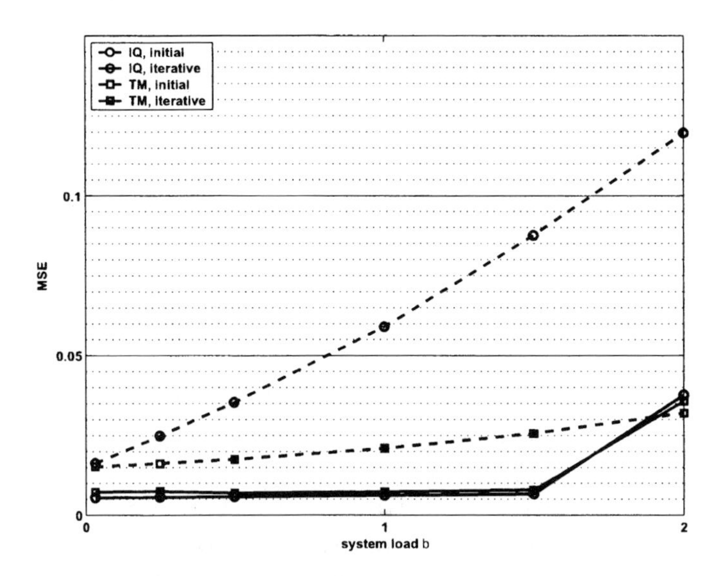

Figure 3. Mean squared error of initial channel estimation for different loads β for $E_b/N_0 =$ 10 dB

3.2 Successive Interference Cancellation

The successive interference cancellation comprises U single user detectors. Users are sorted according to their power level. For the uth user $(1 \leq u \leq U)$ interference for users 1 to $u-1$ determinded at the current iteration step as well as interference for users $u+1$ to U estimated at the previous iteration step are subtracted [10]. The coded bits $\hat{\mathbf{b}}_u$ are then detected by a single user detector and $\hat{\mathbf{d}}_u$ is derived by a FEC decoder processing soft values (SISO decoder). Following this step, $\hat{\mathbf{d}}_u$ is used to soft-reconstruct the received signal in order to determine interference caused by user u. This interference estimation is then used to detect users $u+1$ to U in the same iteration as well as users 1 to $u-1$ in the next iteration step. Due to the decoding gain as well as updated channel estimation the users' bit error rate can be improved in each iteration.
The mean squared error of the iteratively improved channel estimation is as well shown in figure 3.
Detection is only performed on the data part of the signal, i.e. with $\mathbf{r}$ consisting of a training period $\mathbf{r}_T$ and a data period $\mathbf{r}_I$ only $\mathbf{r}_I$ is considered. For the TM-system, this results in neglecting the preceding training part $\mathbf{r}_T$ of $\mathbf{r}$. For

the IQ-system, $\mathbf{r}_T$ contains as well data as training information. The known training information is removed from those symbols leaving the data in the real part thus reducing this part of $\mathbf{r}$ to BPSK symbols. The process of interference cancellation is schematically explained in figure 4.

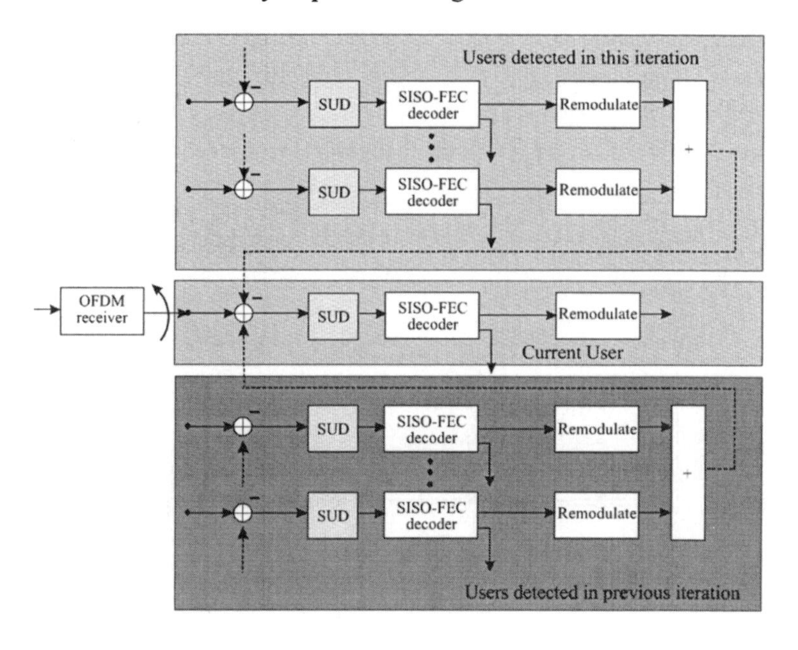

Figure 4. Successive Interference Cancellation

4. Simulation Results

In figure 5 bit error rates for both pilot setups are shown for loads of $\beta = 1$, $\beta = 1.5$ and $\beta = 2$. This corresponds to 32, 48 and 64 users. For $\beta = 1$, the IQ-system performs slightly better than the TM-system in the area between 3 and 8 dB. The gap between IQ- and TM-system is even greater for a load of $\beta = 1.5$. A load of $\beta = 2$ shows the largest performance gain. For a bit error rate of 10^{-3} the IQ-system gains 2 dB compared to the TM-system.
Even though the IQ-system starts with a worse initial channel estimation the iterative update of the channel estimate leads to similar results compared to the TM-system. The average number of iterations required for specific SNRs and different loads β are given in table 1.

However, the IQ-approach removes the training info prior to detection and thus reduces those symbols to BPSK. This applies to 70% of the symbols in an IQ-fading block. A reduction to BPSK corresponds to a reduction of the load by a factor of 2. This is mostly true when the SNR is still conciderably low. For a higher SNR both systems show equal performance since better

Table 1. Average Number of Iterations required for SIC

β	E_b/N_0 in dB	Av. Iter. (TM)	Av. Iter. (IQ)
1	7	13.2	10.4
1.5	9	12.49	11.03
2	13	16	12.5

channel estimation allows better removal of the multi-user interference for the TM-system as well.

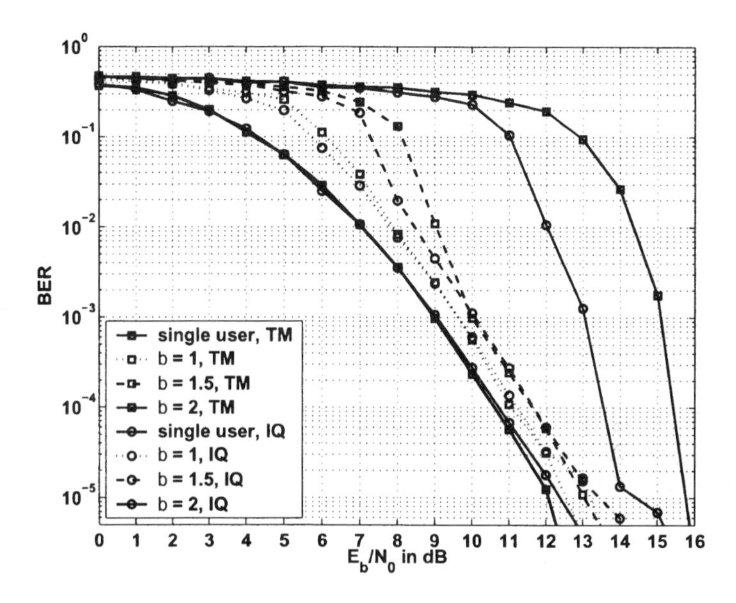

Figure 5. BER performance for the TM- and IQ-system with $L_f = 20$ and different loads β

5. Conclusion

In this paper two systems with different multiplexing schemes for pilot data have been compared. While the TM-system uses preceding QPSK training symbols the IQ-system maps the training information on the imaginary part of a subset of QPSK symbols. Both systems are processed by a combined ML channel estimation and a successive interference cancellation. Even though affected by a worse initial channel estimation the IQ-system shows better bit error performance due to a load reduction since 70% of it's QPSK symbols only carry information in the real part thus suffering by less multi-user interference. The performance gain compared to the TM-system is the better the higher the load β gets. Especially for high loads ($\beta = 2$) the gain of performance is as high as 2 dB.

6. Affiliations

The authors are with the Department of Communications Engineering, Universität Bremen, P.O. Box 33 04 40, D-28334 Bremen, Germany. Email: {feuersaenger,kuehn,kammeyer}@ant.uni-bremen.de

References

[1] Stefan Kaiser, *Multi-Carrier CDMA Mobile Radio Systems – Analysis and Optimization of Detection, Decoding and Channel Estimation*, Ph.D. thesis, German Aerospace Center, VDI, January 1998.

[2] A. Dekorsy and K.D. Kammeyer, "A new OFDM-CDMA Uplink Concept with M-ary Orthogonal Modulation," *European Transactions on Telecommunications (ETT)*, vol. 10, no. 4, pp. 377–390, July/August 1999.

[3] S. Moshavi, "Multi-User Detection for DS-CDMA Communications," *IEEE Communications Magazine*, pp. 124–136, October 1996.

[4] S. Verdu, *Multiuser Detection*, Cambridge University Press, New York, 1998.

[5] L.B. Nelson and V. Poor, "Iterative Multiuser Receivers for CDMA Channels: An EM-based Approach," *IEEE Trans. on Communications*, vol. 44, no. 12, pp. 1700–1710, December 1996.

[6] M.C. Reed, C.B. Schlegel, P.D. Alexander, and J.A. Asenstorfer, "Iterative Multiuser Detection for CDMA with FEC: Near-Single-User Performance," *IEEE Trans. on Communications*, vol. 46, no. 12, pp. 1693–1699, December 1998.

[7] P.D. Alexander, M.C. Reed, J.A. Asenstorfer, and C.B. Schlegel, "Iterative Multiuser Interference Reduction: Turbo CDMA," *IEEE Transactions on Communications*, vol. 47, no. 7, pp. 1008–1014, July 1999.

[8] M. Kobayashi, J. Boutros, and G. Caire, "Successive Interference Cancellation with SISO Decoding and EM Channel Estimation," *IEEE Journal on Selectes Areas in Commununications*, vol. 19, no. 8, pp. 1450–1460, August 2001.

[9] A. Lampe and J. Huber, "Iterative Interference Cancellation for DS-CDMA Systems with High System Loads Using Reliability-Dependent Feedback," *IEEE Trans. on Vehicular Technology*, vol. 51, no. 3, pp. 445–452, May 2002.

[10] V. Kühn, "Iterative Interference Cancellation and Channel Estimation for Coded OFDM-CDMA," in *IEEE International Conference on Communications (ICC)*, Anchorage, Alaska, USA, May 2003.

[11] 3GPP, "Physical channels and mapping of transport channels onto physical channels (FDD)," Tech. Rep. 3GPP TS 25.211 V5.3.0, 3rd Generation Partnership Project, Sophia Antipolis, France, December 2002.

HONGLEI MIAO AND MARKKU J. JUNTTI

DATA AIDED CHANNEL ESTIMATION FOR WIRELESS MIMO–OFDM SYSTEMS

Abstract. Channel estimator for multiple–input multiple–output (MIMO) orthogonal frequency division multiplexing (OFDM) system of the HIPERLAN/2 type with a moderate complexity is proposed. Two types of channel estimators are considered. One uses only time domain pilot symbols, and the other also frequency domain pilot tones. The mean square error performance of the algorithms is analyzed. Their performance is illustrated based on the analysis and computer simulations. The derived algorithms provide good performance with a moderate complexity.

1. INTRODUCTION

Due to complexity constraints, virtually all of today's digital wireless communication systems follow the principle of synchronized detection [1]. It means that a channel estimate is formed and subsequently used for detection and decoding as if it were the true known channel. Channel estimation is known to be a challenging task in wireless communication system in general, and with significant mobility in particular. Channel estimation in single–input single–output (SISO) OFDM systems has been discussed in [2]. Channel estimation for single–input multiple–output (SIMO) OFDM systems has been considered in [3], and MISO–OFDM systems in [4], where also the MIMO–OFDM was briefly addressed. However, a comprehensive treatment of channel estimation a MIMO–OFDM system is not available in the existing open literature. The work on channel estimation for OFDM has mainly concentrated on either frequency [2] or time domain [4] filtering.

In this paper, we study the channel estimation in an MIMO–OFDM system of the HIPERLAN/2 type. We do not restrict our consideration to any particular MIMO signal structure or space-time code, but we formulate a generic MIMO–OFDM system model with pilot symbols and pilot tones, as is the case, e.g., in HIPERLAN/2 standard. This is particularly reasonable, since the MIMO techniques are under intensive study, and only very few standards include specified MIMO features yet; this is the case of the HIPERLAN/2 standard as well. A temporal channel estimator is derived first. It is then extended to utilize also the correlation between OFDM subcarriers. The mean square error (MSE) performance of the derived algorithms is analyzed, and performance is illustrated in several interesting example cases.

2. SYSTEM MODEL

Notations used in this paper are as follows. Upper- and lower-case boldface letters denote matrices and vectors, respectively, $(.)^{\mathrm{T}}$ denotes the transpose, $(.)^{\mathrm{H}}$ denotes the Hermitian transpose. The Kronecker product of matrix $\boldsymbol{A}$ and $\boldsymbol{B}$ is denoted as $\boldsymbol{A} \otimes \boldsymbol{B}$.

K. Fazel and S. Kaiser (eds.), Multi-Carrier Spread-Spectrum, 211–218.

Matrix $\boldsymbol{I}_n$ stands for identity matrix of order n, $\mathrm{diag}(\boldsymbol{a})$ denotes the diagonal matrix whose diagonal is composed of vector $\boldsymbol{a}$. $\boldsymbol{\Sigma}_{\boldsymbol{a}} = \mathrm{E}(\boldsymbol{a}\boldsymbol{a}^{\mathrm{H}})$ denotes the covariance matrix of the random vector $\boldsymbol{a}$, and $\boldsymbol{\Sigma}_{\boldsymbol{a},\boldsymbol{b}} = \mathrm{E}(\boldsymbol{a}\boldsymbol{b}^{\mathrm{H}})$ denotes the covariance matrix of two random vectors $\boldsymbol{a}$ and $\boldsymbol{b}$.

2.1 Channel Model

The channel is assumed to be block-fading with a finite impulse response. It is further assumed that the successive channel snapshots correlate mutually. We also assume that the cyclic prefix (CP) duration is larger than the delay spread of the channels. The channel impulse response (CIR) vector at time n can be expressed

$$\boldsymbol{h}[n] = \Big(h[n,0], h[n,1], \ldots, h[n,L-1]\Big)^{\mathrm{T}} \tag{1}$$

where $L-1$ is the number of the CP samples. The channel frequency response (CFR) at time n can be expressed as

$$\boldsymbol{\theta}[n] = \boldsymbol{F}_K^L \boldsymbol{h}[n] \tag{2}$$

where $\boldsymbol{F}_K^L \in \mathbb{C}^{K\times L}$ is composed of the first left L columns of K-point FFT matrix, and K is the number of the subcarriers in one OFDM symbol.

The CIR correlation matrix $\boldsymbol{\Sigma}_{\boldsymbol{h}}[m] = \mathrm{E}\Big(\boldsymbol{h}[n+m]\boldsymbol{h}^{\mathrm{H}}[n]\Big)$ can be described as

$$\boldsymbol{\Sigma}_{\boldsymbol{h}}[m] = \rho[m]\boldsymbol{\Sigma}_{\boldsymbol{h}} \tag{3}$$

where $\boldsymbol{\Sigma}_{\boldsymbol{h}}$ is termed inter-tap correlation function related to the power delay profile and $\rho[m]$ is the channel autocorrelation function (ACF) depending on the Doppler spectrum. Similarly, the CFR correlation matrix $\boldsymbol{\Sigma}_{\boldsymbol{\theta}}[m] = \mathrm{E}\Big(\boldsymbol{\theta}[n+m]\boldsymbol{\theta}^{\mathrm{H}}[n]\Big)$ can be expressed as

$$\boldsymbol{\Sigma}_{\boldsymbol{\theta}}[m] = \rho[m] \underbrace{\boldsymbol{F}_K^L \boldsymbol{\Sigma}_{\boldsymbol{h}} (\boldsymbol{F}_K^L)^{\mathrm{H}}}_{\boldsymbol{\Sigma}_{\boldsymbol{\theta}}} . \tag{4}$$

Equations (3) and (4) demonstrate the separation properties [5] corresponding to the CIR correlation function and CFR correlation function, respectively.

2.2 MIMO–OFDM System

The block diagram of a MIMO–OFDM system is shown in Fig. 1. There are N_{T} transmit antennas and N_{R} receive antennas. Any channel between one transmitter and one receive antenna is assumed to be independent of the other channels. We denote by $\boldsymbol{h}_{i,j}[n]$ the CIR vector and by $\boldsymbol{\theta}_{i,j}[n]$ the CFR vector of the forms (1) and (2) between the ith transmit antenna and the jth receive antenna during block n. Vector $\boldsymbol{h}_j[n] = \Big(\boldsymbol{h}_{1,j}^{\mathrm{T}}[n], \boldsymbol{h}_{2,j}^{\mathrm{T}}[n], \ldots, \boldsymbol{h}_{N_{\mathrm{T}},j}^{\mathrm{T}}[n]\Big)^{\mathrm{T}} \in \mathbb{C}^{N_{\mathrm{T}}L}$ denotes a concatenation of CIR vectors from all the transmit antennas to the jth receive antenna at time n. Similarly, vector $\boldsymbol{\theta}_j[n] = \Big(\boldsymbol{\theta}_{1,j}^{\mathrm{T}}[n], \boldsymbol{\theta}_{2,j}^{\mathrm{T}}[n], \ldots, \boldsymbol{\theta}_{N_{\mathrm{T}},j}^{\mathrm{T}}[n]\Big)^{\mathrm{T}} \in \mathbb{C}^{N_{\mathrm{T}}K}$ is the

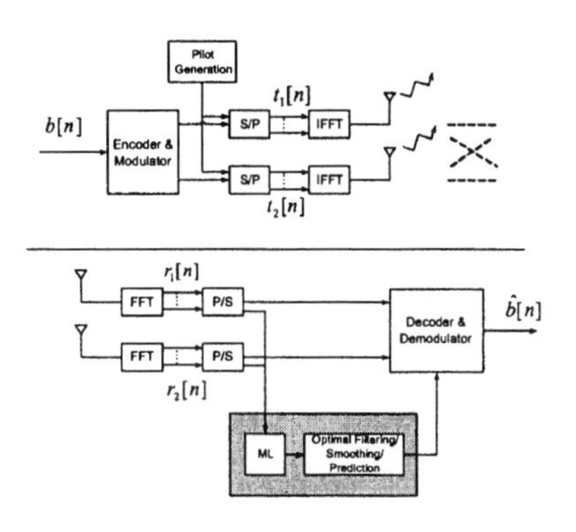

Figure 1. MIMO–OFDM system.

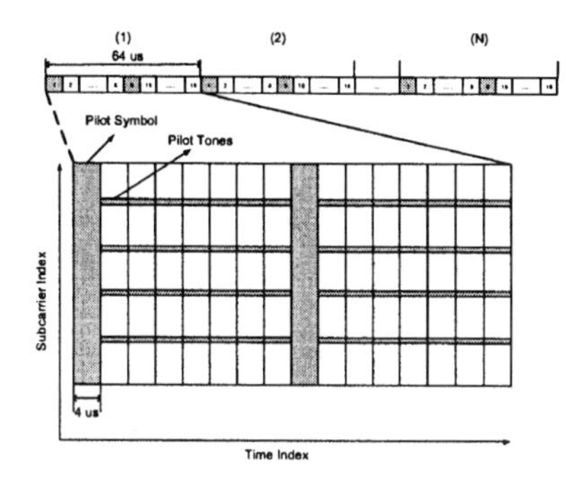

Figure 2. OFDM frame structure, the pilot symbols and tones are shadowed.

corresponding CFR vector concatenation. The transmitted OFDM symbol from the ith transmit antenna at time n is $\boldsymbol{a}_i[n] = \left(a_i[n,1], a_i[n,2], \ldots, a_i[n,K]\right)^{\mathrm{T}} \in \Xi^K$ where Ξ is the modulation symbol alphabet. Thus, at time n, the received signal block $\boldsymbol{r}_j[n] = \left(r_j[n,0], r_j[n,1], \ldots, r_j[n,K-1]\right)^{\mathrm{T}} \in \mathbb{C}^K$ of the jth receive antenna after FFT can be expressed as

$$\boldsymbol{r}_j[n] = \boldsymbol{A}[n]\boldsymbol{\theta}_j[n] + \boldsymbol{\eta}_j[n], \tag{5}$$

where

$$\begin{aligned} \boldsymbol{A}[n] &= \begin{pmatrix} \boldsymbol{A}_1[n] & \boldsymbol{A}_2[n] & \ldots & \boldsymbol{A}_{N_{\mathrm{T}}}[n] \end{pmatrix} \in \Xi^{K \times N_{\mathrm{T}} K}, \\ \boldsymbol{A}_i[n] &= \mathrm{diag}(\boldsymbol{a}_i[n]) \in \Xi^{K \times K}, \end{aligned}$$

$\boldsymbol{\eta}_j[n] = \left(\eta_j[n,0], \eta_j[n,1], \ldots, \eta_j[n,K-1]\right)^{\mathrm{T}} \in \mathbb{C}^K$ is the noise vector at the jth receive antenna at time n. Noise is assumed to be white in both the time and space domains so that the noise covariance matrix is $\boldsymbol{\Sigma}_{\boldsymbol{\eta}_j} = \sigma_{\mathrm{N}}^2 \boldsymbol{I}_K$.

Replacing CFR $\boldsymbol{\theta}_j[n]$ by CIR $\boldsymbol{h}_j[n]$ in (5) and combining the partial FFT matrix $\boldsymbol{F}_K^L$ into the data matrix $\boldsymbol{A}[n]$, (5) can be described in terms of CIR as

$$\boldsymbol{r}_j[n] = \mathcal{A}[n]\boldsymbol{h}_j[n] + \boldsymbol{\eta}_j[n], \tag{6}$$

where

$$\mathcal{A}[n] = \begin{pmatrix} \boldsymbol{A}_1[n]\boldsymbol{F}_K^L & \boldsymbol{A}_2[n]\boldsymbol{F}_K^L & \ldots & \boldsymbol{A}_{N_{\mathrm{T}}}[n]\boldsymbol{F}_K^L \end{pmatrix} \in \mathbb{C}^{K \times N_{\mathrm{T}} L}.$$

2.3 OFDM Frame Structure

We consider an OFDM frame structure similar to that defined in HIPERLAN/2 standard [6]. It is illustrated in Fig. 2. The sample rate is 20 MHz, each OFDM symbol has $K = 64$ subcarriers, the symbol duration is $T_{\text{symb}} = 3.2\ \mu$s, the burst duration is $T_{\text{B}} = T_{\text{symb}} + T_{\text{CP}} = 4\ \mu$s, the number of CP samples is $L - 1 = 16$. It is shown in Fig. 2 that in one protocol data unit (PDU) there are 16 OFDM symbols. The 1st and the 9th symbol are pilot symbols termed as preamble and midamble, respectively. In each data symbol, four pilot tones are assigned to aid the channel estimation. In total, N consecutive PDUs are assumed to form one receive antenna processing block. Thus, there are $16N$ blocks in one receive antenna processing window which are time-indexed by $n \in \mathcal{S} = \{1, 2, \ldots, 16N\}$, the pilot block time-index set is denoted by $\mathcal{S}_{\text{P}} = \{1, 9, 16+1, 16+9, \ldots, (N-1)16+1, (N-1)16+9\}$, and data block time-index set is denoted by $\mathcal{S}_{\text{D}} = \mathcal{S} \setminus \mathcal{S}_{\text{P}}$. The cardinality of $\mathcal{S}_{\text{P}}$ is denoted by N_{P}, and the cardinality of $\mathcal{S}_{\text{D}}$ by N_{D}. For each data symbol, the tone frequency-index set is denoted as $\mathcal{T} = \{0, 1, \ldots, K-1\}$, $\mathcal{T}_{\text{PT}} = \{k_1, k_2, \ldots, k_{N_{\text{PT}}}\}$ stands for the pilot tone index set, and the data tone set is denoted as $\mathcal{T}_{\text{D}} = \mathcal{T} \setminus \mathcal{T}_{\text{PT}}$.

3. TIME DOMAIN CHANNEL ESTIMATION

MMSE estimation contains two steps, i.e, the maximum likelihood (ML) based unbiased estimation and optimal filtering. The ML estimation requires that the number of observations be no less than the number of parameters to be estimated [1].

3.1 Algorithm Derivation

Since the matrix $\mathcal{A}[n], n \in \mathcal{S}_{\text{P}}$, in (6) contains the pilot symbols independent of the block index $n \in \mathcal{S}_{\text{P}}$, it is denoted as $\mathcal{A}$, and (6) can be represented as

$$\boldsymbol{r}_j[n] = \mathcal{A}\boldsymbol{h}_j[n] + \boldsymbol{\eta}_j[n], j \in \{1, 2, \ldots, N_{\text{R}}\}, n \in \mathcal{S}_{\text{P}}. \tag{7}$$

From (7), the ML estimate of CIR $\boldsymbol{h}_j[n]$ related to the jth receive antenna at time $n \in \mathcal{S}_{\text{P}}$ becomes

$$\hat{\boldsymbol{h}}_{\text{ML}j}[n] \quad = \quad (\mathcal{A}^{\text{H}}\boldsymbol{\Sigma}_{\boldsymbol{\eta}_j}^{-1}\mathcal{A})^{-1}\mathcal{A}^{\text{H}}\boldsymbol{\Sigma}_{\boldsymbol{\eta}_j}^{-1}\boldsymbol{r}_j[n], n \in \mathcal{S}_{\text{P}}. \tag{8}$$

When the noise is white over subcarriers, the ML estimate becomes

$$\hat{\boldsymbol{h}}_{\text{ML}j}[n] = (\mathcal{A}^{\text{H}}\mathcal{A})^{-1}\mathcal{A}^{\text{H}}\boldsymbol{r}_j[n], n \in \mathcal{S}_{\text{P}}. \tag{9}$$

We define the ML channel estimate vector as $\hat{\boldsymbol{h}}_{\text{ML}j} = \left(\hat{\boldsymbol{h}}_{\text{ML}j}^{\text{T}}[n_1], \ldots, \hat{\boldsymbol{h}}_{\text{ML}j}^{\text{T}}[n_{N_{\text{P}}}]\right)^{\text{T}}, n_1, n_2, \ldots, n_{N_{\text{P}}} \in \mathcal{S}_{\text{P}}$. It is shown in [1] that the MMSE CIR estimate $\hat{\boldsymbol{h}}_{\text{MMSE}j}[n]$ related to the jth receive antenna at time $n \in S$ can be expressed as

$$\hat{\boldsymbol{h}}_{\text{MMSE}j}[n] = \boldsymbol{W}_j{}^{\text{H}}[n]\hat{\boldsymbol{h}}_{\text{ML}j}, \tag{10}$$

where

$$\boldsymbol{W}_j[n] = \boldsymbol{\Sigma}_{\hat{\boldsymbol{h}}_{\text{ML}j}}^{-1}\boldsymbol{\Sigma}_{\boldsymbol{h}_j[n],\hat{\boldsymbol{h}}_{\text{ML}j}}^{\text{H}}. \tag{11}$$

Since the channels in a MIMO–OFDM system share the common power delay profile and Doppler spectrum, the subscript j of $\boldsymbol{W}_j[n]$ in (11) can be dropped off.

It can be argued from (10) and (11) that by capitalizing on time domain and frequency domain correlation, the linear optimal filtering, smoothing and prediction are performed when $n \in \mathcal{S}_{\mathrm{P}}$, $n \in \mathcal{S}_{\mathrm{D}} \bigcap \{i \mid i < 16N - 8\}$ and $n \in \{i \mid i \geq 16N - 8\}$ respectively.

3.2 Mean Square Error Analysis

At time n, the estimation error vector is $\boldsymbol{e}_{\boldsymbol{h}}^{j}[n] = \boldsymbol{h}_j[n] - \hat{\boldsymbol{h}}_{\mathrm{MMSE}j}[n]$. From [1], the error covariance matrix $\boldsymbol{\Sigma}_{\boldsymbol{e}_{\boldsymbol{h}}^{j}}[n]$ of $\hat{\boldsymbol{h}}_{\mathrm{MMSE}j}[n]$ can be described as

$$\boldsymbol{\Sigma}_{\boldsymbol{e}_{\boldsymbol{h}}^{j}}[n] = \boldsymbol{\Sigma}_{\boldsymbol{h}_j} - \boldsymbol{\Sigma}_{\boldsymbol{h}_j[n],\hat{\boldsymbol{h}}_{\mathrm{ML}j}} \boldsymbol{\Sigma}_{\hat{\boldsymbol{h}}_{\mathrm{ML}j}}^{-1} \boldsymbol{\Sigma}_{\boldsymbol{h}_j[n],\hat{\boldsymbol{h}}_{\mathrm{ML}j}}^{\mathrm{H}}, n \in \mathcal{S}. \tag{12}$$

Due to the edge effect [1], within the processing window, the mean square error (MSE) of the parameter estimates for the blocks located close to the middle of the processing window is less than the MSE for blocks located near the two ends. To evaluate the time domain MSE (TMSE) of channel tap estimate for the whole processing window, we denote the average estimation error variance for each channel tap by $\sigma_{\bar{e}_{\boldsymbol{h}}^{j}}^{2}, j = \{1, 2, \ldots, N_{\mathrm{R}}\}$ as follows

$$\sigma_{\bar{e}_{\boldsymbol{h}}^{j}}^{2} = \sum_{n \in \mathcal{S}} \mathrm{tr}\left(\boldsymbol{\Sigma}_{\boldsymbol{e}_{\boldsymbol{h}}^{j}}[n]\right) / (16 N N_{\mathrm{T}} L). \tag{13}$$

By substituting (12) into (13), the average estimation error variance for each channel tap can be computed.

4. TIME–FREQUENCY DOMAIN CHANNEL ESTIMATION

The channel estimation algorithm presented in this section enhances the algorithm in Section 3 by utilizing the channel information provided by the pilot tones in addition to the preambles and midambles.

4.1 Algorithm Derivation

After performing the ML CIR estimation of the jth receive antenna, we obtain $\hat{\boldsymbol{h}}_{\mathrm{ML}j}[n]$, $n \in \mathcal{S}_{\mathrm{P}}$ as shown in (8). The correspongding ML CFR estimate $\hat{\boldsymbol{\theta}}_{\mathrm{ML}j}[n] \in \mathbb{C}^{N_{\mathrm{T}} K}$ can be obtained by performing FFT to $\hat{\boldsymbol{h}}_{\mathrm{ML}j}[n]$ as follows

$$\hat{\boldsymbol{\theta}}_{\mathrm{ML}j}[n] = \left(\boldsymbol{I}_{N_{\mathrm{T}}} \otimes \boldsymbol{F}_{K}^{L}\right) \hat{\boldsymbol{h}}_{\mathrm{ML}j}[n], \; n \in \mathcal{S}_{\mathrm{P}}. \tag{14}$$

The received pilot tone sample vector $\boldsymbol{r}_{\mathrm{PT}j}[n] = \left(\boldsymbol{r}_j[n, k_1], \ldots, \boldsymbol{r}_j[n, k_{N_{\mathrm{PT}}}]\right)^{\mathrm{T}} \in \mathbb{C}^{N_{\mathrm{PT}}}$, $n \in \mathcal{S}_{\mathrm{D}}$ of the nth received symbol at the jth receive antenna is expressed as

$$\begin{aligned} \boldsymbol{r}_{\mathrm{PT}j}[n] &= \boldsymbol{A}_{\mathrm{PT}}[n] \boldsymbol{\theta}_{\mathrm{PT}j}[n] + \boldsymbol{\eta}_{\mathrm{PT}j}[n], \; n \in \mathcal{S}_{\mathrm{D}}, \qquad (15) \\ \boldsymbol{A}_{\mathrm{PT}}[n] &= \begin{pmatrix} \mathrm{diag}(\boldsymbol{a}_{\mathrm{PT}1}[n]) & \ldots & \mathrm{diag}(\boldsymbol{a}_{\mathrm{PT}N_{\mathrm{T}}}[n]) \end{pmatrix} \in \Xi^{N_{\mathrm{PT}} \times N_{\mathrm{T}} N_{\mathrm{PT}}}, \end{aligned}$$

where $\boldsymbol{a}_{\text{PT}i}[n] = \left(\boldsymbol{a}_i[n,k_1], \boldsymbol{a}_i[n,k_2], \ldots, \boldsymbol{a}_i[n,k_{N_{\text{PT}}}]\right)^{\text{T}}$, $i \in \{1,2,\ldots,N_{\text{T}}\} \in \Xi^{N_{\text{PT}}}$, $\boldsymbol{\theta}_{\text{PT}(i,j)}[n] = \left(\boldsymbol{\theta}_{\text{PT}(i,j)}[n,k_1], \boldsymbol{\theta}_{\text{PT}(i,j)}[n,k_2], \ldots, \boldsymbol{\theta}_{\text{PT}(i,j)}[n,k_{N_{\text{PT}}}]\right)^{\text{T}}$, the combined pilot tone frequency response vector $\boldsymbol{\theta}_{\text{PT}j}[n] = \left(\boldsymbol{\theta}^{\text{T}}_{\text{PT}(1,j)}[n], \boldsymbol{\theta}^{\text{T}}_{\text{PT}(2,j)}[n], \ldots, \boldsymbol{\theta}^{\text{T}}_{\text{PT}(N_{\text{T}},j)}[n]\right)^{\text{T}}$ $\in \mathbb{C}^{N_{\text{T}}N_{\text{PT}}}$ and $\boldsymbol{\eta}_{\text{PT}j}[n] = \left(\boldsymbol{\eta}_j[n,k_1], \boldsymbol{\eta}_j[n,k_2], \ldots, \boldsymbol{\eta}_j[n,k_{N_{\text{PT}}}]\right)^{\text{T}}$.

Let

$$\boldsymbol{\psi}_j = \left(\hat{\boldsymbol{\theta}}^{\text{T}}_{\text{ML}j}, \boldsymbol{r}^{\text{T}}_{\text{PT}j}\right)^{\text{T}} \in \mathbb{C}^{N_{\text{T}}KN_{\text{P}}+N_{\text{PT}}N_{\text{D}}}, \tag{16}$$

where $\hat{\boldsymbol{\theta}}_{\text{ML}j} = \left(\hat{\boldsymbol{\theta}}^{\text{T}}_{\text{ML}j}[n_1], \hat{\boldsymbol{\theta}}^{\text{T}}_{\text{ML}j}[n_2], \ldots, \hat{\boldsymbol{\theta}}^{\text{T}}_{\text{ML}j}[n_{N_{\text{P}}}]\right)^{\text{T}}$, $n_i \in \mathcal{S}_{\text{P}}$ and $\boldsymbol{r}_{\text{PT}j} = \left(\boldsymbol{r}^{\text{T}}_{\text{PT}j}[n_1],\right.$ $\left.\boldsymbol{r}^{\text{T}}_{\text{PT}j}[n_2], \ldots, \boldsymbol{r}^{\text{T}}_{\text{PT}j}[n_{N_{\text{D}}}]\right)^{\text{T}}$, $n_i \in \mathcal{S}_{\text{D}}$. Based on the vector $\boldsymbol{\psi}_j$ in (16), the MMSE estimate $\hat{\boldsymbol{\theta}}_{\text{MMSE}j}[n] \in \mathbb{C}^{N_{\text{T}}K}$ is

$$\hat{\boldsymbol{\theta}}_{\text{MMSE}j}[n] = \boldsymbol{W}^{\text{H}}_{\boldsymbol{\theta}}[n]\boldsymbol{\psi}_j, n \in \mathcal{S}, \tag{17}$$

where

$$\boldsymbol{W}_{\boldsymbol{\theta}}[n] = \boldsymbol{\Sigma}^{-1}_{\boldsymbol{\psi}_j}\boldsymbol{\Sigma}^{\text{H}}_{\boldsymbol{\theta}_j[n],\boldsymbol{\psi}_j} \tag{18}$$

4.2 Mean Square Error Analysis

Similarly to Subsection 3.2, the frequency domain MSE (FMSE) of the MMSE CFR estimate $\hat{\boldsymbol{\theta}}_{\text{MMSE}j}[n]$ for the jth receive antenna at time n can be obtained from the error covariance matrix

$$\boldsymbol{\Sigma}_{\boldsymbol{e}^j_{\boldsymbol{\theta}}}[n] = \boldsymbol{\Sigma}_{\boldsymbol{\theta}_j[n]} - \boldsymbol{\Sigma}_{\boldsymbol{\theta}_j[n],\boldsymbol{\psi}_j}\boldsymbol{\Sigma}^{-1}_{\boldsymbol{\psi}_j}\boldsymbol{\Sigma}^{\text{H}}_{\boldsymbol{\theta}_j[n],\boldsymbol{\psi}_j}, n \in \mathcal{S}. \tag{19}$$

The average FMSE $\sigma^2_{\bar{e}^j_{\boldsymbol{\theta}}}$ is

$$\sigma^2_{\bar{e}^j_{\boldsymbol{\theta}}} = \sum_{n \in S} \text{tr}\left(\boldsymbol{\Sigma}_{\boldsymbol{e}^j_{\boldsymbol{\theta}}}[n]\right)/(16NN_{\text{T}}K). \tag{20}$$

By substituting (19) into (20), the MSE for each tone can be obtained.

5. NUMERICAL EXAMPLES

The normalized time domain MSE (NTMSE) corresponding to $\sigma^2_{\bar{e}^j_{\boldsymbol{h}}}$ in (13) and the normalized frequency-domain MSE (NFMSE) corresponding to $\sigma^2_{\bar{e}^j_{\boldsymbol{\theta}}}$ in (20) have been obtained both by analysis and running Monte Carlo computer simulations. The channel model considered is the Channel Model A provided by the HIPERLAN/2 standard [6]. It corresponds to a typical indoor environment, where the classical Jakes's Doppler spectrum and Rayleigh fading are assumed for all taps. Two values of Doppler spreads, i.e., 40Hz and 200Hz, have been considered. The different processing window lengths correspond to the different Doppler frequencies.

Fig. 3(a) compares the NTMSEs of the time domain MMSE channel estimation with 40Hz Doppler spread, 10 PDUs per processing window. Fig. 3(b) shows results similar to those in Fig. 3(a) except with 200Hz Doppler spread and 5 PDUs per processing window. The MMSE channel estimation has gains of about 16 dB and 12 dB over the ML estimation for a given NTMSE in Figs. 3(a) and 3(b), respectively. When the ideal correlation matrix Σ_h is used, the performance of the MMSE estimation is almost invariant when either Jakes's or ideal lowpass Dopper sprectrum is employed. Thus, it is proved that the MMSE channel estimation is not sensitive to the Doppler spectrum shape. When the estimated correlation matrix Σ_h is used, there are about 7 dB and 5 dB losses for a given NTMSE in Figs. 3(a) and 3(b), respectively. So the estimation of Σ_h plays an important role in the real MMSE estimation. The performance loss due to the estimated Σ_h can be improved by enlarging the window length.

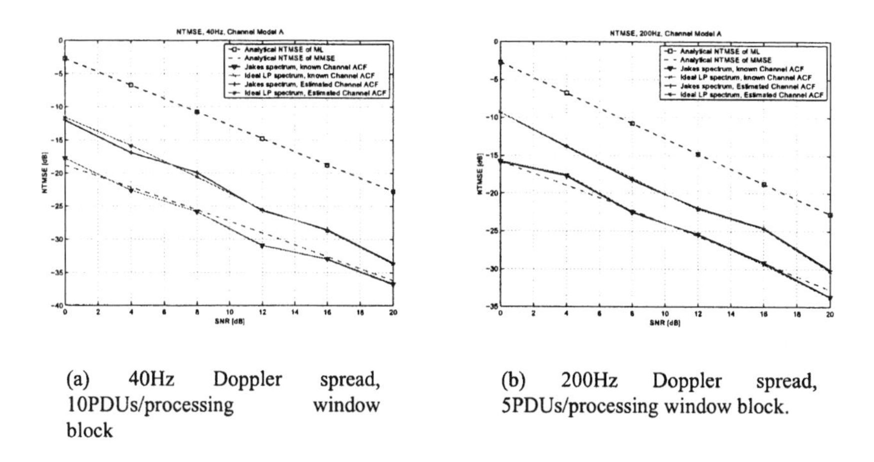

(a) 40Hz Doppler spread, 10PDUs/processing window block

(b) 200Hz Doppler spread, 5PDUs/processing window block.

Figure 3. NTMSE of time domain MMSE channel estimation with known and estimated channel ACF.

Fig. 4 compares the NFMSEs of the time domain MMSE channel estimation and time–frequency domain MMSE channel estimation with 40Hz/200Hz Doppler spread and 2/4 PDUs per processing window. The results show that the performance of the time–frequency domain channel estimation is very close to that of the time domain channel estimation.

6. CONCLUSIONS

Space–time–frequency channel estimation problem in a MIMO–OFDM system of the HIPERLAN/2 type was studied. A generic MIMO–OFDM channel estimation problem was formulated, MMSE solutions were derived, and their MSE performance was analyzed. The derived temporal channel estimator was shown to improve the performance of the well-known ML estimates significantly with a moderate complexity. The performance of the time-frequency domain channel estimator is very close to that of

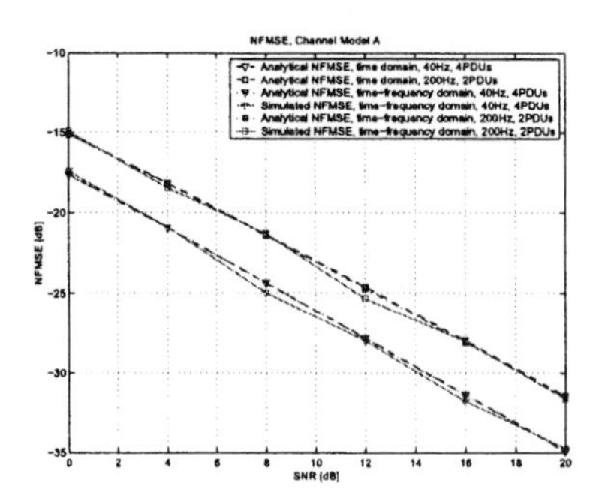

Figure 4. NFMSE of time-frequency domain channel estimation, 40Hz and 200Hz Doppler spread, 2PDUs and 4PDUs/processing window block

the time domain channel estimator. As such, the time domain channel estimation is clearly very promising for practical MIMO–OFDM systems.

7. ACKNOWLEDGEMENTS

This research was supported by TEKES, the National Technology Agency of Finland, Nokia, the Finnish Defence Forces, Elektrobit and Instrumentointi.

8. AFFILIATIONS

Centre for Wireless Communications, P.O.Box 4500, FIN-90014, University of Oulu, Finland.

9. REFERENCES

[1] H. Meyr, M. Moeneclaey, and S. A. Fechtel, *Digital Communication Receivers: Synchronization, Channel Estimation and Signal Processing*, John Wiley and Sons, New York, USA, 1998.

[2] J.-J. van de Beek, O. Edfors, and M. Sandell, "On channel estimation in OFDM systems," in *Proc. IEEE Veh. Technol. Conf.*, 1995, vol. 2, pp. 815–819.

[3] Y. Li and N. R. Sollenberger, "Adaptive antenna arrays for OFDM systems with cochannel interference," *IEEE Trans. Commun.*, vol. 47, no. 2, pp. 217–229, Feb. 1999.

[4] Y. Li, N. Seshadri, and S. Ariyavisitakul, "Channel estimation for OFDM systems with transmitter diversity in mobile wireless channels," *IEEE J. Select. Areas Commun.*, vol. 17, no. 3, pp. 461–471, Mar. 1999.

[5] Y. Li, L. J. Cimini Jr., and N. R. Sollenberger, "Robust channel estimation for OFDM systems with rapid dispersive fading channels," *IEEE Trans. Commun.*, vol. 46, no. 7, pp. 902–915, July 1998.

[6] ETSI, "Broadband radio access networks (BRAN); HIPERLAN type 2; system overview," Tech. Rep., European Telecommunications Standards Institute (ETSI), TS 101 683 V1.1.1 (2000-02), 2000.

A.RENOULT[1,2], M.CHENU-TOURNIER[1] AND I.FIJALKOW[2]

MULTI-USER TRANSMISSIONS FOR OFDM: CHANNEL ESTIMATION AND PERFORMANCES

[1]*Thales communication and* [2]*ETIS/ENSEA – Univ. de Cergy-pontoise - CNRS*

Abstract. To increase the throughput of transmission systems, MIMO transmission have become a natural path. Two different MIMO schemes can be envisioned: point to point communications and multi-point to point communications. We propose in this paper to study the performances of multi-point to point communication (also referred to as multi-user transmissions) for different receivers and with or without channel estimation. The proposed channel estimator is a multi-user maximum likelihood estimator achieving both synchronization and the channel estimates. This work is partially supported by the IST project STRIKE (FP5 IST-2001-38354)

1. INTRODUCTION

To increase the throughput of transmissions systems, MIMO (multiple input multiple output) transmissions have gathered a lot of attention this last decade. These transmission systems either increase the cell throughput using multiple transmission from different users separated at the receiver (using SDMA, joint detection,...) or, more recently, increase a specific link throughput using STC (space-time coding) or multiplexing techniques (BLAST). Many references have proposed the performances for point to point links but fewer have studied the potential for multi-point to point links using joint detection techniques on non spreaded modulations. Moreover, the performance of such systems taking into account the impact of channel estimation have scarcely been studied. In this paper we propose to present the performance of two joint detection receivers the MMSE and the MLSE applied to 802.11a (HiperLAN) like waveforms with a maximum likelihood channel estimator. The channel estimation performs both the synchronization and the channel estimation.
The performance studies of multi-user receivers for OFDM modulations based on spatial filtering with estimated channels have been proposed by [**3**]. In this reference, the channel estimation does not include the synchronization issue and the estimation of the channel coefficients is performed in the frequency domain. A it will be exposed in the following, this techniques requires the knowledge of many preamble symbols to perform accurately. Moreover, a interpolation is also required to provide the response over all the frequency sub-carriers. In [1], the performance of spatio-temporal codes are evaluated with a similar channel estimator. In both cases, the error due to bad the synchronization are not related as the synchronization is assumed perfect. In [2] synchronization techniques for MIMO OFDM systems are proposed but they rely on chirps and not on OFDM symbols. The theoretical

K. Fazel and S. Kaiser (eds.), Multi-Carrier Spread-Spectrum, 219–226.

performances of the MLSE in MIMO propagation contexts have been proposed in [4].
In the next section, the signal model for a multi-user transmission is proposed. In section 3 the maximum likelihood channel estimator is proposed and in section 4 the performance of the receivers with the channel estimator are presented. At last, in section 5 we conclude and present some perspectives.

2. SIGNAL MODEL AND RECEIVERS

In this section we describe the signal model. First a matrix representation of the received signal is proposed in the SISO case and is then extended in the MIMO case. The proposed signal model is given for a single transmitted OFDM symbol. We then briefly describe the studied receivers.
In a classical OFDM system, the received samples, after DFT, on the N_{DFT} sub-carriers are stacked in a vector $\boldsymbol{y}$. This vector can be written as:

$$\mathbf{y}=\mathbf{F}_2\mathbf{I}_{\overline{CP}}\mathbf{H}\mathbf{I}_{CP}\mathbf{F}_1\mathbf{a}+\mathbf{n} \tag{1}$$

where $\mathbf{F}_1$ is a matrix of dimension $N_{DFT}\times N_{DFT}$ processing the inverse DFT (discrete Fourier transform), $\mathbf{F}_1^*=\mathbf{F}_2$ (processes the DFT)

$\mathbf{I}_{CP}$ is a matrix inserting the cyclic prefix (CP) and is of dimension $(N_{DFT}+N_{CP})\times N_{DFT}$

$\mathbf{I}_{\overline{CP}}$ is the matrix that does the synchronization and suppresses the CP of dimension $N_{DFT}\times(N_{DFT}+N_{CP}+N_H)$

H is the channel matrix of dimension

a is the vector containing the N_{DFT} symbols to be transmitted over the sub-carriers (only N_{SC} sub-carriers are non null)

n is the vector of dimension N_{DFT} containing the AWGN samples.

In the multi-user case, (N_u users) and multi-sensor receiver case (N_r receivers), the signal model structure is identical to the one proposed in equation (1):

$$\tilde{\mathbf{y}}=\tilde{\mathbf{F}}_2\tilde{\mathbf{I}}_{\overline{CP}}\tilde{\mathbf{H}}\tilde{\mathbf{I}}_{CP}\tilde{\mathbf{F}}_1\tilde{\mathbf{a}}+\tilde{\mathbf{n}} \tag{2}$$

with $\tilde{\mathbf{y}}$ the vector of size $N_r \cdot N_{SC}$ containing the samples of the sub-carriers for all the receivers , $\tilde{\mathbf{F}}_2=\mathbf{F}_2\otimes\mathbf{I}_{Nr}$ where $\mathbf{I}_{Nr}$ is the identity matrix of size N_r, $\tilde{\mathbf{F}}_1=\mathbf{F}_1\otimes\mathbf{I}_{Nr}$, $\tilde{\mathbf{I}}_{\overline{CP}}=\mathbf{I}_{\overline{CP}}\otimes\mathbf{I}_{Nr}$, $\tilde{\mathbf{H}}$ is the channel matrix, $\tilde{\mathbf{I}}_{CP}=\mathbf{I}_{CP}\otimes\mathbf{I}_{Nr}$, $\tilde{\mathbf{a}}$ the vector of size $N_{DFT} \cdot N_u$ containing the transmitted symbols and $\tilde{\mathbf{n}}$ is the vector containg the samples of the noise supposed to be white in the spatial and frequency domains of variance σ^2.

When the system is synchronized the matrix $\mathbf{H}=\widetilde{\mathbf{I}}_{CP}\widetilde{\mathbf{H}}\widetilde{\mathbf{I}}_{CP}\widetilde{\mathbf{F}}_I$ is bloc circulant and thus it can be written as $\mathbf{H}=\mathbf{F}^*\mathbf{\Lambda}\,\mathbf{F}$ where $\mathbf{F}$ is a DFT matrix and $\mathbf{\Lambda}$ is a bloc diagonal.
Thus the expression of the observation is given by:

$$\widetilde{\mathbf{y}}=\mathbf{\Lambda}\widetilde{\mathbf{a}}+\widetilde{\mathbf{n}}$$

This signal model is valid when the different received signals do not have frequency shifts and the time delay between first received path and the last received path does not exceed the length of the cyclic prefix. Thus for the sub-carrier i the observation reduces to:

$$\mathbf{y}_i=\mathbf{H}_i\mathbf{a}_i+\mathbf{n}_i$$

were $\mathbf{y}_i$ is the vector of size N_r containing the samples of the received signal on the array of sensors for the sub-carrier i ($\widetilde{\mathbf{y}}=\left[\mathbf{y}_i^T\cdots\mathbf{y}_{NDFT}^T\right]^T$, $\widetilde{\mathbf{n}}=\left[\mathbf{n}_i^T\cdots\mathbf{n}_{NDFT}^T\right]^T$)
$\mathbf{H}_i$ is the matrix of dimension $N_r\times N_u$ containing the channel coefficients for the sub carrier **i** ($\Lambda=diag\left(\{\mathbf{H}_i\}_{1\le i\le NDFT}\right)$)
$\mathbf{a}_i$ contains the transmitted symbols on the sub-carrier i by the different. From this linear model, the joint detection techniques can be applied. We will focus on the performances of the MMSE and the MLSE detectors.
The MMSE calculates the matrix filter **M** for each sub-carrier i that minimizes the mean square error on the detected vector of symbols.

$$\mathbf{M}=\underset{\mathbf{M}}{\operatorname{argmin}}\left\|\mathbf{M}\mathbf{y}_i-\mathbf{a}_i\right\|^2$$

The MLSE detects the most likely transmitted symbol vector:

$$\widetilde{\mathbf{a}}_i=\underset{\mathbf{a}_i}{\operatorname{argmax}}P(\mathbf{y}_i|\mathbf{H}_i,\sigma^2)$$

The complexity of the MLSE receiver depends on the number of symbols in the modulation N_s and is given by $N_s^{N_u}$ per sub-carrier. The complexity of the MMSE receiver only depends on the number of transmitters and the number of receivers.

3. CHANNEL ESTIMATOR

In this multi-user scheme it is necessary to propose ad-hoc channel estimators. Classically, in OFDM systems, the synchronization is done temporally using the knowledge and the structure of preamble symbols (or using the CP) and the estimation of the channel coefficients is done in the frequency domain for each sub-carrier. A interpolation can also be applied to increase the reliability of the channel and provide information on the correlation between the sub-carriers. As OFDM systems are generally used in multi-path channels, the propagation channel is generally Rayleigh fading and the demodulation provides the same performances than a single path Rayleigh fading channel with no diversity exploitation. The channel decoder and the interleaver recover these poor performances. Thus generally, OFDM systems are used in rich scattering channels and are designed to work at relatively high SNRs. This makes the channel estimation procedure simple. Indeed, as the SNR is generally high, even with few known symbols, the estimation variance of the propagation channel is low. The variance is given by $\sigma_{hi}^2=\sigma^2/(s^\dagger s)$ with s the vector containing the known samples. Typically, in the 802.11a and HiperLAN/2 systems 2 to 4 symbols are used.
In the multi-user case, the number of coefficients to estimate per sub-carrier is much larger. Indeed, in a N_u transmitter system with N_r reception antennas, the number of coefficients to estimate per sub-carrier is $N_u N_r$ compared to 1 in the SISO case. Thus when estimating the propagation channel in the frequency domain, it is necessary to provide many more known symbols than in the SISO case. We thus propose to estimate the propagation channel temporally. In this case, the number of known samples is much larger than when estimating the propagation channel in the frequency domain.
Thus, in this paper we propose a maximum likelihood synchronization and channel estimator for multi-user OFDM transmissions. The received signal on the sensors before removing the CP and applying the DFT is given by $\tilde{\mathbf{z}}=\tilde{\mathbf{H}}\tilde{\mathbf{I}}_{CP}\tilde{\mathbf{F}}_I\tilde{\mathbf{a}}+\tilde{\mathbf{n}}$ contains the concatenation for all the sensors of the different transmitted signals. The observation on the known received samples can be written as :

$$\mathbf{X}=\mathbf{S}(\tau)\Gamma+\mathbf{N}$$

were the $N_s \times N_r$ matrix $\mathbf{X}$ contains the received samples, the $N_s \times N_{CP} N_u$ matrix $\mathbf{S}(\tau)$ contains the known samples of the preamble shifted by the unknown delay and the $N_u \times N_r$ matrix Γ contains the the channel impulse responses. The matrix $\mathbf{N}$ of dimension $N_s \times N_r$ contains the samples of the AWGN. The matrix $\boldsymbol{S}(\tau)$ is organized as follows:

$$\mathbf{S}(\tau)=\left[\mathbf{S}_1(\tau)\cdots\mathbf{S}_{Nu}(\tau)\right]$$

with the $N_s \times N_{CP}$ matrix $\mathbf{S}_\mathrm{u}(\tau)$ organized as follows:

$$\mathbf{S}_u(\tau)=\begin{bmatrix} s_u(\tau)\cdots s_u((N_s-1)T_e-\tau)\,0 & 0 \\ 0 \quad 0 \quad s_u(\tau) & \cdots s_u((N_s-1)T_e-\tau) \end{bmatrix}$$

were $s_u(t)$ is the known preamble for user u.
With the previous modelization of the received signal, the likelihood of the observation can be derived:

$$P(\mathbf{X}|\tau,\mathbf{\Gamma},\sigma^2)=\frac{1}{\pi^{N_u N_r}\sigma^{2N_u N_r}}e^{-\frac{1}{\sigma^2}\|\mathbf{X}-\mathbf{S}(\tau)\mathbf{\Gamma}\|^2}$$

Thus the log-likelihood is given by $L(\mathbf{X}|\tau,\mathbf{\Gamma},\sigma^2)\approx -N_u N_r\log(\sigma^2)-\frac{1}{\sigma^2}\|\mathbf{X}-\mathbf{S}(\tau)\mathbf{\Gamma}\|^2$. We consider that the noise and the impulse response are nuisance parameters By derivating the criterion by these parameters and by nulling the derivate, we obtain the following estimators:

$$\hat{\sigma}^2=\frac{1}{N_u N_r}\|\mathbf{X}-\mathbf{S}(\tau)\mathbf{\Gamma}\|^2$$

$$\Gamma=(\boldsymbol{S}^{\dagger}(\tau)\boldsymbol{S}(\tau))^{-1}\boldsymbol{S}^{\dagger}(\tau)\boldsymbol{X}$$

Replacing in the log-likelihood the nuisance parameters by their estimates leads to the estimator of the global synchronization:

$$\hat{\tau}=\underset{\tau}{\operatorname{argmin}}\|\mathbf{\Pi}_s(\tau)\mathbf{X}\|^2$$

were $\mathbf{\Pi}_s(\tau)$ is the projector on the noise sub-space given by:

$$\mathbf{\Pi}_s(\tau)=\mathbf{I}-\mathbf{S}(\tau)\,(\mathbf{S}(\tau)^*\mathbf{S}(\tau)\,)^{-1}\mathbf{S}(\tau)^*$$

3.1. Performance of the channel estimate

Supposing accurate synchronization, the variance of the estimated channel is given by:

$$\sigma_{\Gamma}= \quad E[\|\mathbf{\Gamma}-\hat{\mathbf{\Gamma}}\|]$$
$$=E[tr(\mathbf{N}^*\mathbf{S}\,(\mathbf{S}^*\mathbf{S}\,)^{-2}\mathbf{S}^*\mathbf{N}\,)]=\sigma^2(\frac{\mathbf{S}^*\mathbf{S}}{N_s N_r})^{-1}$$

As classically observed with such estimators, the variance degrades linearly when en number of parameters increase. When using the MMSE channel estimator in the frequency domain to estimate the channel coefficients, the variance per sub-carrier is given by $\sigma_{hi}^2=\sigma^2/(s^\dagger s)$ with s containing 2 uncorrelated samples. In following examples used to drive the performances, the S matrix contains either 128 or 256 (correlated) samples. It is straightforward that to achieve the performances of the proposed MIMO maximum likelihood channel estimator using the per sub-carrier frequency domain approach would require many more known OFDM symbols. Moreover, as it will be seen in the following performances, the channel estimator proposes close to optimal performances when using 4 OFDM symbols.

4. PERFORMANCES

The performances presented hereafter have been realized with the 802.11a parameters. The mapping is Grey mapping and the number of emitters and receivers vary. The propagation channels have as maximum length the number of samples in the CP (16 samples). The N_H channel samples are random Rayleigh realizations and the channels are normalized. The signals arrive at the receiver sufficiently synchronized i.e. $\tau_{max}-\tau_{min}\leq N_{CP}$ with τ_{max} the last received path and τ_{min} the first received path. Nevertheless, the received signals arrive in a 20 sample uncertainty range thus requiring a synchronization.
The preamble symbols (known symbols) are randomly chosen at each burst. Each burst contains 100 OFDM symbols and the propagation channel is constant on this burst.

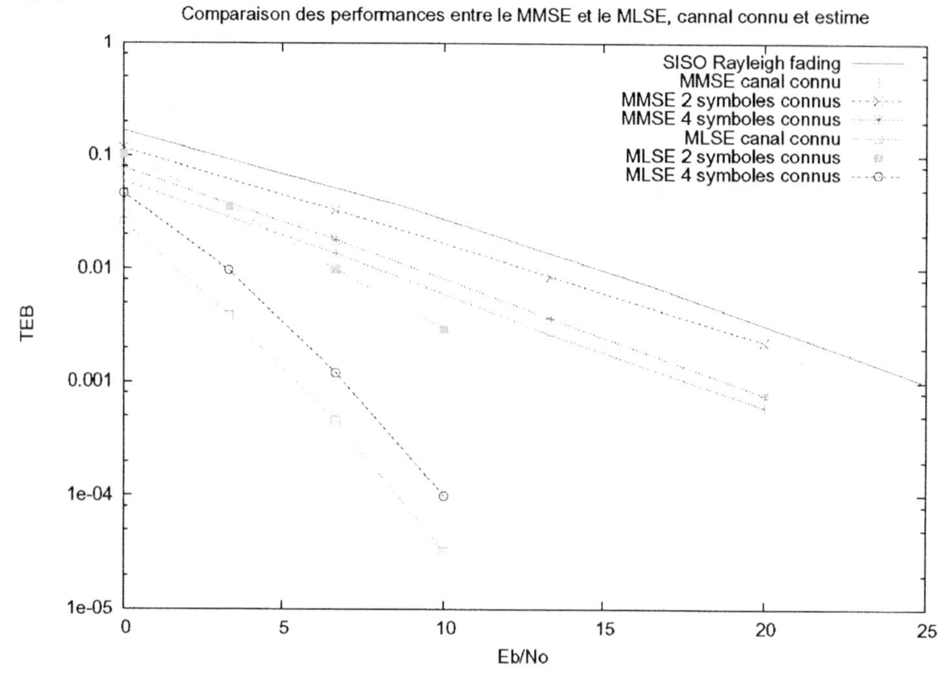

Figure 1. Performances comparaison between the SISO OFDM, the MIMO MMSE and the MIMO MLSE with know and estimated channels for BPSK modulations

On figure 1 the performance of the MMSE and MLSE detectors are compared in the case of known and estimated channels for BPSK modulation. In this simulation, their are 4 transmitters and 4 receivers. We also compare the performance of the system when the number of known OFDM symbols is set to 2 and 4. For comparison issues, the performance of the OFDM SISO case are also depicted for known Rayleigh fading channels.

The presented performances are the mean BER over the different users BER. The abscissa is given in E_b/N_0 and is the per user E_b/N_0. Thus when the number of transmitters increase, the total radiated power also increases. If one wants to compare the performances at constant radiated power within the cell, the BER curve of the MIMO system should be translated of $6dB$, lowering the transmitted power by each user.

On the figures, we notice a $6dB$ gain for the MMSE with known channel compared to the SISO case. Thus at constant power it is possible to have 4 transmitters rather than one for the same demodulation performance or it is possible to dramatically increase the reliability of the transmission. Also notice that the performances of the MLSE are much larger than the performances of the MMSE. Of course the complexity of the MLSE is larger but in the case of small modulations sizes, it is still practical. When 2 OFDM symbols are used in conjunction with the MLSE, for BER in the regions of interest ($5.10^{-2} \rightarrow 10^{-3}$), the performances are limited by the large variance of the estimated channel. Indeed, the BER diverges from the BER for known channel. This is not the case anymore when 4 OFDM symbols are used to estimate the channel.

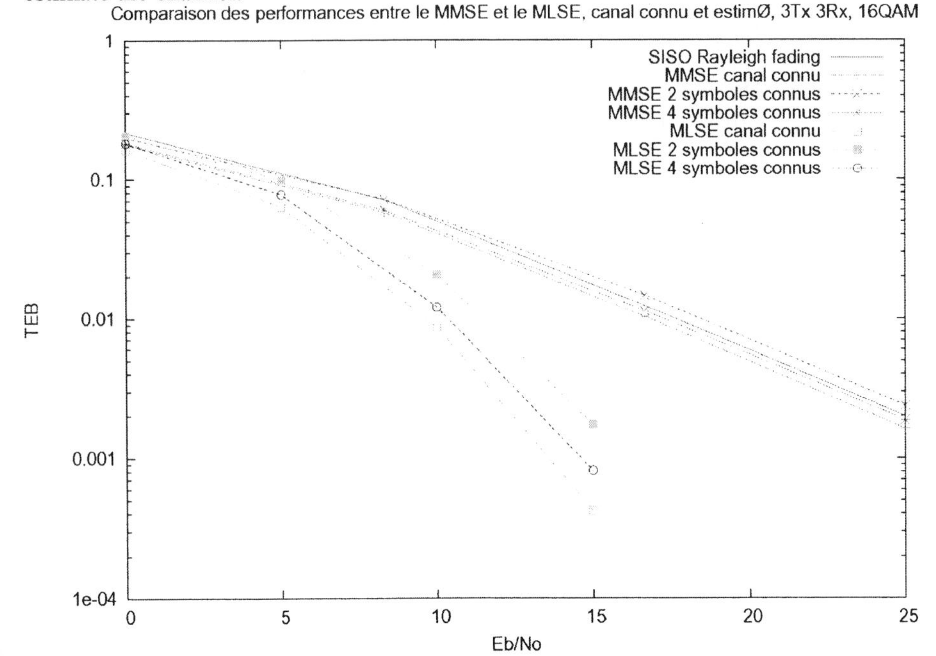

Figure 2. Performances comparaison between the SISO OFDM, the MIMO MMSE and the MIMO MLSE with known and estimated channels for 16 QAM modulations.

On figure 2 the performance of the two previous receivers are compared for 16QAM modulations. In this case, the multiple antenna receiver uses 3 sensors and 3 users transmit simultaneously. Notice that in this case, the performances of the MLSE are not limited by the channel estimation when 2 OFDM symbols are used. This is due to the smaller number of parameters to be estimated due to the lower number of transmitters and receivers. It is also due to the fact that the channel estimation is little sensitive to the modulation size.

5. CONCLUSION

In this paper we have presented the performances of two joint detectors combined to an ad hoc maximum likelihood channel estimator. The proposed channel estimator in the MIMO context provides good results when used with the parameters of the SISO system thus not requiring deep changes. We have also shown that the MLSE receiver can greatly increase the capacity of the system compared to the SISO case at the cost of a complexity increase. On the other hand, the performances of the MMSE show that with little complexity increase, it is possible to multiply the number of simultaneous transmitters. Thus it is possible to provide higher performances, to an increased number of subscribers without increasing the total power in the cell.
In the future studies, iterative receivers will be analyzed as well as the downlink optimization. The statistical performance of the channel estimator will be analyzed and frequency offset will be included and compensated. Also the receivers will be tested on real signals using a multiple sensor receiver.

REFERENCES

[1] Y. Li, N. Seshadri, and S. Ariyavisitakul. Channel estimation for ofdm systems with transmitter diversity in mobile wireless channels. *IEEE JSAC*, Vol. 17(No. 3): pp. 461–471, March 1999.
[2] A.N. Mody and G.L. Stüber. Synchronization for mimo ofdm. In *GLOBECOM 2001*, volume vol. 1, pages pp. 509–513.
[3] F.W. Vook and T.A. Thomas. Mmse multi-user channel estimation for broadband wireless communications. In *GLOBECOM 2001*, volume Vol. 1, pages pp. 470–474.
[4] X. Zhu and R.D. Murch. Performance analysis of maximum likelihood detection in a mimo antenna system. *IEEE Trans. On Communications*, Vol. 50(No. 2): pp. 187–191, Febuary 2002.

STEPHAN SAND, ARMIN DAMMANN, AND GUNTHER AUER

Adaptive Pilot Symbol Aided Channel Estimation for OFDM Systems

Abstract. In this paper we propose a novel scheme for adaptive pilot symbol aided channel estimation. The optimum Wiener filter for pilot symbol aided channel estimation requires the perfect knowledge of the channel correlation function. Since in practice the channel correlation function is unknown, a robust Wiener filter with model mismatch is employed for pilot symbol aided channel estimation [1]. Due to the model mismatch, the robust Wiener filter can perform significantly worse for certain channel models than the optimum Wiener filter. Therefore, we propose an adaptive filter in the frequency domain to reduce the performance loss with a moderate increase in complexity. The received pilot symbols are first interpolated in frequency and time direction. Then, a block frequency normalized least-mean-square algorithm adapts the filter coefficients without making any assumptions about the channel and noise statistics.

1. INTRODUCTION

Modern wireless communications require high data rate transmission over mobile channels. Orthogonal frequency division multiplexing (OFDM) [2], [3] is a suitable technique for broadband transmission in multipath fading environments and is implemented in some new broadcast standards like digital audio broadcasting (DAB) [4] or terrestrial digital video broadcasting (DVB-T) [5] as well as in wireless local area network (WLAN) standards such as IEEE 802.11a/g or HIPERLAN/2 [6]. The successful deployment of OFDM in these standards and its efficient implementation to combat multipath fading make it a promising candidate for the radio air interface of a fourth generation mobile radio system.

Coherent OFDM detection requires information about the channel state that has to be estimated by the receiver. To this purpose, known pilot symbols are periodically multiplexed in the data. Channel estimation is performed by interpolating the time-frequency pilot grid and exploiting the correlations of the received OFDM signal in time and frequency [7]. These correlations are introduced by the time- and frequency-selective mobile radio channel, which is modeled as a wide-sense stationary uncorrelated scattering (WSSUS) channel [8]. The behavior of a WSSUS channel can be described by its delay and Doppler power spectral densities (PSDs).

Since in practice the delay and Doppler PSDs are not perfectly known in the receiver, a robust design is chosen by assuming rectangular delay and Doppler PSDs for a worst-case Doppler-shift and channel delay spread. Often, the transmission channel behaves much better than the assumed worst case [9]. In those situations, Wiener-filters with a smaller bandwidth would yield better estimation results by reducing the noise on the channel estimate. Moreover, the model mismatch between the actual delay and Doppler PSDs of the channel and the assumed ones causes further performance losses.

K. Fazel and S. Kaiser (eds.), Multi-Carrier Spread-Spectrum, 227–234.

In this paper, we will investigate pilot symbol aided channel estimation (PACE) with an adaptive filter in the frequency domain to reduce the performance loss with a moderate increase in complexity. The received pilot symbols are first interpolated in frequency and time direction. Then, a block frequency normalized least-mean-square (BFNLMS) algorithm adapts the filter coefficients without making any assumptions about the channel and noise statistics.

The paper is organized as follows: in Section 2 the investigated system model is presented; Section 3 gives a brief summary of PACE and introduces the adaptive channel estimator (CE); finally, section 4 presents the simulated BER for different channels.

2. System Model

In this section we briefly introduce the considered OFDM system model. All considerations are carried out in the equivalent baseband domain. At the transmitter, the signal from a binary source of a user is encoded by a rate $\frac{1}{2}$ convolutional encoder, interleaved, mapped to the symbol alphabet (QPSK, Gray mapping), multiplexed with pilot symbols, OFDM modulated onto N_c orthogonal subcarriers with a FFT of size N_{fft}, and cyclically extended by the guard interval T_g (see Fig. 1). The subcarrier-spacing is denoted by F_s.

The wide-sense stationary uncorrelated scattering (WSSUS) channel model introduced in [8] was used for the time- and frequency-selective mobile channel. Therefore, the simulations are computed in the frequency domain.

At the receiver, before the OFDM transformation, additive white Gaussian noise (AWGN) is added. After removing the guard interval and OFDM demodulation, the received signal is demultiplexed into pilots and data symbols. The received pilots serve as reference for the channel estimator that estimates the channel state information. The received data symbols are symbolwise demapped into bits, which are soft-decision maximum-likelihood decoded.

3. Pilot Symbol Aided Channel Estimation

The received symbols of an OFDM frame are given by

$$R_{n,k} = H_{n,k} S_{n,k} + N_{n,k}, \quad n = 1, \ldots N_c, \; k = 1, \ldots N_s, \tag{1}$$

where $S_{n,k}$, $N_{n,k}$, N_c, and N_s are the transmitted symbols, the AWGN component, the number of subcarriers per OFDM symbol, and the number of OFDM symbols per frame. The set of pilot positions in an OFDM frame is $\mathcal{P}$ and the number of pilot symbols is $N_{grid} = \| \mathcal{P} \|$.

The first step in the channel estimation stage is to obtain an initial estimate $\breve{H}_{n',k'}$ of the channel transfer function (CTF), i.e.,

$$\breve{H}_{n',k'} = \frac{R_{n',k'}}{S_{n',k'}} = H_{n',k'} + \frac{N_{n',k'}}{S_{n',k'}}, \qquad \forall \{n', k'\} \subset \mathcal{P}. \tag{2}$$

In a second step, the final estimates of the complete CTF are obtained from the

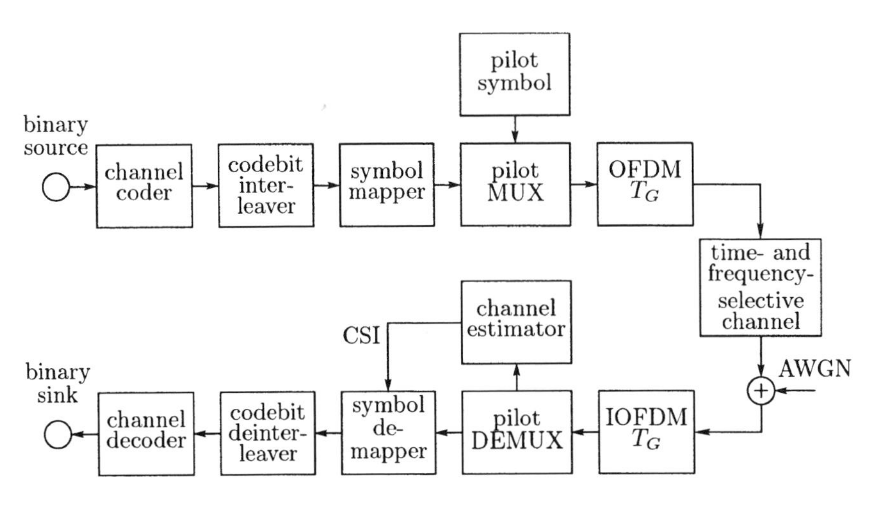

Fig. 1. OFDM transmission channel model

initial estimates $\check{H}_{n',k'}$ by two-dimensional filtering

$$\hat{H}_{n,k} = \sum_{\{n',k'\} \in \mathcal{T}_{n,k}} \omega_{n',k',n,k} \check{H}_{n',k'}, \tag{3}$$

where $\omega_{n',k',n,k}$ is the shift-variant 2-D FIR impulse response of the filter, $n = 1, \ldots, N_c$, and $k = 1, \ldots, N_s$ [7]. The subset $\mathcal{T}_{n,k} \subset \mathcal{P}$ is the set of initial estimates $\check{H}_{n',k'}$, which are actually used to estimate $\hat{H}_{n,k}$. The FIR filter coefficients are based on the Wiener design criterion. The optimal Wiener filter has N_{grid} filter coefficients, in which case the subset $\mathcal{T}_{n,k}$ is identical to the set $\mathcal{P}$. The filter coefficients depend on the discrete time-frequency correlation function (CF) of the CTF $\theta_{n-n'',k-k''} = \mathrm{E}\{H_{n,k} H^*_{n'',k''}\}, \forall\{n'', i''\} \in \mathcal{T}_{n,k}$ and the noise variance σ^2.

Due to the WSSUS assumption of the channel, the CF $\theta_{n-n'',k-k''}$ can be separated into two independent parts

$$\theta_{n-n'',k-k''} = \theta_{n-n''} \cdot \theta_{k-k''}, \tag{4}$$

with $\theta_{n-n''}$ and $\theta_{k-k''}$ representing the discrete frequency and time CF. This allows to replace the 2-D filter by two 1-D filters, one for filtering in frequency direction and the other one for filtering in time direction.

3.1. Robust Wiener Filter

Since in practice the CF $\theta_{n-n'',k-k''}$ is not perfectly known at the receiver, the filters of the channel estimator have to be designed so that they cover a great variety of delay power spectral densities (PSD) and Doppler PSDs. According to [1], a uniform delay PSD ranging from 0 to τ_{max} and a uniform Doppler PSD ranging from $-f_{D,max}$ to $f_{D,max}$ fulfill these requirements. Then, the discrete frequency correlation function results in

$$\theta_{n-n''} = \frac{\sin(\pi\tau_{max}(n-n'')F_s)}{\pi\tau_{max}(n-n'')F_s} e^{-j\pi\tau_{max}(n-n'')F_s}, \tag{5}$$

and the discrete time correlation function yields

$$\theta_{k-k''} = \frac{\sin(2\pi f_{D,max}(k-k'')T_s')}{2\pi f_{D,max}(k-k'')T_s'}, \tag{6}$$

where T_s' denotes the duration of one OFDM symbol including the guard interval T_g.

Due to the model mismatch in the CF $\theta_{n-n'',k-k''}$, the robust Wiener filter performs significantly worse for certain channel models than the optimum Wiener filter. One possible solution to reduce the performance loss is to estimate the CF $\theta_{n-n'',k-k''}$ of the channel as proposed in [9]. Though some performance gains can be achieved, the complexity triples compared to a robust Wiener filter. Therefore, we propose to use an adaptive filter in the frequency domain to reduce the performance loss with a moderate increase in complexity.

3.2. Adaptive Filter

The adaptive channel estimator presented in this section performs a continual update of the filter coefficients and does not assume knowledge of the channel and noise statistics.

In the following sections, we assume that the pilot symbols are rectangularly distributed [1]. Before the adaptive filter, the initial estimates $\check{H}_{n',k'}$ (Eq. 2) are interpolated with two lowpass filters to obtain

$$\check{H}_{n,k} = \sum_{n'=1}^{\frac{N_c}{N_f}} \sum_{k'=1}^{\frac{N_c}{N_t}} \check{H}_{n',k'} \frac{\sin\left(\frac{\pi}{N_f}(n-n')\right)}{\frac{N_c}{N_f}\sin\left(\frac{\pi}{N_c}(n-n')\right)} e^{-j\frac{\pi}{N_c}\left(\frac{N_c}{N_f}-1\right)(n-n')} \frac{\sin\left(\frac{\pi}{N_t}(k-k')\right)}{\frac{\pi}{N_t}(k-k')} \tag{7}$$

for $n = 1, \ldots, N_c$ and $k = 1, \ldots, N_s$. N_f denotes the distance of pilot symbols in the frequency direction and N_t in time direction.

We apply the BFNLMS algorithm to adapt the filter coefficients [10]. The adaptation starts at OFDM symbol $k = 2$, yielding the estimated CTF

$$\hat{H}_{n,k} = w_{n,k}^* \check{H}_{n,k} \qquad n = 1, \ldots, N_c,\ k = 2, \ldots, N_s, \tag{8}$$

where $w_{n,k}$ is the filter coefficient for the n-th subcarrier of the k-th OFDM symbol.

The filter coefficients are updated according to

$$w_{n,k+1} = w_{n,k} + \frac{\mu}{\left\|\check{\boldsymbol{H}}_k\right\|^2} E_{n,k} \check{H}_{n,k} \qquad n = 1, \ldots, N_c,\ k = 1, \ldots, N_s - 1, \tag{9}$$

where μ is the adaptation constant. Stable operation requires $0 < \mu < 2$. $\left\|\check{\boldsymbol{H}}_{n,k}\right\|^2 = \sum_{n=1}^{N_c} \check{H}_{n,k}^* \check{H}_{n,k}$ is the power of the estimator input vector, and $E_{n,k}$ is the estimation error given by

$$E_{n,k} = H_{n,k} - \hat{H}_{n,k}. \tag{10}$$

Since the true channel fading coefficient $H_{n,k}$ is unavailable, we approximate it by $\check{H}_{n,k}$ in Eq. 7 and thus replace Eq. 10 with

$$E_{n,k} \approx \check{H}_{n,k} - \hat{H}_{n,k}. \tag{11}$$

We initialize the filter coefficients to

$$w_{n,1} = 1, \qquad n = 1, \ldots, N_c, \tag{12}$$

as the BFNLMS recursion in Eq. 9 starts with $k = 2$. Thus, $\hat{H}_{n,k} = \check{H}_{n,k}$ for $n = 1$.

In the proposed adaptive filter algorithm, the number of complex additions and multiplications is $O\left(N_c \cdot N_s\right)$. If the interpolation before the adaptive filtering is implemented efficiently by an FFT in frequency direction and a discrete Fourier transform in time direction, additional $O\left(N_c(N_s^2 + N_c \log_2(N_c))\right)$ complex operations are required. In contrast, the adaptive Wiener filter [9] needs approximately $O\left(N_c^3 N_s\right)$ complex operations.

4. Simulation Results

The following BER curves display performance results for a perfect channel estimator, an optimal, model matched Wiener filter, a robust Wiener filter with model mismatch and $\mathcal{T}_{n,k} = \mathcal{P}$, and an adaptive BFNLMS filter. The simulation results were obtained by simulating an OFDM system with 512 subcarriers and 64 OFDM symbols in one frame. The remaining system parameters and the channel model are chosen according to [11] and are displayed in Table 1 and Fig. 2. If not mentioned otherwise, the channel estimators employ all available pilots for channel estimation. The robust Wiener filter assumes a rectangular delay PSD with $\tau_{max} = T_g$ and a rectangular Doppler PSD with $f_{D,max} = 1500\,\text{Hz}$. Since the pilots are spaced on a rectangular grid and the final channel estimate is obtained by interpolation, the sampling theorem in frequency and time direction must be fullfilled [1], i.e.,

$$\tau_{max} \Delta f \cdot N_f \leq 1 \text{ and } f_{D,max} T_s' \cdot N_t \leq 1/2. \tag{13}$$

In frequency direction we exactly fullfill the sampling theorem of Eq. 13 and in time direction we oversample by a factor of 4.

From Fig. 3, we infer that the optimal Wiener filter almost attains the uncoded BER performance of a perfect channel estimator for this scenario, i.e., the performance difference is approximately 0.2 dB. On contrary, the robust Wiener filter with model mismatch shows an error floor above a BER of 10^{-2}. Additionally, a matched and a robust Wiener filter with 15 filter coefficients in frequency and 4 in time direction are

Table 1. OFDM system and channel parameters

Parameter	Symbol	Value
Bandwidth	B	67.5 MHz
Subcarriers	N_c	512
Subcarrier spacing	Δf	131.836 kHz
FFT length	N_{FFT}	512
Guard interval	T_g	128
Sampling duration	T_{spl}	14.8 ns
OFDM symbols per frame	N_s	64
Modulation	4-QAM	
Channel coding	r	1/2
Pilot spacing frequency	N_f	4
Pilot spacing time	N_t	9
Maximum Doppler frequency	$f_{D,max}$	1500 Hz
Maximum delay spread	τ_{max}	T_g
Channel taps	N_p	8
Power decrement	ΔP	1 dB
Tap spacing	$\Delta\tau$	$16\,T_{spl}$
Sampling Offset	ΔT_{spl}	0 or $0.5\,T_{spl}$

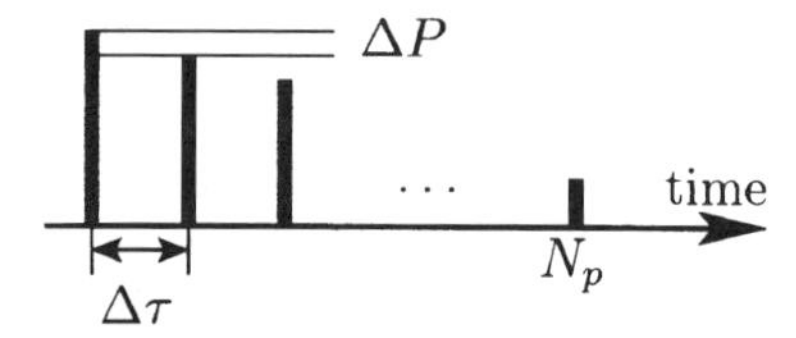

Fig. 2. Power delay profile of the simulated channel model

plotted in Fig. 3. The matched Wiener filter with 15×4 filter coefficients shows a performance loss of 1.7 dB compared to the optimum Wiener filter while the 15×4 robust Wiener filter again exhibits an error floor. The error floor of the robust Wiener filters is caused by the model mismatch and the low sampling rate, which is not sufficient to achieve good performance [1]. The adaptive BFNLMS filter is able to reduce the performance loss of the robust Wiener filter. However, it still performs 3 dB worse than the optimum Wiener filter. According to Fig. 3, the difference between ideal training and the interpolated channel estimates as reference is negligible for the adaptive filter.

Shifting the channel impulse response by $\Delta T_{spl} = 0.5\,T_{spl}$ yields a non-sample spaced channel [12]. Due to the leakage, the channel impulse response is not anymore time limited. Fig. 4 depicts the BER curves for the non-sample spaced channel. Apparently, the robust Wiener filter outperforms the adaptive channel estimator this time, but still loses 3 dB in performance compared to the optimum matched Wiener filter at a BER of 10^{-3}. It attains a lower error floor, now below a BER of 10^{-4}, as the leakage causes a more uniform distribution of the delay PSD. In contrast to Fig. 3, the BER curves of the adaptive channel estimator result in an error floor above a BER

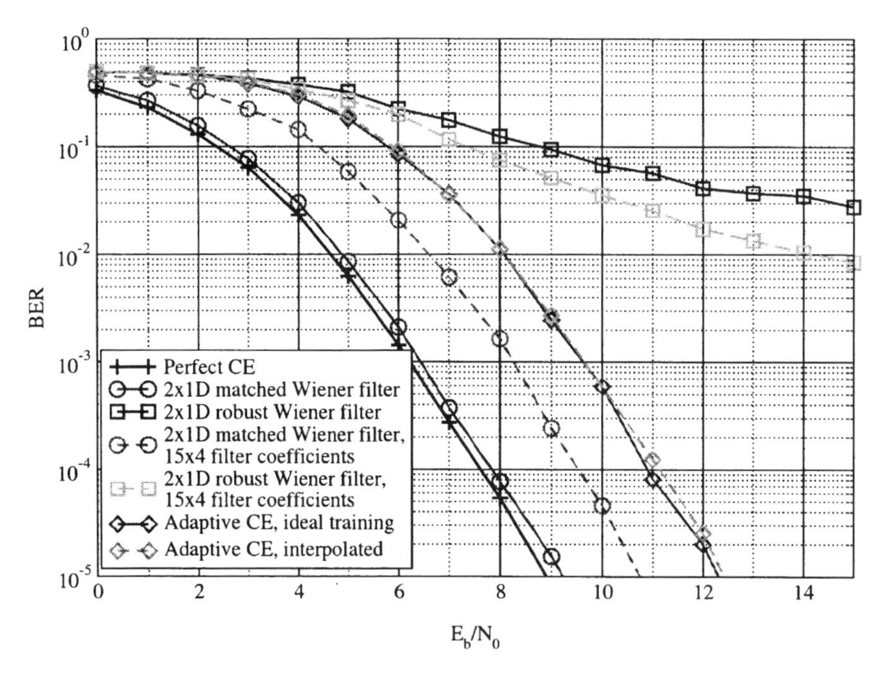

Fig. 3. BER performance for a sample spaced channel

of 10^{-3}. Below a E_b/N_0 of 10 dB, the adaptive filter performs similar to the 15×4 robust Wiener filter. Since the pilot symbols are first interpolated before the adaptive filter is applied, the interpolation error causes an irreducible error floor, which is due to the aliasing of the non-time limited channel impulse response.

5. Conclusions

We have presented a new adaptive PACE. The received pilot symbols are first interpolated in frequency and time direction. Then, a BFNLMS algorithm adapts the filter coefficients without making any assumptions about the channel and noise statistics. Compared to the robust Wiener filter, the adaptive filter reduces the performance loss significantly for a sample spaced channel. However, the robust Wiener filter outperforms the adaptive filter in the case of a non-sample spaced channel. Hence, it is necessary to further investigate adaptive filter algorithms that outperform the robust Wiener filter for both channel scenarios.

6. Affiliations

Stephan Sand and Armin Dammann
German Aerospace Center (DLR),
Institute of Communications and Navigation,
82234 Oberpfaffenhofen, Germany
E–mail: Stephan.Sand@dlr.de
http://www.dlr.de/KN/KN-S/

Gunther Auer
DoCoMo Euro-Labs,
Landsberger Straße 312,
80687 München, Germany

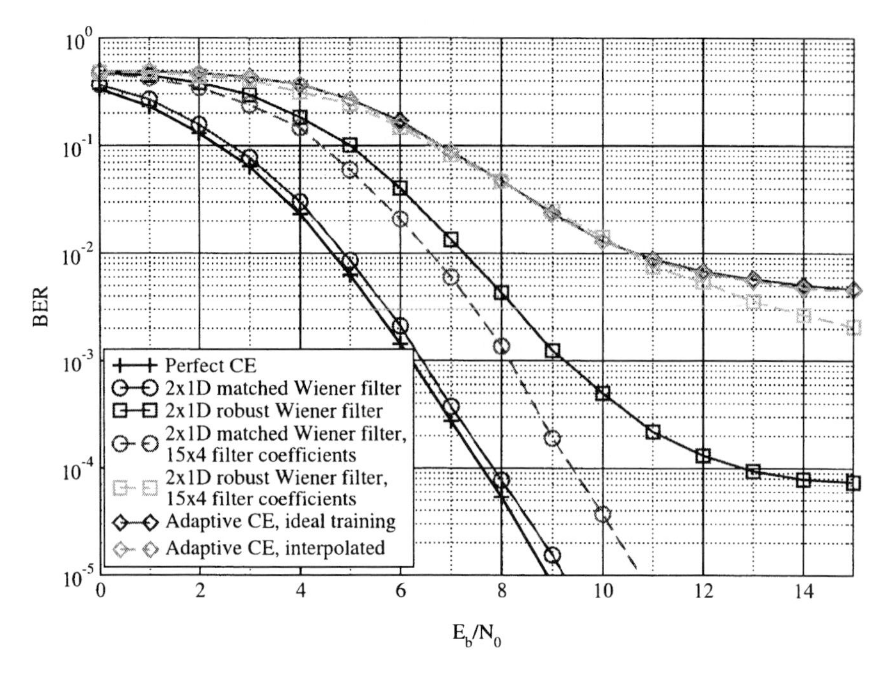

Fig. 4. BER performance for a non-sample spaced channel

REFERENCES

[1] S. Kaiser, "Multi-carrier cdma mobile radio systems - analysis and optimization of detection," Ph.D. dissertation, 1998.

[2] S. Weinstein and P. Ebert, "Data transmission by frequency-division multiplexing using the discrete Fourier transform," *IEEE Trans. Commun.*, vol. 19, pp. 628–634, Oct. 1971.

[3] J. Bingham, "Multicarrier modulation for data transmission: An idea whose time has come," *IEEE Commun. Mag.*, vol. 28, pp. 5–14, May 1991.

[4] ETSI EN 300 401, *Radio Broadcasting Systems; Digital Audio Broadcasting (DAB) to mobile, portable and fixed receivers*, ser. Eurpean Stadard (Telecommunications series), Valbonne, France, February 1995.

[5] ETSI EN 300 744, *Digital Video Broadcasting (DVB); Framing structure, channel coding and modulation for digital terrestrial television*, ser. Eurpean Stadard (Telecommunications series), Valbonne, France, July 1999.

[6] ETSI TS 101 475, *Broadband Radio Access Networks (BRAN); HIPERLAN Type 2; Physical (PHY) layer*, ser. Eurpean Stadard (Telecommunications series), Valbonne, France, February 1995.

[7] P. Hoeher, S. Kaiser, and P. Robertson, "Pilot-symbol-aided channel estimation in time and frequency," in *Proceedings IEEE GLOBECOM*, Phoenix, USA, Apr. 1997, pp. 90–96.

[8] P. Hoeher, "A statistical discrete-time model for the WSSUS multipath channel," *IEEE Trans. Veh. Technol.*, pp. 461–468, Nov. 1992.

[9] M. Necker, F. Sanzi, and J. Speidel, "An adaptive wiener-filter for improved channel estimation in mobile OFDM-systems," in *Proceedings IEEE ISSPIT*, Kairo, Egypt, Dec. 2001.

[10] C. Cowan and P. Grant, *Adaptive Filters*. Englewood Cliffs, NJ: Prentice Hall, 1985.

[11] H. Atarashi, N. Maeda, S. Abeta, and M. Swahashi, "Broadband packet wireless access based on VSF-OFCDM and MC/DS-CDMA," in *PIMRC*, Lisbon, Portugal, Sep. 2002, pp. 992–997.

[12] G. Auer, A. Dammann, and S. Sand, "Channel estimation for OFDM systems with multiple transmit antennas by exploiting the properties of the discrete Fourier transform," in *PIMRC*, Beijing, China, Sep. 2003.

I. MANIATIS, T. WEBER, M. WECKERLE

EXPLOITING A-PRIORI INFORMATION FOR CHANNEL ESTIMATION IN MULTIUSER OFDM MOBILE RADIO SYSTEMS

Abstract. This paper investigates the performance gain for pilot-aided channel estimation in multiuser OFDM mobile radio systems when a-priori information is included in the estimation process. Array antennas are applied at the receivers allowing exploitation of directional information based on the array geometry and the directions of arrival of the considered signals. Furthermore, a-priori channel state information is exploited by the considered channel estimator. Based on an exemplary OFDM air interface described in the paper, the performance of multiantenna channel estimation is compared to the performance of single antenna channel estimation.

1. INTRODUCTION

The suppression of multiuser interference in mobile radio systems is a crucial step towards the improvement of the system performance. The performance of interference mitigation techniques like joint detection (JD) [1] and joint transmission (JT) [2], [3] is limited by the available channel knowledge. This channel knowledge can be obtained by training signal based channel estimation. Significant performance improvements can be achieved for channel estimation by exploiting additional a-priori information such as frequency correlations of the radio channels, directional information about the received signals, correlations of the noise signals at the receiver and channel state information.

This paper describes the pilot-aided technique termed joint channel estimation (JCE), which provides the required channel knowledge in the uplink of a multiuser OFDM mobile radio system [5]. As an exemplary application field for JCE, the system concept termed Joint Transmission and Detection Integrated Network (JOINT) described in [6] is considered in the paper. In [7] JCE is introduced for the case of single transmit and single receive antenna elements in JOINT and in [8] appropriate pilots for JCE are designed. Based on [6], [7] and [8] this paper extends the investigations to the case, where array antennas are applied at the receivers of JOINT.

The considered system model is described in Section 2. In Section 3 the exploitation of directional information concerning the received desired signals and the noise signals is addressed, while Section 4 tackles the inclusion of a-priori channel state information in the JCE process. Along with the definition of the JCE performance criterion and the simulation scenario, simulation results are presented in Section 5. Finally, Section 6 concludes the paper.

K. Fazel and S. Kaiser (eds.), Multi-Carrier Spread-Spectrum, 235–242.

2. SYSTEM MODEL

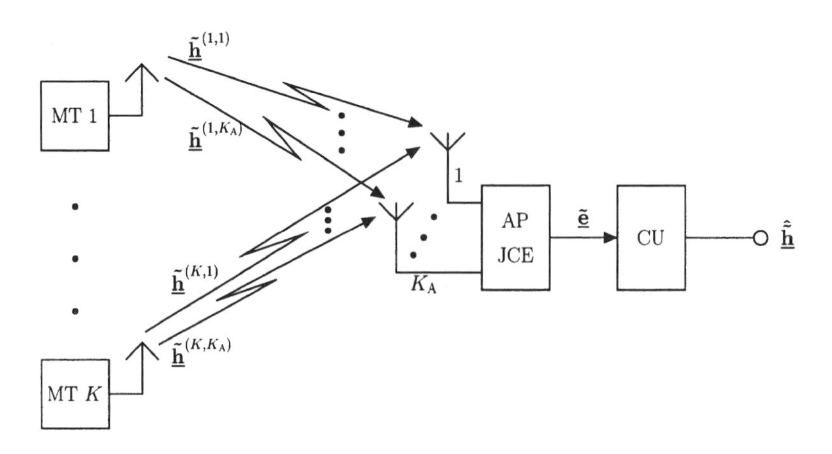

Fig. 1: Uplink scenario in a SA of JOINT

In [6] the system concept of JOINT is introduced. The coverage area of JOINT is divided in service areas (SAs). Each SA contains several access points (APs) and one central unit (CU), which is responsible for signal processing [6]. As stated in [7] it is sufficient to concentrate channel estimation investigations on a single AP. Fig. 1 illustrates the considered uplink scenario in one SA, where K mobile terminals (MTs) are simultaneously active, each of them utilizing a single omnidirectional antenna. In contrast to this, the AP is equipped with a circular array antenna consisting of K_A omnidirectional antenna elements. Each MT $k, k = 1\ldots K$, transmits its pilots using all N_F available subcarriers [7]. The signals reach the AP over the point-to-multipoint mobile radio channels between each MT $k, k = 1\ldots K$, and all K_A antenna elements characterized by the MT specific vectors $\underline{\tilde{\mathbf{h}}}^{(k)}$ of the channel transfer functions. Introducing the total pilot matrix $\underline{\tilde{\mathbf{P}}}$ containing all transmitted pilots, the total vector $\underline{\tilde{\mathbf{h}}}$ including all vectors $\underline{\tilde{\mathbf{h}}}^{(k)}$ and the total vector $\underline{\tilde{\mathbf{n}}}$ representing the noise at the AP, the received signal is expressed by the vector

$$\underline{\tilde{\mathbf{e}}} = \underline{\tilde{\mathbf{P}}}\,\underline{\tilde{\mathbf{h}}} + \underline{\tilde{\mathbf{n}}}. \tag{1}$$

Based on the knowledge of the received signal of (1) and the pilots it is the task of JCE to deliver estimates of the channel transfer functions. For the single receive antenna case described in [7] correlations between transfer function values on adjacent subcarriers are exploited by JCE along with the knowledge of the pilots to produce ML estimates of the radio channels. In the multiple receive antenna case additional a-priori information is exploited by JCE as described in the following two sections.

3. EXPLOITATION OF DIRECTIONAL INFORMATION AND SPATIAL CORRELATIONS

3.1 Desired signals

The application of array antennas at the receiver allows the inclusion of directional information in the JCE process. As mentioned in Section 2, a circular array antenna is utilized at the AP. The K_{A} omnidirectional antenna elements of the array are equidistantly placed on the perimeter of a circle, the centre of which is the reference point (RP) of the array [9]. For simplicity it is assumed in the paper that each signal transmitted by a MT impinges at the AP array from a single direction of arrival (DOA). Furthermore, all K DOAs of the received signals related to the K MTs are assumed to be perfectly known at the CU.

With the DOA knowledge and the given array geometry it is possible to relate the point-to-multipoint radio channels represented by $\underline{\tilde{\mathbf{h}}}$ to the point-to-point radio channels represented by the total vector $\underline{\tilde{\mathbf{h}}}_{\mathrm{d}}$ of the directional channel transfer function [9]. $\underline{\tilde{\mathbf{h}}}_{\mathrm{d}}$ characterizes the radio channel between MT k and an imaginary omnidirectional antenna placed at the RP of the array [9]. Introducing the total blockdiagonal steering matrix $\underline{\tilde{\mathbf{A}}}$, which contains the aforementioned information about the DOAs and the array geometry, the relation

$$\underline{\tilde{\mathbf{h}}} = \underline{\tilde{\mathbf{A}}}\,\underline{\tilde{\mathbf{h}}}_{\mathrm{d}} \tag{2}$$

holds and with (2) the received signal vector of (1) is written as

$$\underline{\tilde{\mathbf{e}}} = \underline{\tilde{\mathbf{P}}}\,\underline{\tilde{\mathbf{A}}}\,\underline{\tilde{\mathbf{h}}}_{\mathrm{d}} + \underline{\tilde{\mathbf{n}}}\,. \tag{3}$$

Exploiting the frequency correlations [7] of the directional point-to-point channel transfer function on adjacent subcarriers and introducing the blockdiagonal matrix $\underline{\tilde{\mathcal{F}}}_{\mathrm{W,tot}}$ containing Fourier coefficients [7] and the total vector $\underline{\mathbf{h}}_{\mathrm{d}}$ of the directional channel impulse response [9], (3) is finally written as

$$\underline{\tilde{\mathbf{e}}} = \underline{\tilde{\mathbf{P}}}\,\underline{\tilde{\mathbf{A}}}\,\underline{\tilde{\mathbf{h}}}_{\mathrm{d}} + \underline{\tilde{\mathbf{n}}} = \underline{\tilde{\mathbf{P}}}\,\underline{\tilde{\mathbf{A}}}\,\underline{\tilde{\mathcal{F}}}_{\mathrm{W,tot}}\,\underline{\mathbf{h}}_{\mathrm{d}} + \underline{\tilde{\mathbf{n}}} = \underline{\tilde{\mathcal{G}}}_{\mathrm{d}}\,\underline{\mathbf{h}}_{\mathrm{d}} + \underline{\tilde{\mathbf{n}}}\,. \tag{4}$$

The inclusion of the frequency correlations together with the DOA knowledge and the array geometry allows the reduction of the number of values to be estimated from $K K_A N_F$ for $\underline{\tilde{\mathbf{h}}}$ to $K W$ for $\underline{\mathbf{h}}_d$. On the other hand, there are still $K_A N_F$ known values available to perform channel estimation. Applying the ML estimation principle for JCE [7] and introducing the total noise covariance matrix $\underline{\mathbf{R}}_{\tilde{n}}$ the vector

$$\hat{\underline{\tilde{\mathbf{h}}}} = \underline{\tilde{\mathbf{A}}} \underline{\tilde{\mathcal{F}}}_{W,tot} \left(\underline{\tilde{\mathcal{G}}}_d^{*T} \underline{\mathbf{R}}_{\tilde{n}}^{-1} \underline{\tilde{\mathcal{G}}}_d \right)^{-1} \underline{\tilde{\mathcal{G}}}_d^{*T} \underline{\mathbf{R}}_{\tilde{n}}^{-1} \underline{\tilde{\mathbf{e}}} = \underline{\mathbf{Z}} \underline{\tilde{\mathbf{e}}} \tag{5}$$

of the estimated channel transfer function results.

3.2 *Noise signals*

The interference originating from adjacent SAs and corrupting the performance of JCE is modelled as temporally white, zero mean Gaussian noise. The omnidirectional noise model [10] is assumed. The spatial covariance matrix $\underline{\mathbf{R}}_s$ describes the spatial correlations of the noise signals among the array antenna elements. With $J_0(\cdot)$ representing the Bessel function of the 1st kind and 0th order, $l^{(i,j)}, i = 1 \ldots K_A, j = 1 \ldots K_A$, representing the distance between antenna elements i and j and λ standing for the carrier wavelength, the elements of the spatial noise covariance matrix $\underline{\mathbf{R}}_s$ are obtained by [10]

$$[\underline{\mathbf{R}}_s]_{i,j} = \sigma^2 \cdot J_0\left(2\pi \frac{l^{(i,j)}}{\lambda}\right), i = 1 \ldots K_A, j = 1 \ldots K_A, \tag{6}$$

for the omnidirectional noise model. Together with the temporal noise covariance matrix

$$\underline{\mathbf{R}}_t = \sigma^2 \cdot \mathbf{I}, \tag{7}$$

which describes the temporal correlations of the noise signals and the Kronecker product $\otimes$, the total noise covariance matrix is given by [10], [11]

$$\underline{\mathbf{R}}_{\tilde{n}} = \underline{\mathbf{R}}_s \otimes \underline{\mathbf{R}}_t . \tag{8}$$

4. EXPLOITATION OF A-PRIORI CHANNEL STATE INFORMATION

In the previous section ML estimation is applied for JCE exploiting a-priori information about the DOAs of the desired and the noise signals. The application of the minimum mean square error (MMSE) estimation principle for JCE allows the inclusion of a-priori channel state information in the estimation process in the face of the power delay profile (PDP) of the considered channel model. This information is included in the channel covariance matrix

$$\underline{\mathbf{R}}_{\tilde{\mathbf{h}}} = \mathrm{E}\left\{\underline{\tilde{\mathbf{h}}}\,\underline{\tilde{\mathbf{h}}}^{*\mathrm{T}}\right\}, \tag{9}$$

of the channel transfer function $\underline{\tilde{\mathbf{h}}}$, which with (2) and (4) is rewritten as

$$\underline{\mathbf{R}}_{\tilde{\mathbf{h}}} = \underline{\tilde{\mathbf{A}}}\,\underline{\tilde{\mathcal{F}}}_{\mathrm{w,tot}}\mathrm{E}\left\{\underline{\mathbf{h}}_{\mathrm{d}}\underline{\mathbf{h}}_{\mathrm{d}}^{*\mathrm{T}}\right\}\underline{\tilde{\mathcal{F}}}_{\mathrm{w,tot}}^{*\mathrm{T}}\underline{\tilde{\mathbf{A}}}^{*\mathrm{T}} = \underline{\tilde{\mathbf{A}}}\,\underline{\tilde{\mathcal{F}}}_{\mathrm{w,tot}}\underline{\mathbf{R}}_{\mathbf{h}_{\mathrm{d}}}\underline{\tilde{\mathcal{F}}}_{\mathrm{w,tot}}^{*\mathrm{T}}\underline{\tilde{\mathbf{A}}}^{*\mathrm{T}}. \tag{10}$$

It can be seen from (10) that $\underline{\mathbf{R}}_{\tilde{\mathbf{h}}}$ includes the knowledge of the array geometry, the DOA information of the desired signals, the frequency correlations and information about the PDP of the point-to-point radio channels included in the covariance matrix $\underline{\mathbf{R}}_{\mathbf{h}_{\mathrm{d}}}$ of the directional channel impulse response $\underline{\mathbf{h}}_{\mathrm{d}}$. With (10) the MMSE estimates of the channel transfer function are obtained by [12]

$$\hat{\underline{\tilde{\mathbf{h}}}} = \underline{\mathbf{R}}_{\tilde{\mathbf{h}}}\underline{\tilde{\mathbf{P}}}^{*\mathrm{T}}\left(\underline{\mathbf{R}}_{\tilde{\mathbf{n}}} + \underline{\tilde{\mathbf{P}}}\,\underline{\mathbf{R}}_{\tilde{\mathbf{h}}}\underline{\tilde{\mathbf{P}}}^{*\mathrm{T}}\right)^{-1}\underline{\tilde{\mathbf{e}}} = \underline{\mathbf{Z}}\,\underline{\tilde{\mathbf{e}}}. \tag{11}$$

5. PERFORMANCE RESULTS

5.1 Performance criterion

In this section the signal-to-noise(-and-interference) ratio (SN(I)R) degradation is defined as the criterion for the JCE performance assessment [7], [8]. The SN(I)R degradation $\delta^{(k,k_{\mathrm{A}},n_{\mathrm{F}})}$ compares the SN(I)R $\gamma_{\mathrm{ref}}^{(k,k_{\mathrm{A}},n_{\mathrm{F}})}$ at the output of a channel estimator in a reference system to the SN(I)R $\gamma_{\mathrm{jce}}^{(k,k_{\mathrm{A}},n_{\mathrm{F}})}$ at the output of the joint channel estimator in the multiantenna case of JOINT. As the reference system the single MT single receive antenna case of JOINT is chosen [7]. With the energy E_{P} of the pilot symbols, the length W of the point-to-point channel impulse response in taps and the variance σ^2 of the noise

$$\delta^{(k,k_{\mathrm{A}},n_{\mathrm{F}})} = \frac{\gamma_{\mathrm{ref}}^{(k,k_{\mathrm{A}},n_{\mathrm{F}})}}{\gamma_{\mathrm{jce}}^{(k,k_{\mathrm{A}},n_{\mathrm{F}})}} = \frac{2E_{\mathrm{P}}}{W\sigma^2} \cdot \left[\underline{\mathbf{Z}}\underline{\mathbf{R}}_{\tilde{\mathbf{n}}}\underline{\mathbf{Z}}^{*\mathrm{T}} \right]_{i,i}, \qquad (12)$$
$$i = (k-1)K_{\mathrm{A}}N_{\mathrm{F}} + (k_{\mathrm{A}}-1)N_{\mathrm{F}} + n_{\mathrm{F}},$$

follows for the SN(I)R degradation. At first glance, the impression is created from (12) that $\delta^{(k,k_{\mathrm{A}},n_{\mathrm{F}})}$ depends on E_{P}, W and σ^2. For the case of ML-JCE presented in Section 3.1 no such dependencies exist [7] and for the case of MMSE-JCE presented in Section 4 the SN(I)R degradation depends on the noise variance σ^2. Of course, in both cases the SN(I)R degradation also depends on the choice of pilots, the array geometry, the MT specific DOAs and the considered noise model.

5.2 *Simulation scenario and results*

In order to assess the performance of JCE with the SN(I)R degradation of (12) computer simulations are conducted based on the following scenario: pilots based on Walsh codes as described in [8] are applied by the MTs. A fully loaded system is assumed , i.e., $N_{\mathrm{F}} = K \cdot W$ holds. As already mentioned in Section 3.1, single DOA is assumed for each received signal and all K DOAs are perfectly known at the CU. The omnidirectional noise case described in Section 3.2 is considered. The MTs are equidistantly located on a circle around the AP. The behaviour of the SNR degradation depending on the MT specific DOA $\varphi^{(k)} = \frac{2\pi}{K}(k-1), k = 1 \ldots K$, of the received signals is investigated. For MMSE-JCE, five different values of the reference SNR

$$10\log_{10}\left(\gamma_{\mathrm{ref}}^{(k,k_{\mathrm{A}},n_{\mathrm{F}})}\right) / \mathrm{dB} = -10, -5, 0, 5, 10 \qquad (13)$$

depending on the noise variance σ^2 are considered.

Fig. 2 shows the obtained results for the case of K_{A}=4 antenna elements at the AP. The figure illustrates the SN(I)R degradation $\delta^{(k,k_{\mathrm{A}},n_{\mathrm{F}})}$ in dB versus the DOA $\varphi^{(k)}$. The upper curve represents the results for ML-JCE and the remaining give the performance of MMSE-JCE for the five different values of the reference SN(I)R $\gamma_{\mathrm{ref}}^{(k,k_{\mathrm{A}},n_{\mathrm{F}})}$ listed next to the figure. Compared to the single antenna case where the SN(I)R degradation equals 0 dB due to the applied pilots [8], the application of array antennas at the receiver decreases the SN(I)R degradation further. Including a priori channel state information in the estimation process by MMSE-JCE results to an additional decrease of the SN(I)R degradation. It is seen that for the considered values of the reference SN(I)R $\gamma_{\text{ref}}^{(k,k_A,n_F)}$ the SN(I)R degradation is decreased up to a factor of 3 compared to ML-JCE.

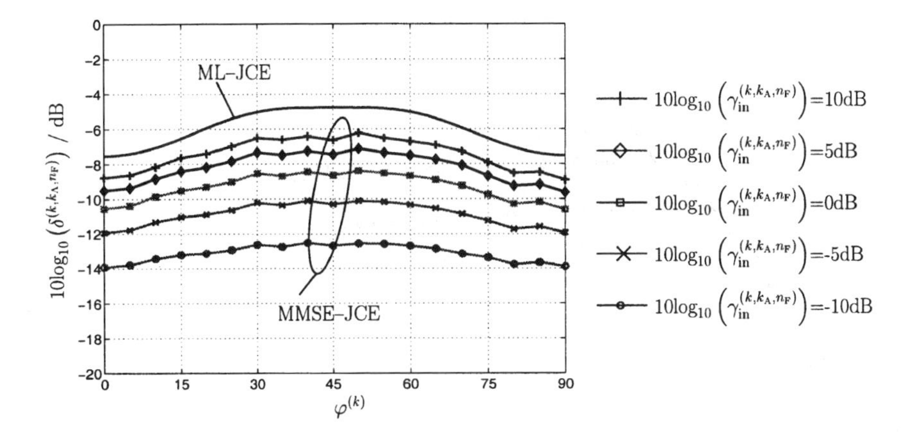

Fig. 2: SN(I)R degradation $\delta^{(k,k_A,n_F)}$ in dB versus the MT specific DOA $\varphi^{(k)}$ for K_A=4 antenna elements

6. CONCLUSIONS

In this paper the performance improvement of the channel estimation technique JCE for OFDM based mobile radio systems is shown for the exemplary system concept termed JOINT. The cases of ML-JCE and MMSE-JCE are compared. The application of array antennas at the receivers of JOINT allows the inclusion of a priori directional information of desired and noise signals in the estimation process and together with the exploitation of the known array geometry the performance of ML-JCE is improved compared to the single antenna case. Exploiting further a priori information in terms of the PDP of the radio channel, MMSE-JCE outperforms ML-JCE, thus introducing a further improvement.

7. ACKNOWLEDGEMENTS

The authors gratefully appreciate the stimulating discussions and the fruitful exchange of ideas with their colleagues at the Research Group for RF Communications, and P.W. Baier, Technical University of Kaiserslautern. The invaluable support of the supercomputing staff at the Central Computer Facility (RHRK), of the Technical University of Kaiserslautern is highly acknowledged.

8. REFERENCES

[1] Verdu, S. (1998). *Multiuser detection.* Cambridge University Press.

[2] Meurer, M., Baier, P.W., Weber, T., Lu, Y. and Papathanassiou, A. (2000): *Joint transmission: advantageous downlink concept for CDMA mobile radio systems using time division duplexing.* IEE Electronics Letters, vol. 36, pp. 900 – 901

[3] Fischer, R.F.H. (2002). *Precoding and Signal Shaping for Digital Transmission.* New York: John Wiley & Sons

[4] Steiner, B. (1995): *Ein Beitrag zur Mobilfunk-Kanalschätzung unter besonderer Berücksichtigung synchroner CDMA-Mobilfunksysteme mit Joint Detection.* Fortschrittberichte VDI, Reihe 10, Nr. 337, VDI-Verlag

[5] Van Nee, R. and Prasad, R. (1984): *OFDM for wireless multimedia communications.* London: Artech House

[6] Weber, T., Maniatis, I., Sklavos, A., Liu, Y., Costa, E., Hass, H. and Schulz, E. (2002): *Joint transmission and detection integrated network (JOINT), a generic proposal for beyond 3G systems.* Proc. 9th International Conference on Telecommunications (ICT '02), vol. 3, pp. 479 – 483

[7] Sklavos, A., Maniatis, I., Weber, T., Baier, P.W., Costa, E., Hass, H. and Schulz, E. (2001): *Joint channel estimation in multi-user OFDM mobile radio systems.* Proc. 6th International OFDM – Workshop (InOWo '01), pp. 3-1 – 3-4

[8] Maniatis, I., Weber, T., Sklavos, A., Liu, Y., Costa, E., Hass, H. and Schulz, E. (2002): *Pilots for joint channel estimation in multi-user OFDM mobile radio systems.* Proc. 7th International Symposium on Spread Spectrum Techniques & Applications (ISSSTA '02), vol. 1, pp. 44 – 48

[9] Papathanassiou, A. (2000): *Adaptive antennas for mobile radio systems using Time Division CDMA and joint detection.* Fortschrittberichte Mobilkommunikation, Band 6, Universität Kaiserslautern

[10] Weckerle, M., Papathanassiou, A., Schmalenberger, R. (1999): *Spatial interference covariance matrix estimation in multiantenna TD-CDMA systems.* COST 259 TD(99) 18

[11] Weckerle, M., Papathanassiou, A., Emmer, D. (1998): *The benefits of intelligent antenna arrays in TD-CDMA – a study based on measured channel impulse responses.* Proc. IEEE 9th International Symposium on Personal, Indoor and Mobile Radio Communications (PIMRC '98), vol. 2, pp. 962-966

[12] Whalen, A.D. (1971): *Detection of signals in noise.* New York: Academic Press

9. AFFILIATIONS

I. Maniatis, T. Weber: Research Group for RF Communications
Technical University of Kaiserslautern
P.O. Box 3049,
67653 Kaiserslautern, Germany
Tel: +49 631 2052920
Fax: +49 631 2053612
e-mail: maniatis@rhrk.uni-kl.de

M. Weckerle: SIEMENS AG
Information & Communication Mobile
Gustav-Heinemann-Ring 115
81739 Munich, Germany

A. KOPPLER, M. HUEMER, A. SPRINGER, R. WEIGEL

TIMING OF THE FFT-WINDOW IN SC/FDE SYSTEMS

Abstract. In this work the positioning of the FFT-window in a single carrier system with frequency domain equalization (SC/FDE) is analyzed. Similar to OFDM (orthogonal frequency division multiplexing), SC/FDE systems are based on blockwise transmission and FFT processing, which enables channel equalization in the frequency domain with low complexity. Comparable to OFDM, the interblock interference (IBI) caused by multipath propagation and by time domain pulse shaping puts some constraints on the FFT-window positioning for SC/FDE. We describe the distortions which are induced by the time domain pulse shaping in combination with blockwise processing. Since in SC/FDE systems information is carried by short pulses in time domain, the effects differ from the corresponding impacts in OFDM, where information is carried by subcarriers in frequency domain. We present error vector magnitude (EVM) measurements and error rate curves obtained from computer simulations. Our results indicate that the performance loss caused by IBI, which becomes worse for higher order modulation schemes, can be minimized by an optimization of the FFT-window position.

1. INTRODUCTION

The Wireless LAN systems described in the IEEE 802.11a and the ETSI Hiperlan/2 standards significantly reduce intersymbol interference (ISI) caused by multipath propagation by the utilization of OFDM modulation [1]. In OFDM systems, equalization is performed by simple multiplication operations in the frequency domain. Consequently, in terms of complexity, OFDM is more attractive for broadband wireless transmission than conventional single carrier systems using time domain equalizer structures. An alternative solution for broadband communication systems has been provided by the concept of SC/FDE [2]-[5]. This approach combines the properties of OFDM and single carrier transmission advantageously.

The paper is organized as follows: In the next section, the main characteristics of an SC/FDE transmission system are resumed briefly. Section 3 describes the pulse shaping and FFT-window positioning and their effects on the system performance. Simulation results are presented in section 4.

2. SC/FDE

The main advantage of SC/FDE over conventional single carrier systems is the efficient frequency domain equalization scheme, where simple multiplication operations are applied to the complex-valued frequency domain samples. Since the samples of the received signal are transformed to the frequency domain utilizing FFT operations, the sample stream has to be divided into blocks determined by the FFT size. Samples are taken for a duration of T_{FFT}, the so-called FFT-window, for further (FFT) processing. In [2] a blockwise transmission scheme similar to OFDM, with the insertion of a cyclic prefix (CP), functioning as a guard interval (GI) between successive blocks, has been described. This GI mitigates interblock interference (IBI) rising from multipath radio channels, in the same way as ISI between OFDM symbols is combated. The duration T_G of the guard interval has to

K. Fazel and S. Kaiser (eds.), Multi-Carrier Spread-Spectrum, 243–250.

be chosen to be longer than T_h, the duration of the channel impulse response $h(t)$. The task of the frequency domain equalizer is to eliminate ISI between the individual symbols of the actual block.

Figure 1 shows exemplarily the structure of one transmitted block, which consists of the original sequence of N symbols with duration $T_{FFT}=NT$ and the cyclic extension with duration T_G. The main physical layer (PHY) parameters for the investigated system are presented in Table 1. Furthermore, Table 1 also contains the parameters of an SC/FDE scheme with similar parameters, which employs a fixed pilot sequence (PS) in the guard interval instead of the cyclic prefix (see also [3]).

The CP lets the linear convolution of the signal with the channel appear to be a circular convolution. In other words, the linear convolution of one cyclically extended transmit block $\hat{s}(t)$, having a length T_G+T_{FFT}, with the channel $h(t)$ can be described by a circular convolution of the original signal $s(t)$ (having only the length T_{FFT}) with the channel $h(t)$. The corresponding frequency domain relation is

$$R(nf_0) = H(nf_0)S(nf_0) + N(nf_0) \tag{1}$$

for $n \in \mathbf{Z}$ and $f_0=1/T_{FFT}$. $R(f)$, $S(f)$ and $H(f)$ are the frequency domain representations of the time domain signals $r(t)$, denoting one period of the received data block, $s(t)$, and $h(t)$, respectively. $N(f)$ is the Fourier transform of the additive noise.

Table 1. Main PHY Parameters of the investigated SC/FDE Systems.

Parameter	SC/FDE w/ cyclix prefix	SC/FDE w/ guard pilots
Net symbol rate	12 Msps	12 Msps
Modulation schemes	BPSK, QPSK, 16/64QAM	BPSK, QPSK, 16/64QAM
Coding rates	1/2, 2/3, 3/4	1/2, 2/3, 3/4
Total symbols per block	76	64
Data symbols per block N_S	64	52
Guard symbols per block N_G	12	12
Block duration T_B	5.33 µs	4.33 µs
Guard time T_G	842 ns	812.5 ns
FFT integration time T_{FFT}	4.49 µs	4.33 µs
Symbol duration T	70 ns	68 ns
Pulse shaping	RRC (α=0.25, α=0.5)	RRC (α=0.25, α=0.5)

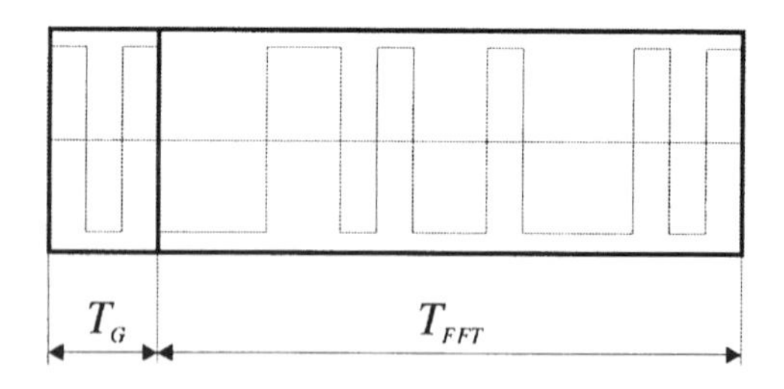

Figure 1. SC/FDE transmit block with cyclic prefix.

If the overall channel frequency response from the input of the transmit pulse shaping filter to the equalizer input, i.e. including transmitter pulse shaping, the physical radio channel and receiver filters like anti-aliasing and matched filter, is given by $C(f)$, the equalizer frequency response is given as

$$E_{ZF}(nf_0) = \frac{1}{C(nf_0)}, \tag{2}$$

if the zero forcing (ZF) criterion is to be fulfilled, or for the minimum mean square error (MMSE) criterion, taking into account the signal-to-noise ratio SNR, as

$$E_{MMSE}(nf_0) = \frac{1}{C(nf_0) + 1/\mathrm{SNR}}. \tag{3}$$

3. FILTER DESIGN AND FFT-WINDOW POSITIONING

Several parameters have to be synchronized in an SC/FDE system [5]. Proposals for time and frequency synchronization techniques based on pilot symbols have been made e.g. in [6].

The exact synchronization in time is a difficult task in many communication systems. However, due to the cyclic extended blockwise transmission, it is sufficient to start the FFT-window somewhere in that fraction of the guard interval, that is not (or only little) distorted by the channel impulse response (see Figure 2). As a consequence, a coarse synchronization of the FFT-window position is usually assumed to be sufficient. From the multipath/IBI problem point of view, similar to OFDM, the optimum FFT-window start seems to be at $t_{start}=0$ at first sight, since then the system is able to combat channel echos with delays up to T_G, see Figure 2.

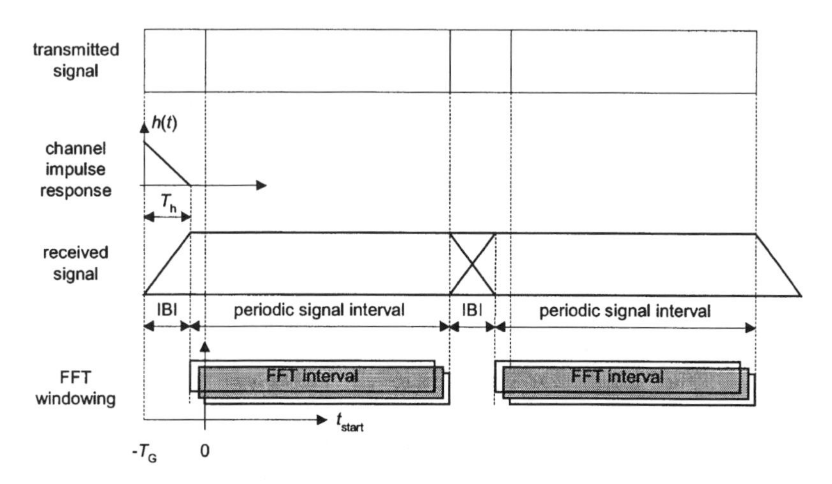

Figure 2. Prevention of interblock interference by the guard interval.

3.1 Influence of the Transmit Pulse Shape

In contrast to OFDM, SC/FDE systems have to utilize time domain pulse shaping. IBI is introduced at the beginning (depending on the pulse length) and at the end of each FFT-block because of the time dilatation of the pulses. Hence the optimal FFT-window position varies, mainly depending on the transmit pulse shape and the multipath channel. In the following, root-raised-cosine (RRC) pulses are considered, where IBI is already produced in the transmitter. The distortions mainly stem from precursors of the (guard) symbols of the subsequent block, but can also result from postcursors from the last symbols of the previous FFT-block. As another (minor) effect, chopping some precursor part of the first few symbols of the data block (because of FFT-windowing) causes a matching loss for these symbols. Matching is however no problem for the end symbols of a block, as their postcursors are included at the beginning of the block due to the CP.

3.2 Shifting of the FFT-Window

Our proposed solution, similar as proposed for OFDM systems, where the subject matter of filtering and optimum FFT-window positioning is for instance addressed in [7], is to shift the FFT-window starting position towards the guard interval, which of course reduces the allowed range for the window position. With this FFT-window shift, the amount of sampled 'noisy' precursors at the block end decreases. At the same time, the distortion from postcursors of the previous block increases slightly, but the matching loss for the symbols at the block begin is reduced. These influences are depicted in Figure 3.

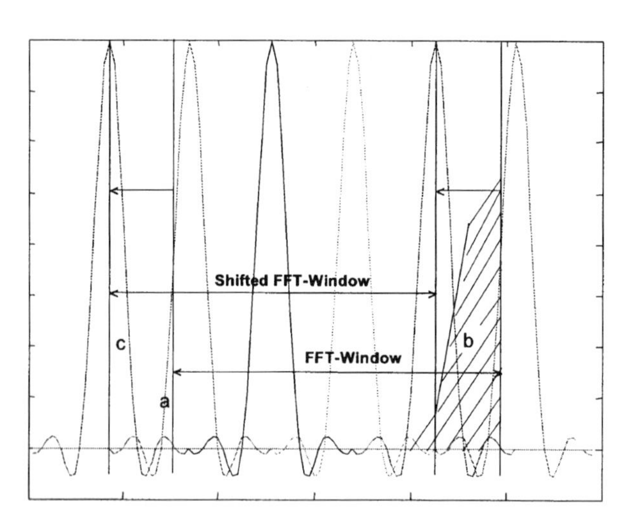

Figure 3. FFT-window shifting. Reduced matching loss for begin symbols (a) and reduction of precursors from subsequent block (b). No matching loss for end symbols which are repeated in the guard interval (c).

3.3 Error Vector Magnitude Measurements

To optimize the FFT-window shift, the influence of the synchronization location shall be evaluated by means of error vector magnitude measurements, as described in the IEEE 802.11a standard. The RMS (root mean square) EVM value is thus given by

$$EVM = \frac{\sum_{i=1}^{N_b} \sqrt{\frac{\sum_{j=1}^{L_b} \left[\sum_{k=1}^{N_S} \{(I_{i,j,k} - I_{i,j,k}^0)^2 + (Q_{i,j,k} - Q_{i,j,k}^0)^2\} \right]}{N_S L_b P_0}}}{N_b}, \tag{4}$$

where L_b is the number of blocks per burst, N_b is the number of simulated bursts, (I, Q) denotes the in-phase and quadrature-phase parts of the actual observed constellation point, (I^0, Q^0) is the corresponding ideal constellation point, and P_0 is the average power of the constellation.

4. SIMULATION RESULTS

All of the following simulation results have been obtained for 64-QAM modulation, since the effects of FFT-window position variations can be observed best for this constellation. The EVM in dB over the synchronization location, i.e. t_{start}, is depicted in Figure 4 for ideal multipath-free and noise-free transmission, for RRC pulses with a duration of L=24 symbol intervals and a roll-off factor of α=0.25 and α=0.5, respectively. The results for the CP and the PS schemes are shown. While the distortions are very weak for synchronization locations in the middle of the GI, they grow towards the edges.

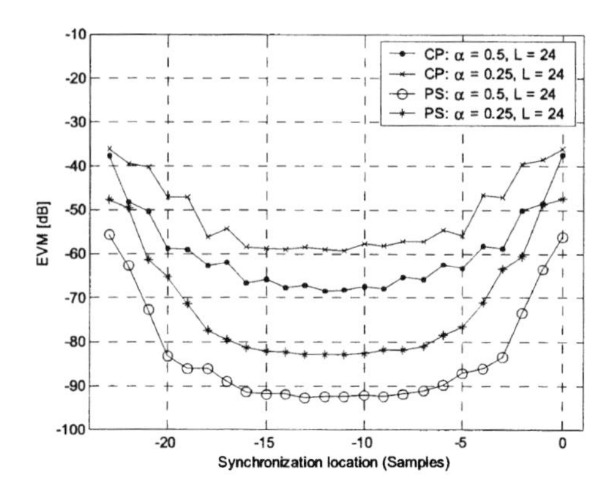

Figure 4. EVM over synchronization position. The sampling rate is 2/T, therefore e.g. –2 describes a shift of one symbol interval T towards the guard interval.

Better insight into the different influences is possible with Figure 5, which shows how each of the 64 block symbols is affected by the synchronization variations. Here, the EVM has been individually calculated by dropping the summation over k in (4). Starting the sampling for the FFT-window at the end of the GI (t_{start} near zero), the distortions due to precursors of the subsequent block and the mismatching of the first symbols in the block can be observed. If the window starts near the beginning of the GI, the influence of the postcursors of the previous block increases. This effects the guard symbols, which (due to circular convolution) correspond to the end symbols of the block after equalization. This explains the EVM characteristics plotted in Figure 5.

The influence on the system performance in form of BER is shown in Figure 6 for the AWGN channel. FFT-window position $t_{start} = 0$ causes a saturation in the error behavior for uncoded 64-QAM modulation. This saturation is exclusively caused by the weak performance of the last symbol of the block, which is heavily distorted by precursors of the subsequent block. A small shift of the window towards the GI eliminates these distortions, the influence on the BER is then negligible.

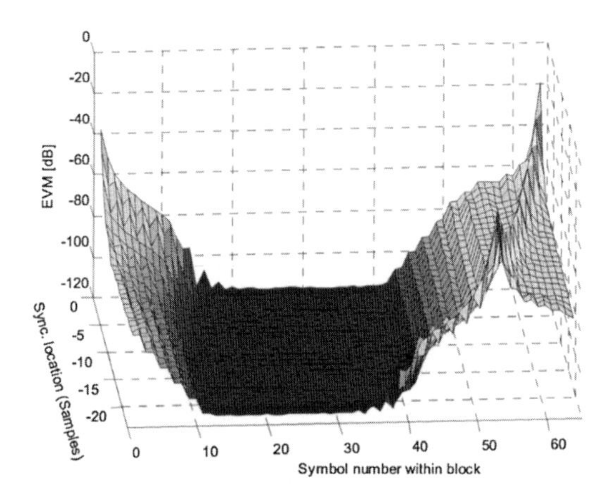

Figure 5. EVM over block symbols and synchronization position (α=0.5, L=24).

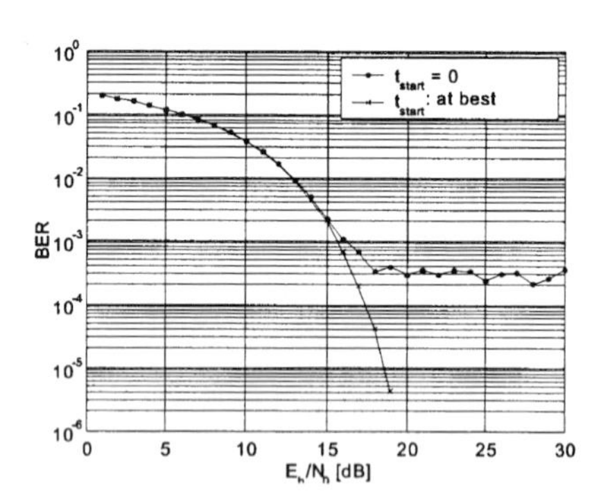

Figure 6. BER for the AWGN channel (α=0.5, L=24).

Practical results for multipath conditions are obtained with channel snapshots that have been generated with the model described in [8]. Two snapshots with a delay spread of 100ns, respectively, but with different frequency selectivity have been used for simulation. For bandwidth efficiency reasons, RRC pulses with α=0.25, L=8 are employed. EVM measurements for both snapshots, and BER curves for the minor frequency selective channel are presented in Figure 7 and Figure 8, respectively, for the CP scheme. Since the channel impulse response distorts the GI, the optimum synchronization location is displaced from the center towards the right end of the GI. The EVM curves clearly indicate that also for multipath conditions the optimum synchronization location is not at t_{start}=0. Simulations applying a large number of different multipath impulse responses have shown that the optimum position is typically located within the interval $-2T \le t_{start} \le T/2$.

It must be mentioned that the BER behavior in multipath environment is mainly dictated by the actual frequency selectivity of the channel. Optimizing the window position does not guaranty low error rates, but optimizes the BER behavior for the particular channel.

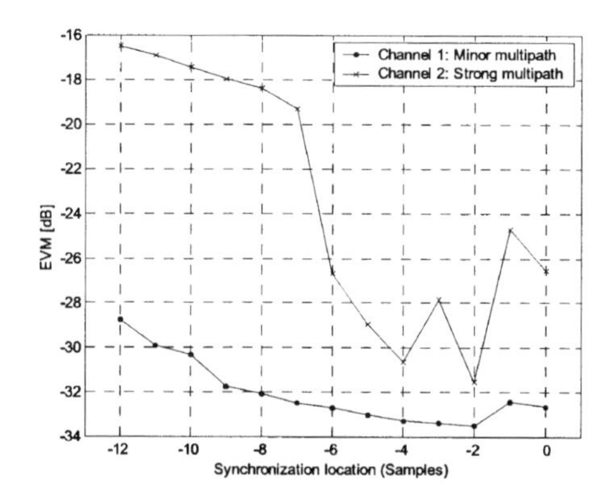

Figure 7. EVM for multipath conditions (α=0.25, L=8).

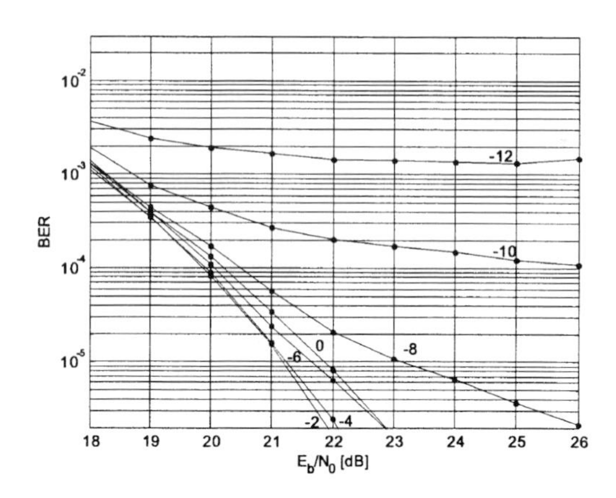

Figure 8. BER for channel 1 and different FFT-window offsets (64-QAM uncoded).

We note that simulation runs using the PS scheme (parameters see Table 1) instead of the CP scheme show improved EVM characteristics for the data symbols (see e.g. Figure 4 for the AWGN case). This is mainly due to the fact that here the last N_G symbols of an equalized block, which suffer most from the IBI effects, represent the known pilot sequence, which may be used for decision feedback equalization (DFE) detection or for synchronization purposes, but which carries no data. However, in the multipath case the same conclusion as for the CP scheme, namely, that the optimum synchronization location is typically located within the interval $-2T \leq t_{start} \leq T/2$, proves to be true also for the PS scheme.

5. CONCLUSION

The aspect of FFT-window positioning in SC/FDE systems has been discussed and the diverse impacts have been described. The optimum window position is mainly influenced by the multipath impulse response and by time domain pulse shaping, which causes IBI. We conclude, that the optimum synchronization location, i.e., the starting point of the FFT-window, is typically located within the interval $-2T \leq t_{start} \leq T/2$. To optimize the BER behavior, we propose to shift the FFT-window 2-4 samples from the end of the GI towards the GI.

6. REFERENCES

[1] R. van Nee and R. Prasad, *OFDM for Wireless Multimedia Communications*. Boston: Artech House Publishers, 2000.

[2] H. Sari, G. Karam, and I. Jeanclaude, "Frequency-domain equalization of mobile and terrestrial broadcast channels," in *Proc. International Conference on Global Communications (Globecom '94)*, San Francisco, CA, Nov.-Dec. 1994, pp. 1-5.

[3] D. Falconer, S. L. Ariyavisitakul, A. Benyamin-Seeyar, and B. Eidson, "Frequency domain equalization for single-carrier broadband wireless systems," *IEEE Communications Magazine*, vol. 40, no. 4, pp. 58-66, April 2002.

[4] M. Huemer, A. Koppler, L. Reindl, and R. Weigel, "A review of cyclically extended single carrier transmission with frequency domain equalization for broadband wireless transmission," *European Transactions on Telecommunications (ETT)*, vol. 14, no. 4, July/August 2003.

[5] M. Huemer, *Frequenzbereichsentzerrung für hochratige Einträger-Übertragungssysteme in Umgebungen mit ausgeprägter Mehrwegeausbreitung*, Ph.D. dissertation, University of Linz, Austria, 1999 (in German).

[6] A. Czylwik, "Low overhead pilot-aided synchronization for single carrier modulation with Frequency Domain Equalization," in *Proc. IEEE Global Telecommunications Conference (GLOBECOM'98)*, Sydney, Australia, 1998.

[7] M. Engels, *Wireless OFDM Systems*. Boston: Kluwer Academic Publishers, 2002.

[8] J. Fakatselis, "Criteria for 2.4 GHz PHY comparison of modulation methods," Document IEEE P802.11-97/157r1, Nov. 1997.

A. Koppler and A. Springer are with the Institute of Communications and Information Engineering, Johannes Kepler University, Linz, Austria.
Mario Huemer is with the University of Applied Sciences, Hagenberg, Austria.
R. Weigel is with the Institute for Technical Electronics, University of Erlangen-Nuremberg, Germany.

SNJEZANA GLIGOREVIC, RAINER BOTT,
ULRICH SORGER

EQUALIZATION FOR MULTI-CARRIER SYSTEMS IN TIME-VARYING CHANNELS

1. INTRODUCTION

Orthogonal Frequency Division Multiplexing (OFDM) systems generally require estimation and tracking of the channel parameters to perform coherent demodulation. A pilot pattern for channel estimation in OFDM systems mostly consists of equidistant pilot symbols in time and frequency direction. The channel estimation can be done by exploiting the correlations of the channel frequency response at different frequencies and times [1, 2, 3].

In [4] a parameter based equalization scheme is used, where equalization is performed in the frequency domain using model functions, which are optimized in the Maximum Likelihood (ML) sense based on the received signal. However, this approach leads to lower data rates since smaller pilot spacing in time is required due to large time-variations of the channel. Moreover, Inter-Carrier-Interference (ICI) cannot be avoided, and hence an equalizer is needed. The method in [5] for estimation and equalization in the time domain is considered in this letter for application in the frequency domain. Here, the channel is described by the discrete spreading function (DSF) [1], which is constant within one OFDM symbol. No a-priori knowledge about the channel is available, except the maximum delay, τ_{max}, and the maximum possible Doppler frequency, f_{Dmax}, of the considered mobile radio communication.

However, there are some disadvantages of the frequency domain approach in comparison with the time domain approach, which are discussed later in this letter. In Section II, a description of a time-variant channel is given. Section III describes channel estimation along a known data sequence while in Section IV joint channel and data estimation is considered. Finally, in Section V simulation results are presented and conclusions are given in Section VI.

[1] This function corresponds to the sampled Doppler-Delay-Spread Function defined in [6]

K. Fazel and S. Kaiser (eds.), Multi-Carrier Spread-Spectrum, 251–258.

2. TIME–VARYING CHANNEL

We assume the l-th OFDM symbol, $l = -\infty, ..., +\infty$,

$$d^{(l)}(t) = \sum_{n=0}^{N-1} D_n^{(l)} \cdot \hat{g}(t - l(T_s + T_g)) \cdot e^{j2\pi f_n (t - lT_s)} \quad (1)$$

to be sent over a frequency–selective and time–varying channel $h(\tau, t)$. The data signal is given in the frequency domain by

$$D(f) = \sum_{n=0}^{N-1} D_n \cdot \delta(f - f_n) \text{ with } f_n = \left(n - \frac{N}{2} + 1\right) \cdot \Delta f,$$

where f_n denotes the carrier frequency, $n = 0, ..., N-1$, and $\Delta f = 1/T_s$ the carrier spacing.

The transmit pulse is assumed to have a rectangular form, i.e. $\hat{g}(t) = \text{rect}^{+}_{T_s+T_g}(t + T_g)$, where $\text{rect}^{+}_{T}(t) = 1$ for $0 \leq t < T$ and zero otherwise. Moreover, it is assumed that Inter-Symbol-Interference (ISI) is avoided by a guard interval using the 'cyclic prefix' (CP) technique [7]. At the receiver the filter $\breve{g}(t) = \text{rect}^{+}_{T_s}(t)$ and a sampling with frequency $f_a = \frac{1}{T} = N \cdot \Delta f$ are used. This gives the received signal

$$r_i = \sum_{m=0}^{M-1} d_{i-m} \cdot h_m(i) + w_i \quad (2)$$

with $h_m(t)$ an equivalent channel. Moreover, $\mathbf{w}$ are uncorrelated samples of the filtered additive white Gaussian noise and $M = \lfloor \frac{\tau_{max}}{T} \rfloor + 1$ with τ_{max} the maximum delay introduced by the channel. For the mobile channel the time-variation of $h_m(t)$ is induced by the relative movements of transmitter, scatterer and receiver. These motions introduce a Doppler shift that describes the time–variations.

As the channel is physically limited in delay and Doppler frequency, it is convenient to describe the channel by the spreading function which depends only on these two variables.

This function is linked to the channel impulse response (CIR) by the Fourier transform over the absolute time t:

$$S_m(f_D) = \int_{-\infty}^{\infty} h_m(t) \cdot e^{-j2\pi f_D t} dt. \quad (3)$$

According to the sampling theorem, if the observation time is limited, the values $S_{m,k}$, obtained by sampling in the Doppler frequency domain, are sufficient to describe $h_m(t)$ completely.

Here, the observation interval is one OFDM symbol of length NT. This gives

$$h_m(t) = \sum_{k=-K}^{K} S_{m,k} \cdot e^{j\frac{2\pi k}{NT}t}, \tag{4}$$

where $K = \lfloor f_{Dmax} NT \rfloor + 1$ if the time-variations of the CIR are periodical in t within one OFDM symbol [5].

Applying the discrete Fourier transform (DFT) with respect to the delay samples yields the discrete Doppler Resolved Transfer Function

$$H_{n,k} = \frac{1}{N} \sum_{m=0}^{M-1} S_{m,k} \cdot e^{-j\frac{2\pi m}{N}n}, \tag{5}$$

where n, $0 \leq n \leq N-1$, denotes the normalized sub-carriers frequencies. The received signal in the frequency domain is given by

$$R_n = \frac{1}{N} \sum_{i=0}^{N-1} \left(\sum_{m=0}^{M-1} d_{i-m} \sum_{k=-K}^{K} S_{m,k} e^{j\frac{2\pi i}{N}k} \right) e^{-j\frac{2\pi n}{N}i} + W_n$$

$$= \sum_{k=-K}^{K} \sum_{m=0}^{M-1} S_{m,k} e^{-j\frac{2\pi(n-k)}{N}m} \cdot \underbrace{\frac{1}{N} \sum_{i=0}^{N-1} d_{i-m} e^{-j\frac{2\pi(n-k)}{N}(i-m)}}_{D_{n-k}} + W_n$$

$$= \sum_{k=-K}^{K} \sum_{m=0}^{M-1} S_{m,k} \cdot D_{n-k} e^{-\frac{j2\pi m(n-k)}{N}} + W_n. \tag{6}$$

As the W_n are obtained from a DFT of the uncorrelated and white w_n the W_n are uncorrelated AWGN samples in the frequency domain.

For a time-invariant channel $K = 0$ and thus

$$R_n = \sum_{m=0}^{M-1} S_{m,0} \cdot D_{n-0} e^{-\frac{j2\pi m(n-0)}{N}} + W_n = N \cdot D_n \cdot H_n + W_n. \tag{7}$$

Hence, the time-variation introduces Inter-Carrier-Interference (ICI) in the Multi-Carrier system, leading to the same equalization problem as in the time domain.

On the other hand for channel estimation only a relatively small number of coefficients $S_{m,k}$ need to be computed as the spreading function is bounded in delay and Doppler frequency. Hereby the $S_{m,k}$ are uncorrelated under the wide sense stationary uncorrelated scattering (WSSUS) [6] assumption. If, moreover, the $S_{m,k}$ are Gaussian distributed, minimizing the mean square error (MMSE) between received signal and its estimate yields the optimum solution.

3. CHANNEL ESTIMATION

Obviously, $S_{m,k} = 0$ for $|k| > K$ is only valid if $h_m(t)$ is periodical in t on the observation interval. Otherwise, $h_m(t)$ exhibits a jump at the borders leading to other nonzero $S_{m,k}$. This problem is mitigated by oversampling the spreading function with respect to the Doppler frequency f_D which leads to a virtually longer observation interval. Due to oversampling, the spreading matrix is expanded to $M \times (2KG_f + 1)$ coefficients, where the grid-factor, $\frac{1}{G_f} < 1$, describes the step width in f_D-direction. This, however, implies that interpolated values between received data symbols R_n and transmitted data symbols D_n, $\bar{R}_n$ and $\bar{D}_n$ respectively, have to be calculated. One obtains

$$\bar{R}_n = \sum_{k=-\bar{K}}^{\bar{K}} \sum_{m=0}^{M-1} S_{m,k} \cdot \bar{D}_{n-k} \cdot e^{-\frac{j2\pi m(n-k)}{\bar{N}}} + W_n, \tag{8}$$

where $\bar{N} = \lceil N \cdot G_f \rceil$, $\bar{K} = \lceil K \cdot G_f \rceil$ and $n = 0, ..., \bar{N} - 1$, is given in a matrix form by

$$\bar{\mathbf{R}} = \bar{\mathcal{D}} \cdot \mathbf{S} + \mathbf{W}. \tag{9}$$

Here, $\mathbf{S} = (S_{0,-\bar{K}}, ..., S_{m,k}, ..., S_{(M-1),\bar{K}})^T$ is the vector of spreading coefficients, $\bar{\mathbf{R}} = (\bar{R}_0, \bar{R}_1, ..., \bar{R}_{\bar{N}-1})$, $\mathbf{W} = (W_0, W_1, ..., W_{\bar{N}-1})$ and $\bar{\mathcal{D}}$ is a $\bar{N} \times M(2\bar{K}+1)$ data matrix with rows $\underline{\bar{D}}(n) = (\bar{D}_{n-k} e^{-j\frac{2\pi m \cdot k}{\bar{N}}})_{k=-\bar{K},..,\bar{K}}^{m=0,..,M-1}$, $n = 0, .., \bar{N} - 1$. Note that due to the cyclic convolution $\bar{D}_{-k} = \bar{D}_{\bar{N}-k}$ is valid for $k > 0$. Parseval's Theorem [8] now allows us to apply the MMSE approach in the frequency domain. Assuming that the N data symbols of one transmitted OFDM symbol to be known and uncorrelated W_n, we obtain from $\frac{\partial}{\partial S_{i,k}} (\bar{\mathbf{R}} - \bar{\mathcal{D}} \cdot \mathbf{S})^H (\bar{\mathbf{R}} - \bar{\mathcal{D}} \cdot \mathbf{S}) = 0$ the normal equation

$$\bar{\mathcal{D}}^H \bar{\mathcal{D}} \tilde{\mathbf{S}} = \bar{\mathcal{D}}^H \bar{\mathbf{R}}, \tag{10}$$

[8] where $\tilde{\mathbf{S}}$ are the spreading coefficients to be estimated. This estimation problem has an unique solution if

$$N \cdot G_f > M \cdot (2K \cdot G_f + 1) \tag{11}$$

and $\bar{\mathcal{D}}^H \bar{\mathcal{D}}$ is a regular matrix.

For the estimation one needs to rely on the transmitted OFDM symbol as one cannot predict a fast time varying channel over a full OFDM symbol. Furthermore, one should consider only N_P known symbols within one OFDM symbol, where $N_P << N$ and N_P is not fulfilling (11). Therefore, in Section IV the algorithm proposed in [9] for combined channel estimation and equalization in the time domain is considered for application in the frequency domain.

4. JOINT CHANNEL ESTIMATION AND EQUALIZATION (JCE)

Here, we propose the JCE algorithm, which applies a continuously adapted channel estimation in a reduced state equalizer. For simplicity, we first consider the N_P sub-carriers at the beginning of the OFDM symbol to be modulated with known data symbols. After transforming the received signal in the frequency domain, the first N_P data symbols are used for the initialization of the algorithm. Then, the recursive least square (RLS) algorithm renews the channel estimation for each following sub-carrier n by extending this 'training sequence' with the hypothesis of the corresponding (interpolated) data symbol $\bar{D}_n$. Table 1 summarizes the filter algorithm.

Table 1: RLS algorithm for estimating the spreading coefficients.

Initialization:	$\tilde{\mathbf{S}}(0) = 0$
	$\mathcal{P}(0) = \delta^{-1}\mathbf{I}, \quad \delta = \text{small positive constant}$
Loop	$\mathbf{K}(n) = \frac{\mathcal{P}(n-1)\cdot\underline{\bar{D}}(n)}{1+\underline{\bar{D}}^H\cdot\mathcal{P}(n-1)\cdot\underline{\bar{D}}(n)}$
$n = 1, 2, ...$	$e(n\|n-1) = \bar{R}(n) - \tilde{\mathbf{S}}^H(n-1)\cdot\underline{\bar{D}}(n)$
compute:	$\tilde{\mathbf{S}}(n) = \tilde{\mathbf{S}}(n-1) + \mathbf{K}(n)\cdot e^*(n\|n-1)$
	$\mathcal{P}(n) = \mathcal{P}(n-1) - \mathbf{K}(n)\cdot\underline{\bar{D}}^H(n)\cdot\mathcal{P}(n-1)$
	$e(n) = \bar{R}(n) - \tilde{\mathbf{S}}^H(n)\cdot\underline{\bar{D}}(n)$

A separate channel estimation is calculated for every path, assuming the hypothetical data sequence to be the correct one. As the spreading function is 'non-causal' in f_D direction, the hypotheses for the data symbols $\bar{D}_{n-K}$, .., $\bar{D}_{n+K}$ are needed. All possible hypotheses of the data sequence build a data tree. To keep the complexity low only a limited number of paths in the tree is traced to the end of the transmitted sequence. The traced paths are chosen according to the metric Δ, which can be calculated recursively [10] by $\Delta(n) = \Delta(n-1) + e(n)e(n|n-1)$. At the end of the tree the path with the smallest metric corresponds to the estimated data sequence. By decreasing the number of known symbols at the beginning of one OFDM symbol, the number of paths retained at the beginning of the tree should be increased for reliable equalization. OFDM symbols with equidistant pilots are also considered in the simulations, assuming the first one at the beginning of the symbol. In this case higher complexity is required for the initialization, because more paths have to be considered at the beginning, to ensure the true path is kept till the end.

5. SIMULATION RESULTS

To demonstrate the performance of the proposed JCE approach, a transmission over a time–variant channel with [2] $|f_{D_{max}}NT| \geq 1$ is considered in the simulations. No coding is applied and therefore we assume an automatic gain control in the simulations, i.e. there is no information loss in the received signal due to fading. The variance of the additive white Gaussian noise is $\sigma^2 = \frac{N_0}{2}$. One OFDM symbol consists of $N = 64$ sub-carriers, modulated by using binary phase shift keying (BPSK) modulation, i.e. $D_n \in \{-1, +1\}$. The length of the training sequence or number of pilot symbols is $N_P = 12$.

In the following figures bit error rate (BER) over signal to noise ratio (SNR) is given, where the superposition of $N_{int} = 3$ neighboring data symbols is considered for each interpolation value. In Fig. 1(a,b) the result of simulations in frequency domain for a different number of paths P retained in each step of the equalization tree are presented for a periodical (a) and for a non-periodical CIR (b) and compared with the time domain with $P = 8$. The results for a non distorting channel (AWGN) are given for comparison. With $G_f = 1.4$ in Fig.1(b) the Fourier series expansion error can be significantly reduced only by additional increasing the number of paths P. However, the time domain approach shows for the same number of paths a better performance if oversampling is needed.

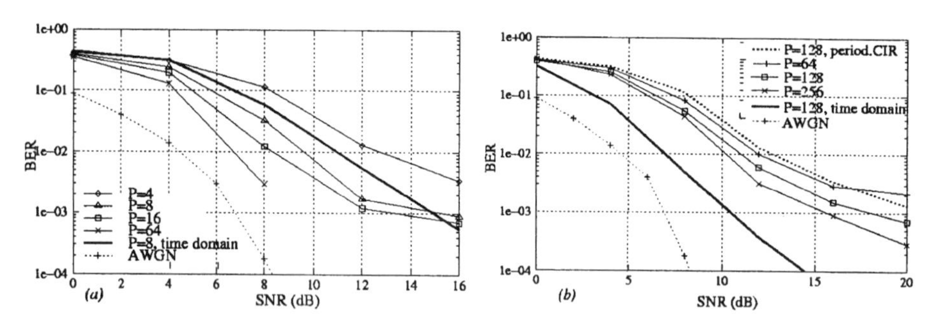

Figure 1. Performance of the JCE method in the frequency domain for (a) a periodical CIR with $|\tau_{max} \cdot f_{D_{max}}| = 0.0156$ *and (b) a non-periodical CIR with* $|\tau_{max} \cdot f_{D_{max}}| = 0.04$. *The number of estimated spreading coefficients is (a)* $N_S = 9$ *and (b)* $N_S = 21$.

The simulations with an initial training sequence and equidistant pilot symbols are compared in Fig. 2a. The degradation of the performance for the simulation with equidistant pilot symbols is due to

[2] We characterize the time–variation of the channel by the number $|f_{D_{max}}NT|$ of periods of a cosine signal in an observation interval of length NT.

a poor estimate of the spreading coefficients at the beginning of the block, which often causes the loss of the true path. Fig. 2b shows the mean square identification error $MSIE = E\{||\tilde{h}_m(i) - h_m(i)||^2\}$ of the channel estimation over the time for different number of neighboring symbols N_{int}, which are considered for the interpolation. By increasing N_{int}, more hypotheses and therefore more paths should be considered in the equalization tree.

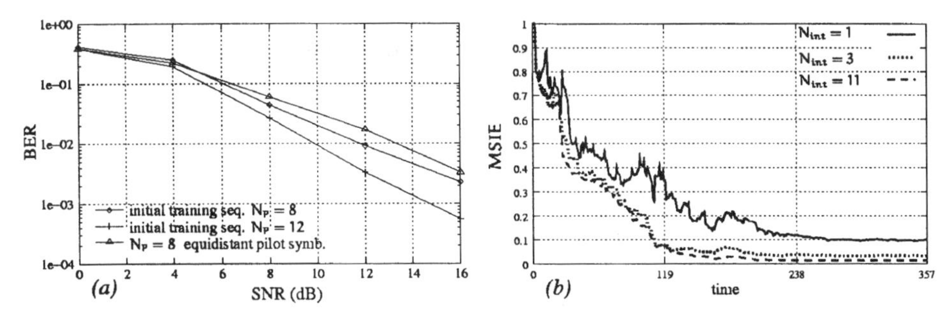

Figure 2. (a) Performance of the JCE method in the frequency domain for a non-periodical CIR with an initial training sequence and equidistant pilot symbols. (b) Mean squared error of the estimated CIR over the time, i.a. iterations step of the RLS algorithm.

6. SUMMARY AND CONCLUSIONS

In this paper, an algorithm for joint channel estimation and equalization for OFDM systems has been derived. In case of a large time-variation of the CIR, it is advantageous to describe the channel by its spreading function, which is time-invariant within one OFDM symbol. With the proposed method a tracking of the channel is feasible, and the velocity of the mobile station movement affects only the complexity but not the system performance. Still in comparison with the equivalent approach in the time domain [5], the JCE approach in the frequency domain shows some disadvantages:

- Due to the DFT and the cyclic extension, the information from the actual OFDM symbol can not be used for the estimation of the DSF for the subsequent OFDM symbol.
- Higher complexity is required for the initialization because the spreading function is non-causal in f_D and the CIR is not continuous after cutting the guard interval.
- Oversampling in the frequency domain increases the complexity and affects the performance.
- The JCE algorithm benefits from known data symbols at the beginning of the transmission block, i.e. OFDM symbol, which is in general

not convenient for OFDM transmission.
Anyhow, crucial for the complexity in time or frequency domain is the size of the frequency selectivity or respectively the time-variation of the channel. Apart from the complexity, the proposed method shows the same performance as in the time domain, i.e. an effective data estimation is possible despite the fast time–variation of the channel impulse response. The accuracy of the estimation can be improved by increasing the number of spreading coefficients to be estimated or the number of paths retained in each state of the equalization tree. Significant improvements are to be expected by limiting the number of possible data sequences, for example by considering coding. Furthermore, potential assumptions about the channel model could also be considered in this algorithm to reduce the dimension of the estimation problem.

REFERENCES

1. Hoeher,P., Kaiser,S., Robertson,I. (1997). Two-dimensional pilot-symbol-aided channel estimation by Wiener filtering. *in Proc. ICASSP'97*,1845-1848.
2. Li,Y., Cimini,L.J, Sollenberger,N.R. (1998, July). Robust channel estimation for OFDM systems with rapid dispersive fading channels. *IEEE Tran. Commun.*, 46, 902-915.
3. Yang,B., Letaief,K.B., Cheng,R.S., Cao,Z. (2001, March). Channel Estimation for OFDM Transmission in Multipath Fading Channels Based on Parametric Channel Modeling. *IEEE Tran. Commun.*, 49(5).
4. Chang,M.X., Su,Y.T. (2002, April). Model-Based Channel Estimation for OFDM Signals in Rayleigh Fading. *IEEE Tran. Commun.*, 30(4).
5. Gligorevic,S. (2002, September). Joint Channel Estimation and Equalization for Fast Time-Variant Multipath Channels. *In Proc. PIMRC'02.*
6. Bello,P.A. (1963). Characterization of Randomly Time-Variant Linear Channels. *IEEE Tran. on Communications Systems*, 11, 360-393.
7. Nee,R.van, Prasad,R. (2000). OFDM for Wireless Multimedia Communications. Artech House.
8. Haykin,S. (1986). Adaptive Filter Theory. Prentice-Hall.
9. Broeck,I.De, Kullmann,M., Sorger,U. (1999, September). Reduced State Joint Channel Estimation and Equalization. *Sec. Int. Workshop on Multi-Carrier Spread-Spectrum, Oberpfaffenhofen.* Kluwer Academic Publishers.
10. Traeger,J. (1998). Kombinierte Kanalschtzung und Decodierung fuer Mobilfunkkanle. Dissertation TU Darmstadt. Aachen: Berichte aus der Kommunikationstechnik, Shaker Verlag.

S. Gligorevic is with the Institute for Telecommunication,Darmstadt University of Technology, Germany. R. Bott is with Rohde & Schwarz GmbH, Munich, Germany. U. Sorger is with the Institute Supérieur de Technologie, Luxembourg.

Wei Zhang, Markus A. Dangl and Jürgen Lindner

PERFORMANCE ANALYSIS OF THE DOWNLINK AND UPLINK OF MC-CDMA WITH CARRIER FREQUENCY OFFSET

Abstract: In this paper, we investigate the influence of carrier frequency offset (CFO) on a MC-CDMA system and evaluate the system performance. At the receiver either a multiuser detector based on a recurrent neural network structure or a conventional linear MMSE multiuser detector is employed. The evaluation is done for the additive white Gaussian noise (AWGN) channel and a time-invariant multipath channel, respectively.

1. INTRODUCTION

MC-CDMA, *Multi-Carrier Code Division Multiple Access*, is one of the most promising candidates for future communication systems due to its high power and bandwidth efficiency. MC-CDMA employs OFDM modulation such that a high bandwidth efficiency can be achieved. However, one drawback of OFDM is its sensitivity to Doppler shift and carrier frequency offset (CFO) caused by the frequency difference between oscillators at transmitter and receiver. CFO gives rise to inter-channel interference (ICI) as well as multiuser interference (MUI) in a multiuser system, and thus results in severe performance degradation even if the CFO is relatively small compared to the subcarrier frequency spacing [5].

In this contribution, we first analyze the effects of CFOs on the downlink and uplink of MC-CDMA on the basis of a general vector-valued transmission model [6, 7], where a transmission with full load is inherently assumed. For simplicity we suppose a time-invariant channel. The effect of CFO on OFDM may be formulated by means of an equivalent discrete-time matrix, by which we can analyze CFO independent of the channel characteristics. However, in a multipath channel, MUI caused by CFO can be increased or reduced depending on the channel characteristics.

A multiuser detector based on a recurrent neural network structure (RNN-MUD) [8] will be employed at the receiver, which takes advantage of the principle of an iterated nonlinear feedback of tentative decisions (soft feedback) and provides a competitive solution for interference limited multiuser communications. RNN-MUD shows good performance for the cancellation of the MUI caused by CFO for the AWGN channel and also for some multipath channels. Furthermore, a conventional linear MMSE detector (MMSE-MUD) is used for comparison.

The paper is organized as follows. Section 2 describes the MC-CDMA system model. In Section 3, the effects of CFOs on the downlink and uplink of an MC-CDMA system are studied. Section 4 gives a brief introduction to the detection

K. Fazel and S. Kaiser (eds.), Multi-Carrier Spread-Spectrum, 259–268.

method based on the structure of a recurrent neural network. Finally, in Sections 5 and 6, we show simulation results and conclude the paper, respectively.

2. SYSTEM MODEL

2.1 *Downlink*

As a first step, a downlink vector transmission at a MC-CDMA system with N users (fully loaded system) is considered. At the transmitter, at any time instant k, a single source symbol $x_l[k]$ of user l is replicated into N parallel copies. Each branch of the parallel stream is multiplied by an associated element $u_{i,l}$ of a given spreading code $\underline{u}_l$ (in our case an orthogonal spreading code is utilized, and $\underline{u}_l = [u_{i,l}]_{N\times 1}$) and modulated onto the corresponding OFDM-subcarrier i. For simplicity we ignore k when only one symbol block transmission is investigated. The transmit symbol on subcarrier i is the sum of spread symbols from all users:

$$s_i = \frac{1}{\sqrt{N}} \sum_{l=1}^{N} u_{i,l} x_l \;. \tag{1}$$

The transmit symbols on all the subcarriers can be expressed as a vector $\underline{s}$:

$$\underline{s} = \underline{U}\,\underline{x}\,, \tag{2}$$

where $\underline{x} = [x_l]_{N\times 1}$, $\underline{s} = [s_i]_{N\times 1}$, and $\underline{U}$ is a spreading matrix with $\underline{U} = (1/\sqrt{N})[\underline{u}_l]_{1\times N}$. The $\underline{s}$ is then transmitted over the OFDM subchannels where an N-point DFT is used. In the downlink case, the system is synchronous such that the received vector of user l is:

$$\underline{\tilde{s}} = \underline{R}_{O,l}\underline{s} + \underline{n}_0\,, \tag{3}$$

where $\underline{R}_{O,l}$ represents the equivalent discrete-time OFDM channel matrix for user l. Since signals arriving at the receiver of user l go through a common channel and user l could be any user, we replace $\underline{R}_{O,l}$ with $\underline{R}_O$. $\underline{n}_0$ denotes the noise vector with covariance matrix $2N_0\underline{R}_O$, where N_0 is the component variance of the complex AWGN process prior to matched filtering. The elements in $\underline{n}_0$ are uncorrelated. Assuming that the length of the cyclic prefix N_G is not less than the channel delay spread L, each subcarrier signal experiences flat fading. Hence, $\underline{R}_O$ is a diagonal matrix, and

$$\underline{R}_O = \underline{H}^H \underline{H}\,, \tag{4}$$

where $\underline{H} = diag\{\underline{h}\}$ with main diagonal elements $\underline{h} = [h_1, \cdots h_i \cdots h_N]$, and h_i represent the transfer function of subchannel i. $(\cdot)^H$ denotes the conjugate transpose operation. After despreading, the despread vector $\underline{\tilde{x}}$ will be:

$$\underline{\tilde{x}} = \underline{U}^H \underline{\tilde{s}} = \underline{R}_{M,DL} \underline{x} + \underline{\tilde{n}}\,, \tag{5}$$

where $\underline{R}_{M,DL}$ stands for the equivalent discrete-time channel matrix of the downlink of an MC-CDMA system and $\underline{R}_{M,DL} = \underline{U}^H \underline{R}_O \underline{U}$. In general, $\underline{\tilde{n}}$ is the colored noise vector after despreading. In the case of AWGN only, since $h_i = 1$ for any sub-channel i and the usage of orthonormal spreading codes, both $\underline{R}_O$ and $\underline{R}_{M,DL}$ are identity matrices. Therefore the transmission of signals is just disturbed by the white noise $\underline{n}$. On the other hand, in the case of a multipath channel where $\underline{R}_O$ is a diagonal matrix but $\underline{R}_O \neq \underline{I}$ ($\underline{I}$ denotes the identity matrix), the orthogonality between spreading codes will be destroyed and thus multiuser interference (MUI) will occur.

2.2 *Uplink*

For the uplink of an MC-CDMA system, we consider a synchronous scenario where perfect symbol timing is assumed. Note that signals arriving at the base station receiver are from different users and may undergo different fading, hence index l in $\underline{R}_{O,l}$ cannot be ignored. The despread vector is obtained by:

$$\underline{\tilde{x}} = \underline{U}^H \sum_{l=1}^{N} \underline{R}_{O,l} \underline{s}_l + \underline{\tilde{n}} = \underline{U}^H \sum_{l=1}^{N} \underline{R}_{O,l} \underline{c}_l x_l + \underline{\tilde{n}}\,, \tag{6}$$

A general equivalent channel matrix can also be obtained by combining the equivalent channels of the different users, which we denote as $\underline{R}_{M,UL}$, and

$$\underline{\tilde{x}} = \sum_{l=1}^{N} (\underline{U}^H \underline{R}_{O,l} \underline{c}_l)\, x_l = \underline{R}_{M,UL} \underline{x} + \underline{\tilde{n}}\,. \tag{7}$$

3. EFFECT OF CARRIER FREQUENCY OFFSET

3.1 *Downlink*

Now we consider the system with carrier frequency offset. In the time domain, the influence of a carrier frequency offset can be viewed as a multiplication with $\exp(j2\pi\Delta f t + \theta_0)$, where Δf denotes the absolute carrier frequency offset and θ_0 is a constant phase shift. For simplicity, we assume Δf to be constant and θ_0 to be zero. Consequently, the equivalent OFDM channel matrix $\underline{R}_O$ can be rewritten as:

$$\underline{R}_O = \underline{F}\, \underline{E}\, \underline{F}^H \underline{H}^H \underline{H}\,, \tag{8}$$

where $\underline{F}$ represents the Fourier matrix and $\underline{F}^{-1} = \underline{F}^H$ the inverse of $\underline{F}$. $\underline{E}$ is a diagonal matrix with $\underline{E} = diag\{\underline{e}\}$,where $\underline{e}$ is a vector consisting of sampled phase errors caused by CFO in the time domain. Taking into account the guard time, we define $\underline{e} = [\exp(j2\pi\varepsilon N_G/(N+N_G)), \cdots, \exp(j2\pi\varepsilon(N+N_G-1)/(N+N_G))]$, where $\varepsilon = \Delta f T_s$

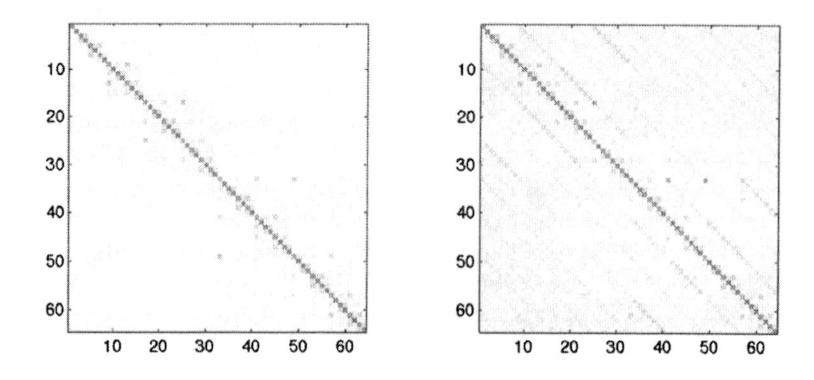

Figure 1. *The equivalent channel matrix of an MC-CDMA system with CFO*

denotes the carrier frequency offset normalized to subchannel spacing. T_s is the OFDM symbol duration. For the downlink of MC-CDMA with CFO, the equivalent channel matrix $\underline{R}_{M,DL}$ can also be expressed as $\underline{R}_{M,DL} = \underline{U}^H \underline{R}_O \underline{U}$, but now $\underline{R}_O$ is defined by Eq. (8).

To concentrate on the effect of carrier frequency offset, we assume that $\underline{H}^H \underline{H} = \underline{I}$ (AWGN channel). If there is no CFO, namely $\varepsilon = 0$, $\underline{R}_O = \underline{F}\,\underline{E}\,\underline{F}^H = \underline{I}$ and thus the orthogonality between OFDM-subcarriers as well as the orthogonality between spreading codes is preserved. If $\varepsilon \neq 0$, then $\underline{F}\,\underline{E}\,\underline{F}^H \neq \underline{I}$, therefore both orthogonalities will be destroyed, resulting in inter-channel interference (ICI) for OFDM and multiuser interference (MUI) for MC-CDMA. The despread vector in Eq. (5) can be rewritten as:

$$\underline{\tilde{x}} = \underbrace{\mathrm{diag}\{\underline{R}_{M,DL}\}\underline{x}}_{\text{desired part}} + \underbrace{(\underline{R}_{M,DL} - \mathrm{diag}\{\underline{R}_{M,DL}\})\underline{x}}_{\text{MUI}} + \underline{\tilde{n}}\,, \tag{9}$$

where the main diagonal elements of $\underline{R}_{M,DL}$ correspond to the desired part of the despread symbols, while off-diagonal elements correspond to MUI. Note that each main diagonal element of $\underline{R}_{M,DL}$ consists of two parts: one is the summation of the diagonal elements of the $\underline{R}_O$ matrix and the other is the combination of the given spreading code and off-diagonal elements of $\underline{R}_O$, which corresponds to ICI for OFDM. The left picture in Fig. 1 depicts the magnitudes of $\underline{R}_{M,DL}$ for such a case, where normalized CFO is $\varepsilon = 0.33$.

For a multipath channel where $\underline{H} \neq \underline{I}$, as discussed in Subsection 2.1, the orthogonality may also be destroyed by multipath propagation giving rise to even stronger MUI. Meanwhile, the mutual effects of multipath propagation and frequency offset result in much worse system performance. The right picture in Fig. 1 illustrates such a scenario.

It must be emphasized that increasing the transmit power will not help to

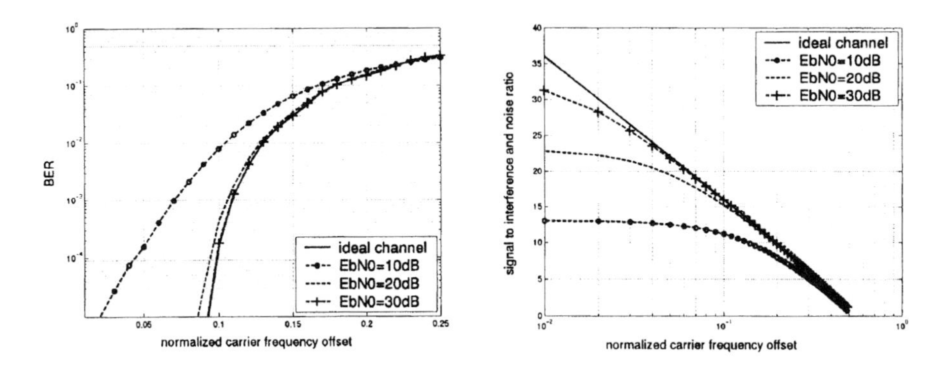

Figure 2. BER (left) performance and SINR (right) degrade with CFO

suppress the MUI caused by CFO, since the power of interference is proportional to transmit power and this proportion depends on the value of ε. Therefore, accurate knowledge of CFO (optimal estimate) and compensation at the receiver are necessary. To reveal this relation, we calculate the signal to interference and noise ratio (SINR) for a given E_b/N_0. The SINR is obtained by:

$$SINR = \frac{\sigma_u^2}{\sigma_{MUI}^2 + \sigma_n^2}, \tag{10}$$

where σ_u^2, σ_{MUI}^2 and σ_n^2 denote the average power of desired part of despread signal, MUI and the power of noise, respectively. For the special case of $\sigma_n^2 \to 0$, only the signal to interference ratio (SIR) is considered. Fig. 2 illustrates how the SINR decreases with increase of CFO (left) and therefore the bit error rate (BER) performance degrades (right). In order to maintain an BER less than 10^{-4} at an $E_b/N_0 = 20\,\text{dB}$, the normalized CFO must be less than 0.09.

3.2 Up*link*

As mentioned above, we consider a synchronous scenario for the uplink, assuming that the signals from distinct users are accompanied by different carrier frequency offsets. Therefore, for user l the equivalent OFDM channel matrix $\underline{R}_{O,l}$ is obtained by:

$$\underline{R}_{O,l} = \underline{F}\,\underline{E}_l \underline{F}^H\,\underline{H}_l^H\,\underline{H}_l \tag{11}$$

The despread vector at the receiver is also calculated by Eq. (6). Apparently, mutual effects of the different $\underline{E}_l$ and relevant channel characteristics will make the entries of the equivalent channel matrix $\underline{R}_{M,UL}$ more complicated. Thus, for simplicity the same channel characteristics are assumed for all the users.

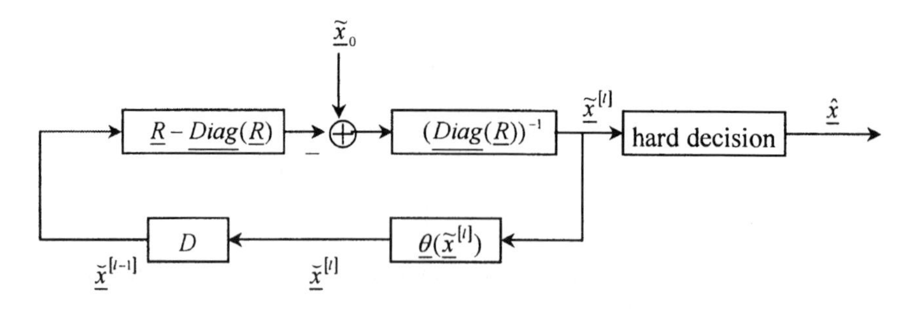

Figure 3. Structure of multiuser detector based on RNN

4. MULTIUSER INTERFERENCE CANCELLATION

A multiuser detector based on a recurrent neural network structure (RNN-MUD) is employed at the receiver for the suppression of multiuser interference. The advantage of RNN-MUD is its lower complexity compared with Minimum Mean Square Error multiuser detector (MMSE-MUD).

4.1 *Structure of RNN-MUD*

Fig. 3 depicts the block diagram structure of the RNN-MUD. The fundamental principle is a symbol-based iterative (partial) interference subtraction. The iteration process can be performed in different ways. In this paper serial updating is employed for the better performance. In Fig. 3, $\underline{\tilde{x}}_0$ denotes the initial value of the received vector, $\underline{\tilde{x}}^{[l]}$ represents an estimate of the transmit vector in the lth iteration, and $\underline{R}$ is the equivalent channel matrix with $\underline{R} = [r_{ij}]_{N \times N}$. The equalized symbol $\tilde{x}_i^{[l]}$ in iteration l is thus obtained by:

$$\tilde{x}_i^{[l]} = \frac{\tilde{x}_{0i}}{r_{ii}} - \sum_{j=1}^{i-1} \frac{r_{ij}}{r_{ii}} \tilde{x}_j^{[l]} - \sum_{j=i+1}^{N} \frac{r_{ij}}{r_{ii}} \tilde{x}_j^{[l-1]}, \tag{12}$$

Details can be found in [8, 10].

4.2 *RNN-MUD for MUI cancellation*

We first employ RNN-MUD under the assumption that perfect knowledge of the channel as well as CFO is available. Then a CFO estimation approach introduced by Moose [3] is used. It takes advantage of the fact that CFO does not vary with the time. The estimation method is derived for AWGN channel, but shows also good performance in multipath channel. We assume that the estimation of CFOs in the uplink is possible. In addition, a conventional linear MMSE-MUD is utilized for comparison.

5. SIMULATION RESULTS

5.1 *System parameters*

The physical layer of the WLAN standard IEEE 802.11a has been partly employed:

1) Total number of the OFDM-subcarriers: $N = 64$.
2) The cyclic prefix are 16 samples.
3) QPSK constellation is used for transmit symbols.
4) Possible dynamical range of normalized CFO: [-0.33, 0.33], according to the tolerable accuracy of the carrier frequency.
5) In the uplink perfect time synchronization is assumed.
6) 10 iterations are used in the RNN-MUD.
7) For multipath propagation, the channel impulse responses $h(n)$ have length of 5 and the power of $h(n)$ decays exponentially with n.

5.2 *Performance analysis*

Fig. 4 depicts the average BER versus the average E_b / N_0 (Signal to Noise Ratio per bit) for the downlink of an AWGN channel (left) and a multipath channel (right), where we set the normalized CFO to $\varepsilon = 0.33$. Compared to the detection without equalization, both RNN-MUD and MMSE-MUD show good performance compensating for the degradation caused by CFOs. At high E_b / N_0, the performance of MMSE-MUD is better than that of RNN-MUD. If Moose method is utilized and the estimate of the CFO is available, several dB SNR loss will occur compared to the perfect case for a given BER. In addition, it is evident that better performance is obtained for the AWGN channel. It must be pointed out that for the downlink, the frequency compensation can also be realized in the time domain by adjusting the receiver oscillator, and for some multipath channels RNN-MUD will show an error floor.

Fig. 5 illustrates the uplink scenario, where ε_l is distributed uniformly in the range of [-0.33, 0.33]. The simulation results show that the MUI caused by CFOs can be suppressed if the exact knowledge of CFOs is available and RNN-MUD shows better performance than MMSE-MUD. However, estimates from Moose method give no help for the interference cancellation due to the complexity of the system and the large range of the frequency offsets. If the CFOs are limited to a small range, e.g. in the range of [-0.1, 0.1], a tolerable BER performance can be obtained. In addition, if the number of users transmitting data simultaneously is small compared to full load case, a better performance can also be achieved.

6. CONCLUSION

In this paper, the effects of carrier frequency offsets on a fully loaded MC-CDMA system have been investigated. The RNN-MUD and MMSE-MUD have been employed at the receiver to compensate the impairment caused by CFOs. It was shown that MUI caused by CFOs alone can be compensated for by RNN-MUD and MMSE-MUD in the downlink as well as in the uplink if perfect knowledge of carrier frequency offsets is available. The analysis and simulation were done for an uncoded transmission. Future work has to include coding.

REFERENCES

[1] J. Armstrong, "Analysis of new and existing methods of reducing intercarrier interference due to carrier frequency offset in OFDM", *IEEE Trans. Commun.*, vol. 47, No. 3, pp 365-369, March 1999.

[2] Y. Zhao and S. Häggman, "Intercarrier interference self-cancellation scheme for OFDM mobile communi-cation systems," *IEEE Trans. Commun.*, vol. 49, No. 7, July 2001.

[3] P. H. Moose, "A technique for orthogonal frequency division multiplexing frequency offset correction," *IEEE Trans. Commun.*, vol. 42, pp 2908-2913, Oct. 1994.

[4] T. M. Schmidl and D. C. Cox, "Robust frequency and timing synchronization of OFDM," *IEEE Trans. Commun.*, vol. 45, No. 12, Dec. 1997.

[5] H. Steendam and M. Moeneclaey, "The effect of carrier frequency offsets on downlink and uplink MC-DS-CDMA", *IEEE Trans. Journal. on select. Areas*, Vol. 19, No. 12, pp 2528-2536, Dec. 2001.

[6] J. Lindner, M. Nold, and W. G. Teich, "Future mobile communication systems: signal processing in time, space, and user direction," *DSPCS'99, February 1999, Australia,* pp 281-288, 1999.

[7] J. Lindner, "MC-CDMA in the context of general multiuser/multisubchannel transmission methods," *Europen Trans.Telecomm.*, vol. 10, No. 5, pp 351-367, July/Aug. 1999.

[8] C. Sgraja, W. G. Teich, A. Engelhart, J. Lindner, "Multiuser/multisubchannel detection based on recurrent neural network structures for linear modulation schemes with general complex-valued symbol alphabet," *Proceedings COST 262 Workshop Multiuser Detection in Spread Spectrum Communications,* Schloss Reisensburg, pp. 45-52, Jan. 17-18, 2001.

[9] S. Verdu, "Multiuser detection" *Cambridge, Cambridge University Press*, 1998.

[10] W. G. Teich, and M. Seidl, "Code division multiple access communications : multiuser detection based on a recurrent neural network structure," *Proceedings IEEE ISSSTA 1996, Mainz/Germany,* pp. 979-984.

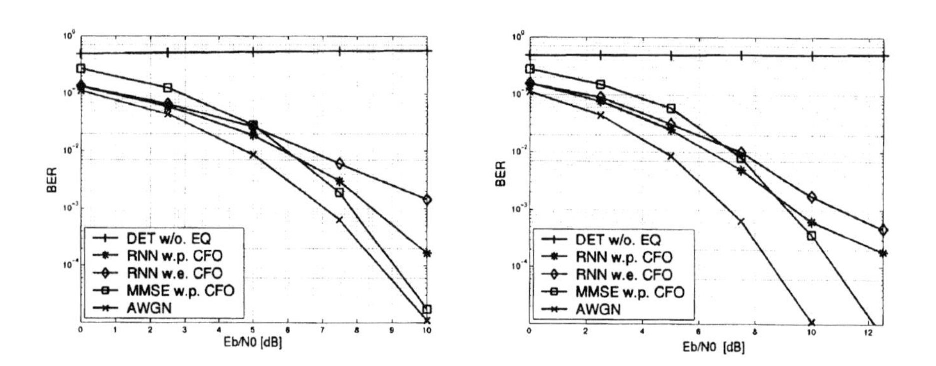

Figure 4. Downlink performance for the AWGN channel (left) and a multipath channel(right)

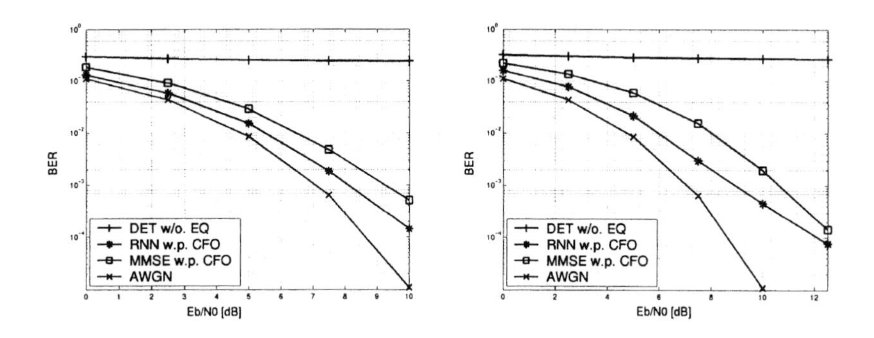

Figure 5. Uplink performance for the AWGN channel (left) and a multipath channel (right) (with 64 users)

L. SANGUINETTI, M. MORELLI, U. MENGALI

SPACE-TIME MULTI-USER DETECTION FOR MC-CDMA SYSTEMS IN THE PRESENCE OF CHANNEL ESTIMATION ERRORS

Dept. of Information Engineering, University of Pisa - Italy

Abstract. We assess the impact of channel estimation errors in the uplink of a multicarrier code-division multiple-access (MC-CDMA) system. Channel estimates are computed by means of the least-mean-square (LMS) algorithm and are passed to either a linear minimum mean square error (MMSE) detector or to a non-linear parallel interference cancellation (PIC) receiver. Both single-antenna and multiple-antenna receivers are considered. It is shown that the system performance depends heavily on the quality of the channel estimates, confirming that channel estimation plays a crucial role in the uplink of the network.

1. INTRODUCTION

Multi-Carrier Code-Division Multiple-Access (MC-CDMA) is a multiplexing technique that combines orthogonal frequency division multiplexing (OFDM) with direct sequence CDMA [1]. It has been proposed as a viable candidate for future generation broadband communications due to its advantages over other conventional multiplexing techniques, which include higher spectral efficiency, increased flexibility and robustness to frequency selective fading [2].

In an MC-CDMA system the data of different users are spread in the frequency domain using orthogonal signature sequences. In the presence of multipath propagation, however, signals undergo frequency-selective fading and the spreading codes loose their orthogonality. This results in multiuser interference (MAI) at the receiver, which strongly limits the system performance. Some advanced signal-processing techniques are available to mitigate interference and multipath distortion. They are largely categorized into space-time processing with antenna array and multi-user detection [3]-[4]. Linear multiuser receivers in the form of decorrelating detectors [5] or minimum mean square error (MMSE) detectors are usually proposed to achieve a reasonable trade-off between performance and complexity. Alternatively, non-linear techniques can be adopted in the form of parallel interference cancellation (PIC) receivers, that are very promising for applications on the uplink channel.

All the above techniques require explicit knowledge of the channel impulse response of each user. While several channel estimation schemes have been proposed for the downlink [6]-[7], few results are available for the uplink [8]-[9]. The main problem here is that the channel responses of the active users are different from one another and the base station (BS) must estimate a large number of

K. Fazel and S. Kaiser (eds.), Multi-Carrier Spread-Spectrum, 269–278.

parameters. This is expected to degrade the quality of the estimates with respect to the downlink, where only a single channel response is involved.

In the following we investigate the effect of channel estimation errors in the uplink of an MC-CDMA network. We consider a quasi-synchronous system in which each user is time-aligned to the BS reference in a way similar to that discussed in [10]. A least-squares (LS) approach is employed to perform channel acquisition while channel tracking is pursued by means of the LMS algorithm. Both linear and non-linear space-time multiuser receivers are considered to perform data detection.

The rest of the paper is organized as follows. Next section describes the signal model and introduces basic notation. In Sect. 3 we discuss linear and non-linear multiuser data detection while channel acquisition and tracking is addressed in Sect. 4. Simulation results are discussed in Sect. 5 and some conclusions are offered in Sect. 6.

2. SIGNAL MODEL

2.1. MC-CDMA system

We consider the uplink of an MC-CDMA network in which the total number of subcarriers, N, is divided into smaller groups of Q elements. Several users within a group are simultaneously active and are separated by their specific spreading codes. Without loss of generality we concentrate on a single group with K different users ($K \leq Q$). The BS is equipped with P antennas and the Q subcarriers are uniformly spread over the signal bandwidth in order to better exploit the channel frequency diversity. The channel is assumed static over each OFDM block (slow fading) and a cyclic prefix is employed to eliminate inter-block interference.

At the receiver side, the incoming waveform is first filtered and then sampled with period T_s. Next, the cyclic prefix is removed and the remaining samples are passed to an N-point discrete Fourier transform (DFT) unit. We concentrate on the m-th OFDM block and denote $\boldsymbol{X}_i(m) = [X_i(m,1), X_i(m,2), \ldots, X_i(m,Q)]^T$ the DFT outputs at the i-th antenna corresponding to the Q subcarriers of the considered group. Thus, we have

$$\boldsymbol{X}_i(m) = \sum_{k=1}^{K} a_k(m)\boldsymbol{d}_{i,k}(m) + \boldsymbol{w}_i(m) \qquad i = 1,2,\ldots,P \tag{1}$$

where $a_k(m)$ is the symbol transmitted by the k-th user, $\boldsymbol{w}_i(m)$ is thermal noise and $\boldsymbol{d}_{i,k}(m)$ is a Q-dimensional vector with entries

$$[\boldsymbol{d}_{i,k}(m)]_n = H_{i,k}(m,n)c_k(n) \qquad 1 \leq n \leq Q. \tag{2}$$

In the above formula, $c_k(n) \in \{\pm 1/\sqrt{Q}\}$ is the (unit-energy) spreading code of the k-th user and $H_{i,k}(m,n)$ is the channel frequency response over the n-th subcarrier at the i-th antenna.

Inspection of (1) reveals that $X_i(m)$ may also be written as

$$X_i(m) = D_i(m)a(m) + w_i(m) \tag{3}$$

where $D_i(m) = [d_{i,1}(m)\ d_{i,2}(m) \cdots d_{i,K}(m)]^T$, $a(m) = [a_1(m), a_2(m), \ldots, a_K(m)]^T$ and the superscript $(\cdot)^T$ denotes the transpose operator.

2.2. Channel model

The signal transmitted by each user propagates through a multipath channel with N_p distinct paths and the P receive antennas are arranged in a uniform linear array with inter-element spacing δ. The k-th baseband channel impulse response at the i-th antenna (during the m-th OFDM block) takes the form

$$h_{i,k}(m,t) = \sum_{\ell=1}^{N_p} a_{\ell,k}(m) e^{j(i-1)\omega_{\ell,k}(m)} g(t - \tau_{\ell,k}(m)) \tag{4}$$

where $g(t)$ is the convolution between the impulse responses of the transmit and receive filters, $\tau_{\ell,k}(m)$ is the delay of the ℓ-path and $a_{\ell,k}(m)$ is the corresponding complex amplitude. Finally, $\omega_{\ell,k}(m)$ is defined as

$$\omega_{\ell,k}(m) = \frac{2\pi}{\lambda} \delta \sin[\varphi_{\ell,k}(m)] \tag{5}$$

where λ is the free-space wavelength and $\varphi_{\ell,k}(m)$ is the direction-of-arrival (DOA) of the ℓ-path. The path gains $\{a_{\ell,k}(m)\}$ are modelled as narrow-band independent Gaussian random processes with zero-mean and average power $\sigma_\ell^2 = \mathrm{E}\{|a_{\ell,k}(m)|^2\}$.

The channel frequency response $H_{i,k}(m,n)$ is computed by taking the DFT of $h_{i,k}(m, pT_s)$ and reads

$$H_{i,k}(m,n) = \sum_{p=1}^{L} h_{i,k}(m, pT_s) e^{-j2\pi\, np\Delta/N}. \tag{6}$$

where $\Delta = N/Q$ is the distance between adjacent subcarriers of the same group and L is the duration of $h_{i,k}(m,t)$ in sampling periods.

3. MULTIUSER DATA DETECTION

Stacking the $\{X_i(m); i = 1, 2, \ldots, P\}$ into a single PQ-dimensional vector $X(m) = [X_1^T(m)\ X_2^T(m) \cdots X_P^T(m)]^T$ yields

$$X(m) = D(m)a(m) + w(m) \tag{7}$$

where $\boldsymbol{D}(m) = [\boldsymbol{D}_1^T(m)\ \boldsymbol{D}_2^T(m) \cdots \boldsymbol{D}_P^T(m)]^T$ and $\boldsymbol{w}(m) = [\boldsymbol{w}_1^T(m)\ \boldsymbol{w}_2^T(m) \cdots \boldsymbol{w}_P^T(m)]^T$ is a Gaussian vector with zero mean and covariance matrix $\sigma^2 \boldsymbol{I}_{PQ}$ ($\boldsymbol{I}_{PQ}$ denotes the identity matrix of order PQ).

To detect the data $\boldsymbol{a}(m)$ we employ either a linear MMSE multiuser detector or a PIC-based receiver.

3.1. MMSE multiuser detector

The decision statistic in the MMSE multiuser detector during the m-th OFDM block is

$$\boldsymbol{Y}(m) = [\boldsymbol{D}^H(m)\boldsymbol{D}(m) + \sigma^2 \boldsymbol{I}_{PQ}]^{-1} \boldsymbol{D}^H(m)\boldsymbol{X}(m) \tag{8}$$

where $(\cdot)^H$ denotes Hermitian transposition. The entries of $\boldsymbol{Y}(m)$ are then fed to a threshold device to produce an estimate $\hat{\boldsymbol{a}}(m)$ of the transmitted symbols.

3.2. PIC detector

The PIC detector is a multistage receiver in which MAI is estimated using tentative data decisions and subtracted out in parallel for each user. Without loss of generality, we concentrate on the k-th user. At the ℓ-th stage the PIC detector computes the following vectors

$$\boldsymbol{Z}_{i,k}^{(\ell)}(m) = \boldsymbol{X}_i(m) - \sum_{\substack{j=1 \\ j \neq k}}^{K} \hat{a}_j^{(\ell-1)}(m)\boldsymbol{d}_{i,j}(m) \qquad i = 1, 2, \ldots, P \tag{9}$$

where $\{\hat{a}_j^{(\ell-1)}(m)\}$ are data decisions from the previous stage. A decision statistic is obtained from $\{\boldsymbol{Z}_{i,k}^{(\ell)}(m); i = 1, 2, \ldots, P\}$ using maximum-ratio combining

$$Y_k^{(\ell)}(m) = \sum_{i=1}^{P} \boldsymbol{d}_{i,k}^H(m)\boldsymbol{Z}_{i,k}^{(\ell)}(m). \tag{10}$$

Finally, passing $Y_k^{(\ell)}(m)$ to a threshold device produces the estimate $\hat{a}_k^{(\ell)}(m)$ at the ℓ-stage. The performance of the PIC detector depends heavily on the quality of the initial estimate $\hat{\boldsymbol{a}}^{(0)}(m)$. In our simulations, $\hat{\boldsymbol{a}}^{(0)}(m)$ is taken as the output of the MMSE detector.

4. CHANNEL ESTIMATION

From (8)-(10) we see that data detection requires knowledge of the PK vectors $\{\boldsymbol{d}_{i,k}(m);\ 1 \le i \le P,\ 1 \le k \le K\}$ which are related to the channel frequency responses $\boldsymbol{H}_{i,k}(m) = [H_{i,k}(m,1), H_{i,k}(m,2), \ldots, H_{i,k}(m,Q)]^T$ as indicated in (2). This means that channel estimation is necessary to perform multiuser detection. In the following, we aim directly at the estimation of $\boldsymbol{d}_{i,k}(m)$ rather than $\boldsymbol{H}_{i,k}(m)$. To this purpose we assume that the OFDM blocks are organized in frames and each frame is preceded by a suitable number of training blocks that are exploited to get initial estimates of $\boldsymbol{d}_{i,k}(m)$ (*acquisition*). These estimates are then updated as a function of time during the data section of the frame (*tracking*).

4.1. Acquisition

We denote by N_T the number of training blocks and we assume that the channel variations are negligible over the entire training sequence, i.e., we set $\boldsymbol{d}_{i,k}(m)=\boldsymbol{d}_{i,k}$ for $m=1,2,\ldots,N_T$. Then, collecting $\{\boldsymbol{d}_{i,k};k=1,2,\ldots,K\}$ into a single KQ-dimensional vector $\boldsymbol{d}_i=[\boldsymbol{d}_{i,1}^T\ \boldsymbol{d}_{i,2}^T\ \cdots\ \boldsymbol{d}_{i,K}^T]^T$, from (1) we have

$$\boldsymbol{X}_i(m)=\boldsymbol{B}(m)\boldsymbol{d}_i+\boldsymbol{w}_i(m) \qquad m=1,2,\ldots,N_T \tag{11}$$

where $\boldsymbol{B}(m)=[a_1(m)\boldsymbol{I}_Q\ a_2(m)\boldsymbol{I}_Q\ \cdots\ a_K(m)\boldsymbol{I}_Q]$ and $\{a_k(m);m=1,2,\ldots,N_T\}$ is the training sequence of the k-th user. The observations $\{\boldsymbol{X}_i(m);m=1,2,\ldots,N_T\}$ are exploited to get a least-squares (LS) estimate of $\boldsymbol{d}_i$ in the form

$$\hat{\boldsymbol{d}}_i=\boldsymbol{R}^{-1}\sum_{m=1}^{N_T}\boldsymbol{B}^H(m)\boldsymbol{X}_i(m) \tag{12}$$

with

$$\boldsymbol{R}=\sum_{m=1}^{N_T}\boldsymbol{B}^H(m)\boldsymbol{B}(m). \tag{13}$$

The complexity of the above estimator can be greatly reduced by employing orthogonal training sequences. In these circumstances $\boldsymbol{R}$ becomes diagonal and (12) reduces to

$$\hat{\boldsymbol{d}}_{i,k}=\frac{1}{E_T}\sum_{m=1}^{N_T}a_k^*(m)\boldsymbol{X}_i(m) \qquad k=1,2,\ldots,K \tag{14}$$

where E_T is the energy of the training sequences.

4.2. Tracking

The variations of $\boldsymbol{d}_{i,k}(m)$ during the data section of the frame are tracked with the least-mean-square (LMS) algorithm. This leads to the following recursion

$$\hat{\boldsymbol{d}}_{i,k}(m+1)=\hat{\boldsymbol{d}}_{i,k}(m)+\mu\,\hat{a}_k^*(m)\left[\boldsymbol{X}_i(m)-\sum_{\ell=1}^{K}\hat{a}_\ell(m)\hat{\boldsymbol{d}}_{i,\ell}(m)\right] \qquad m\geq N_T \tag{15}$$

in which $\{\hat{a}_\ell(m)\}$ are data decisions provided by either the MMSE or PIC detectors and the initial estimate $\hat{\boldsymbol{d}}_{i,k}(N_T)$ is computed from (14). The step-size μ controls the convergence properties of the algorithm and is chosen as a trade-off between steady-state performance and tracking capabilities.

5. PERFORMANCE EVALUATION

Computer simulations have been run to assess the impact of channel estimation errors on the system performance. The transmitted symbols belong to a QPSK constellation and are obtained from the information bits through a Gray map. The total number of subcarriers is $N = 128$ and Walsh-Hadamard codes of length $Q = 16$ are used for spreading purposes. The signal bandwidth is $B = 20$ MHz so that the useful part of each OFDM block has length $T = N / B = 6.4$ μs. A cyclic prefix of length $T_G = 1.6$ μs is employed to eliminate inter-block interference. The users are synchronous within the cyclic prefix and have the same power. The carrier frequency is $f_0 = 2$ GHz (corresponding to a wavelength $\lambda = 15$ cm) and the inter-element spacing in the antenna array is $\delta = \lambda / 2$. The channel impulse response of the k-th user at the i-th antenna is generated as indicated in (4) with eight paths ($N_p = 8$). Pulse $g(t)$ is a raised-cosine function with roll-off 0.22 and duration $10T_s$. The path delays and DOAs are uniformly distributed within [0, 1.0 μs] and $[-60^\circ, 60^\circ]$ respectively and are kept constant over a frame. The path gains vary independently of each other within a frame with power $\sigma_\ell^2 = \exp(-\ell)$ for $0 \le \ell \le 7$. They are generated by filtering statistically independent white Gaussian processes in a third-order low-pass Butterworth filter. The 3-dB bandwidth of the filter is taken as a measure of the Doppler rate $f_D = f_0 v / c$, where v denotes the mobile speed and $c = 3 \times 10^8$ m/s is the speed of light.

A simulation run begins with the generation of the channel responses of each user. Channel acquisition is then performed exploiting Walsh-Hadamard training sequences of length $N_T = 16$ while the LMS algorithm tracks the channel variations during the data section of the frame. The optimal value of the step-size μ depends on the mobile velocity v and on the number K of active users. Simulations indicate that a good choice for mobile speeds between 40 and 160 Km/h is given by the rule-of-thumb formula $\mu = 0.2 / \sqrt{K}$.

Figure 1 illustrates the BER performance of the MMSE detector vs. E_b / N_0, where E_b is the energy per bit and $N_0 / 2$ is the two-sided noise spectral density. The mobile velocity is 80 Km/h (corresponding to $f_D = 150$ Hz) and the number of users is either $K = 4$ or $K = 8$. A single receiving antenna is employed ($P = 1$). For comparison, the single user bound (SUB) and the performance with ideal channel information (ICI) are also shown. We see that the MMSE receiver with ICI approaches the SUB when $K = 4$ while it looses 2 dB with $K = 8$ (half-loaded system). The loss due to channel estimation errors is approximately 2 dB with 4 users but grows to 3dB when 8 users are simultaneously active.

Figure 2 shows results obtained in the operating conditions of Fig. 1, except that data detection is performed using the PIC receiver. It is seen that the PIC and the

MMSE detectors have comparable performance with 4 users, but the PIC is superior with $K=8$. The impact of the channel estimation errors is comparable with both detectors.

BER performance with 2 receiving antennas ($P=2$) is illustrated in Figs 3-4. The number of active users is either $K=4$ or $K=8$ and the mobile speed is $v=80$ Km/h. MMSE data detection is performed in Fig. 3 while PIC detection is employed in Fig. 4. Comparing with Figs. 1-2, we see that using two receiving antennas instead of one entails a gain of approximately 3 dB. The loss due to imperfect channel knowledge is still 2 dB with $K=4$ and 3 dB with $K=8$.

Figures 5-6 show the performance of the PIC detector with $K=8$ and different mobile speeds. A single receive antenna is employed in Fig. 5 while two antennas are used in Fig. 6. Note that simulation results with ICI do not depend on the fading rate as the channel is assumed constant within each OFDM block. We see that the BER deteriorates as the mobile speed increases. For an error probability of 10^{-2}, the loss with respect to ICI is approximately 3 dB with $v=40$ Km/h and becomes 7 dB with $v=160$ Km/h.

6. CONCLUSIONS

We have discussed a simple scheme to perform channel estimation in the uplink of an MC-CDMA system equipped with multiple receiving antennas. Orthogonal training sequences are exploited to perform LS channel acquisition. The LMS algorithm is employed to track the channel variations. Either MMSE or PIC receivers are used to detect the transmitted data symbols.

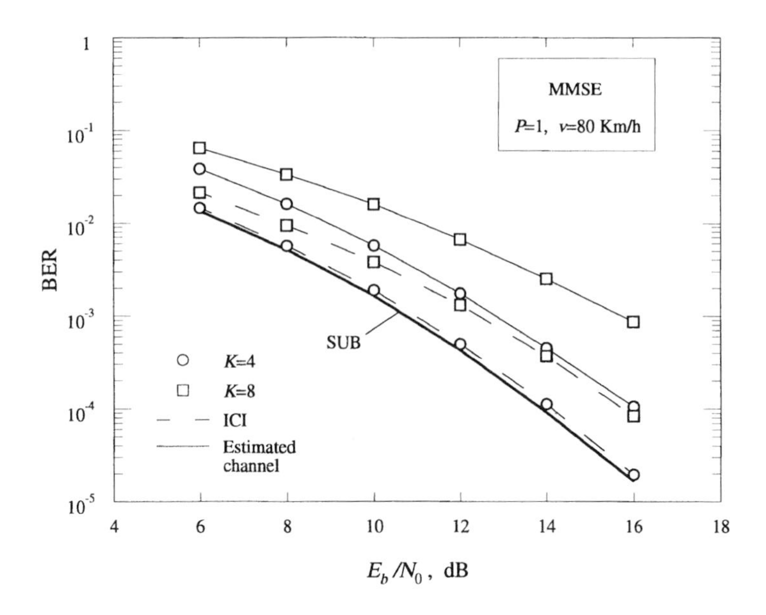

Figure 1. Performance of the MMSE detector with $P=1$ and $v=80$ Km/h

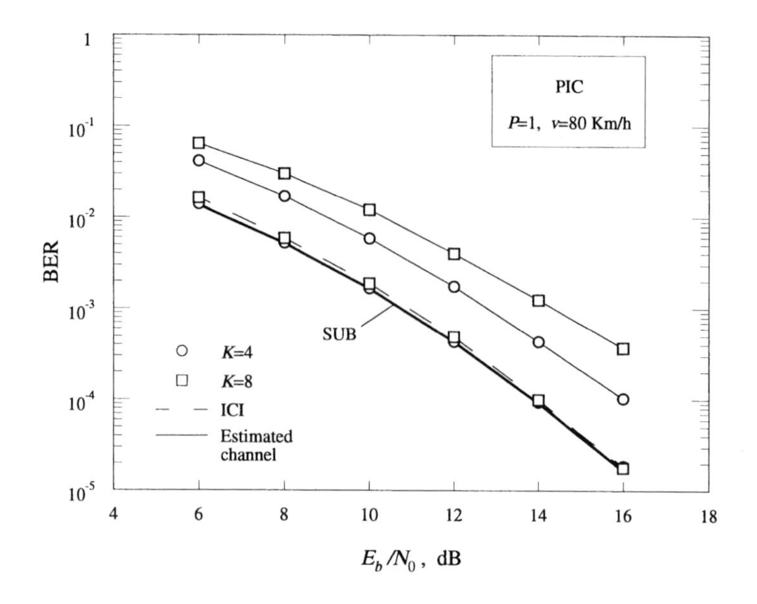

Figure 2. Performance of the PIC detector with $P = 1$ *and* $v = 80$ *Km/h*

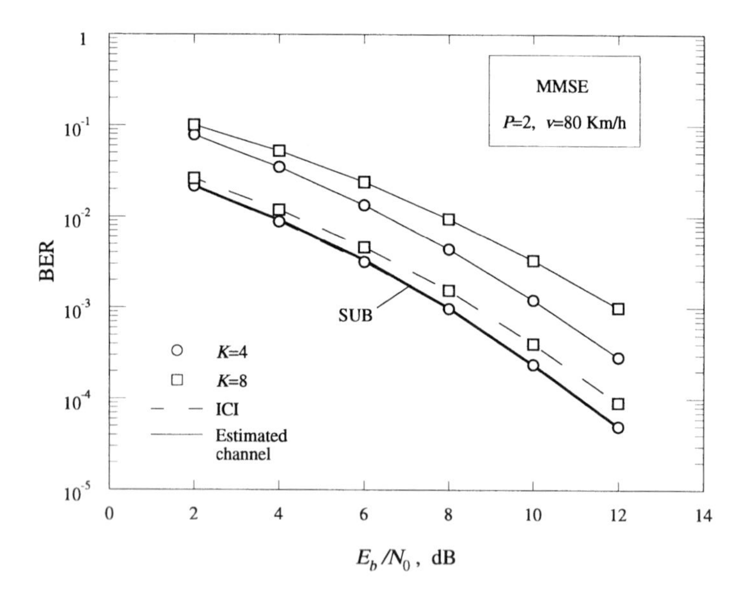

Figure 3. Performance of the MMSE detector with $P = 2$ *and* $v = 80$ *Km/h*

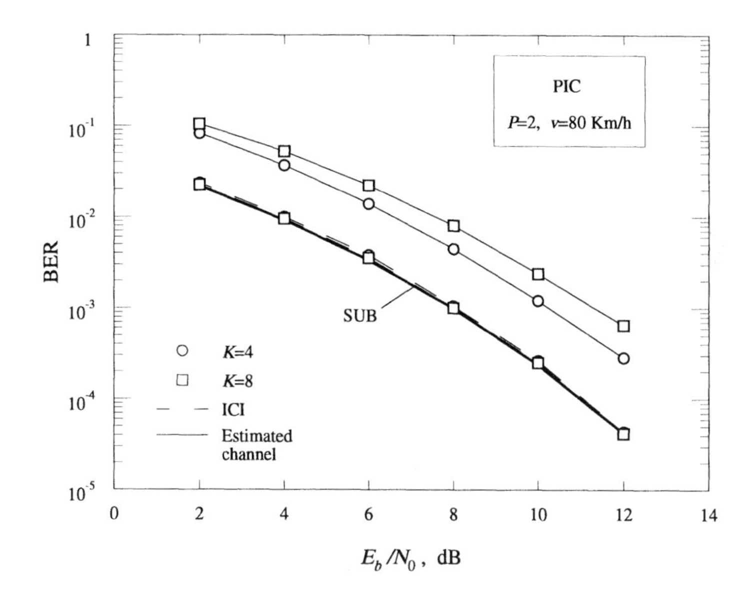

Figure 4. Performance of the PIC detector with $P = 2$ *and* $v = 80$ *Km/h*

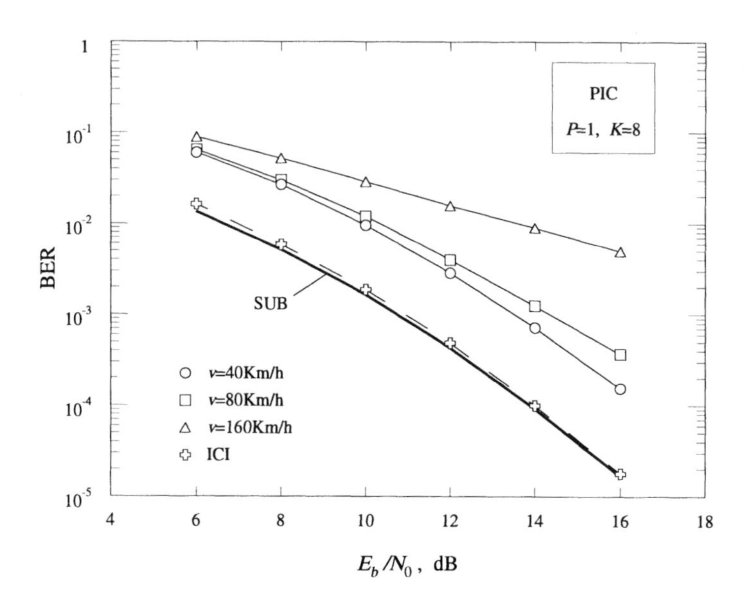

Figure 5. Performance of the PIC detector with $P = 1$ *and various mobile speeds*

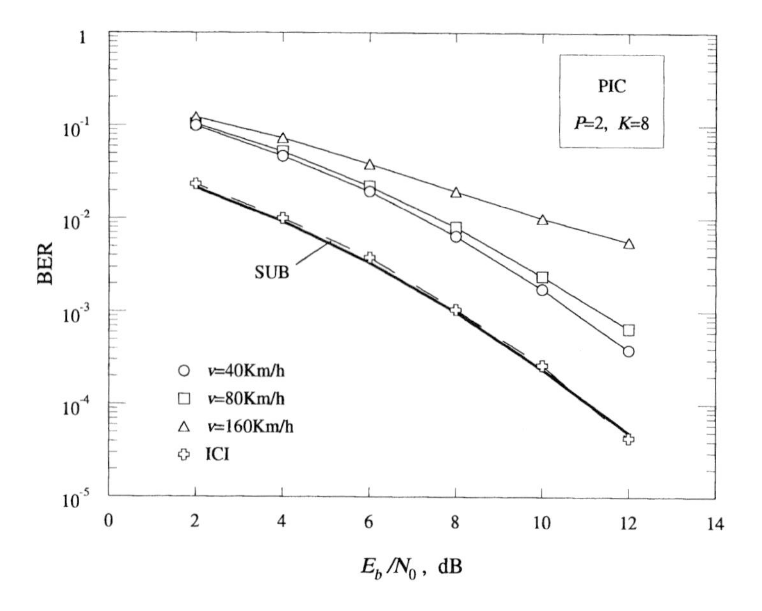

Figure 6. Performance of the PIC detector with $P = 2$ *and various mobile speeds*

Computer simulations have been run to evaluate the impact of channel estimation errors on the system performance. It is shown that the loss due to imperfect channel knowledge increases with the number of active users and with the fading rate while it depends weakly on the number of receiving antennas. For a half-loaded system and a mobile speed of 80 Km/h, a loss of 3 dB is incurred with respect to a system with ideal channel information.

7. REFERENCES

[1] K.Fazel, "Performance of CDMA/OFDM for Mobile Communication Systems", *Proc. 2nd IEEE Int. Con[illegible] Universal Personal Commun. (ICUPC)*, 1993, pp. 975-979.

[2] S.Hara, P.Prasad, "Overview of Multicarrier CDMA", *IEEE Comm. Magazine*, pp. 126-133, Dec. 1997.

[3] X.Wang, V.H.Poor, "Space-Time Multiuser Detection in Multipath CDMA Channels", *IEEE Tran[illegible] Signal Processing*, pp. 2356-2374, Sept. 1999.

[4] S.Moshavi, "Multiuser Detection for DS-CDMA Communications", *IEEE Comm. Magazine*, pp. 12[illegible] 136, Oct. 1996.

[5] R.Lupas, S.Verdù, "Linear Multiuser Detectors for Synchronous Code-Division Multiple-Acce[illegible] Systems", *IEEE Trans. Information Theory*, pp. 123-136, Jan. 1989.

[6] S.Kaiser, P.Hoeher, "Performance of Multi-Carrier CDMA Systems with Channel Estimation in Tw[illegible] Dimensions", *Proc. of PIMRC'97*, pp. 115-119, Sept. 1997.

[7] C.J.Escudero, D.I.Iglesia, L.Castedo, "A Novel Channel Identification Method for Downlink Multicarri[illegible] CDMA Systems", *Proc. of PIMRC 2000*, pp. 103-107, Sept. 2000.

[8] Z.Li, M.Latva-aho, "Analysis of MRC Receivers for Asynchronous MC-CDMA with Channel Estimati[illegible] Errors", *Proc. of 7th Int. Symp. On Spread-Spectrum Techn. & Appl.*, Prague, pp. 343-347, Sept. 2002.

[9] J.G.Andrews, T.H.Y.Meng, "Performance of Multicarrier CDMA with Successive Interferen[illegible] Cancellation with Estimation Error in a Multipath Fading Channel", *Proc. of 7th Int. Symp. On Sprea[illegible] Spectrum Techn. & Appl.*, Prague, pp. 150-154, Sept. 2002.

[10] M.Morelli, U.Mengali, "Timing Synchronization for the Uplink of an OFDMA System", *Proc. of MC[illegible] 2001*, pp.23-34, Kluwer Academic Publisher, Sept. 2001.

R. LEGOUABLE, D. CALLONNEC AND M. HELARD

SYNCHRONIZATION AND POWER CONTROL PROCESSES FOR UPLINK MULTICARRIER SYSTEMS BASED ON MC-CDMA TECHNIQUE

Abstract. In this paper, a new synchronization and power control process for uplink multicarrier systems based on MC-CDMA technique is presented. In fact, a main challenge for multicarrier systems is the time synchronization of several users, who transmit data on the same OFDM symbol by using OFDMA or MC-CDMA techniques. The aim is to apply in an established phasis at the reception side (the Base Station), a global and unique fast Fourier transform to determine the data information of all users. To allow this unique operation, we need to estimate all the propagation delays between each user and the BS. This paper presents the principles of this ranging method based on the MC-CDMA technique and the new algorithm processed at the reception side. This new algorithm is based on the differential demodulation technique and is compared to classical correlation methods applied either in the frequential or in the temporal domain in terms of performance and complexity.

1. INTRODUCTION

In 1990, the Orthogonal Frequency Division Multiplexing (OFDM) modulation was retained for the first time in a standard; it was for the Digital Audio Broadcasting (DAB) [1]. Since this date, the broadcasting systems, like Digital Video Broadcasting – Terrestrial (DVB-T) in 1993 [2] or the Digital Radio Mondiale (DRM) in 2001, using OFDM modulation scheme, have been standardised too. In parallel to these broadcasting systems, several high data rate OFDM wireless standards such as: Hiperlan/2, IEEE 802.11a and more recently Hiperman have emerged. The growing deployment of OFDM systems is due to the fact that this modulation has the good capacity to fight against frequency selective channels (multipath channels) by providing a flat fading channel on each transmitted subcarrier. In order to share the bandwidth between several users, different multiple access techniques are combined to the OFDM modulation. In Hiperlan/2 and IEEE 802.11a standards a Time Division Multiple Access (TDMA) technique is used, where each user transmits its data informations during one specific time slot. The Hiperman standard uses the Frequency Division Multiple Access (FDMA), providing the OFDMA technique. This technique consists in allocating one or several carriers during one or several OFDM symbols to a specific user and has been introduced in interactive channel for digital terrestrial television standard (DVB-RCT [3]). The third multiple access technique that can be combined to the OFDM modulation is the Coded Division Multiple Access (CDMA) providing the technique called MC-CDMA. It has been proposed for multimedia services in high data rate wireless networks [4][5] and is likely to be also a candidate for 4G mobile radio

K. Fazel and S. Kaiser (eds.), Multi-Carrier Spread-Spectrum, 279–286.

systems. The principle of this technique is to allocate to each user one or several specific sequences (called codes) and to spread, in the frequency domain, the data symbols on the modulated carriers.
One of the challenge when considering an OFDMA or a MC-CDMA system is to be able to synchronize in time all the users transmitting simultaneously in the uplink in order to carry out one single OFDM demodulator (one FFT) at the BS side. The frequency synchronization is also a hard point but it can be compensated in part by channel estimation. Frequency synchronization is not studied in this paper.
This paper is organised as follows: In Section 2, we briefly describe the process implemented at the transmission side (User Equipment UE) corresponding to the principle detailed in the DVB-RCT standard [3]. In Section 3, we describe the methods carried out at the reception side and we give some complexity figures of each of them. In Section 4, performance results are described, considering several scenarii. In addition, the processes are compared in terms of performance /complexity. Finally, Section 5 summarises the results and draws conclusions.

2. TX SIDE PRINCIPLES

2.1. Definitions and principles

The UL transmission is considered and the goal is to synchronize in time all the UE, in order to apply at the BS side one unique FFT. The principle is to allocate to each user one specific sequence, called code in the following, and to transmit it into a ranging sub-channel. The ranging sub-channels enable the user terminal to synchronize to the DVB-RCT RF channel, to maintain their connection with the BS and to request additional bandwidth if necessary. In our case, one ranging sub-channel is composed of 145 (or 116) scattered carriers (the carriers are allocated following a specific law and are no-adjacent), on which until 32 users can coexist. Also, the MC-CDMA technique is applied to allow the synchronization procedure in the DVB-RCT standard. In the burst structures BS1 and BS2, when considering the 1K mode (841 modulated carriers), we have 6 ranging sub-channels (5 of 145 carriers and 1 of 116 carriers) and when considering the 2K mode (1711 modulated carriers), we have 12 ranging sub-channels (11 of 145 carriers and 1 of 116 carriers). For each of these 2 burst structures, we can have 6, 12, 24 or 48 OFDM ranging symbols at the beginning of the frame, dedicated to the synchronization procedure; this variable OFDM symbol number is depending on the system load and on the occurred synchronization false alarms.
Two ranging procedures can be applied and are represented in Figure 1:

- The long ranging transmission procedure is used by any terminal that wants to synchronize for the first time at the BS. In that case the 2 first OFDM symbols are used and the phase continuity between the 2 symbols is maintained;
- The short ranging transmission procedure is used by only terminal that has already synchronized to the BS or that is asking for bandwidth allocation request.

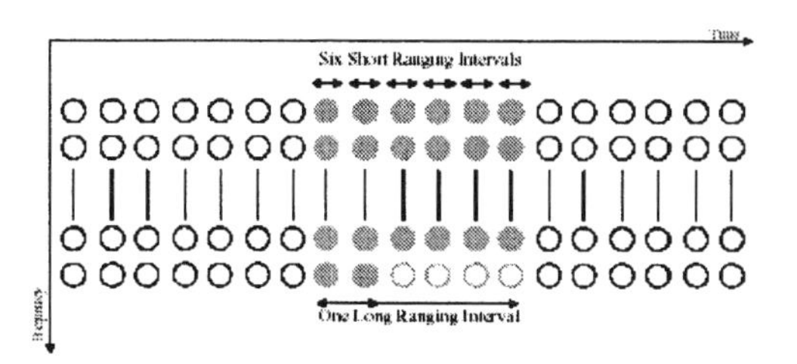

Figure 1: RangingProcedures.

2.2. Ranging Codes

The ranging codes allocated at each user are formed by several series of 145 (or 116) bits produced by the following PRBS generator: $1+X+X^4+X^7+X^{15}$. Only the first series of 96 codes are used as ranging codes. The first 32 codes are used for long ranging, the next 32 codes are used for short ranging and the last 32 codes are used by a terminal already connected to the system and asking for additional transmission resources.
We can have until 32 simultaneously asynchronous users transmitting on the same ranging sub-channel. However, as the users are asynchronous, a high Multiple Access Interference (MAI) appears due to the no perfect inter-correlation between the codes. The inter-correlation formula is depending on the respective delays between the users arriving at the BS, introducing a phase mismatch in the frequency domain between them. The use of spreading codes to synchronize in time the mobile terminals is the principle already implemented in UMTS and often leads to the detection of false alarms.

3. RX SIDE TECHNIQUES

Having the transmission concept, we have to define at the BS side a process, as less complex as possible, allowing us to synchronize in time, the different users transmitting asynchronously. Three methods have been compared, in terms of performance and complexity. Their goal is to determine the user time shift, characterizing by a peak, corresponding to the relative user delay, with the maximum of dynamic. What we call dynamic is the magnitude between the main peak and the second most important peak. The more important the dynamic is, the less probable the false alarm detection is. In addition, the peak magnitude provides power information that can be used to adjust the power control of the detected users. The three methods are: the correlation in the time domain, the correlation in the frequency domain and the proposed method called "differential method".

3.1 Correlation method in the time domain

This technique imposes to translate the code applied in the frequency domain in the time domain and to process chip by chip the correlation according to all the possible shifts. This method is very complex since for each ranging sub-channel, the number of complex multiplications is equal to: 32*fft_size*fft_size, where fft_size is the size of the FFT. So we have: 33.55 E+06 complex multiplications for the 1K mode and 134.21 E+06 for the 2K mode.

3.2 Correlation in the frequency domain

This method is less complex as the one above. The principle is to extract the 145 carriers of the ranging sub-channel and to multiply them by the code of each user i; then we can insert null points to carry out one 1K or 2K IFFT points, in order to obtain a peak in the time domain relating to the signal delay arriving at the BS. So, the numbers of complex multiplications are equal to:

$$\underbrace{512\log_2 512}_{\text{FFTreception}} + \underbrace{145*32}_{\text{Multiplication by code}} + \underbrace{32*512\log_2 512}_{\text{crossing in temporal domain}} = 156704 \text{ for 1K mode}$$

$$\underbrace{1024\log_2 1024}_{\text{FFT reception}} + \underbrace{145*32}_{\text{Multiplication by code}} + \underbrace{32*1024\log_2 1024}_{\text{crossing in temporal domain}} = 342560 \text{ for 2K mode}$$

This method still stays relatively complex because we have to do, for each ranging sub-channel, 32 complex IFFT of size 1024 (1K mode) or 2048 (2K mode). The choice of so big IFFT sizes is due to the use of scattered carriers for the ranging sub-channel. It is possible to decrease the IFFT size in order to decrease the complexity but at the price of degraded results.

3.3 "Differential method"

The main idea is that a phase variation in the frequency domain corresponds to a temporal shift in the time domain. So the proposed technique relies on the frequential differential demodulation that allows to obtain the average phase variation between the set of carrier couples, with ranging carrier indexes taken into the sub-channel, separated of $k\Delta f$. As the set of carriers, on which the code is applied are scattered, the goal is to look for the set of carrier couples spaced out of $k\Delta f$, where Δf is the inter-carrier spacing ($\Delta f = 1/t_s$, where t_s is the useful OFDM symbol duration) and k=1, 2, 3, ...Nfft. The principle of the method is described in Figure 2 and summarized below:

- We apply the conjugate complex multiplication by the code relating to each user on the extracted ranging sub-channel;
- Knowing the carrier couple indexes, we apply the frequential differential demodulation for each couple of carriers relating to the k value. As we have several couples of carrier indexes checking one k value, we can obtain an average phase variation relating to k (the number of points that are used for the average can be adjusted in order to limit the complexity);

- Then, we apply Kuser (Kuser is the number of users transmitting at the same time) IFFT of Nfft points to obtain the temporal shift of the Kuser, characterized by the peak with the highest magnitude.

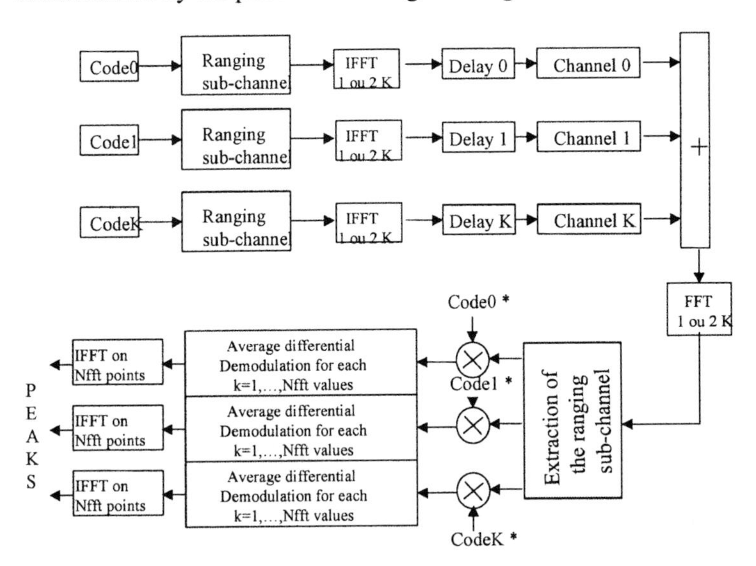

Figure 2: "Differential method" Procedure.

To mathematically illustrate the differential demodulation method, we consider the case where only one user transmits over the ranging sub-channel 0 of the 1K mode, with one time delay equal to $\tau 0$ and with the code vector $\mathbf{C}=(C_0, C_1, C_2, \ldots C_{31})$. We denote by Ω the set of modulated carriers used. The indexes of the 8 first carriers defined into the sub-channel 0 of the 1K mode are: 8 10 13 14 24 37 42 43. We denote by H_n the channel perturbation on the *nth* carrier and C_i* denotes the complex conjugate of C_i. For the detection of the C code, if Y_k is the received signal in the *kth* sub-carrier, we process:

$$for\ k=1: C_3^* Y_{14} * C_2 Y_{13}^* + C_7^* Y_{43} C_6 Y_{42}^* \,;\ for\ k=2: C_1^* Y_{10} * C_0 Y_8^* \,;\ for\ k=3: C_2^* Y_{13} * C_1 Y_{10}^*$$

$$for\ k=4: C_3^* Y_{14} * C_1 Y_{10}^* \,;\ for\ k=5: C_2^* Y_{13} * C_0 Y_8^* \ \ldots.$$

Then if C was really transmitted, we obtain:

$$for\ k=1: C_3 C_3^* H_{14} \exp(-2i\pi f_{14}\tau_0)(C_2 C_2^* H_{13} \exp(-2i\pi f_{13}\tau_0))^* + C_7 C_7^* H_{43} \exp(-2i\pi f_{43}\tau_0)(C_6 C_6^* H_{42} \exp(-2i\pi f_{42}\tau_0))^*$$

$$for\ k=2: C_1 C_1^* H_{10} \exp(-2i\pi f_{10}\tau_0)(C_0 C_0^* H_8 \exp(-2i\pi f_8\tau_0))^* ; for\ k=3: C_2 C_2^* H_{13} \exp(-2i\pi f_{13}\tau_0)(C_1 C_1^* H_{10} \exp(-2i\pi f_{10}\tau_0))^*$$

$$for\ k=4: C_3 C_3^* H_{14} \exp(-2i\pi f_{14}\tau_0)(C_1 C_1^* H_{10} \exp(-2i\pi f_{10}\tau_0))^* ; for\ k=5: C_2 C_2^* H_{13} \exp(-2i\pi f_{13}\tau_0)(C_0 C_0^* H_8 \exp(-2i\pi f_8\tau_0))^* ;.$$

This series of terms leads to:

$$(H_{14}\,H_{13}{}^{*}+H_{43}\,H_{42}{}^{*})\exp(-2i\pi\,\Delta f\,\tau_0)\,;H_{10}\,H_8{}^{*}\exp(-4i\pi\,\Delta f\,\tau_0)\,;H_{13}\,H_{10}{}^{*}\exp(-6i\pi\,\Delta f\,\tau_0)\,;$$
$$H_{14}\,H_{10}{}^{*}\exp(-8i\pi\,\Delta f\,\tau_0);\,H_{13}\,H_8{}^{*}\exp(-10i\pi\,\Delta f\,\tau_0)\,;...$$

and can be generalized by:

$$\sum_{\substack{k\in\Omega\\k-1\in\Omega}} H_k H_{k-1}{}^{*}\exp(-2i\pi\,\Delta f\,\tau_0);\ \sum_{\substack{k\in\Omega\\k-2\in\Omega}} H_k H_{k-2}{}^{*}\exp(-4i\pi\,\Delta f\,\tau_0);...;\ \sum_{\substack{k\in\Omega\\k-Nfft\in\Omega}} H_k H_{k-Nfft}{}^{*}\exp(-2^{*}Nfft^{*}i\pi\,\Delta f\,\tau_0)$$

In the case of a channel with one tap, we have $H_k=H_{k'}{}^*=1$ and we obtain after normalization and after an IFFT of the serie, a delayed dirac function:

$$\exp(-2i\pi\,k\Delta f\tau_0)\ \xrightarrow{IFFT}\delta(t-\tau_0)$$

If C was not the transmitted code, the detection of C yield a noise response. When considering one single user not affected by MAI and noise, we can thus easy find the time shift corresponding to the delay between the transmitted signal by the UE and the OFDM demodulation performed at the BS side.
About complexity, this method leads to apply respectively for the 1K and 2K modes:

$$\underbrace{512\log_2 512}_{FFT\text{ of reception}} + \underbrace{32*145}_{\text{multiplications by codes}} + \underbrace{32*\text{Nfft}*\text{pt\_aver}}_{\text{differential method}} + \underbrace{32*\text{Nfft}}_{\text{division for average}} + \underbrace{32*\frac{Nfft}{2}\log_2\frac{Nfft}{2}}_{\text{Crossing in temporal domain by IFFT}}$$

$$\underbrace{1024\log_2 1024}_{FFT\text{ of reception}} + \underbrace{32*145}_{\text{multiplications by codes}} + \underbrace{32*\text{Nfft}*\text{pt\_aver}}_{\text{differential method}} + \underbrace{32*\text{Nfft}}_{\text{division for average}} + \underbrace{32*\frac{Nfft}{2}\log_2\frac{Nfft}{2}}_{\text{Crossing in temporal domain by IFFT}}$$

complex multiplications, where Nfft corresponds to the IFFT size for crossing in the temporal domain and pt_aver corresponds to the number of average points. The higher this parameter is, the better the phase channel variation determination between sub-carrier couples is. When considering Nfft=256 and pt_aver=32 in 1K mode, the frequential correlation and the differential method have around the same complexity (~160 000 complex multiplications per ranging sub-channel and for 32 users). When considering 2K mode, then the frequential method has a complexity two times higher than the differential method. The temporal correlation is two hundred times more complex.

4. PERFORMANCE RESULTS

All the results presented have been found via statistics. Numerous simulations have been launched according to the number of users and for various time delays for each of them whatever the ranging sub-channel index and the mode.

4.1 Influence of the MAI term

The performance results given in figures 2a, 2b and 2c correspond respectively to the temporal correlation, the frequential correlation and the "differential method".

The represented scenario deals with 2 users delayed of 0.3847*ts (390 samples) and 0.1835*ts (188 samples) without channel and gaussian noise (only the interference effect is evaluated). The results show that we can determine the both time shifts for each method. However, the highest dynamics are obtained with the differential demodulation method leading to less of false alarms. We have dynamics close to 25dB with the differential method against 5dB with the temporal method and 10dB with the frequential method. In addition, the differential method is less complex in that case. The results have shown that until 10 users the differential method allows to recognize all the delays with good dynamics compared to the others where more false alarms are recorded.

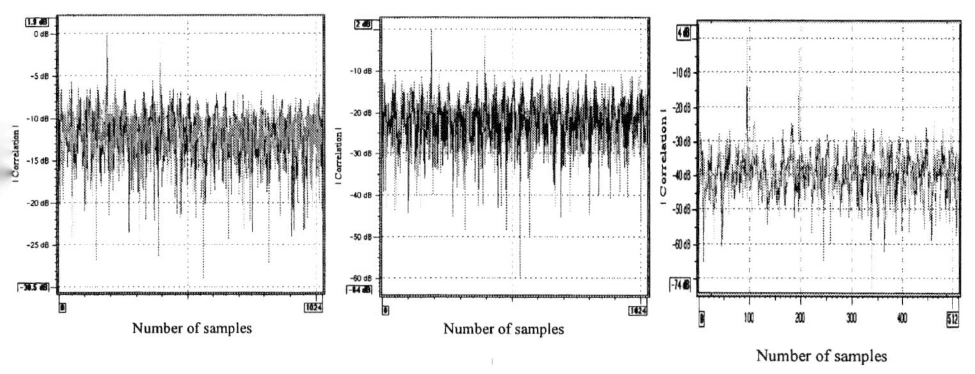

Fig 2a: temporal correlation *Fig.2b: frequential correlation* *Fig. 2c: differential method*

4.2 Performance results with MAI and gaussian noise

We have found that a Signal to Noise Ratio (SNR) per sub-carrier close to -5dBs, corresponding to a global SNR near 16dBs, allows recognizing without false alarms until 8 users. When this number of users increases, the number of false alarms increases too, whatever the method. With gaussian noise, the performances are quasi-similar for the 3 methods but the differential technique is less complex and the statistical results have shown that it most often obtained the best dynamics.

4.3 Performance results in multipath channel

The 3 methods have been tested in multipath channels with a SNR per sub-carrier equals to –5dB. The channels are the "F1" and "P1" models described in [2]. "F1" is a Rice channel (LOS) and P1 is a NLOS Rayleigh channel. At first, we have simulated one user transmitting in a propagation channel with 2 taps having the same magnitude and delayed of 0.01 and 0.04*ts respectively (10 and 40 samples). Figure 3 shows the performance results of the 3 methods. We remark that the best results are obtained with the differential method, where the highest dynamics are obtained and the temporal correlation gives the worst performance results. In F1 channel, we obtain the same results as for Gaussian channel while in P1 channel, the best results are obtained with the differential method that can recognize until 4 users (when 4 users transmit) whereas the temporal correlation method recognize 3 users and the frequential correlation method can recognize 2 users. So we see that the differential method gives the best results in terms of performance/complexity.

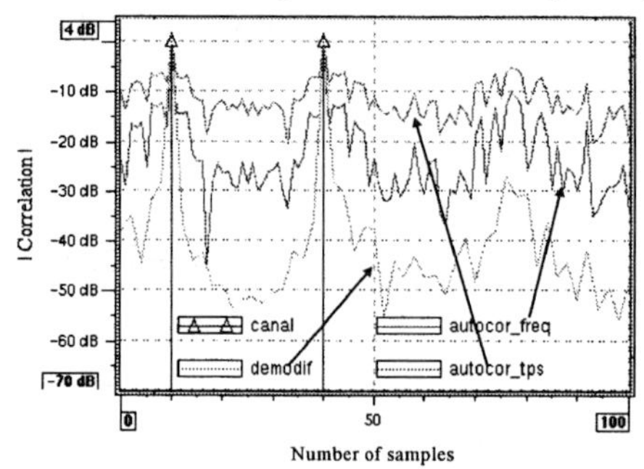

Figure 3: Performance in 2-taps channel (only 1 user + SNR=-5dB)

5. CONCLUSION

This paper presents a new method, based on the MC-CDMA technique, allowing to determine the time shifts between several users, transmitting asynchronously in the uplink. The goal is to be able to implement a single OFDM demodulator. In addition, the proposed method can inform the BS about the signal power of each user via the peak magnitude and so, to carry out a power control mechanism. It can be applied when OFDMA or MC-CDMA technique is implemented. The proposed method is based on the frequential differential demodulation principle and has been compared to more classical correlation techniques. The results have shown that the differential method gives the best trade-off between performance and complexity. Further studies based on code research, interference canceling processes are carryied out into the IST-MATRICE project [6], in order to still improve the performance.

6. REFERENCES

[1]. ETSI "Radio Broadcasting Systems; Digital Audio Broadcasting (DAB) to mobile , portable and fixed receivers", april 2000, EN 300 401 V1.3.1

[2]. ETSI "Digital Video Broadcasting (DVB); Framing structure, channel coding and modulation for digital terrestrial television", July 1999, EN 300 744 V1.2.1

[3]. ETSI EN 301.958V1.1.1, Digital Video Broadcasting (DVB), Interaction channel for digital terrestrial television (RCT) incorporating multiple access OFDM; March 2002.

[4]. YEE (N.), LINNARTZ (J.P.), FETTWEIS (G.). Multicarrier CDMA in indoor wireless radio networks. *Proceedings of IEEE PIMRC'93*, pp 109-113, Yokohama, Japan, (1993).

[5]. FAZEL (K.), PAPKE (L.). On the performance of convolutionnally-coded CDMA/OFDM for mobile communication system. *Proceedings of IEEE PIMRC'93*, pp 468-472, Yokohama, Japan, (1993).

[6]. IST MATRICE project, web site http://www.ist-matrice.org

7. AFFILIATIONS

The authors work at France Télécom R&D in DMR/DDH lab. For more detailed informations, you can contact them at the following email addresses:
rodolphe.legouable/denis.callonnec/maryline.helard@francetelecom.com

Section IV

MIMO, DIVERSITY AND SPACE TIME CODING

Kyesan Lee* and Masao Nakagawa ‡

A Novel Soft Handoff Technique using STTD for MC-CDMA in a Frequency Selective Fading Channel

* Graduate School of Information and Communication, Kyunghee University,
1, Seochun-ri, Kihung-eup, Yongin-shi, Kyunggido, Korea.

‡ Dept. of Information and Computer Science, Faculty of Science and Technology, Keio University
3-14-1 Hiyoshi, Kohoku-ku, Yokohama-shi, Kanagawa, Japan

Abstract

Antennas are commonly located in near each other and there is high correlations among antennas. In such cases, there exists correlated fading among antennas, which results in worse performance for this system. This paper presents an effective soft handoff technique using Space Time Transmitter Diversity (STTD) with two base station for Multi-Carrier (MC)-CDMA to achieve high diversity gain in a wireless broadband channel. The conventional antenna The proposed system is a dual step diversity scheme in frequency selective fading channel environments combining the frequency diversity effect achieved by multi-carriers and the antenna diversity gain by STTD. This is an effective diversity scheme, which improves the signal quality at the receiver by simple processing across two transmit antennas. No feedback is required in this system from the receiver to the transmitter and the computation complexity is low.

1 Introduction

Mobile communication systems are required to be sufficiently flexible to support a variety of multimedia services such as video, image, picture and data services with high quality[1],[2],[3]. A multi-carrier modulation scheme providing high data rate transmission with high frequency utilization efficiency has been proposed for the DS/CDMA system based on orthogonal frequency division multiplexing (OFDM), which is a parallel data transmission technique. It is crucial for multi-carrier transmission to have a non-frequency selective fading channel over each sub-carrier[2],[4],[5],[6],[7]. The Orthogonal Frequency Division Multiplexing-Code Division Multiple Access (OFDM-CDMA) systems is effective in a frequency selective fading channel environment to provide frequency diversity gain avoiding ISI (Inter Symbol Interference)[5],[6],[8]. Because the MC-CDMA(OFDM-CDMA) method spreads the original data in a frequency domain using spreading code, it can obtain a frequency diversity effect through de-spreading since the fading of each subcarrier is different[5],[6],[9],[10].

The transmitter scheme proposed by S.M. Alamount improves the signal quality at the receiver on one side of the link by simple processing with two transmitter antennas[11]. However, for broadband wireless systems, additional techniques need to be used to avoid performance deterioration in such a multiple frequency selective fading channel.

Antennas are commonly located near each other and there is high correlation among antennas. In such cases, the system performance using STTD diversity cannot be improved sufficiently, because there exists correlated fading among antennas, which results in worse performance for this system. The signals from one base station is abruptly deteriorted due to shadowing by obstacles.

K. Fazel and S. Kaiser (eds.), Multi-Carrier Spread-Spectrum, 289–298.

In addition, the conventional STTD system deteriorates the system quality from multipath fading, which results in worse performance in a frequency selective fading channel. A new diversity scheme is therefore required to solve the above problems.

We propose a novel soft handoff system using the space time transmitter diversity technique for MC-CDMA in a frequency selective fading channel, which is a dual step diversity scheme. The proposed system provides a combined diversity effect created by not only the frequency diversity achieved from multi-carriers but also by the antenna space diversity from STTD of two base station in frequency selective fading channel environments. This proposed STTD system is effective in a wireless broadband channel avoiding the distortion of signal occurring by multipath fading. This proposed system improves the signal quality at the receiver by simple processing across two transmit antennas.

2 Proposed System Model

Antennas are commonly located in near each other and there is high correlations among antennas. In such cases, the system performance using STTD diversity cannot be improved sufficiently, because there exists correlated fading among antennas, which results in worse performance for this system. A new type of soft handoff system for MC-CDMA scheme using STTD is proposed, which provides a combined diversity effect created by not only the antenna space diversity by STTD from two base stations but also by the frequency diversity from multicarrier in frequency selective fading channel environments. The signals from multiple base station are combined by the same frequency, which is called SFN(Single Frequency Network). This is the dual step diversity scheme, which improves the signal quality at the receiver by simple processing across two base station with transmit antennas. No feedback is required in this system from the receiver to the transmitter and the computation complexity is low.

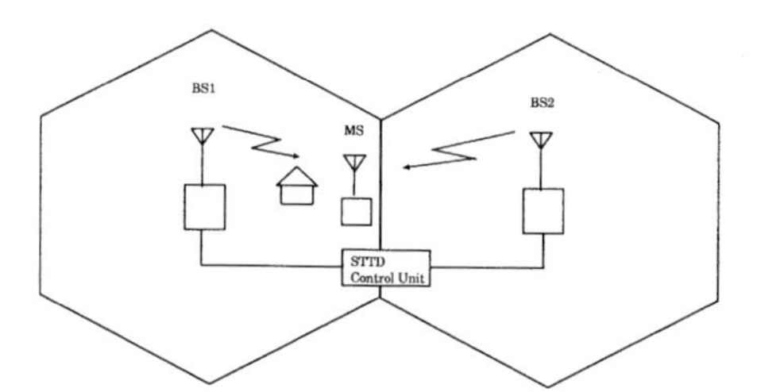

Figure 1: System configuration.

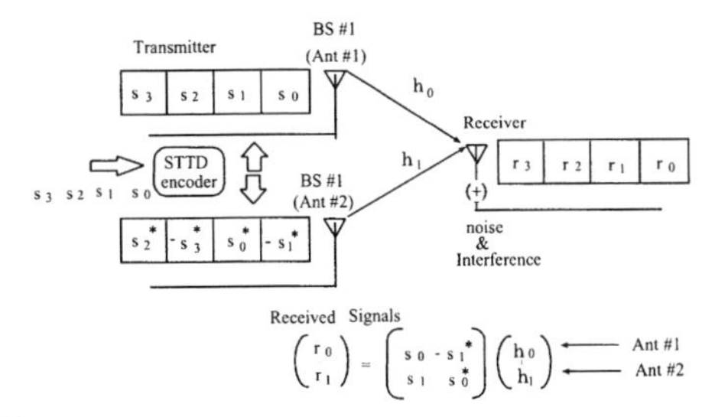

Figure 3: STTD scheme with two base station and one mobile stations.

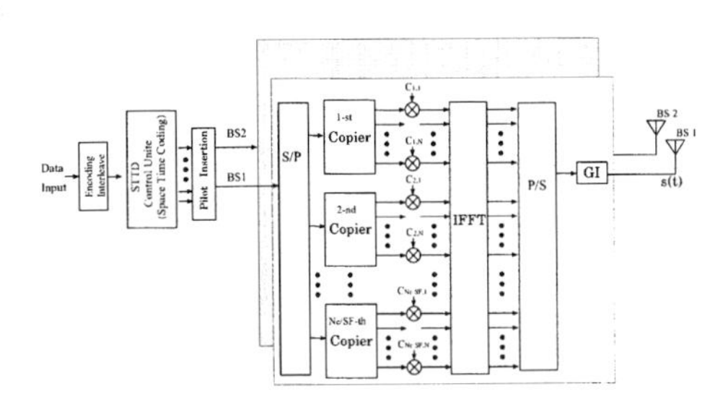

Figure 2: Transmitter configuration of site diversity using STTD.

Fig.1 shows the model consisting of 2 Base Station(BS) and one Mobile Station (MS), and STTD control unite controls the transmit signal with space time coding.

Fig.2 shows the configuration of the transmitter consisting of two base stations. Input information data are encoded in space and time (space-time coding), serial to parallel converted, and then spread in the frequency domain by PN codes in each carrier. Inverse Fast Fourier Transform (IFFT) is utilized for multi-carrier modulation. The output signals of IFFT are parallel to serial converted. A guard interval is inserted between the output signals to prevent multipath fading. The proposed system transmits two space-time coded symbols on different antennas simultaneously. The signal transmitted from the first antenna is denoted by s_0, and the next symbol is donoted by s_1, while, that from the second antenna is denoted by $-s_1^*$, and the next symbol by s_0^* as shown in Fig.3. Table 1 shows the transmitted signal pattern for space-time transmission diversity spreading by PN codes. $C_{i,j}$ is the j th chip of the ith carrier.

The channel at time t between the transmit antenna and the receive antenna 1 is denoted by h_0 and between the transmit antenna and the receive antenna 2 is denoted by h_1.

$$h_0(t) = h_0(t+T) = h_0 = \alpha_0 \exp(j\theta_0) \quad (1)$$
$$h_1(t) = h_1(t+T) = h_1 = \alpha_1 \exp(j\theta_1) \quad (2)$$

where α means the fading channel gain and θ denotes angle.
The receiver receives the space-time coded signals, which are serial to parallel converted, FFT is

Table 1: Transmission sequences pattern for STTD

	Antenna:1	Antenna:2
Time t	$\sum_{j=1}^{j=N} s_0 \cdot C_{i,j}$	$\sum_{j=1}^{j=N} -s_1^* \cdot C_{i,j}$
Time t+T	$\sum_{j=1}^{j=N} s_1 \cdot C_{i,j}$	$\sum_{j=1}^{j=N} s_0^* \cdot C_{i,j}$

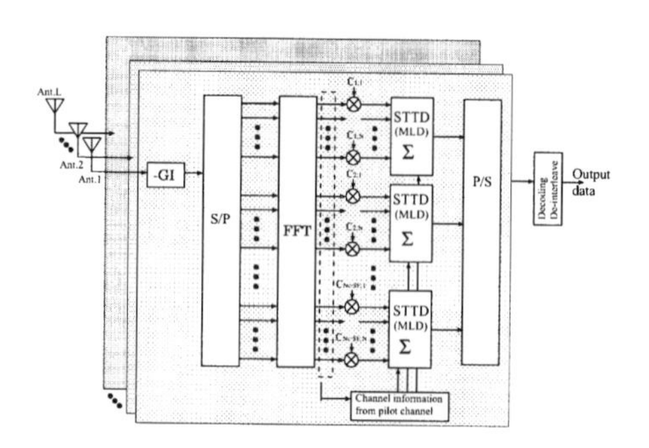

Figure 4: Receiver configuration of site diversity with STTD.

used to demodulate all the carriers, then these signals are de-spread by PN codes. The signals of all the carriers are combined in order to achieve the frequency diversity gain. The received signals at time t and time $t+T$ after de-spreading by PN codes are given by

$$r_0 = r(t) = h_0 s_0 - h_1 s_1^* + n_0 \tag{3}$$

$$r_1 = r(t+T) = h_0 s_1 + h_1 s_0^* + n_1 \tag{4}$$

where n_0 and n_1 represent complex random variables denoting noise and interference. The combined received signals are given by

$$\begin{aligned} s_0' &= h_0^* r_0 + h_1 r_1^* \\ &= h_0 h_0^* s_0 + h_1 h_1^* s_0 + h_0^* n_0 + h_1 n_1^* \\ &= (|h_0|^2 + |h_1|^2) s_0 + h_0^* n_0 + h_1 n_1^* \end{aligned} \tag{5}$$

$$\begin{aligned} s_1' &= h_0^* r_1 - h_1 r_0^* \\ &= h_0 h_0^* s_1 + h_1 h_1^* s_1 + h_0^* n_1 - h_1 n_0^* \\ &= (|h_0|^2 + |h_1|^2) s_1 + h_0^* n_1 - h_1 n_0^* \end{aligned} \tag{6}$$

The received signals with the two base base station antennas and the two receive antennas after de-spreading by PN codes are given by

$$r_0 = h_0 s_0 - h_1 s_1^* + n_0 \tag{7}$$
$$r_1 = h_0 s_1 + h_1 s_0^* + n_1 \tag{8}$$
$$r_2 = h_2 s_0 - h_3 s_1^* + n_2 \tag{9}$$
$$r_3 = h_2 s_1 + h_3 s_0^* + n_3 \tag{10}$$

The combined signals received of the two transmit antennas and the two receive antennas are expressed as

$$\begin{aligned} s_0' &= h_0^* r_0 + h_1 r_1^* + h_2^* r_2 + h_3 r_3^* \\ &= (|h_0|^2 + |h_1|^2 + |h_2|^2 + |h_3|^2) s_0 \\ &+ h_0^* n_0 + h_1 n_1^* + h_2^* n_2 + h_3 n_3^* \end{aligned} \tag{11}$$

$$\begin{aligned} s_1' &= h_0^* r_1 - h_1 r_0^* + h_2^* r_3 - h_3 r_2^* \\ &= (|h_0|^2 + |h_1|^2 + |h_2|^2 + |h_3|^2) s_1 \\ &+ h_0^* n_1 - h_1 n_0^* + h_2^* n_3 - h_3 n_2^* \end{aligned} \tag{12}$$

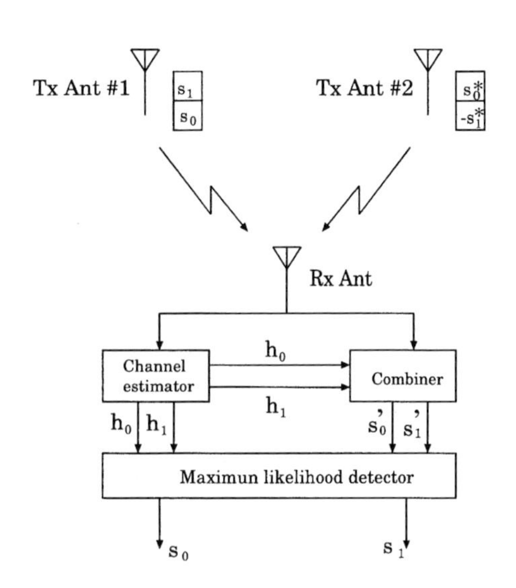

Figure 5: STTD scheme with two transmitter antennas with one receiver antenna.

Fig.4 shows a model of the receiver of the proposed system. A receiver receives the transmitted signals, which are converted from serial to parallel, and FFT is used to demodulate all the carriers.

Table 2: Simulation Parameters

Parameter	Value
Carrier Frequency	5[GHz]
Number of Subcarrier	128
Modulation	DQPSK, QPSK
Bandwidth	20MHz
Spreading Code	Walsh Hadamard code
Process Gain	4, 16, 32
Doppler Frequency	400[Hz]
Symbol Duration	5.2 [μsec]
FFT Length	128
GI	32
Number of Tx Antenna	1, 2
Number of Rx Antenna	1 , 2, 3, 4
Fading Model	Flat fading model
Fading Model	2 path fading model 6 path fading model

The signals of all the carriers are combined in order to achieve the frequency diversity gain. These combined signals sent to the Maximum Likelihood Detector(MLD) in Fig.5 provide a diversity order of $2L$ of two base station with one transmit antenna and L receive antennas. MLD is implemented to recover the information data signals (s_0,s_1)by space time decoding.

3 Results

3.1 Simulation Conditions

The performance of the proposed system was demonstrated by computer simulation in a frequency selective Rayleigh fading channel as well as in a frequency non-selective (flat) fading channel. In particular, a 2 path fading model and 6 path fading model in the frequency selective fading channel as mentioned in Section 2 are used. Ideal channel estimation is assumed in this paper. Carrier frequency is assumed at 5[GHz], and there are 128 sub-carriers. More detailed parameters used in the simulation are shown in Table 2.

3.2 Simulation Results

The Bit Error Rate(BER) performance of STTD for MC-CDMA is evaluated as a function of E_b/N_0. Fig.6 shows BER performance of site diversity using STTD in a non-frequency selective fading channel. Tx number indicates the number of transmitter antennas, while, Rx number means the number of receiver antennas. QPSK and DQPSK is the modulation scheme, and STTD indicates Space Time Transmitter Diversity in figures. These results of Fig.6 show that site diversity using STTD consising of two base stations using MC-CDMA can achieve better performance compared to a single antenna by combined antenna diversity and frequency diversity effect. The proposed system effectively achieves a better diversity gain by increasing the number of antennas by the antenna diversity effect.

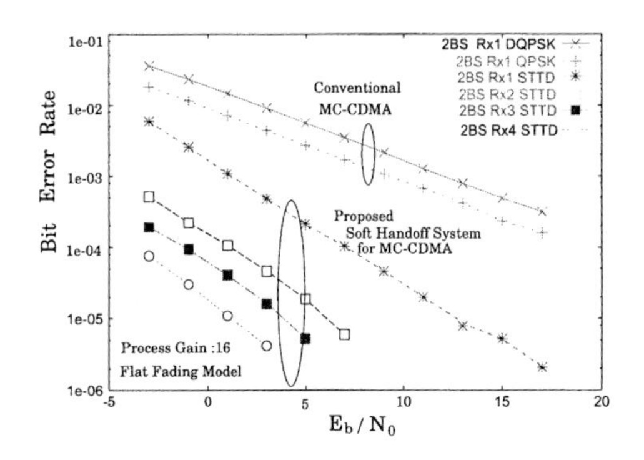

Figure 6: BER performance of site diversity using STTD scheme. (Flat Fading model, Doppler frequency: 400 [Hz])

Fig.7 shows the results of the 2-path frequency selective fading channel. Here, the delay time of the 1st path is 300[ns] relative to the 0th path(0[ns]). It is an exponential decay model (5[dB] decay). BER performance of the proposed system is also improved by antenna diversity gain from the site diversity using STTD in the 2-path frequency selective fading channel. By increasing receiver antenna number, the diversity gain provided by the site diversity using STTD of two base station for MC-CDMA can be improved compared to a system with no diversity. This proposed the softh handoff system using using STTD system has highly desirable features of Guard Interval inserted in MC-CDMA compared to the conventional STTD system, which is robust for the mutipath fading. While a single carrier STTD is worse in performance deteriorated by the multipath fading. The proposed system is effective under multipath fading.

Fig.8 shows the results of the 6-path frequency selective fading channel. The delay time between paths is 50[ns] as explained in Section 2. It is an exponential decay model of 5 [dB] decay with 6 paths. The results of Fig.8 are similar those of Fig.7. The proposed system is effective for improving the diversity gain provided by the site diversity using STTD system for the MC-CDMA method in 6-path frequency selective fading channel. For these reasons, The proposed soft handoff system using STTD can be considered effective for the frequency selective fading channel providing the large frequency diversity gain achieved by MC-CDMA with increase of multipath. This differs from the conventional STTD scheme, which deteriorates abruptly in the muti-path fading channel.

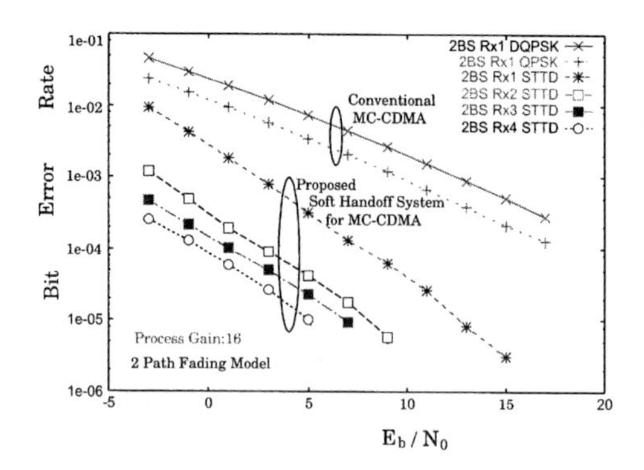

Figure 7: BER performance of site diversity using STTD scheme with 2 path fading model.(Doppler frequency:400 [Hz])

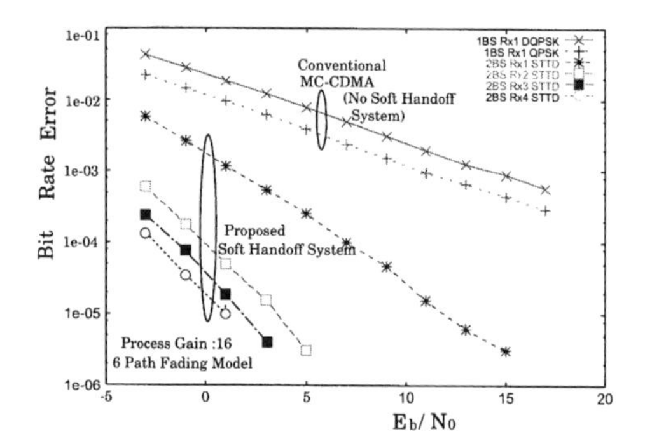

Figure 8: BER performance of Site diversity using STTD scheme with 6 Path fading model. (Doppler frequency: 400 [Hz])

Fig.9 shows the frequency diversity effect of MC-CDMA as a function of Process Gain. In this figure, one is the proposed site diversity using STTD with two base station (2 BS) and two receive antenna(Rx 2), while, another shows no diversity system with one transmit antenna of one BS(1 BS) and one receive antenna(Rx 1). The BER performance of the proposed systems is much improved by increasing the process gain due to the frequency diversity gain of MC-CDMA. The proposed system also achieves better BER performance than a system with no softh handoff system due to the antenna diversity gain from STTD.

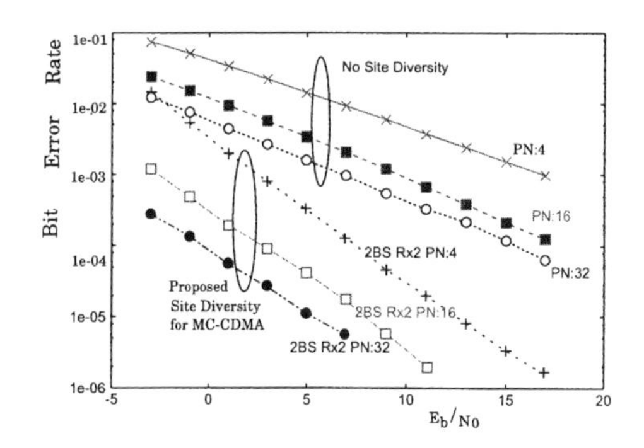

Figure 9: BER performance of Site diversity using STTD scheme with variable process gains.

4 Conclusion

An soft handoff system using STTD scheme for MC-CDMA was proposed to achieve a combined diversity effect between the antenna diversity by STTD and the frequency diversity achieved by multi-carriers in frequency selective fading channel environments. This method achieves the effective diversity gain created by the antenna diversity order of 2L in the case of two base station, Rx L, and the frequency diversity. The simulation results showed sufficiently better performance by the proposed system can be obtained compared to conventional systems due to site diversity and the frequency diversity. It was verified by simulation results that the proposed soft handoff system using STTD technique is effective and practical in the frequency selective fading channel. Therefore, the proposed system is appropriate for supporting a variety of multimedia services with high quality in the broadband wireless channel.

References

[1] R.V. Nee and R. Prasad, "OFDM Wireless multimedia communications" Artech House.

[2] K. Lee, and M. Nakagawa, "Multiple Antenna Transmission System Using RAKE Combining Diversity for a Multicarrier DS/CDMA in a Frequency Selective Fading Channel," IEICE Tran. on Comm., Vo. E84-B, No.4, pp. 739-746, Apr. 2001

[3] L.M. Correia and R. Prasad, "An overview of wireless broadband communications", IEEE Comm. Mag., Vo.35, No.1, pp. 28-33, Jan. 1997

[4] K. LEE, M. Nakagawa, "Distributed Antenna System using RAKE Combining Diversity for a Multi-Carrier DS/CDMA in a Frequency Selective Fading Channel," in proc. IEEE PIMRC'2000, London, pp.1490-1494, Sep. 2000.

[5] R. Prasad, and S. Hara, "An Overview of Multi-carrier CDMA," IEEE ISSSTA'96, pp. 107-114, Aug. 1996

[6] S. Hara, and R. Prasad, "Design and performance of multicarrier CDMA system in frequency selective rayleigh fading channel," IEEE Tran. on Vehicular Tech., pp. 1584-1595, May. 1999

[7] E. A Sourour and M. Nakagawa,"Performance of othogonal multicarrier CDMA in a multipath fading channel," IEEE Tran. on Commu., pp. 356-367, March. 1996

[8] N. Yee, J. P. Linartz, and G. Fettweis,"Multicarrier CDMA in indoor wireless networks," IEEE pr. PIMRC'93, pp.109-113, Yokohama, Japan, Sep. 1993

[9] S. Abeta, H. Atarashi, M. Sawahashi, and F. Adachi, "Performance of Coherent Multi-carrier/DS-CDMA and MC-CDMA for Broadband Packet Wireless Access," IEICE Tran. on Comm., Vo. E84-B, No.3, pp. 406-413, March 2001

[10] X. Gui and T. S. Ng,"Performance of asynchronous othogonal multicarrier CDMA system in frequency selective fading channel," IEEE Tran. Veh. Tech., Vol. 47, No.7, pp. 1084-1091, July 1999

[11] S.M. Alamount,"A Simple Transmit Diversity Technique for Wireless Communications,"IEEE Jou. on Select Areas in Comm., Vol.16, No.8, pp.1451-1457 1998.

N. NAGAI, M. OKADA, M. SAITO, and H. YAMAMOTO

ARRAY ANTENNA ASSISTED DOPPLER SPREAD COMPENSATOR WITH VEHICLE SPEED ESTIMATOR FOR OFDM RECEIVER

abstract. This paper proposes a new array antenna assisted Doppler spread compensator having a vehicle speed estimation function. The array antenna assisted Doppler spread compensator compensates for the Doppler spread by estimating the received signal at the stationary point with respect to the ground. However, the compensator requires the vehicle speed for compensation. The proposed vehicle speed estimator first observes the residual Doppler spread compensation error and then adjusts the estimated speed which minimises the error. Computer simulation results show that the speed estimator allows us to obtain the accurate estimate of the vehicle speed. The proposed compensator can considerably improve the bit error rate in the mobile reception of DTTB.

1. INTRODUCTION

OFDM (Orthogonal Frequency Division Multiplexing) is a multicarrier modulation technique capable of establishing wideband digital transmission over multipath environment. It has been adopted in DTTB (Digital Terrestrial Television Broadcasting) systems and wireless local area networks (W-LANs).

Although OFDM is robust to the frequency selective fading due to multipath propagation, it is vulnerable to the channel time-variation caused by Doppler spread. This weak point in OFDM causes the degradation of the reception quality in the mobile reception of DTTB.

The array antenna assisted Doppler spread compensator, which is described in , estimates the received signal at the stationary point with respect to the ground. Although the compensation mechanism is simple, this operation requires the vehicle speed.

The previously proposed compensator [1, 2] assumed that the vehicle speed was given from the speedometer that is equiped with vehicles. However, considering the convenience of the system, it is desired that the vehicle speed is obtained automatically in the system.

In this paper, we propose a new array antenna assisted Doppler spread compensator with vehicle speed estimation function. The proposed compensator first assumes a certain vehicle speed, and estimates the residual Doppler spread compensation error using the assumed vehicle speed. The assumed vehicle speed is then changed toward the direction that the residual error is reduced. By repeating this, the proposed compensator can estimates the vehicle speed as well as compensating for the degradation due to Doppler spread.

K. Fazel and S. Kaiser (eds.), Multi-Carrier Spread-Spectrum, 299–306.

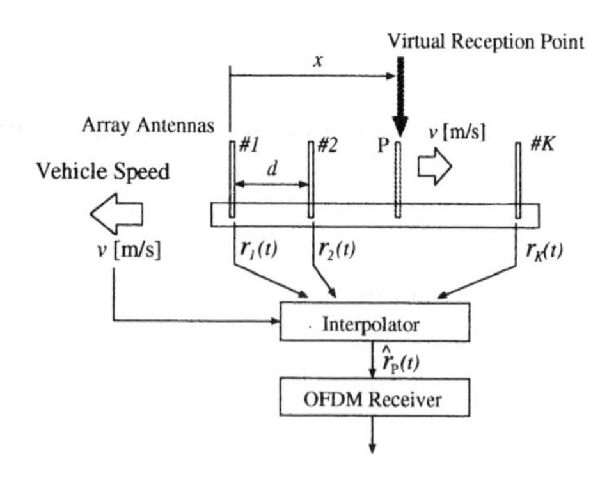

Figure 1: Array antenna assisted Doppler spread compensator

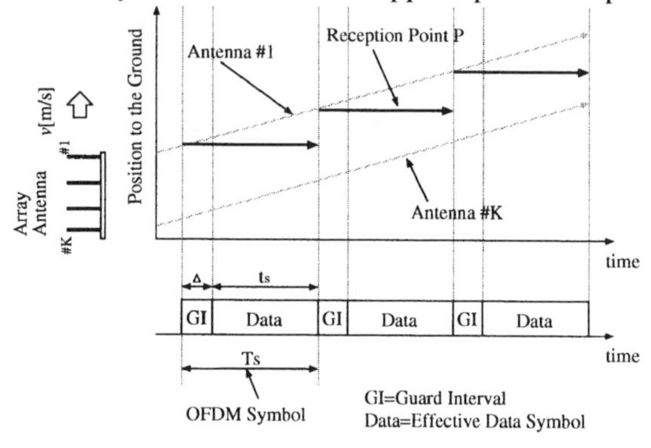

Figure 2: Positions of antennas and reception point P with respect to the groud

The rest of the paper is organized as follows. The section 2 briefly describes the principle of the Doppler spread compensator followed by the proposed vehicle speed estimator in section 3. Computer simulation results are then shown in section 4. Finally, section 4 summarizes concluding remarks.

2. DOPPLER SPREAD COMPENSATOR

In the beginning, we describe the principle of the Doppler spread compensator that has been proposed previously in [1, 2]. It compensates for Doppler spread in the mobile reception of DTTB by moving the reception antenna at the same speed as the receiver in the oppsite direction of the movement of the receiver. This operation is equal to making the reception antenna immovable to the ground. If it is possible to make the reception antenna stop to the ground in spite of the movement of the receiver, Doppler frequency shift does not occur and the reception quality is not influenced by Doppler spread.

However, it is not actual to move the real antenna in accordance with the movement of the receiver. Instead of moving the antenna, we introduce a virtual reception point and move it in accordance with the receiver. By doing this, we obtain the received signals at the static points to the ground.

In order to get the received signal at the virtual reception point (we will call this point 'P'), we use an array antenna. We estimate the received signal at the point P by interpolating the signals that are received at the elements of the array antenna.

The block diagram of the array antenna assisted Doppler spread compensator is shown in Fig.1. The antenna elements are lined up in the direction of the vehicle movements.

In order to make the reception point P immovable to the ground, we move P from the position of the antenna element #1 toward the position of the antenna element #K and we estimates the reception signal at P by an interpolator.

Let us assume that the vehicle speed is v (m/s) and $T_s = \Delta + t_s$ is the length of a OFDM symbol. In the reception of k-th OFDM symbol, the distance from antenna #1 to the point P is given by $x = vt - kT_s$, where t represents time.

The positions of the array antenna and reception point P are shown in Fig.2. We make the reception point P immovable to the ground while reciving every OFDM symbol. We can obtain the received signals that are not affected by Doppler spread.

3. VEHICLE SPEED ESTIMSTING DOPPLER SPREAD COMPENSATOR

The array antenna assisted Doppler spread compensator described in the previous section requires the vehicle speed for compensation process. In the following, we propose a new vehicle speed estimator that estimates the vehicle speed in the Doppler spread compensator. The proposed vehicle speed estimator first calculate the residual compensation error. In order to estimate the vehicle speed, a convergence algorithm is used for minimizing the residual error.

Let us begin with the definition of the residual compensation error. In the case of 16QAM, for example, the constaration of the transmitted signals are as shown in Fig.3. Let the received signal of subcarrier k be $\mathbf{r}_k$, and the decided signal of subcarrier k be $\mathbf{d}_k$. Then, we define the error for subcarrier k as

$$\mathbf{e}_k = \mathbf{r}_k - \mathbf{d}_k \tag{1}$$

Furthermore, we define the mean square of $\mathbf{e}_k$ as

$$\frac{1}{N}\sum_{k=1}^{N} |\mathbf{e}_k|^2 = \frac{1}{N}\sum_{k=1}^{N} |\mathbf{r}_k - \mathbf{d}_k|^2 \equiv E\left[|\mathbf{e}|^2\right] \tag{2}$$

where N is the number of effective carriers of OFDM.

Now, let the vehicle speed v and the wave length λ. The maximal Doppler frequency f_d is given by $f_d = v/\lambda$. For simplicity, we use the notation $f_d T_s \equiv \zeta$. Then, we can get

$$\zeta = f_d T_s = \frac{v}{\lambda} T_s \tag{3}$$

Since λ and T_s are determined by the system configurations, we can assume that ζ represents the vehicle speed itself.

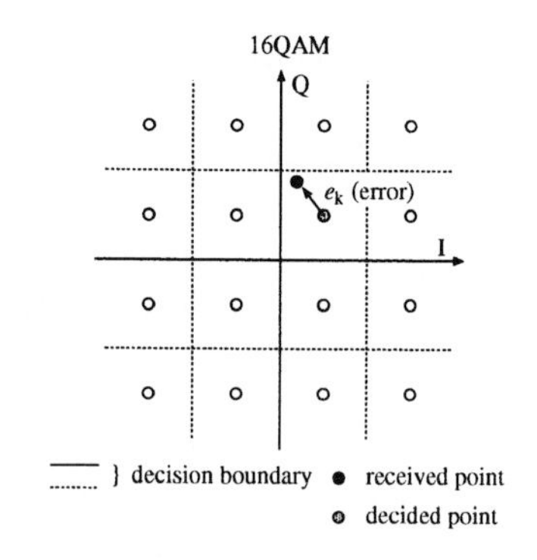

Figure 3: Calculation of residual compensation error in case of 16QAM

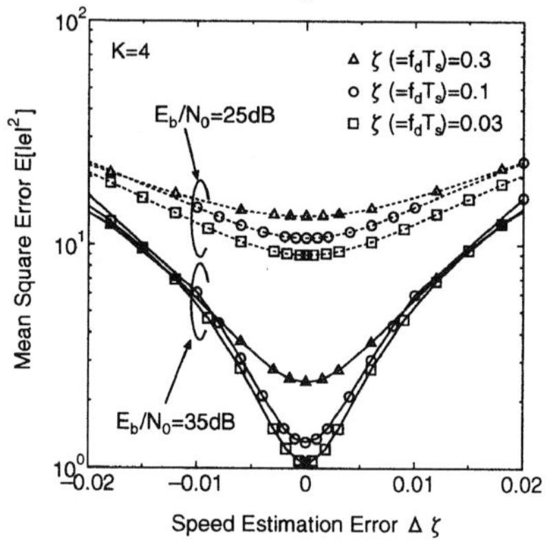

Figure 4: Mean square error against estimation error of vehicle speed. Number of antenna elements K=4.

Next, we examine the relation between the speed estimation error $\Delta\zeta = \hat{\zeta} - \zeta$ and the mean square error $E\left[|\mathrm{e}|^2\right]$, where $\hat{\zeta}$ is the estimated speed and ζ is the real speed. In Fig. 4, we calculate $E\left[|\mathrm{e}|^2\right]$ using the parameters shown in Table 1.

According to Fig. 4, the mean square error curve is in the shape of convex downward. And, independently of E_b/N_0 or $\zeta(= f_d T_s)$, both the value of the mean square error and the absolute value of gradient of the mean square error curve increase in proportion to the absolute value of the speed estimation error.

The proposed algorithm seeks the vehicle speed $\hat{\zeta}$ that minimizes the above-described mean square error.

Let the estimated speed at OFDM symbol i be $\hat{\zeta}_i$, the mean square error at $\hat{\zeta}_i$ be $E\left[|e(\hat{\zeta}_i)|^2\right]$ and the gradient of $E\left[|e(\zeta)|^2\right]$ at $\zeta = \hat{\zeta}_i$ be $\nabla E\left[|e(\hat{\zeta}_i)|^2\right]$, then we will

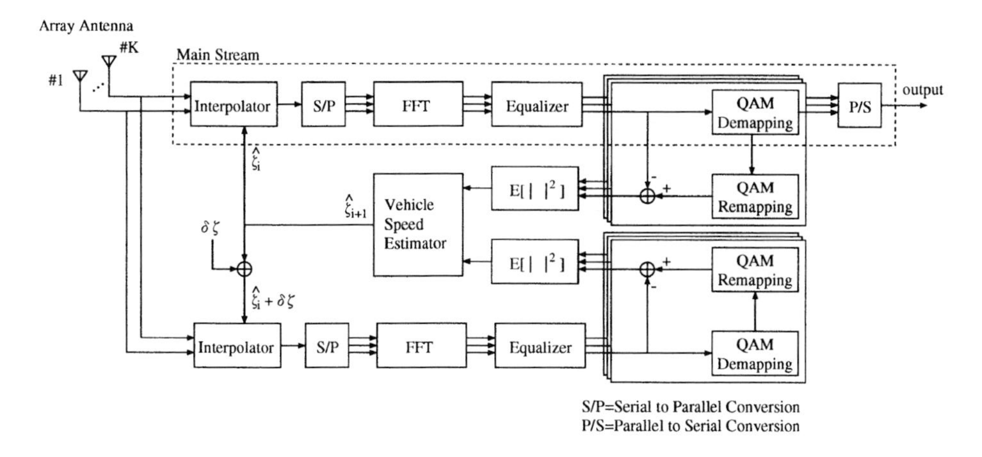

Figure 5: Block diagram of the proposed array antenna assisted Doppler spread compensator with vehicle speed estimator for OFDM receiver

get the estimated speed at OFDM symbol $i+1$ as

$$\hat{\zeta}_{i+1} = \hat{\zeta}_i - \mu \cdot \nabla E\big[|e(\hat{\zeta}_i)|^2\big] \tag{4}$$

$$= \hat{\zeta}_i - \mu \cdot \left| \frac{\partial E\big[|e(\zeta)|^2\big]}{\partial \zeta} \right|_{\zeta=\hat{\zeta}_i} \tag{5}$$

$$= \hat{\zeta}_i - \mu \cdot \left| \frac{E\big[|e(\zeta+\delta\zeta)|^2\big] - E\big[|e(\zeta)|^2\big]}{\delta\zeta} \right|_{\zeta=\hat{\zeta}_i} \tag{6}$$

where μ is a step size parameter and $\delta\zeta$ is a minute constant for acquiring the gradient.

One of the significant features of this algorithm is that it uses the gradient of the mean square error curve to renew the estimated vehicle speed. By doing this, the estimated vehicle speed is renewed drastically when the speed estimation error is comparatively large, and is renewed to a nicety when the speed estimation error is comparatively small.

And, another feature is that the estimated speed will not go into wrong extremal value because the mean square error curve is in the shape of convex downward.

Fig. 5 illustrates the block diagram of the proposed array antenna assisted Doppler spread compensator with vehicle speed estimator. The proposed one is composed of two Doppler spread compensators. One of the compensators compensates assumes the maximum Doppler frequency $\hat{\zeta}_i$ and the other assumes $\hat{\zeta}_i+\delta\zeta$ for performing compensation. The residual compensation errors for both compensators are then calculated. The estimated vehicle speed $\hat{\zeta}_i$ is changed according to the difference between two errors. By repeating this, the estimator estimates the vehicle speed accurately.

Table 1: System configurations

Transmission method	OFDM
Bandwidth	5.572 MHz
Carrier spacing	0.992 kHz
FFT size	8192
Number of carriers	5617
Carrier modulation	64QAM
Effective symbol duration	1.008 ms
Guard interval	126 μs (1/8)
Propagation model	Two-ray Rayleigh fading
D/U	0 dB
Delay	4 μs
Arrival directions	Uniform distribution
Number of antenna elements and antenna spacing	$K = 2, d = 0.1\lambda$ $K = 4, d = 0.2\lambda$

NUMERICAL RESULTS

Computer simulation is carried out in order to verify the performance of the system. The system configuration is shown in Table 1. The parameters used in this section is based on Japanese terrestrial digital television standard (ISDB-T).

Fig. 6 shows the estimated vehicle speed against the number of OFDM symbols. The estimated speed converges within 100 OFDM symbols at $\mu = 5 \times 10^{-15}$. The convergence speed can be reduced to 10 OFDM symbols when $\mu = 5 \times 10^{-14}$.

The bit error rate performances against E_b/N_0 and f_dT_s are respectively shown in Figs. 7 and 8. The solid lines indicate the bit error rate in case of perfect vehicle speed estimation while dashed lines corresponds to the proposed vehicle speed estimator. In both the figures, the performance of proposed estimator agrees well with the one with perfect vehicle speed. This implies that the proposed estimator estimates the vehicle speed accurate enough for the Doppler spread compensator.

CONCLUSION

In this paper we have proposed an array antenna assisted Doppler spread compensator with vehicle speed estimator. Computer simulation results show that the proposed vehicle speed estimator can estimate the speed accurate enough for Doppler spread compensation.

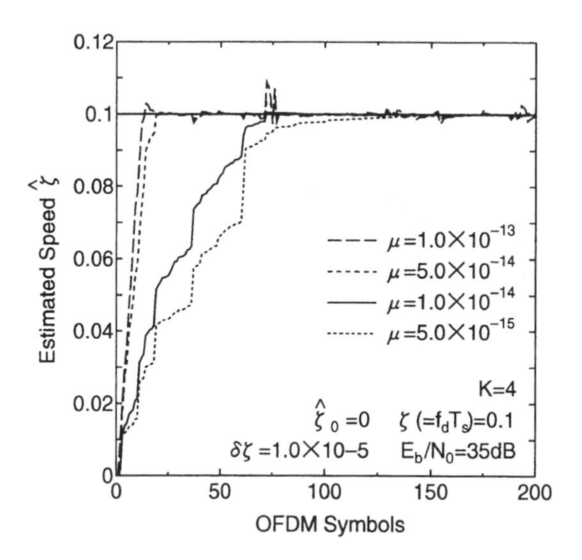

Figure 6: Convergence performance of estimated vehicle speed. Number of antenna elements $K = 4$. $\zeta(= f_d T_s) = 0.1$. E_b/N_0=35dB.

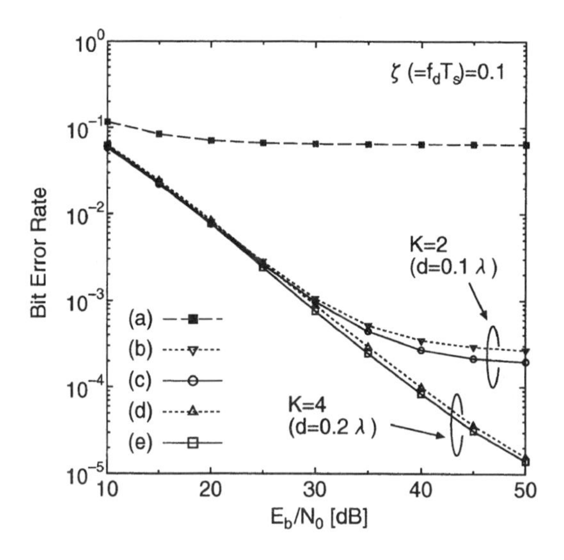

Figure 7: Bit error rate performance against E_b/N_0. Number of antenna elements $K = 2, 4$. $\zeta(= f_d T_s) = 0.1$. (a) is without Doppler spread compensation. (c) and (e) are with Doppler spread compensation where the vehicle speed is known. (b) and (d) are with Doppler spread compensation where the vehicle speed is estimated using the proposed method.

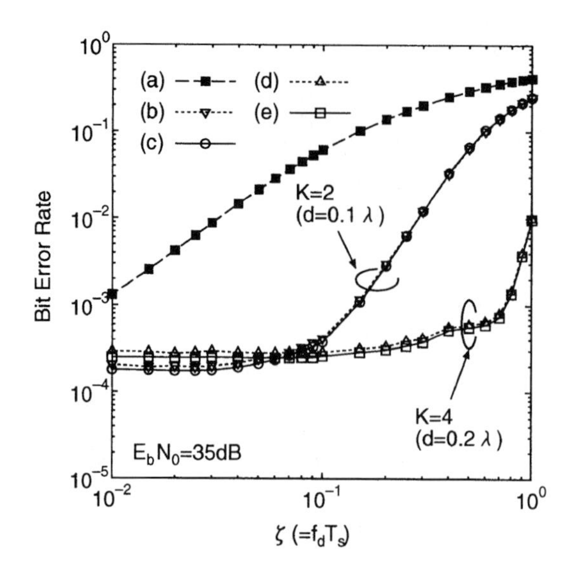

Figure 8: Bit error rate performance against $\zeta(= f_d T_s)$. Number of antenna elements $K = 2, 4$. $E_b/N_0 = 35$dB. (a) is without Doppler spread compensation. (c) and (e) are with Doppler spread compensation where the vehicle speed is known. (b) and (d) are with Doppler spread compensation where the vehicle speed is estimated using the proposed method.

References

[1] M.Okada and S.Komaki. Random FM Noise Compensation Scheme for OFDM. *IEE Electron. Lett.*, 36(19):1653–1654, Sep. 2000.

[2] M.Okada, H.Takayanagi, and H.Yamamoto. Array Antenna Assisted Doppler Spread Compensator for OFDM. *European Transactions on Telecommunications (ETT)*, 13(5):507–512, Sep.-Oct. 2002.

Nara Institute of Science and Technology (NAIST)

ADÃO SILVA, ATÍLIO GAMEIRO

PRE-FILTERING TECHNIQUES USING ANTENNA ARRAYS FOR DOWNLINK TDD MC-CDMA SYSTEMS

Abstract. This paper deals with downlink time division duplex MC-CDMA system, and presents a space-frequency pre-filtering technique designed for two different receivers: a simple despread receiver without channel equalization and an EGC conventional receiver, where the base station is equipped with an antenna array. We show that the space-frequency pre-filtering approach proposed allows to format the transmitted signals so that the multiple access interference at mobile terminals is reduced allowing to transfer the most computational burden to the base station. Simulations results are carried out to demonstrate the effectiveness of the proposed pre-filtering schemes.

1. INTRODUCTION

MC-CDMA is a very promising multiple access scheme for achieving high data rate transmission in a mobile cellular environment [1]. This technique combines Orthogonal Frequency Division Multiplex (OFDM) and CDMA. In [2] it is shown that MC-CDMA significantly increases the capacity of the mobile cellular system compared with third generation technologies based on DS-CDMA.

It is weel known that CDMA based systems are limited by the Multiple Access Interference (MAI) caused by the loss of orthogonality among users in multipath propagation. Usually, in conventional MC-CDMA downlink, i.e from base station (BS) to mobile terminals (MT), the MAI is mitigated by frequency domain equalization techniques at the receiver. Considering time division duplex (TDD) another solution consists in performing pre-filtering at the transmitter side using the TDD channel reciprocity between alternative uplink and downlink transmission period [3]. The principle is to use the same channel estimates obtained during a reception slot, in order to pre-compensate the signal during the following transmission slot. The aim of this solution is to move the main MAI mitigation task from the MT to BS, allowing the use of simple low-cost, low power-consuming receiver at the MT.

Another different approach to further increase the system capacity without allocating additional frequency spectrum is the use of spatial processing technique with antenna arrays at the BS [4]. In most scattering environments, antenna diversity is a practical, effective and, hence, a widely applied technique for reducing the effect of frequency selective fading and improve the spectral efficiency [4]. Combining antenna arrays with MC-CDMA systems is very advantageous in cellular communications[5].

This paper proposes a space-frequency pre-filtering technique designed for two different receivers, for a downlink TDD MC-CDMA system using antenna arrays at transmission. Both schemes are performed in frequency domain and optimization is

K. Fazel and S. Kaiser (eds.), Multi-Carrier Spread-Spectrum, 307–314.

done jointly in space and frequency. Moreover, these algorithms are designed using as criterion the minimization of the transmitted power at BS.

The remaining paper is organized as follows: In section 2 we present the proposed downlink MC-CDMA system. In section 3, we analytically derive the space-frequency CZF and CZF-EGC algorithms. In section 4, we present some simulation results obtained with the CZF and CZF-EGC pre-filtering techniques in two different scenarios (beamforming and diversity). We also compare the pre-filtering algorithms with conventional equalizer techniques such as Maximal Ratio Combining (MRC), Equal Gain Combining (EGC). Finally, the main conclusions are presented in section V.

2. SYSTEM DESCRIPTION

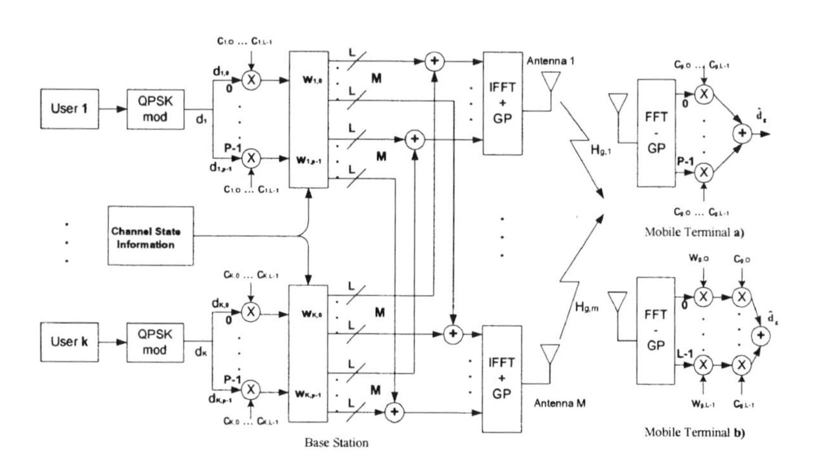

Figure 1 .Proposed Transmitter and Receivers schemes for MC-CDMA downlink using antenna arrays.

The generic block diagram of the proposed transmitter and receivers is shown in Figure 1. From this figure we can see that for each user K, a complex QPSK data symbol d_k ($k=1, ..., K$) is converted from serial-to-parallel to produce p symbols, $d_{k,p}$ ($k=1, ..., K$ and $p=0, ...,P-1$), where P denotes the number of data symbols transmitted per OFDM symbol. The data symbols are spread into L chips using the orthogonal Walsh-Hadamard code set and scrambled by a pseudo-random code. We denote the code vector of user k as $C_{k,l}=[c_{k,0}, ..., c_{k,L-1}]^T$, where $(.)^T$ is the transpose operator. Then, the chips of the data symbols are copied M times in order to obtain $L.M$ versions of the original symbols which are weighted and transmitted over M antenna branches. A vector weight $W_{k,p}$ of length $L.M$ is computed for each user and data symbol. These weights are calculated using the CSI according to the criteria to be presented in section 3. Then, the signals of all users on each sub-carrier and antenna branch are added to form the multi-user transmitted signal. Finally, a guard period (GP) longer than the channel multipath spread is inserted in the transmitted signal, on each antenna, to avoid inter symbol interference (ISI).

The transmitted signal, in frequency domain, for a generic data symbol p is given by,

$$Y_p = \sum_{k=1}^{K} d_{k,p} \widehat{C}_k \circ W_{k,p} \tag{1}$$

where $\widehat{C}_k = \left[C_k^T, \ldots, C_k^T\right]$ is size of $L.M$ and represents the spreading operation, once the same code is used for all antenna branches, $W_{k,p}$ is the weight vector of size $L.Mx1$ and $\circ$ is the element wise vector product. This vector signal of length $L.M$ is mapped to the antenna branch so that the first L elements are transmitted over the L subcarriers of the OFDM modulation on the first antenna branch, the second L elements to the second branch and so on.

The input signal at the generic mobile g, for symbol p, is obtained multiplying (1) for the channel frequency response of the desired user and adding AWGN noise,

$$X_{g,p} = \sum_{m=1}^{M} \sum_{k=1}^{K} d_{k,p} C_k \circ W_{k,p,m} \circ H_{g,p,m} + n_g \tag{2}$$

where $W_{g,p,m}$ is the vector weight for antenna m and $H_{g,p,m}$ of size of $Lx1$ is the channel frequency response between antenna m and mobile terminal.

At the receiver side we propose two different receivers: receiver a) which is composed just by a single antenna, a FFT, despreading and descrambling operations and a conventional EGC single user receiver, b).

For receiver a), the decision variable at the input of the QPSK demodulator, is for the desired user g and symbol p given by,

$$\hat{d}_{g,p} = \underbrace{d_{g,p}.C_g(\sum_{m=1}^{M} W_{g,p,m} \circ H_{g,p,m})C_g^H}_{Desired\ Signal} + \underbrace{\sum_{k=1,k\neq g}^{K} d_{k,p}.C_k \sum_{m=1}^{M} (W_{k,p,m} \circ H_{g,p,m})C_g^H}_{MAI} + \underbrace{N_g}_{Noise} \tag{3}$$

For receiver b), the decision is also given by equation (3), but the vector H is replaced by,

$$H = \frac{(|H_m|^2 + H_m \sum_{i=1,i\neq m}^{M} H_i^*)}{\sum_{m=1}^{M} |H_m|} \tag{4}$$

where (.)* denotes the complex conjugate.

The vector N_g represents the noise samples of ML sub carriers. The signal of (3) involves the three terms: the desired signal, the MAI caused by the loss of code orthogonality among the users, and the noise after despreading.

3. PRE-FILTERING ALGORITHMS

In this section we analytically derive a space-frequency pre-filtering algorithm for the two receivers: a) and b). In the latter case the weights are computed taking into account that at the receiver we have the EGC combiner. However, we use the same criterion, zero forcing, in both pre-filtering schemes.

The use of pre-filtering algorithms has two main advantages: reduce the MAI at mobile terminals by pre-formatting the signal so that the received signal at the decision point is free from interferences and allow to move the most computational burden from MT to BS, keeping the MT at a low complexity level. When we use an antenna array at the BS, the pre-filtering can be done in both dimensions, space and frequency. We propose to jointly optimize the user separation in space and frequency by the use of criteria based on the decision variable after despreading at the MT. This optimization task is performed taking into account the power minimization at the transmitter side.

2.1. Space-Frequency CZF Algorithm

The CZF pre-filtering algorithm is based on zero forcing criterion. This algorithm is designed in order to remove the MAI term of (3) at all MTs. Furthermore, it takes into account the transmitted power at BS, reason we call this algorithm the constrained zero forcing.

Applying the zero-forcing criterion to equation (3), we obtain the following conditions:

$$\begin{cases} C_g \circ (\sum_{m=1}^{M} W_{g,p,m} \circ H_{g,p,m}) C_g^H = 1 \\ \sum_{k=1,k\neq g}^{K} C_k \circ (\sum_{m=1}^{M} W_{k,p,m} \circ H_{g,p,m}) C_g^H = 0 \end{cases} \tag{5}$$

ensuring that each user receives a signal that after despreading is free of MAI. The first term of the right side of (3) is the desired signal, and has been made, for normalization purposes, equal to 1, while the second term represents the interference caused by other K-1 users and according to the criterion used should be equal to 0.

The interference that the signal of a given user g produces at an other MT k is obtained for a generic data symbol according to (3),

$$MAI(g \rightarrow k) = C_g (\sum_{m=1}^{M} W_{g,p,m} \circ H_{k,p,m}) C_k^H \tag{6}$$

The weight vector for user g is then obtained by constraining the desired signal part of its own decision variable to one while cancelling its MAI contribution all other mobile terminals at same time. This leads to the following set of conditions,

$$\begin{cases} C_g \circ (\sum_{m=1}^{M} W_{g,p,m} \circ H_{g,p,m}) C_g^H = 1 \\ C_g \circ (\sum_{m=1}^{M} W_{g,p,m} \circ H_{k,p,m}) C_k^H = 0 \quad \forall k \neq g \end{cases} \tag{7}$$

Hence, to compute the weights for user g we have to solve a linear system of K equations (constraints) and $L.M$ variables (degrees of freedom) given by,

$$A_p W_p^H = B \tag{8}$$

where A_p is a channel coefficients and codes matrix of size $KxML$, $W_p=[W_{p,1}, \ldots, W_{p,m}]$ is a vector weight size of $L.M$ and B is a constraint vector of size $Kx1$.

As pointed above the pre-filtering algorithms should take into account the minimization of the transmitted power. Therefore, the transmitted power must be minimized under $A_p W_p=B$ constraint. When the number of constraints equals the number of degrees of freedom, a single solution exists provided there are no singularities. If however we have more degrees of freedom than constraints ($ML>K$) then signal design can be done to optimize some cost function, normally the total transmitted power. This optimization can be solved with the Lagrange multipliers method.

After some mathematical manipulations, we obtain the CZF based pre-filtering vector,

$$W_p = A_P^H (A_p A_p^H)^{-1} B = A_p^H \psi_p^{-1} B \tag{9}$$

where $\psi_p = \left[A_p A_p^H \right]$ is a square and Hermitian matrix of size KxK.

3.2. Space-Frequency CZF-EGC algorithm.

As referred, for this scheme we also use the zero forcing criterion, but now to compute the weights we should take into account that we have the EGC combiner at receiver side, reason we call this algorithm CZF-EGC. Thus, using equations (3), (4) and (6) we obtain the following set of conditions,

$$\begin{cases} C_g (\sum_{m=1}^{M} W_{g,pm} \circ \dfrac{(|H_{g,p,m}|^2 + H_{g,p,m} \cdot \sum_{i=1, i \neq m}^{M} H_{g,p,i}^*)}{\sum_{m=1}^{M} |H_{g,p,m}|}) C_g^H = 1 \\ C_g (\sum_{m=1}^{M} W_{g,p,m} \circ \dfrac{(|H_{k,p,m}|^2 + H_{k,p,m} \cdot \sum_{i=1, i \neq m}^{M} H_{k,p,i}^*)}{\sum_{m=1}^{M} |H_{k,p,m}|}) C_k^H = 0 \quad \forall g \neq k \end{cases} \tag{10}$$

As the CZF algorithm, the CZF-EGC based pre-filtering vector is given by equation (9). However, the matrix A is fulfilled according equation (10) and not according equation (7) as the CZF algorithm. From equation (10) we can see that for the case $M=1$ (single antenna) we obtain an expression very similar to equation (7), given by,

$$\begin{cases} C_g(\sum_{m=1}^{M} W_{g,p,m} \mid H_{g,p,m} \mid) C_g^H = 1 \\ C_g(\sum_{m=1}^{M} W_{g,p,m} \mid H_{k,p,m} \mid) C_k^H = 0 \quad \forall k \neq g \end{cases} \tag{11}$$

In this case the weights are real, because we use the modulus of the channel frequency response, thus we just equalize the amplitude at the transmitter whereas the phase is equalized at the receiver side.

4. NUMERICAL RESULTS

To evaluate the performance of the proposed pre-filtering algorithm, we used a pedestrian Rayleigh fading channel, whose system parameters are derived from the European BRAN Hiperlan/2 standardization project [6]. This channel model has 18 taps, multipath spread of 1.76μs and coherence bandwidth approximately equal to 637KHz.

We extended this time model to a space model in two different ways: for the diversity case, we assumed that the distance between antenna elements is large enough, to consider for each user M independents channels, i.e, we assume independent fading process; for the beamforming case we allocated a Direction Of Arrival (DOA) to each path (beamforming), with the DOA's randomly chosen within a 120° sector. In this latter case the BS is equipped with a half wavelength spaced uniform linear array. We considered a DL synchronized transmission using Walsh-Hadamard spreading sequences of length 32 scrambled by a pseudo-random code. We used 1024 carriers, a bandwidth equal 100Mz, a carrier frequency equal 5.0GHz. The duration of the guard period (GP) is 20% of the total OFDM symbol duration. The channel is considered to be constant during an OFDM symbol.

The simulations were carried out to assess the performance of the CZF and CZF-EGC algorithms in the two different scenarios presented above, and to compare against the performance achieved with conventional frequency equalization receivers, such as MRC and EGC. For a better comparison with a variable number of antennas, the results have been normalized, i.e. for the case of multiple transmitting antennas, the figures do not take into account the array gain which is 10.Log(M) in dB.

The simulation results for the diversity case are shown in Figure 2 a). The simulations were run for a number of users K=32 i.e. a full-load system, and the metric used is the average bit error rate (BER) as function of Et/No, the transmitted energy (assuming a normalized channel) per bit over the noise spectral density. The performance of the CZF and CZF-EGC algorithms is illustrated for the cases of M=1, 2, 4 and 8 transmit antennas. With a single antenna at BS, there is no spatial separation and the pre-filtering operation is done only in the frequency dimension. As can be seen from Figure 2 a) the performance of the CZF algorithm for a single antenna is modest. At low values of Et/No the performance is even worse than with all single user conventional detectors, and only for high values of Et/No the CZF outperforms the conventional EGC equalizer. This occurs because, for a single antenna and full load system we do not have enough degrees of freedom to minimize the transmitted power. The number of degrees of freedom is equal to $M.L$ and the number of constraints is K. Thus, for M=1 and K=L (full load system), the number of degrees is equal to the number of constraints. For multiple antennas at BS, it is possible to optimize the pre-filtering algorithm in both dimensions, space and frequency. When we use an array of 2, 4 and 8 antennas the performance of the CZF algorithm is much better than all single user conventional equalizers for any Et/No value. We can see that with 4 and 8 antennas the performance is very close to the one obtained with the Gaussian channel. As it can be seen from Figure 2 a) the performance of the CZF-EGC algorithm for single antenna outperforms the MRC, EGC conventional equalizer and the CZF. This occurs because, as can be seen from equation (11), with single antennas the CZF-EGC weights are real. Thus, we just perform a pre-filtering amplitude operation at the transmission side while the phase equalization is done at the receiver. In the case of CZF we perform the amplitude and phase equalization at transmission side. For two antennas the performance of the CZF-EGC is slightly better than CZF algorithm, while for a number of antenna elements greater than four the performance of both algorithm is nearly identical.

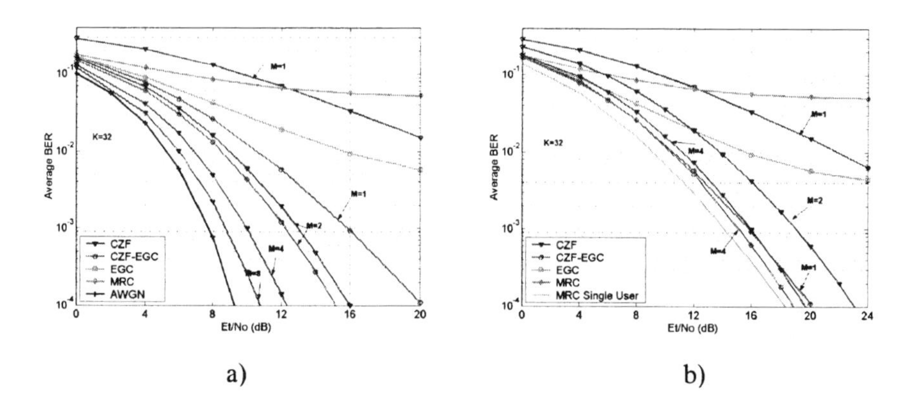

Figure 2. Performance comparison between the CZF, CZF-EGC and conventional receiver for two scenarios: a) diversity b) beamforming

The simulation results for the beamforming case are shown in Figure 2 b), where the simulation metrics and parameters other than the channels correlation are identical to the ones considered for Figure 2 a). The results show that the CZF algorithm outperforms all the conventional equalizers, except for the single antenna case as happened with the diversity scenario. With four antenna elements the CZF performance is very close to the performance of the MRC single user. The performance of the CZF-EGC with single antenna is very good as compared with all conventional equalizers and CZF. For the single antenna case we have the same number of degrees of freedom for both algorithms, but for CZF-EGC case we just perform the amplitude equalization at transmitter side while for CZF we perform amplitude and phase equalization. We can even see that the performance is similar to the one obtained with CZF algorithm for four antennas. The performance of the CZF-EGC for four antennas is very close to the one obtained with CZF.

5. CONCLUSIONS

We presented a space-frequency pre-filtering technique for downlink TDD MC-CDMA, using antenna arrays at BS, for two different receivers: the conventional EGC and a simple despread receiver without channel equalization. We analytically derived the proposed pre-filtering algorithms, based one a constrained zero-forcing criterion. The performance was assessed either for the diversity and beamforming scenarios and compared against the one of conventional receivers. The results have shown that a considerable MAI reduction is obtained with the CZF-EGC technique, and with CZF when an antenna array is used at base station. For a single antenna, the performance of the CZF-EGC outperforms the CZF, while with multiple antennas the performance are very similar. These techniques allows a drastic raise of the user capacity and move the most demanded processing task to the BS, keeping the mobile terminal as simple as possible.

6. ACKNOWLEDGMENT

The work presented in this paper was supported by the European IST-2001-32620 MATRICE project.

7. REFERENCES

[1] IST MATRICE project, web site *http://www.ist-matrice.org*.

[2] S.Hara, R. Prasad, "Overview of multicarrier CDMA", *IEEE Communications Magazine*, vol.35, December 1997.

[3] B.R. Vojcic, W.M. Jang, 'Transmitter Precoding in Synchronous Multiuser Communications,' *IEEE Transact. on Communications*, vol. 46, no. 10, pp. 1346-11355, Oct. 1998.

[4] A.F. Naguib, A.J. Paulraj, T. Kailath, 'Capacity improvement with base-station antenna arrays in cellular CDMA,' *IEEE Transact. on Vehic. Tech.*, vol. 43, no. 3, pp. 691-698, Aug. 1994.

[5] C. K. Kim, Y. S. Cho, 'Performance of a wireless MC-CDMA system with an Antenna Array in a fading channel: reverse link,' *IEEE Transact. on Communications*, vol. 48, no. 8, pp. 1257-1261, Aug. 2000.

[6] J. Medbo, 'Channel Models for Hiperlan/2 in different Indoor Scenarios,' *ETSI BRAN doc.* 3ERI085b, 1998.

Dept. of Electronics and Telec, University of Aveiro/Institute of Telecommunications

T. SÄLZER AND D. MOTTIER

DOWNLINK STRATEGIES USING ANTENNA ARRAYS FOR INTERFERENCE MITIGATION IN MULTI-CARRIER CDMA

Abstract. The paper considers transmitter strategies for interference mitigation in the downlink of a multi-carrier CDMA system using antenna arrays. We describe typical indoor and outdoor scenarios and derive the corresponding transmitter strategy according to the channel knowledge available at the base station for each case. We show the effectiveness of multi-user space-frequency transmit filtering for the indoor and single-user beamforming for the outdoor scenario. Both strategies considerably reduce the multiple access interference and thus allow a low-complexity receiver design for the mobile terminal.

1. INTRODUCTION

Virtues such as robustness to multipath propagation and flexibility of the multi-user access make Multi-Carrier CDMA (MC-CDMA) a transmission scheme that is now considered by a growing community of researchers striving for new efficient air interfaces [1][2]. Recent publications show that this scheme is particularly advantageous for the Down-Link (DL), i.e. from a Base-Station (BS) to Mobile Terminals (MT) [3]. However, like all CDMA-based systems, MC-CDMA suffers from Multiple Access Interference (MAI), which is caused by the loss of orthogonality among the users' signals in multipath propagation. MAI mitigation has, therefore, been a challenging research topic since the very beginning of studies on MC-CDMA [1]. Yet, the frequently considered approach of performing Multi-User Detection (MUD) at the receiver is quite unattractive for the DL, because it entails an increase of complexity and power consumption at the MTs [1][4]. Here, we propose an alternative approach using an antenna array at the BS.

Antenna arrays endow a wireless system with the spatial dimension, which can be exploited in various ways. Striving for a light design of the MT, we only consider an array for transmission at the BS. Here, the strategy to adopt essentially depends on the knowledge about the propagation channel that is available at the BS prior to transmission. Three cases may be distinguished: If the BS has no knowledge at all, we may opt for a transmit diversity or space time coding scheme, e.g. [5]. These systems transmit redundant information over the different antenna branches to benefit from spatial diversity by combining or decoding at the receiver side. In contrast, if the BS has perfect instantaneous knowledge of the channel fading, it is preferable to adapt the transmitted signal to the channel conditions. Such techniques are generally referred to as pre-filtering or pre-coding. In a multi-user context they not only aim at pre-compensating the channel fading but can also reduce the MAI. The general concept of multi-user prefiltering can be found in [6], and we already proposed approaches for prefiltering in space and frequency applied to MC-CDMA in [7] and [8]. When only partial knowledge, e.g. the Directions of Departure (DODs) of the multipath components or any other related long-term spatial channel statistics, is available, Beamforming (BF) may be a good choice, especially when

K. Fazel and S. Kaiser (eds.), Multi-Carrier Spread-Spectrum, 315–326.

the channels at different antennas are correlated. BF exploits the spatial separation of MTs by adapting the antenna pattern so as to illuminate only desired directions and avoid interference at other MTs positions [9]. BF applied to MC-CDMA was studied in [10] for reception in the Up-Link (UL).

In this paper, we focus on the latter two cases where we have either instantaneous knowledge, which may be available at the BS from estimation of the UL in a Time Division Duplex (TDD) system with low mobility, or long-term spatial statistics, which can still be obtained in scenarios with high mobility and arbitrary duplex mode [11]. In both cases, we can use the channel knowledge to transfer at least some of the demanding signal processing tasks for MAI reduction from the MT to the BS.

The paper is organised as follows. We present the system model of a DL MC-CDMA transmission with an antenna array and the corresponding transmit filter at the BS in section 2. The scenarios and system considerations are exposed in section 3. Section 4 is dedicated to the optimisation of the transmit filter in the two mentioned scenarios leading to Space-Frequency Transmit Filtering (SFTF) and Beamforming (BF), respectively. Numerical results for both scenarios including an assessment of the impact of Doppler variations are presented in section 5. Finally, we give concluding remarks in section 6.

2. THE SYSTEM MODEL

We consider the DL MC-CDMA system depicted in figure 1. Like in the conventional system, the data symbols, e.g. QPSK symbols, of users 1 to K are spread into L chips using orthogonal codes $\mathbf{c}_k$ (k=1,..,K), for instance Walsh-Hadamard codes. These chips are then copied M times for the M antenna branches. Both operations are mathematically represented by vector $\tilde{\mathbf{c}}_k=[\mathbf{c}_k^T,\ldots,\mathbf{c}_k^T]^T$ of length ML, which is a repetition of the code vector $\mathbf{c}_k$ of length L. $(.)^T$ denotes vector transposition. These chips are weighted by the transmit filter, whose components are represented by vector $\tilde{\mathbf{w}}_k$ of length ML. The transmit filters are calculated using available channel knowledge, c.f. section 3. Finally, the contributions of all users are summed chip-by-chip, and the result is mapped on L subcarriers of the Orthogonal Frequency Division Multiplex (OFDM) system on each antenna branch.

The underlying OFDM transmission uses a guard interval Δ and is assumed to be ideal in the sense that the channel can be represented in the frequency domain by a single flat fading coefficient on each subcarrier. Hence, we can represent the channel between the M antennas of the BS and the single antenna of MT g by ML complex fading coefficients gathered in vector $\mathbf{h}_g$.

To keep the complexity at the MT as low as possible, we only use a single antenna and Single User Detection (SUD) techniques. Thus, the receiver antenna implicitly recombines the M signals from the transmit array in space. Then, after OFDM demodulation and a potential channel estimation, the receiver equalises and despreads the signals of the L subcarriers using vector $\mathbf{q}_g=[q_g(1),\ldots,q_g(L)]^T$. In analogy to the transmitter description, we use an expanded vector $\tilde{\mathbf{q}}_g=[\mathbf{q}_g^T,\ldots,\mathbf{q}_g^T]^T$

of size ML to mathematically represent the signal recombination in space and frequency. The resulting decision variable for MT g is given by

$$\hat{d}_g = \underbrace{\tilde{\mathbf{q}}_g^H \cdot \left(\mathbf{h}_g \circ \tilde{\mathbf{w}}_g^* \circ \tilde{\mathbf{c}}_g\right) d_g}_{\text{Desired Signal}} + \underbrace{\tilde{\mathbf{q}}_g^H \cdot \left(\mathbf{h}_g \circ \sum_{k=1,k\neq g}^{K} \left(\tilde{\mathbf{w}}_k^* \circ \tilde{\mathbf{c}}_k\right) d_k\right)}_{\text{MAI}} + \underbrace{\mathbf{q}_g^H \cdot \mathbf{n}_g}_{\text{Noise}} \tag{1}$$

where vector $\mathbf{n}_g$ gathers the noise samples on the L subcarriers, $(.)^*$ is the complex conjugate operator, $(.)^H$ denotes the conjugate transposition, and $\circ$ represents the element-wise vector multiplication. The decision variable is the sum of the desired signal, the MAI and the noise after subcarrier combining. We notice that the desired signal and the MAI are functions of the transmit filters. In contrast to a conventional MC-CDMA DL, the MAI, i.e. the loss of the signal orthogonality installed by the spreading codes, arises here not only from the channel fading but from the combination of channel fading and transmit filtering. As this combination is specific to each user, a subcarrier combining aiming at restoring the orthogonality among users' signals is not advantageous. Therefore, we use Equal Gain Combining (EGC), i.e. phase equalisation on each subcarrier and despreading, as low complexity SUD scheme. The corresponding weight vector $\mathbf{q}_g$ depends on the transmit filter, and we derive it in section 3.

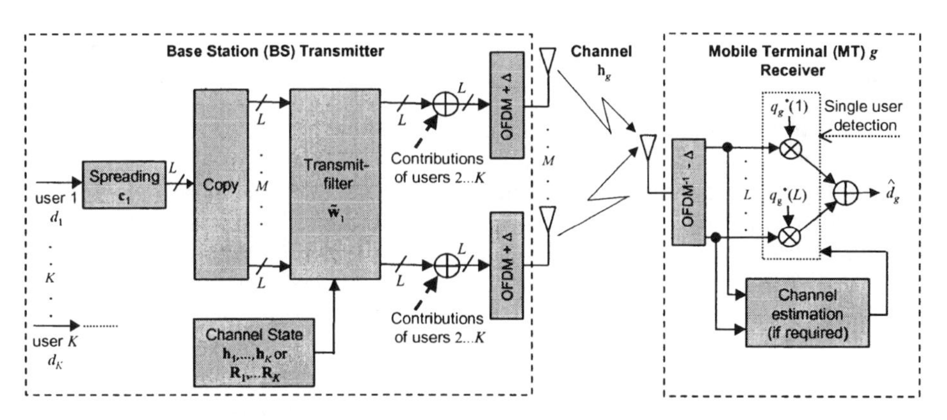

Figure 1: Downlink MC-CDMA system using an antenna array at the BS

3. TRANSMISSION SCENARIOS

The transmit filtering strategy depends on the transmission scenario and the related channel knowledge available at the BS. In the sequel, two different scenarios are identified.

3.1 Indoor scenario

The indoor scenario is characterised by a low mobility of the MT, typically less than 10 km/h. This means that the coherence time of the channel is much greater than the frame duration of the transmission. Hence, in a TDD system, the fading coefficients estimated during the UL transmission slot are still valid for the following DL transmission slot. We can thus assume to have perfect knowledge of the channel coefficient vector $\mathbf{h}_g$. Concerning the spatial properties, the BS may be surrounded by many obstacles so that the Directions of Arrival (DOAs) of the multipath components are spread over a large Angular Sector (AS). For instance, we will assume a uniform distribution of the DOAs of all users within 120°. Finally, due to the stationarity of the channel the DOAs of the UL are equal to the DODs in the DL.

3.2 Outdoor scenario

As a consequence of the potentially high mobility of MTs in an outdoor scenario, we cannot assume that the channel fading is constant during consecutive UL and DL slots. A measure of these channel variations is the correlation of a given element of $\mathbf{h}_g$ taken at two instants separated by a duration τ. Denoting this element by $h_{m,\ell}(t)$ and assuming a Jakes Doppler spectrum [12], the correlation ρ is given by the following expression:

$$\mathrm{E}\left[h_{m,\ell}(t)\cdot h_{m,\ell}^{*}(t+\tau)\right]=\rho=J_0(2\pi\tau v_{MT}/\lambda_c) \tag{2}$$

Here, v_{MT} is the mobile speed, λ_C the carrier wavelength and J_0 denotes the zero order Bessel function of the first kind.
Concerning the spatial channel characteristics, the DOAs of each user generally lay in a smaller AS than in the indoor scenario, since the reflecting obstacles may be located far from the isolated BS. For instance, we will assume a uniform distribution of the DOAs in an AS of 10° around the main DOA, which itself is randomly chosen in a 120° sector for each user. Even in the high mobility case, we can still assume that the DOAs of the UL slot are identical to the DODs of the following DL slot.
As a consequence, we assume that only long term channel knowledge is available at the BS in the outdoor scenario. This means that the BS knows the spatial covariance matrices, $\mathbf{R}_g$. Let $\mathbf{h}'_g(\ell)$ gather the M coefficients of the channel between the BS array and MT g on subcarrier ℓ, then $\mathbf{R}_g$ is given by

$$\mathbf{R}_g=\mathrm{E}\left[\sum_{\ell=1}^{L}\mathbf{h}'_g(\ell)\cdot\mathbf{h}'^{H}_g(\ell)\right] \tag{3}$$

where E[x] denotes the expected value of x. Note that $\mathbf{R}_g$ is simply averaged over the L subcarriers, because we assume that, with frequency interleaving, there is no correlation of the fading on the L subcarriers. It has been shown that these matrices can also be obtained from the UL even in FDD systems [11].

4. TRANSMIT FILTERING STRATEGIES

The transmit filtering strategies use the available channel knowledge at the BS to form the transmitted signal of each user with respect to Single-User (SU) and Multi-User (MU) optimisation criteria. The SU criteria are low complex and simply aim at maximising the desired signal part of the decision variable without taking into account the MAI. However, due to the spatial selectivity of the transmitted signal, SU techniques implicitly reduce the MAI generated at other MTs. The MU criteria explicitly reduce the MAI at the cost of a higher complexity, but this computational burden may be tolerable at the BS. It has to be noted that, in the DL, the spatial dimension can only be used for MAI cancellation at the BS side. Indeed, at a given MT, all signals have passed through the same space-frequency channel.
As it can be seen in figure 1, the transmit filters are placed before the OFDM modulation, which corresponds to filtering in the frequency domain. A time domain approach would imply a distinct OFDM operation for each user and, therefore, would be disadvantageous in terms of computational complexity.
We consider two distinct approaches: A joint optimisation of the transmit filter in space and frequency, called Space-Frequency Transmit Filtering (SFTF), when instantaneous channel knowledge is available, and an optimisation in space only, i.e. Beamforming (BF), if the BS has long-term channel knowledge only.
When modifying the transmit signal, the resulting signal power is likely to vary as well. Since power control is out of the scope of this paper, we ensure that all transmit filters are power normalised:

$$\left|\tilde{\mathbf{w}}_k\right|^2 = \tilde{\mathbf{w}}_k \tilde{\mathbf{w}}_k^H = 1 \quad \forall k = 1 \ldots K \tag{4}$$

Yet, it is worth noting that the presented techniques can readily be used in a joint transmit filtering and power control scheme, e.g. [13].

4.1 Space-Frequency Transmit Filtering (SFTF)

Space Frequency Transmit Filtering (SFTF) is a combination of spatially selective transmission and pre-filtering in the frequency domain. It is intended for the indoor scenario exposed in section 3.1. Hence, we assume that we have perfect knowledge of the channel vectors $\mathbf{h}_g$. We already presented several versions of SFTF in [8]. Here, we will focus on the SU criterion called Maximum Ratio Transmission (MRT) and a MU criterion based on a maximisation of the Signal over Interference plus Noise Ratio (SINR). Since instantaneous channel knowledge is available, we can include pre-equalisation in the transmit filter. This allows to simplify the detection at the MT to a pure despreading, i.e. $\tilde{\mathbf{q}}_g^H = \tilde{\mathbf{c}}_g^H$. So, there is no more need for channel estimation on each subcarrier at the MT side.
For the SU-SFTF criterion, the optimisation of the filter is based on the maximisation of the signal to noise ratio (SNR) after despreading. Since the noise

term in (1) is not affected by the transmit filter, this is equivalent to maximising the desired signal part with a given transmit power. This criterion is well known and analogous to maximum ratio combining at the receiver. Hence, at the transmitter we may call it MRT. The corresponding transmit filtering vector is

$$\text{SU-SFTF:} \qquad \tilde{\mathbf{w}}_g = \kappa_g \mathbf{h}_g \tag{5}$$

where the scalar κ_g is used to meet the power constraint in (4).
The MU-SFTF criterion is based on maximising the SINR. Since the MAI in (1) is a function of the transmit weights of all other users $k \neq g$, direct SINR maximisation would lead to a joint optimisation problem for the transmit filters of all users at once. [8] presents an approach using a modified SINR (m-SINR) and decoupled optimisation, which leads to a closed form solution for the transmit filters of each user. The basic idea of this m-SINR is to replace the interference term of MT g by the sum of the interference that user g creates at the other MTs $k \neq g$. Assuming that the power of the data symbols is unity, the m-SINR is given as:

$$m - SINR_g = \frac{\left|\tilde{\mathbf{w}}_g^H \mathbf{h}_g\right|^2}{\tilde{\mathbf{w}}_g^H \left(\sum_{k=1, k \neq g}^{K} \left(\tilde{\mathbf{c}}_k^H \circ \mathbf{h}_k \circ \tilde{\mathbf{c}}_g\right)\left(\tilde{\mathbf{c}}_k^H \circ \mathbf{h}_k \circ \tilde{\mathbf{c}}_g\right)^H \right) \tilde{\mathbf{w}}_g + \sigma_n^2} \tag{6}$$

with σ_n^2 being the noise variance per subcarrier. (6) has to be maximised under the power constraint in (4). When we include this constraint, we see that the maximisation problem itself gets independent of a scalar factor in $\tilde{\mathbf{w}}_g$. This means that we can add the term $k=g$ to the sum in the denominator without any impact. Defining the $ML \times K$ matrix $\mathbf{A}_g = \left[\tilde{\mathbf{c}}_1^H \circ \mathbf{h}_1 \circ \tilde{\mathbf{c}}_g, \ldots, \tilde{\mathbf{c}}_k^H \circ \mathbf{h}_k \circ \tilde{\mathbf{c}}_g, \ldots, \tilde{\mathbf{c}}_K^H \circ \mathbf{h}_K \circ \tilde{\mathbf{c}}_g\right]$ and a vector $\bar{\mathbf{w}}_g = \tilde{\mathbf{w}}_g / \kappa_g$, where the scalar κ_g is used to ensure (4), the optimum transmit filtering vector is obtained from

$$\max_{\bar{\mathbf{w}}_g} \frac{\left|\bar{\mathbf{w}}_g^H \mathbf{h}_g\right|^2}{\bar{\mathbf{w}}_g^H \left(\mathbf{A}_g \mathbf{A}_g^H + \sigma_n^2 \mathbf{I}_{ML}\right) \bar{\mathbf{w}}_g} \quad \text{subject to } \left|\bar{\mathbf{w}}_g^H \mathbf{h}_g\right| = 1 \tag{7}$$

where $\mathbf{I}_X$ denotes the identity matrix of size X. After solving (7) with the method of Lagrange multipliers and some simplifications we finally get:

$$\text{MU-SFTF:} \quad \tilde{\mathbf{w}}_g = \kappa_g \mathbf{A}_g (\mathbf{A}_g^H \mathbf{A}_g + \sigma_n^2 \mathbf{I}_K)^{-1} \mathbf{b}_g, \text{ with } \mathbf{b}_g = [0, \ldots, \underset{g\text{-th element}}{1}, \ldots, 0]^T \tag{8}$$

In contrast to SU-SFTF, MU-SFTF provides explicit MAI reduction at the price of a higher complexity, since it implies a matrix inversion of size $K \times K$.

4.2 Beamforming (BF)

In the second scenario, where the BS has long term channel knowledge only, spatial transmit filtering, i.e. BF, can be performed. We here assume that the BS has knowledge about the spatial covariance matrices $\mathbf{R}_g$ defined in (3). The optimisation criteria for BF are very close to those described for SFTF, with the only difference that they are averaged over the channel fading. We represent $\tilde{\mathbf{w}}_g$ as a repetition of BF vector $\mathbf{w}_g=[w_g(1),\ldots,w_g(M)]^T$, which is identical for all carriers, i.e. $\tilde{\mathbf{w}}_g=[w_g(1)\xleftrightarrow{L\text{ times}}w_g(1),\ldots,w_g(M)\xleftrightarrow{L\text{ times}}w_g(M)]^T$. After despreading, the power of the desired signal averaged over the channel fading and the subcarriers can be expressed by

$$\mathrm{E}\left[\tilde{\mathbf{w}}_g^H\mathbf{h}_g\mathbf{h}_g^H\tilde{\mathbf{w}}_g\right]=\mathrm{E}\left[\sum_{\ell=1}^{L}\mathbf{w}_g^H\mathbf{h}'_g(\ell)\mathbf{h}'^H_g(\ell)\mathbf{w}_g\right]=\mathbf{w}_g^H\mathbf{R}_g\mathbf{w}_g \tag{9}$$

(9) is maximised by the principal eigenvector (m_eig(.)) [14] of $\mathbf{R}_g$ for BF, i.e.

SU-BF: $$\mathbf{w}_g=\kappa_g\,\mathrm{m\_eig}(\mathbf{R}_g) \tag{10}$$

The scalar κ_g is used to meet the power constraint in (4) and ensure a common phase at a given array element for all users. Like SU-SFTF, SU-BF yields no explicit but implicit interference reduction.

We can also formulate a MU-BF approach and obtain the interference plus noise covariance matrix $\mathbf{R}_{Ig}$ by averaging the denominator of the fraction we had to maximise in (7). Hence, we get

$$\mathbf{R}_{Ig}=\sum_{k,k\neq g}\mathbf{R}_k+\sigma_n^2\mathbf{I}_M \tag{11}$$

The MU-BF vector aims then at maximising the averaged m-SINR:

$$\max_{\mathbf{w}_g}\frac{\mathbf{w}_g^H\mathbf{R}_g\mathbf{w}_g}{\mathbf{w}_g^H\mathbf{R}_{Ig}\mathbf{w}_g}\quad\text{subject to (4)} \tag{12}$$

The solution is the principal generalised eigenvector (gm_eig(.)) of the matrix pair [14] formed by the signal and the interference plus noise covariance matrices:

MU-BF: $$\mathbf{w}_g=\kappa_g\,\mathrm{gm\_eig}(\mathbf{R}_g,\mathbf{R}_{Ig}) \tag{13}$$

As before, κ_g ensures (4) and a common phase at a given array element for all users. Note that the only degree of freedom available for MU-BF is the number of antennas. As, in practice, M may be quite low compared to the number of DOAs

multiplied by the number of users, the MU-BF vector may be sub-optimum and even lead to a loss of desired signal power.
When BF is performed at transmission, channel estimation and frequency domain equalisation is required in addition to despreading at the MT side. As presented in section 2, we chose in this case the EGC SUD technique [1] for its low complexity, where the coefficient on the ℓ-th subcarrier is $q_g(\ell)=c_g(\ell)e^{j\varphi_g(\ell)}$ and $\varphi_g(\ell)$ is the phase of the estimated channel fading on subcarrier ℓ.

5. NUMERICAL RESULTS

The numerical results for the two scenarios exposed in section 3 were obtained with a system configuration similar to the European BRAN Hiperlan/2 standardisation project. The channel model is based on the channel A defined in [15]. This channel model comprises 18 paths with a multipath spread of 390 ns. We extended this time model to a space-time model by allocating a DOA to each of the paths. The DOAs are randomly chosen within an AS of 120° (indoor) and 10° (outdoor). The channel response is recalculated at each OFDM symbol to ensure a good average over the spatial configurations. The BS is equipped with a half wavelength spaced uniform linear array. The system bandwidth is 20 MHz and the OFDM system comprises 64 subcarriers. Walsh-Hadamard spreading codes of size L=8 are considered. We perform subcarrier interleaving to avoid that the chips of a symbol are transmitted on carriers lying within the coherence bandwitdth of the channel for the sake of diversity. All plots show the uncoded average Bit Error Rate (BER) computed over all active users versus the E_t/N_0, where E_t is the energy per bit transmitted over all antennas and N_0 the noise spectral density.

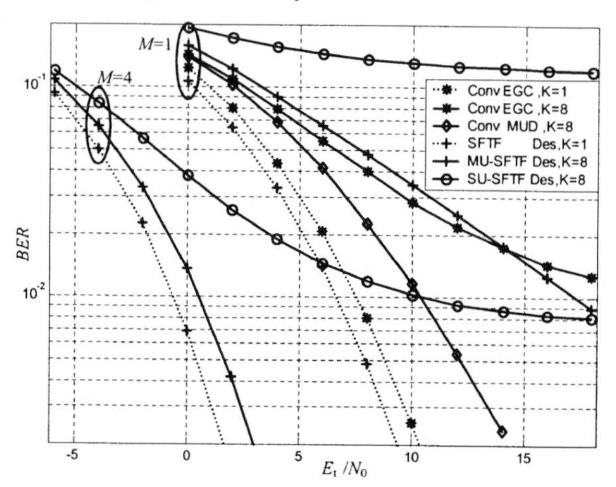

Figure 2: Comparison of SFTF to conventional systems in an indoor scenario (M=4)

Figure 2 compares the performance of SFTF, i.e. SU-SFTF (5) and MU-SFTF (8), in the indoor scenario to the Conventional (Conv) system (no transmit filtering and

M=1) with EGC and optimum linear MUD [4] at the MT receiver. For K=1 (dotted curves) SU- and MU-SFTF vectors are identical. With M=1, MU-SFTF already gains with respect to Conv EGC, because of the power weighting over the subcarriers. For M=4, the gain of MU-SFTF is about 7 dB at BER=10^{-2}, which is higher than the antenna gain itself (10log$M$=6 dB) and shows that MU-SFTF exploits space-frequency diversity in this case. For full load (K=8) and M=1, SU-SFTF has poor performance, because user-specific SU transmit filtering without spatial separation aggravates the loss of orthogonality among users' signals and thus increases the MAI. MU-SFTF has better performance, but shows no significant gain compared to Conv EGC, which means that transmit filtering without spatial separation is almost useless for the considered system. However, for K=8 and M=4, we clearly see the advantage of the antenna gain together with the spatial separation of users. Here, even low-complexity SU-SFTF outperforms Conv EGC. Compared to the Conv. System with MUD, MU-SFTF yields a gain of about 10 dB at BER=10^{-2}. Hence, joint SFTF yields a considerable MAI reduction.

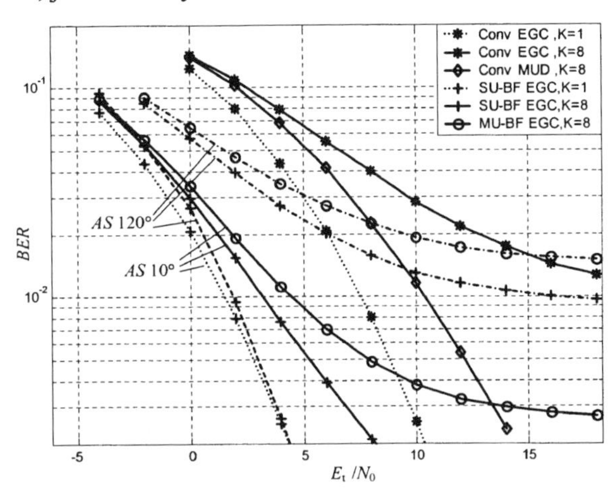

Figure 3: Comparison of BF to conventional systems in an outdoor scenario (M=4)

In figure 3, the outdoor scenario is considered and we compare the proposed BF schemes, i.e. SU-BF (10) and MU-BF (13), with M=4 to the conventional system with EGC or opt. lin. MUD at MTs. For comparison, we also plotted the curves with AS=120°. For K=1 (dotted curves) SU-BF yields the expected antenna gain, i.e. 6 dB, for AS=10°. With AS=120°, the gain is less, because the main lobe of the antenna diagram cannot cover all multipath components. For full load (K=8) and AS=10°, SU-BF outperforms Conv. MUD by about 7 dB at BER=10^{-2}. With AS=120°, SU-BF only leads an advantage at low SNR, as it experiences an error floor already for medium SNR. Thus, separation of users' signals in space only yields no advantage for a wide AS. For MU-BF, the fact that M is smaller than K and that there are multiple DOAs per user degrade the desired signal power as already explained in section 4. Therefore, the MU-BF actually has poorer performance than SU-BF.

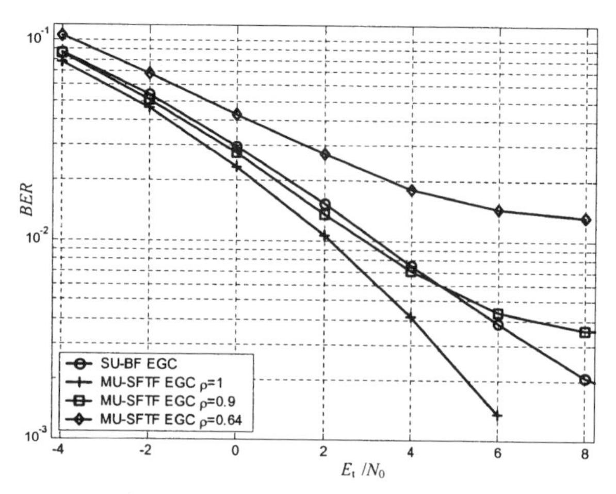

Figure 4: Comparison of MU-SFTF to SU-BF in an outdoor scenario (M=4, K=8)

In figure 4, we compare MU-SFTF and SU-BF in the outdoor scenario, i.e. AS=10° and imperfect channel estimates at the transmitter due to Doppler variations. Here, frequency equalisation is not only required for BF but also for SFTF. Note that the channel is still perfectly known at the receiver. The mobile speed considered here is in a range, where the Doppler effect has no impact on the underlying OFDM transmission. So the only effect is a mismatch in the channel estimates at the BS as quantified by (2). In this case, there is no impact on the performance of BF neither. We assume a worst case delay corresponding to a typical slot duration, i.e. 1 ms, between estimation and usage of the channel coefficients and mobile speeds of 20 km/h and 45 km/h. This leads to a correlation of ρ=0.9 and ρ=0.64, respectively. Since SFTF exploits the channel diversity, its performance is reduced by the smaller AS. Thus, even with perfect channel knowledge (ρ=1) for K=8 and M=4, there is a loss of about 1.5 dB at BER=10^{-2} compared to figure 2. But it remains a gain of 1 dB compared to SU-BF. This gain vanishes at a mobile speed of 22 km/h, where we have similar performance of MU-SFTF and SU-BF. Besides, MU-SFTF experiences an error floor for $E_t/N_0 \geq 5$ dB. For higher speeds, SFTF looses its superiority and BF becomes more appropriate.

6. CONLUSION

We considered MC-CDMA DL transmission systems using an antenna array at the BS in order to mitigate MAI. Two different scenarios have been distinguished: a low-mobility TDD system with perfect instantaneous channel knowledge the BS and a high-mobility TDD or FDD scheme where only spatial information is known in advance. The presented results show that for both cases, there are appropriate transmitter strategies for efficient interference mitigation in the DL, which not only allow the transfer of the computational effort from the MTs to the BS, but also outperform a conventional system with high complex MUD at the MT.

ACKNOWLEDGEMENT

This work was supported by the European IST MATRICE project (MC-CDMA Transmission Techniques for Integrated Broadband Cellular Systems) [2].

REFERENCES

[1] S. Hara, R. Prasad, 'Overview of multicarrier CDMA,' *IEEE Comm. Mag.*, vol. 35, pp. 126-133, Dec. 1997.
[2] IST MATRICE project, web site http://www.ist-matrice.org
[3] H. Atarashi, N. Maeda, S. Abeta, M. Sawahashi, 'Broadband Wireless Access based on VSF-OFCDM and MC/DS-CDMA,' *Proc. of PIMRC,* vol. 3, pp. 992-997, Sept. 2002.
[4] J.F. Hélard, J.Y. Baudais, J. Citerne, 'Linear MMSE detection technique for MC-CDMA,' *Electronics Letters*, vol. 36, Mar. 2000, pp. 665-666.
[5] V. Le Nir, M. Hélard, R. Le Gouable, 'Space-time block coding applied to turbo coded multicarrier CDMA,' *Proc. of Vehic. Tech. Conf. VTC 2003-Spring,* Apr. 2003
[6] B.R. Vojcic, W.M. Jang, 'Transmitter Precoding in Synchronous Multiuser Communications,' *IEEE Trans. on Comm., vol.* 46, no. 10, pp. 1346-11355, Oct. 1998.
[7] T. Sälzer, D. Mottier, D. Castelain, 'Comparison of Antenna Array Techniques for the Downlink of Multi-Carrier CDMA systems,' *Proc. of Vehic. Tech. Conf. VTC 2003-Spring,* Apr. 2003.
[8] T. Sälzer, A. Silva, A. Gameiro, D. Mottier, 'Pre-Filtering Using Antenna Arrays for Multiple Access Interference Mitigation in Multi-Carrier CDMA Downlink,' *Proc. of IST Mobile Comm. Summit*, Aveiro, June 2003.
[9] A.F. Naguib, A.J. Paulraj, T. Kailath, 'Capacity improvement with base-station antenna arrays in cellular CDMA,' *IEEE Trans. on Vehic. Tech.*, vol. 43, no. 3, pp. 691-698, Aug 1994.
[10] C. K. Kim, Y. S. Cho, 'Performance of a wireless MC-CDMA system with an antenna array in a fading channel: reverse link,' *IEEE Trans. on Comm.*, vol. 48, no. 8, pp. 1257-1261, Aug. 2000.
[11] Y.C. Liang, F.P.S Chin, 'Downlink Channel Covariance Matrix (DCCM) Estimation and Its Application in Wireless DS-CDMA Systems,' *IEEE Journal on Sel. Areas in Comm.*, vol. 19, no. 2, pp. 222-232, Feb. 2001.
[12] C.J. Jakes, 'Microwave Mobile Communications,' *IEEE Press*, New Jersey, 1994.
[13] F. Rashid-Farrokhi, K.J.R. Liu, L. Tassiulas, 'Transmit Beamforming and Power Control for Cellular Wireless Systems,' *IEEE Journal on Sel. Areas in Comm.*, vol. 16, no. 8, pp. 1437-1450, Oct 1998.
[14] G. H. Golub, C.F. van Loan, 'Matrix Computations,' Johns Hopkins Univ. Press, 3rd ed., 1996.
[15] J. Medbo, 'Channel Models for Hiperlan/2 in different Indoor Scenarios,' *ETSI BRAN doc.* 3ERI085b, 1998.

T. Sälzer and D. Mottier are with
Mitsubishi Electric ITE, Telecommunications Laboratory, Rennes, France.
Contact: Mitsubishi Electric ITE-TCL
1, Allée de Beaulieu, CS 10806
35708 Rennes cedex 7, France
phone: +33 2 23455858, fax: +33 2 2345585
e-mail: {salzer, mottier}@tcl.ite.mee.com

H. WITSCHNIG, G. STRASSER, K. REICH,
R. WEIGEL, A. SPRINGER

ANTENNA DIVERSITY TECHNIQUES FOR SC/FDE – A SYSTEM ANALYSIS

Abstract. In this work we investigate the possibilities of combining diversity techniques with a Single Carrier System with Frequency Domain Equalization (SC/FDE), as future transmission concepts will also be judged by their possibilities to be combined with multiple antennas. Concepts of multiple antennas allow either a significant performance gain through diversity (and) or significant higher capacity based on spatial multiplexing. While it is topic of many contributions to improve the concepts of antenna diversity itself, it is topic of this paper to point out the advantages when combining SC/FDE with diversity techniques. It will be demonstrated that SC/FDE shows excellent possibilities to be combined with diversity techniques, as the process of equalization itself can be combined with receive diversity as well as with space time block (STBC) decoding, that can be carried out advantageously in the frequency domain. Besides that an overall system approach will be discussed that includes aspects of implementation, equalization, modulation and coding.

1 INTRODUCTION

The ambition for extreme high data rates as well as high quality of service make new concepts for communication systems indispensable. Future physical layer concepts will have to ensure better performance, higher data rates, higher spectral efficiency as well as an acceptable signal processing complexity, especially for mobile applications.

Concepts of multiple antennas allow to fulfill several of the above mentioned criteria in a superior way. From that point of view the question is obvious, if there are transmission concepts (physical layer concepts) that can be combined in a more profitable way with multiple antennas than others

One of the most decisive factors when talking of high data rate mobile communication is the cost for time domain equalization that grows quadratically with the bit rate [1] which prevents the realization of transmission rates beyond several Mb/s. Concepts that implement this equalization in the frequency domain such as OFDM (Orthogonal Frequency Division Multiplexing) and SC/FDE (Single Carrier System with Frequency Domain Equalization) are able to fulfill both criteria, good performance and reduced implementation effort as the implementation effort grows only slightly more than linear [2], [3], [4].

This paper is organized as follows: The first section introduces the concept of SC/FDE. The following chapters deal with transmit- and receive diversity and how they can be combined with the concept of SC/FDE in an advantageous way. Based on that, possible performance gains and the required implementation effort will be discussed, with special reference to equalization concepts, higher order modulation schemes and channel coding. Finally a short conclusion and outlook will be given.

K. Fazel and S. Kaiser (eds.), Multi-Carrier Spread-Spectrum, 327–336.

2 THE CONCEPT OF SC/FDE

In an SC/FDE system the received sequence must be transferred to the frequency domain (by means of efficient FFT operations), where the equalization takes place. Similar to OFDM, a cyclic prefix is inserted between successive blocks in the transmitter in order to mitigate interblock interference (IBI) [2]. To prevent this IBI the duration of the guard period T_G must be longer than the duration of the channel impulse response T_h. Due to the cyclic extension, the linear convolution of one cyclically extended transmitted block and the channel impulse response $h(t)$ appears as a circular convolution corresponding to the frequency domain relation

$$R(nf_0) = H(nf_0)S(nf_0) + N(nf_0) \tag{1}$$

for $n \in \mathbf{Z}$ and $f_0=1/T_{FFT}$. Here the functions $R(f)$, $S(f)$ and $H(f)$ are related to the time domain signals $r(t)$ (one period of the received data block), $s(t)$ (the original, non cyclically extended block) and $h(t)$ by the continuous Fourier transform. $N(f)$ is the Fourier transform of the additive noise. Fig.1 shows the transmitted data structure, which consists of the data sequence containing N symbols and the sequence of the cyclic prefix with N_G symbols. The duration of a processed block is $T_{FFT}=NT$ and the duration of the cyclic prefix is $T_G=N_GT$.

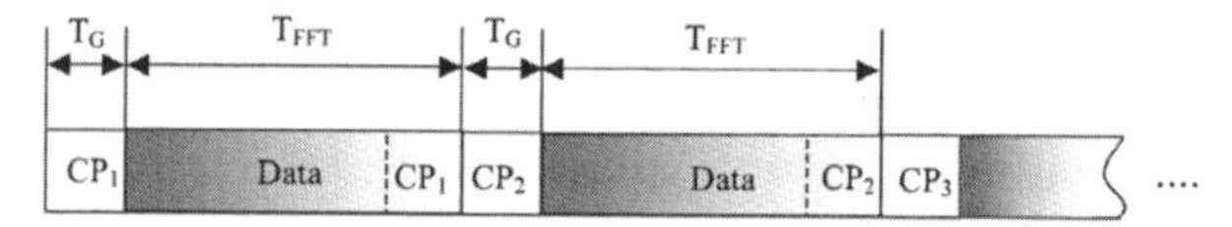

Fig. 1 Transmitted data structure in an SC/FDE system.

It is essential to mention that in the case of a blockwise transmission based on a cyclic extension, an optimal equalizer of infinite length is implemented and the equalization of a received block containing N symbols is carried out with only N complex multiplications in the frequency domain. In this contribution fractionally spaced equalizer structures are implemented that allow a low complexity implementation as well as a compact mathematical description that includes multiple antennas. The received signal is sampled with twice the symbol rate, which is sufficient for band-limited pulses (root-raised cosine pulse shaping is applied at the transmitter) and the sampled sequence of each block is split into its polyphases $r_1(k)$ and $r_2(k)$, were both sequences contain N samples that correspond to symbol rate sequences. The N-point FFT transformed sequences $R_1(nf_0)$ and $R_2(nf_0)$ are then multiplied (equalized) with the equalizer phases $E_1(nf_0)$ and $E_2(nf_0)$. The equalized signal

$$Y(nf_0) = E_1(nf_0)R_1(nf_0) + E_2(nf_0)R_2(nf_0) \tag{2}$$

is finally transformed back to the time domain by an N-point IFFT.

Three equalization criteria/structures are discussed in this contribution: the Zero Forcing (ZF) criterion, the Minimum Mean Square Error (MMSE) criterion and a MMSE decision feedback equalizer (MMSE-DFE). An advantageous combination

of frequency domain feed forward equalization and time domain decision feed back equalization is discussed in [4], [5].
The polyphase representation of the ZF equalizer has been developed in [6] and is given in equation (3)

$$\mathbf{E}_{ZF}(f)=\frac{\hat{\mathbf{F}}^{H}(f)}{\sum_{i=1}^{2}\left|\hat{F}_{i}(f)\right|^{2}}=\frac{\hat{\mathbf{F}}^{H}(f)}{\left\|\hat{\mathbf{F}}(f)\right\|^{2}} \tag{3}$$

where the superscript ^{H} denotes Hermitian transpose. In this equation

$$\hat{\mathbf{F}}(f)=\left[\hat{F}_{1}(f),\hat{F}_{2}(f)\right]' \tag{4}$$

describes the fourier transformed, channel distorted transmitted pulse that has to be derived by pilot aided channel estimation procedures. From the analytical expression of the equalizer it gets obvious that the nominator of equation (3) characterizes a matched filter that maximizes the SNR for the two polyphases, while the denominator describes the ZF equalizer itself that forces an ISI free detection. It is well known that the ZF equalizer suffers from noise amplification if the channel transfer function shows deep spectral fades. The mentioned MMSE equalizer avoids this problem by compromising the noise amplification and the ISI reduction. Additional performance gain is reached by using decision feedback. As it will be demonstrated that the suboptimal ZF-equalizer is sufficient if combined with multiple antennas, MMSE equalizer and DFE are not described in detail.

3 RECEIVE DIVERSITY FOR SC/FDE

Maximum Ratio Combining is the most powerful receive diversity concept but suffers from the highest implementation effort. There, the signal that is received due to multiple antennas (M) is added and additionally weighted in an optimal way to enhance the SNR. The optimal ZF-equalizer for multiple antennas can be defined as

$$\mathbf{E}_{ZF}(f)=\frac{\hat{\mathbf{F}}^{H}(f)}{\left\|\hat{\mathbf{F}}(f)\right\|^{2}}=\frac{\hat{\mathbf{F}}^{H}(f)}{\sum_{j=1}^{M}\sum_{i=1}^{2}\left|\hat{F}_{i,j}(f)\right|^{2}} \tag{5}$$

were

$$\hat{\mathbf{F}}(f)=\left[\hat{F}_{1,1}(f),\hat{F}_{1,2}(f),\hat{F}_{2,1}(f),\hat{F}_{2,2}(f)\right]' \tag{6}$$

for the case of $M=2$. Notice that the equalization term itself is independent of the individual antenna path or polyphase and can therefore be separated from the matched filter part. This allows the following implementation as shown in Fig. 2.

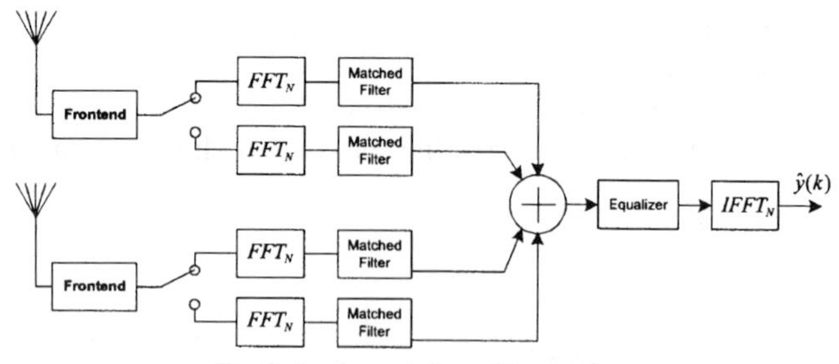

Fig. 2: Implementation of the receiver.

Here, the matched filter of every path implements the optimal weighting and guarantees an optimal performance. The equalization itself has to be carried out only once for every block. Furthermore there is only a simple N-point IFFT necessary to transform the received and equalized blocks back to the time domain.

The performance gain of receive diversity is based on two aspects – antenna gain and diversity gain [7].

Antenna gain is simply based on the reception of the same sent data sequence on several antennas which enhances the received signal power – M receive antennas lead to a gain of a factor of M. In comparison to this, diversity gain is based on the summation of several channel transfer functions. Due to this addition, deep spectral fades in a single channel are canceled under the assumption of uncorrelated channel transfer functions. A high number of receive antennas leads to an almost flat behaviour of the overall channel transfer function and the BER approximates the AWGN case – documented in Fig. 3. There, the possible performance gain based on diversity that is limited by the matched filter bound is shown as a function of the number of antennas (for this diagram, the constant gain of M based on the antenna gain has been subtracted). This possible additional gain is already within *1 dB* for four antennas and would not justify additional antennas. To conclude this consideration, the bit error-behaviour for receive diversity is given in Fig. 4. The simulation results are based on QPSK modulation, and are averaged over 20 randomly chosen indoor radio channel snapshots that are defined by the IEEE 802.11a standardization model.

While the performance loss of a simple ZF equalizer is about 2-3 dB compared to an MMSE equalizer and about 5 dB compared to an MMSE-DFE (using 4 feedback taps) for one antenna, this changes significantly already for two antennas. All three

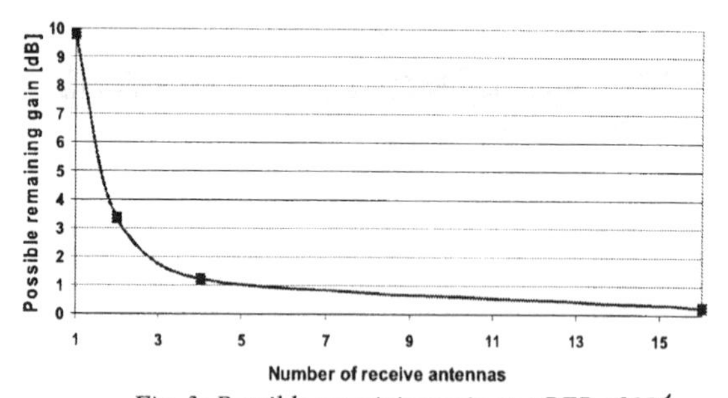

Fig. 3: Possible remaining gain at a BER of 10^{-4}.

equalizers perform within 1 dB – an obvious behaviour as MMSE equalizer or DFE outperform ZF equalization especially in the case of deep spectral fades of the radio channel. For four antennas the performance of the investigated equalizer structures is comparable and would not justify the use of non linear equalizers or even an MMSE equalizer. Comparable results have been reached for the modulation schemes of 16 QAM and 64 QAM.

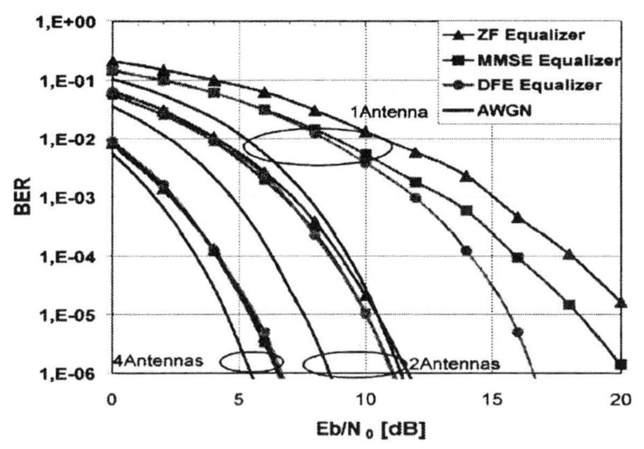

Fig. 4: BER due to different equalizer structures.

Additional performance gain may be reached by using channel coding. It is well known that channel coding is of essential means for OFDM. In comparison to this, it is of reduced importance for SC/FDE, as this system does not suffer from unreliable subcarriers that dominate the overall performance. The used coder is an industry standard rate 1/2, constraint length 7 convolutional encoder with generator polynomials (171-133) (defined by the IEEE 802.11a standard). By puncturing the rate 1/2 code, the coding rate of 3/4 was achieved. At the receiver an MMSE equalizer and a Viterbi decoder was used. Fig. 5 shows the simulation results for the modulation scheme of 16 QAM. It gets obvious that for a high coding rate as 3/4 a performance gain is reached not before a E_b/N_0 of about 20 and the expected performance gain due coding is not more than 2-3 dB at a BER of 10^{-4}. This supports the statement that coding is of reduced meaning for SC/FDE. But nevertheless, an additional performance gain of 5-7 dB is reached (rate 1/2) when combining diversity with channel coding.

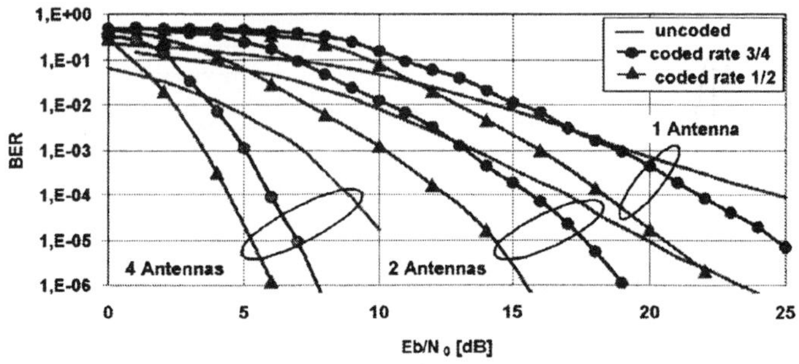

Fig. 5: Receive diversity combined with channel coding.

4 TRANSMIT DIVERSITY BASED ON SPACE-TIME BLOCK CODING

Orthogonal designs of space time codes provide an efficient means to achieve the full diversity gain. For two transmit antennas, the orthogonal design is known as the Alamouti scheme [8], originally formulated for the flat fading case. In [9] this concept was extended for the case of frequency selective fading. The concept of Al-Dahir implements a block by block coding/processing, that is combined with the concept of SC/FDE with almost no additional effort. Fig. 6 shows the block format of Space-Time Block Coding (STBC) for SC/FDE as well as the coding scheme. The expression $b_m^N(i)$ stands for the *ith* of N symbols of the *mth* block and $(-i)_N$ stands for *(-i) modulo N*. The Fourier transformation of the coded blocks are given as follows:

$$
\begin{aligned}
&b_m^N(i) \bullet\!-\!\circ\, B_m^N(k) \\
&b_m^{N*}((-i)_N) \bullet\!-\!\circ\, B_m^{N*}(k)
\end{aligned}
\qquad (7)
$$

Based on the blockwise processing, two received blocks are given by equation (8), where the factor $\sqrt{2}$ ensures that the total transmit energy is the same as for the system with one antenna.

$$
\begin{aligned}
R_1^N(k) &= H_1^N(k)\frac{B_1^N(k)}{\sqrt{2}} + H_2^N(k)\frac{B_2^N(k)}{\sqrt{2}} + N_1 \\
R_2^N(k) &= -H_1^N(k)\frac{B_2^N*(k)}{\sqrt{2}} + H_2^N(k)\frac{B_1^N*(k)}{\sqrt{2}} + N_2
\end{aligned}
\qquad (8)
$$

The received distorted and coded blocks, have to be decoded based on the following scheme:

$$
\begin{aligned}
\hat{B}_1^N(k) &= H_1^N*(k)R_1^N(k) + (H_2^N*(k)R_2^N(k))* \\
\hat{B}_2^N(k) &= H_2^N*(k)R_1^N(k) - (H_1^N*(k)R_2^N(k))*
\end{aligned}
\qquad (9)
$$

From equation (9) the main advantage of a combination of SC/FDE and STBC gets obvious again. The main part of the decoding is the multiplication with $H(k)*$ – which characterizes the matched filter that is implemented anyway.
A decoded block of symbols can finally be calculated as follows:

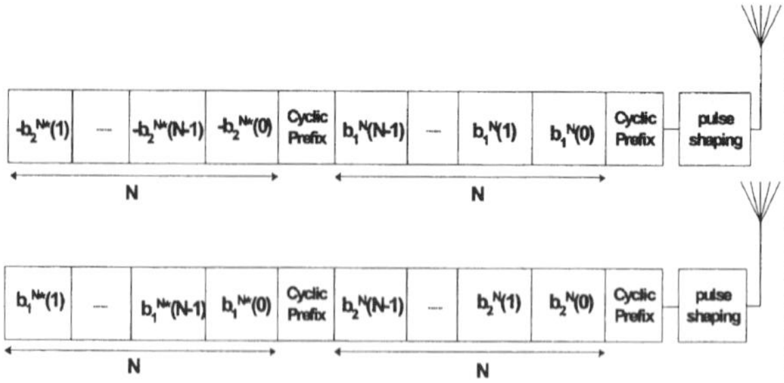

$-b_2^{N*}(1)$	...	$-b_2^{N*}(N-1)$	$-b_2^{N*}(0)$	Cyclic Prefix	$b_1^N(N-1)$	...	$b_1^N(1)$	$b_1^N(0)$	Cyclic Prefix	pulse shaping
N					N					

$b_1^{N*}(1)$	...	$b_1^{N*}(N-1)$	$b_1^{N*}(0)$	Cyclic Prefix	$b_2^N(N-1)$	...	$b_2^N(1)$	$b_2^N(0)$	Cyclic Prefix	pulse shaping
N					N					

Datablock	1. Antenna	2. Antenna
b_1^N	---	---
b_2^N	$b_1^N(i)$	$b_2^N(i)$
b_3^N	$-b_2^{N*}((-i)_N)$	$b_1^{N*}((-i)_N)$
b_4^N	$b_3^N(i)$	$b_4^N(i)$
b_5^N	$-b_4^{N*}((-i)_N)$	$b_3^{N*}((-i)_N)$
...	...	...

Fig. 6: STBC for SC/FDE.

$$\hat{B}_1^N(k)=\frac{\left|H_1^N(k)\right|^2+\left|H_2^N(k)\right|^2}{\sqrt{2}}B_1^N(k)+H_1^{N*}(k)N_1+H_2^N(k)N_2^* \tag{10}$$
$$\hat{B}_2^N(k)=\frac{\left|H_1^N(k)\right|^2+\left|H_2^N(k)\right|^2}{\sqrt{2}}B_2^N(k)+H_2^{N*}(k)N_1-H_1^N(k)N_2^*$$

Comparable coding concepts of STBC exist for four and eight antennas [10],[11]. Notice that the concept of block coding leads to a time delay of one block using two antennas and four blocks using four antennas that might be taken into account for time critical implementations. Additionally it is to mention that STBCs for more than two antennas will lead to a loss of data rate, because only three blocks of information are sent over four antennas in the investigated case.
The performance gain due to STBC is based on diversity but not on antenna gain as described in secton 3. Due to that the BER curve for multiple transmit antennas approximates the AWGN case of *one* receive antenna - in comparison to receive diversity, the possible gain is reduced by *3 dB* for two and *6 dB* for four antennas. An identical discussion in terms of equalizer structures might be carried out as already done for receive diversity.

5 THE COMBINATION OF TRANSMIT AND RECEIVE DIVERSITY

Finally we want to point out the advantages of a combined implementation of transmit and receive diversity. Fig. 7 shows the possible implementation for the investigated system. The concept of coding is identical to a system containing several transmit antennas and only one receive antenna. The difference gets obvious at the receiver, where for every receive antenna an individual decoding unit has to be implemented and after this the equalization is carried out. Notice that matched filtering, optimal weighting and decoding is implemented in one. The *mth* block of the *jth* receive unit can be written as follows

$$\hat{B}_{m,j}^N(k)=B_m^N(k)\frac{1}{\sqrt{t}}\sum_{l=1}^{t}\left|H_{l,j}^N(k)\right|^2+N_{\hat{B}_{m,j}} \tag{11}$$

In this equation $H_{l,j}$ describes the channel transfer function between transmit antenna l and receive antenna j. t describes the total number of transmit antennas. As every

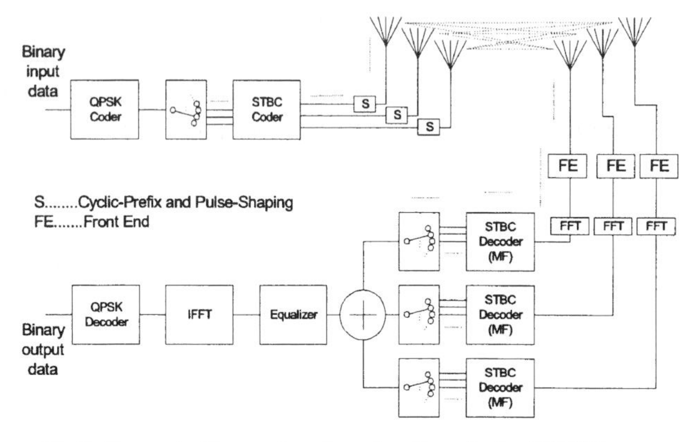

Fig.7: Concept of transmit and receive diversity for SC/FDE.

antenna receives the same information, a simple addition of the r receive paths implements the receive diversity

$$\hat{B}_m^N(k) = \sum_{j=1}^{r} \hat{B}_{m,j}^N(k) = B_m^N(k)\frac{1}{\sqrt{t}}\sum_{j=1}^{r}\sum_{l=1}^{t}\left|H_{l,j}^N(k)\right|^2 + N_{\hat{B}_m} \tag{12}$$

Without going into details we conclude that the SNR of this system is

$$SNR = \frac{1}{t}\sum_{j=1}^{r}\sum_{l=1}^{t} SNR_{l,j} \tag{13}$$

Again it gets obvious that receive diversity results in a summation of the average SNR while transmit diversity results in an averaging of the SNR. Fig. 8 shows the BER due to transmit and receive diversity. For the 4Tx-4Rx system the performance corresponds almost to the case of AWGN using four receive antennas. The additional gain in comparison to simple receive diversity is about *1 dB* – at the cost of four transmit antennas. Focusing an competitive implementation, the most interesting concept is obviously the concept of 2Tx-2Rx. This system combines high performance gain with affordable implementation effort.

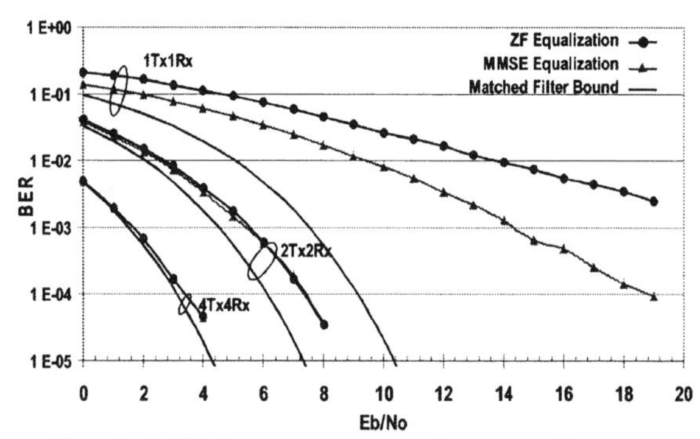

Fig. 8: BER for transmit- and receive diversity.

6 CONCLUSION

It can be concluded that the concept of SC/FDE is not only a powerful candidate for future high rate communication systems, but can be combined with diversity techniques advantageously. The advantages of an SC/FDE system are its reduced implementation complexity combined with high performance. The combination of SC/FDE with diversity techniques enhances these advantages additionally, as the concept of the frequency domain equalization can be combined with the implementation of maximum ratio combining as well as STBC in an easy way. The highest profit due to diversity is reached already when using two receive or transmit antennas - that combines affordable implementation effort with significant better performance. Furthermore it has been demonstrated that a low complexity ZF equalizer is sufficient when using several antennas, although this equalizer is known as a suboptimal solution in the case of multipath conditions. Besides that the concept of multiple antennas was combined with channel coding that allows an addition BER improvement of several dB.

7 REFERENCES

[1] J. G. Proakis, "*Digital Communications*", McGraw-Hill, New York, 1995

[2] H. Sari, G. Karam, I. Jeanclaude. "An Analysis of Orthogonal Frequency-Division Multiplexing for Mobile Radio Applications". *Proc. of the IEEE Vehicular Technology Conference (VTC '94)*, Stockholm, Sweden, pp. 1635-1639, June 1994

[3] H. Witschnig, M. Huemer, A. Koppler, A. Springer, R. Weigel, „A Comparison of an OFDM System and a Single Carrier System Using Frequency Domain Equalization", *ETT* Vol. 13, No 5, *Special Issue on "Multi-Carrier Spread-Spectrum and related topics"*, September 2002

[4] S. Lek Ariyavisitakul, Anader Benyamin-Seeyarm B. Eidison, B. Falconer, "Frequency Domain Equalization for Single-Carrier Broadband Wireless Systems", White Paper – online available at www.sce.carleton.ca/bbw/papers /Ariyavisitakul.pdf

[5] H. Witschnig, T. Mayer, A. Springer, L. Maurer, M. Huemer, R. Weigel „The Advantages of a Known Sequence versus Cyclic Prefix for a SC/FDE System", WPMC 2002, 5th international Symposium on Wireless Personal Multimedia Communications, Honolulu Hawaii, October 2002

[6] C. B. Papadias, D.T. Slock, " Fractionally Spaced Equalization of Linear Polyphase Channels and Related Blind Techniques Based on Multicahnennel Linear Prediction", *IEEE Trans. on Signal Processing*, Vol. 47, no. 3, pp. 641-654, March 1999

[7] M.V. Clark, "Adaptive Frequency-Domain Equalization and Diversity Combining for Broadband Wireless Communications", *IEEE Journal on Selected Areas in Communications*, Vol. 16, No. 8, Oct. 1998

[8] S. M. Alamouti, "A Simple Transmit Diversity Technique fore Wireless Communications", IEEE Journal on Selected Areas in Communications, Vol. 16, No. 8, Oct., 1998

[9] N. Al-Dahir, "Single Carrier Frequency-Domain Equalization for Space-Time Block-Coded Transmission Over Frequency-Selective Fading Channels", *IEEE Communications Letters*, Vol. 5, No. 7, July 2001

[10] V. Tarokh, "Space-Time Block codes from Orthogonal Designs", *IEEE Trans. on Information Theory*, Vol. 45, No. 7, July 1999

[11] O. Tirkkonen, A. Hottinen, „Complex Space-Time Block Codes for four TX Antennas", Nokia Research Center, Finland 2000

H. Witschnig, G.Strasser, K. Reich, A. Springer are with the Institute of Communications and Information Enginieering, University of Linz, Austria.
R. Weigel is with the Institute for Technical Electronics, University of Erlangen, Germany.

TOBIAS HIDALGO STITZ, MIKKO VALKAMA, JUKKA RINNE AND MARKKU RENFORS

PERFORMANCE OF MC-CDMA VS. OFDM IN RAYLEIGH FADING CHANNELS

Abstract. Multicarrier techniques are of high interest in modern communication systems due to their spectral efficiency and simplicity of channel equalization even in the case of heavily frequency selective channels. Further advantages of MC techniques include, e.g., their insensitivity to timing errors as well as to limited narrowband interference. This paper presents a comparison of MC-CDMA and OFDM with diversity from the spectral efficiency point of view in a Rayleigh fading channel. We come to the general conclusion that MC-CDMA provides best performance with moderate (50 ... 75%) loadings. We also conclude that in the synchronous downlink case, MC-CDMA with moderate loading provides clearly better performance than OFDM with comparable diversity. In the quasi-synchronous uplink case, MC-CDMA starts showing better efficiency than OFDM with diversity only in cases with rather low raw bit error rates (10^{-3} or lower).

1. INTRODUCTION

In general, multicarrier modulation seems to be a key ingredient in future beyond 3G and 4G communication system developments. In OFDM, data symbols are serial-to-parallel converted and then modulated onto N orthogonal subcarriers that are separated by the inverse of the symbol duration [1], [2]. This leads to spectral overlapping of the subchannel signals, and hence to high spectral efficiency. If there is no signal distortion in the channel, the original symbols can be perfectly recovered due to the orthogonality of the modulated subchannel signals. A big advantage of OFDM is that the modulation and demodulation operations can in practice be simply and efficiently implemented using IFFT and FFT, respectively [1]. In a multipath environment, the delayed replicas of the signal produce ISI that can be prevented by the insertion of a guard interval (GI) that absorbs the OFDM symbol 'tails' due to multipath dispersion. Typically a cyclic prefix (CP) is used for this purpose, which also causes the linear channel distortion to appear as flat within each subchannel, allowing simple channel equalization [1], [2]. Different diversity techniques (time and/or frequency [2]), can in general be used to improve OFDM system performance. In this paper, frequency domain diversity is obtained by transmitting same information over multiple subcarriers.

MC-CDMA tries to combine the advantages of OFDM and CDMA, allowing several users or code channels to be transmitted over the same set of subchannels [3]. Each code channel signal is spread using a unique spreading code (assumed mutually orthogonal) and the spread signal is transmitted using OFDM, on a one chip per subcarrier basis. Interestingly, sending the same information over several subcarriers provides natural frequency diversity. MC-CDMA can be used as a true user multiplexing technique in which the signals to be transmitted over different code channels arise from different users. As an alternative, it can also be used simply as a modulation method where all the codes in use are allocated to a single information source. In the following, the terms "user" and "code channel" are used

K. Fazel and S. Kaiser (eds.), Multi-Carrier Spread-Spectrum, 337–344.

interchangeably. In general in MC-CDMA, guard interval and cyclic prefix are used in the same way as in OFDM to cope with the multipath channel.

When MC-CDMA is used as a user multiplexing method, two scenarios are possible; either all the code channels are subject to identical channel distortion (referred to as downlink) or the channel responses for different users are different (referred to as uplink). One fundamental difference between the uplink and the downlink is that in the (single-cell) downlink case, the different user signals are perfectly synchronized, whereas in the uplink case accurate synchronization is not possible. In the quasi-synchronous [4] uplink case, which we consider here, the timing differences are small compared to the guard interval, and thus the useful parts of the consecutive symbols of different users are not overlapping.

To increase flexibility, an MC-CDMA multiplex could use only a subset of the subcarriers of the underlying OFDM system. Then, with low data rate, a user would use just a single code in one of the multiplexes, whereas with higher data rates, a single user could use a number of multiplexes and/or several codes in each multiplex. Also the MC-CDMA detection complexity increases heavily with the number of subchannels, so from the implementation point of view, the number of subcarriers should not be too high. It should also be noted that to increase frequency diversity, it is beneficial to allocate to each MC-CDMA multiplex subcarriers that are not very close to each other.

2. SYSTEM MODELS

In general, we consider multicarrier systems with N subcarriers and the subcarrier symbol/chip rate is denoted by R. The size of the symbol alphabet is M and the whole system bandwidth is W [Hz]. In case of OFDM with diversity (OFDM-DIV), a given information symbol is transmitted over $L \leq N$ subcarriers (L-th order diversity). This being the case, the system spectral efficiency E_1 in terms of the total bit rate divided by the system bandwidth is

$$E_1 = \frac{\frac{1}{L} \times \log_2(M) \times R \times N}{W} = \frac{\log_2(M) \times N}{L} \times \frac{R}{W}. \tag{1a}$$

For MC-CDMA with $K \leq N$ simultaneous code channels and spreading factor being equal to the number of subcarriers, the corresponding spectral efficiency is given by

$$E_2 = \frac{\log_2(M) \times R \times K}{W} = \log_2(M) \times K \times \frac{R}{W}. \tag{1b}$$

Both (1a) and (1b) have the common scaling factor of R/W which is ignored in the following for simplicity. Suffice to say that this scaling factor does affect the actual spectral efficiencies but does not change their relations to one another. Furthermore, the following comparisons will be made in terms of the normalized spectral efficiencies *per subchannel*, i.e., $E_1'=\log_2(M)/L$ and $E_2'=\log_2(M)\times K/N$. Naturally, both have the maximum value of $\log_2(M)$ corresponding to $L=1$ (OFDM-DIV) and $K=N$ (MC-CDMA).

Considering the signal detection, perfect channel state knowledge is assumed. Given that the guard interval (GI) is longer than the channel delay spread, intersymbol interference is avoided, and due to the cyclic prefix (CP) the transmission channel is effectively flat within each subchannel. (Quasi-) Synchronicity is also assumed.

In case of OFDM-DIV, the symbol contributions at different subcarriers are detected separately and then combined using maximal ratio combining (MRC) approach. With time-selective fading, each subchannel can be assumed to have Rayleigh-like amplitude variations, and the resulting signal-to-noise ratio (SNR) can be shown to be chi-square distributed when MRC is used [2]. This forms the basis for analytical error probability results.

With MC-CDMA, a general linear detection strategy is deployed and formulated here shortly. With the previous assumptions, only one symbol from each user is contributing to the observed data within one detection interval, and a direct frequency domain signal model can be used. This being the case, the signal within one detection interval after GI removal and FFT appears as (time index omitted)

$$\mathbf{r} = \begin{bmatrix} H_{1,1}c_{1,1} & H_{2,1}c_{2,1} & \cdots & H_{K,1}c_{K,1} \\ H_{1,2}c_{1,2} & H_{2,2}c_{2,2} & \cdots & H_{K,2}c_{K,2} \\ \vdots & \vdots & \ddots & \vdots \\ H_{1,N}c_{1,N} & H_{2,N}c_{2,N} & \cdots & H_{K,N}c_{K,N} \end{bmatrix} \begin{bmatrix} A_1 \\ A_2 \\ \vdots \\ A_K \end{bmatrix} + \begin{bmatrix} n_1 \\ n_2 \\ \vdots \\ n_N \end{bmatrix} = \mathbf{Sa} + \mathbf{n} \tag{2}$$

where $\mathbf{r} = [r(1), r(2), \ldots, r(N)]^T$ with $r(i)$ being the i-th subchannel observation, A_k is the transmitted symbol for user k and $\mathbf{n}$ denotes the noise vector. In general, the code chips of user k are denoted as $c_{k,1}, c_{k,2}, \ldots, c_{k,N}$ and $H_{k,i}$ denotes the i-th subchannel response for user k. In downlink, $H_{1,i} = H_{2,i} = \ldots = H_{K,i}$ for all i, i.e., all the users or code channels experience the same channel response. For the simultaneous detection of all the data symbols, the general minimum mean-squared error (MMSE) estimation principle [6] is deployed here. Assuming (for analysis purposes) that $\mathbf{S}$ is deterministic, it is easy to show that the resulting estimator can be written as

$$\begin{aligned} \mathbf{y} &= \mathbf{D}_{\text{MMSE}}\mathbf{r} = \mathbf{\Sigma}_{\mathbf{a}}\mathbf{S}^H(\mathbf{S}\mathbf{\Sigma}_{\mathbf{a}}\mathbf{S}^H + \mathbf{\Sigma}_{\mathbf{n}})^{-1}\mathbf{r} \\ &= \mathbf{\Sigma}_{\mathbf{a}}\mathbf{S}^H(\mathbf{S}\mathbf{\Sigma}_{\mathbf{a}}\mathbf{S}^H + \mathbf{\Sigma}_{\mathbf{n}})^{-1}\mathbf{Sa} + \mathbf{n}_{\text{D,MMSE}} \end{aligned} \tag{3}$$

where $\mathbf{\Sigma}_{\mathbf{a}}$ is the autocovariance matrix of the symbol vector $\mathbf{a}$, $\mathbf{\Sigma}_{\mathbf{n}}$ is the observation noise autocovariance, and symbols and noise are assumed to be mutually uncorrelated.

3. OBTAINED RESULTS

Some numerical results are presented next. In all the following studies, the subchannels are assumed to be independently Rayleigh fading and non-frequency-selective. Thus the model is valid if the subchannel spacing is greater than the channel coherence bandwidth and proper cyclic prefix is used. With suitable choice of the multicarrier system parameters and suitable interleaving of the subcarriers used in the multiplexes this situation can be approached quite well also in practice.

In MC-CDMA, cases with $N = 4$ and 16 subchannels are studied. The spreading is done by Walsh-Hadamard codes, and different system loads (K/N) are considered. Both the cases of quasi-synchronous uplink with different channels for each user and the downlink with the same channel for all the users are analyzed.

The key target of this study is to compare OFDM-DIV and MC-CDMA from the spectral efficiency point of view. The comparison is based on the needed signal-to-noise ratio (SNR) to obtain a target raw BER (i.e., no error control coding included) at different normalized spectral efficiency values (bits per symbol). The results are obtained by computer simulations. The SNR is represented in the form of the bit energy to noise spectral density ratio (E_b/N_0). Denoting the subcarrier bandwidth by B, the relation between the total subcarrier SNR and the received E_b/N_0 is of the form

$$SNR_1 = \frac{\log_2(M) \times R \times E_b / L}{N_0 \times B} = \frac{E_b}{N_0} \times \frac{\log_2(M)}{L} \times \frac{R}{B} \tag{4a}$$

for OFDM-DIV with L-th order diversity and of the form

$$SNR_2 = \frac{K \times \log_2(M) \times R \times E_b / N}{N_0 \times B} = \frac{E_b}{N_0} \times \frac{\log_2(M)}{N/K} \times \frac{R}{B} \tag{4b}$$

for MC-CDMA. The spectral efficiency values are obtained from the data modulation method and from the degree of diversity (OFDM) or user load (MC-CDMA) as $E_1' = \log_2(M)/L$ and $E_2' = \log_2(M) \times K/N$. As an example, 16-QAM OFDM with diversity order 2 has a relative spectral efficiency of 2. The same data modulation in an MC-CDMA system with four subchannels and three users has an efficiency of 3 bits per symbol.

Figure 1 shows the comparison for a target BER of 10^{-2} between OFDM with diversity orders 1, 2, and 4 and MC-CDMA with $N = 4$ and 16 subchannels and different user loads in the quasi-synchronous uplink direction. With low order modulation (QPSK), MC-CDMA outperforms OFDM-DIV at all the studied spectral efficiency levels. However, when the size of the data constellation is increased, the situation starts to change. With 64-QAM, e.g., OFDM-DIV is always better than MC-CDMA with 4 subchannels. With 16 subchannels, MC-CDMA is still better at low spectral efficiencies. Going even further towards higher-level data modulations,

OFDM-DIV consistently outperforms MC-CDMA with 256-QAM. In general it is also interesting to note that the slope in the spectral efficiency curves is much smaller for MC-CDMA than for corresponding OFDM-DIV at the high spectral efficiency regions. This indicates that much larger increase in SNR is needed for MC-CDMA than for OFDM-DIV to achieve a similar relative increase in the system spectral efficiency. Notice also that for a constant user load, increasing the spreading factor in MC-CDMA improves its performance slightly at moderate system loads, but in case of fully loaded systems, the performance degrades clearly with increasing spreading factor. This is because increasing the spreading factor in a fully or close to a fully loaded system results in a higher increase in multiaccess interference (MAI) per user than the gain obtained due to increased frequency diversity.

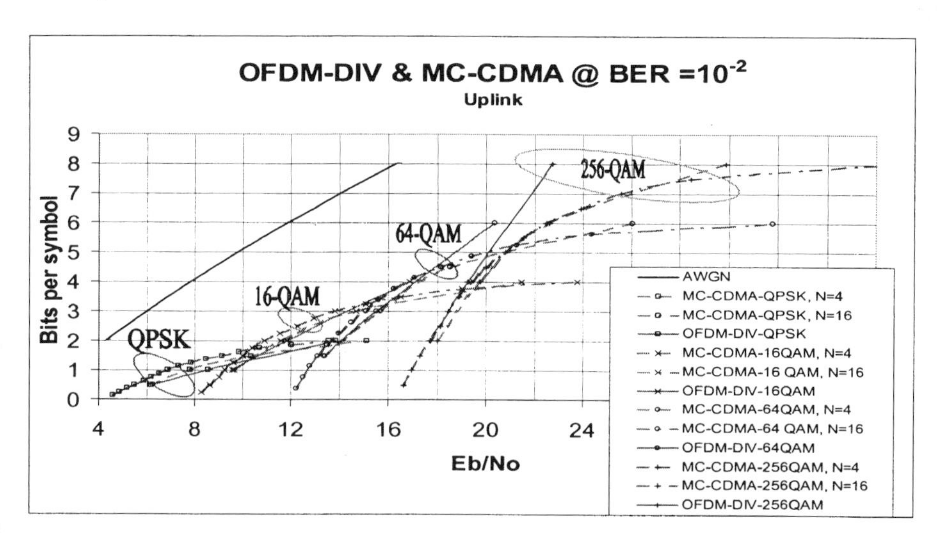

Figure 1. *Spectral efficiency and the needed SNR to achieve a target BER of 10^{-2}. Quasi-synchronous uplink with different number of subchannels (MC-CDMA) and different data modulations.*

The same comparison is presented in Figure 2 for the synchronous downlink case at the target BER of 10^{-2}. The MC-CDMA systems performs now clearly better than OFDM with diversity with data modulations up to and including 64-QAM. This is mainly due to less severe loss of orthogonality, because of identical channel responses, than in the uplink. Otherwise, similar observations hold as in the uplink case.

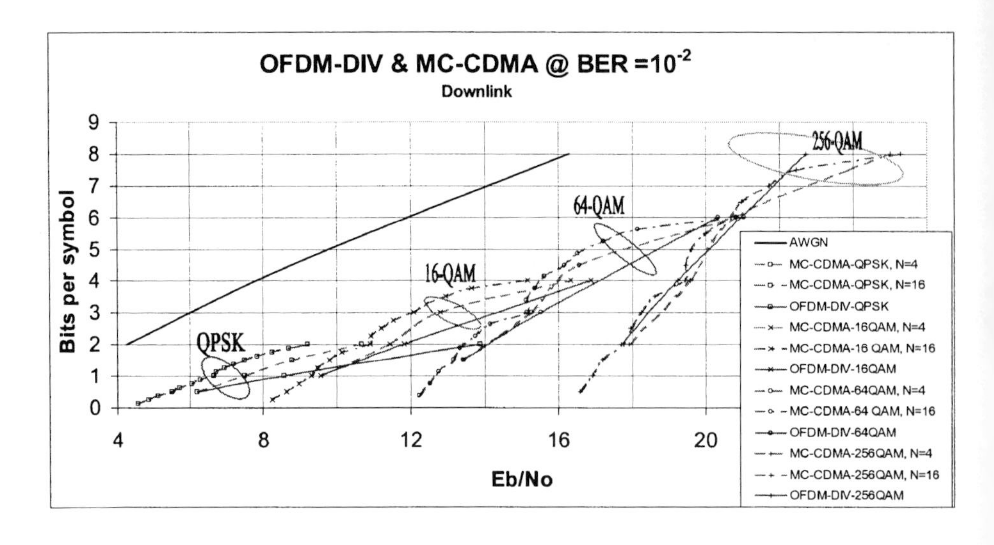

Figure 2. *Spectral efficiency and the needed SNR to achieve a target BER of* 10^{-2}. *Downlink with different number of subchannels (MC-CDMA) and different data modulations.*

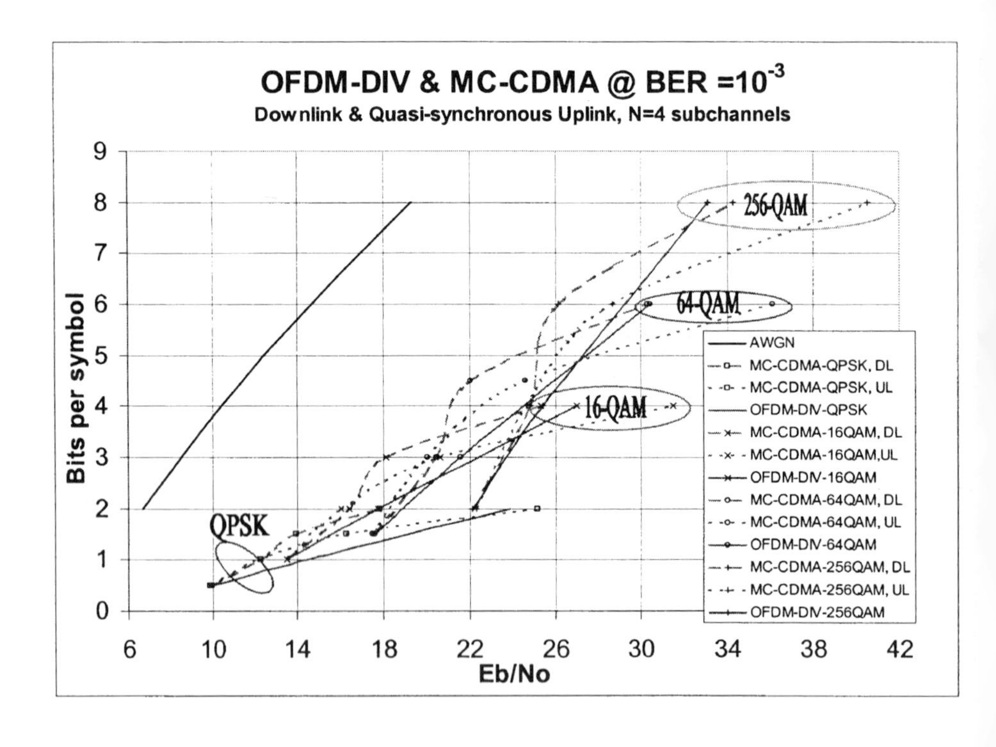

Figure 3. *Spectral efficiency and the needed SNR to achieve a target BER of* 10^{-3}. *Downlink and quasi-synchronous uplink with different data modulations,* $N = 4$.

In Figure 3, the spectral efficiencies are shown at a target BER of 10^{-3} for $N = 4$ subchannels in both quasi-synchronous uplink and downlink directions. The figure shows that the downlink case performs clearly better than the uplink case. This is obvious and again due to more severe loss of orthogonality and thus higher MAI in the uplink. At this target BER (compare to the previous results for 10^{-2}), both MC-CDMA directions perform better than the corresponding OFDM with diversity, except for the case of fully loaded MC-CDMA vs. OFDM with diversity order 1 (no diversity). The trend at still lower BER values is that the MC-CDMA system increasingly outperforms the OFDM system, according to the tendency indicated in this study

4. SUMMARY

MC-CDMA and diversity OFDM systems were compared from the throughput efficiency point of view. In the synchronous downlink case, MC-CDMA provides consistently better results than OFDM with diversity. In the quasi-synchronous uplink case, MC-CDMA starts showing better efficiency than OFDM with diversity only in cases with rather low raw BER (10^{-3} or lower). Therefore, in the uplink case, error control coding should be included to be able to make clear conclusions about the relative performance of the two systems under study. This is a good topic for further studies. We also came to the general conclusion that MC-CDMA provides best performance with moderate (50% ... 75%) loadings. Increasing the spreading factor (i.e., the number of subchannels) improves the MC-CDMA performance clearly in the low BER region (lower than 1%), while on the higher BER levels (10% or higher) the improvement is not that obvious. Interference canceling and multiuser detection (MUD) ideas can be used to improve the MC-CDMA performance, but understanding the limitations of linear detection based system considered here is anyway a corner-stone for all the future developments.

Tobias Hidalgo Stitz, Mikko Valkama, Jukka Rinne & Markku Renfors
Institute of Communications Engineering
Tampere University of Technology
P.O. Box 553, FIN-33101 Tampere
Finland

ACKNOWLEDGMENTS

This research was supported by Nokia, Graduate School in Electronics, Telecommunications, and Automation (GETA), Tampere Graduate School in Information Science and Engineering (TISE), and the Academy of Finland .

REFERENCES

[1] R. van Nee and R. Prasad, *OFDM for Wireless Multimedia Communications*. Boston, MA: Artech House, 2000.

[2] J. G. Proakis, *Digital Communications*, 4th ed. New York, NY: McGraw-Hill, 2001.

[3] S. Hara and R. Prasad, "Overview of multicarrier CDMA," *IEEE Commun. Mag.*, vol. 35, pp. 126-133, Dec. 1997.

[4] S. Tsumura and S. Hara, "Design and performance of quasi-synchronous multi-carrier CDMA system," in *Proc. IEEE Veh. Technol Conf.*, Atlantic City, NJ, Oct. 2001, pp. 843-847.

[5] Z. Li and M. Latva-aho, "Performance comparison of frequency domain equalizers for MC-CDMA systems," in *Proc. IEEE Conf. Mobile Wireless Networks*, Recife, Brazil, Aug. 2001.

[6] S. M. Kay, *Fundamentals of Statistical Signal Processing: Estimation Theory*. Englewood Cliffs, NJ: Prentice-Hall, 1993.

V. LE NIR, M. HELARD, R. LE GOUABLE

EFFICIENT DIVERSITY TECHNIQUES USING LINEAR PRECODING AND STBC FOR MULTI-CARRIER SYSTEMS

1. INTRODUCTION

Since the work of Foschini [1], there has been a huge interest concerning Multiple Input Multiple Output (MIMO) systems in order to exploit the capacity varying linearly with the minimum of transmit N_t and receive antennas N_r and then to exploit the diversity of these systems using Orthogonal Space Time Block Codes (OSTBC) as discovered by Alamouti [2] for $N_t=2$ and then generalized by Tarokh [3] for $2 \leq N_t \leq 4$. Quasi-Orthogonal (QO) STBC were then described in [4][5]. New codes are given in [6] for $N_t=5$ or 6. In parallel, linear precoding was demonstrated to be very efficient in SISO transmission in order to exploit temporal diversity using Maximum Likelihood (ML) detector [7]. Using same type of detectors, linear precoders were adapated to multi-antenna transmissions. The linear precoders used as space-time codes were carried out in [8][9]. The concatenation of linear precoders with QOSTBC was carried out in [10]. In this paper, we combine a particular linear precoder with OSTBC in a specific way allowing a simple linear decoding for various cases of MIMO systems [11]. Several linear precoding matrices based on either Hadamard matrix or Fourier Transform construction are compared. In the second part, we present this linear precoding that has the effect of increasing the overall diversity of the system by scattering the information in the time and/or frequency domains for multi-carrier modulations. We apply our precoder to OFDM and MultiCarrier Code Division Multiplex Access (MC-CDMA) systems, exploiting spatial, temporal and frequency diversities.

2. OSTBC REPRESENTATION

The different channel coefficients are modelled as independent flat fading channels that are quite realistic for OFDM-like modulations. We consider uncorrelated channels from each transmit antenna t to each receive antenna r $h_{tr} = \rho_{tr} e^{i\theta_{tr}}$. Assuming one receive antenna, the Alamouti code can be represented as follows:

$$G_2 = \begin{bmatrix} s_1 & -s_2^* \\ s_2 & s_1^* \end{bmatrix} \tag{1}$$

Assuming fading coefficients constant over two consecutive symbol transmissions, the received signal over two consecutive symbols periods are:

$$\begin{bmatrix} r_1 \\ -r_2^* \end{bmatrix} = \begin{bmatrix} h_1 & h_2 \\ -h_2^* & h_1^* \end{bmatrix} \cdot \begin{bmatrix} s_1 \\ s_2 \end{bmatrix} + \begin{bmatrix} n_1 \\ -n_2^* \end{bmatrix} \tag{2}$$

K. Fazel and S. Kaiser (eds.), Multi-Carrier Spread-Spectrum, 345–352.

where n_1 and n_2 are independent complex variables with zero mean and one-sided power spectral density N_0, representing Additive White Gaussian Noise (AWGN). For this study, perfect channel estimation is assumed. Applying the transpose conjugate of the channel matrix to the equivalent received vector, we obtain:

$$\begin{bmatrix} \hat{s}_1 \\ \hat{s}_2 \end{bmatrix} = \lambda . I_2 . \begin{bmatrix} s_1 \\ s_2 \end{bmatrix} + \begin{bmatrix} h_1^* n_1 + h_2 n_2^* \\ h_2^* n_1 - h_1 n_2^* \end{bmatrix} \tag{3}$$

with $\lambda = |h_1|^2 + |h_2|^2$. This receiving process corresponds to a Maximum Ratio Combining (MRC) equalizer. However, an equalization process can be carried out according to the Zero Forcing or Minimum Mean Square Error criteria. This matrix representation can be extended for other OSTBC schemes [3][6] where we obtain:

$$\lambda = \sum_{j=1}^{N_t} |h_j|^2 \qquad \text{or} \qquad \lambda = \sum_{j=1}^{N_t} |h_j|^2 \Big/ \left(\sum_{j=1}^{N_t} |h_j|^2 + 1/\gamma \right) \tag{4}$$

in the case of a MRC or a MMSE equalizer respectively, where γ is the Signal to Noise Ratio at the receive antenna. We use different equalizers because they lead to different performance when using linear precoding.

3. LINEAR PRECODING

This linear precoding given in [11] is briefly presented. According to the theorem of diagonal decomposition, let A_L be a Hermitian LxL matrix with eigenvalues $\lambda_1 \ldots \lambda_L$. Then A_L can be expressed as:

$$A_L = \Theta_L \Lambda_L \Theta_L^H \tag{5}$$

where $\Lambda_L = diag(\lambda_1, \ldots, \lambda_L)$ and Θ_L is an unitary matrix so that $\Theta_L^{-1} = \Theta_L^H$, where $(.)^H$ stands for transpose conjugate. We propose to use the following linear precoding based on the Hadamard construction matrix such as:

$$\Theta_L = \sqrt{\frac{2}{L}} \begin{bmatrix} \Theta_{L/2} & \Theta_{L/2} \\ \Theta_{L/2} & -\Theta_{L/2} \end{bmatrix} \tag{6}$$

with $L=2^n$, $n \in \mathbf{N}^*$, $n \geq 2$ and:

$$\Theta_2 = \begin{bmatrix} e^{j\theta_1} . \cos\eta & e^{j\theta_2} . \sin\eta \\ -e^{j\theta_2} . \sin\eta & e^{-j\theta_1} . \cos\eta \end{bmatrix} \tag{7}$$

belonging to the Special Unitary group SU(2), therefore $\det(\Theta_2) = 1$. This leads to the following expression:

$$A_L = \frac{2}{L} \begin{bmatrix} A_{L/2}^1 + A_{L/2}^2 & A_{L/2}^1 - A_{L/2}^2 \\ A_{L/2}^1 - A_{L/2}^2 & A_{L/2}^1 + A_{L/2}^2 \end{bmatrix} \tag{8}$$

with $A^1_{L/2} = \Theta_{L/2}\Lambda^1_{L/2}\Theta^H_{L/2}$ and $A^2_{L/2} = \Theta_{L/2}\Lambda^2_{L/2}\Theta^H_{L/2}$, and

$$\Lambda_L = diag\left(\Lambda^1_{L/2}, \Lambda^2_{L/2}\right) \tag{9}$$

where $\Lambda^1_{L/2} = diag(\lambda_1,...,\lambda_{L/2})$ and $\Lambda^2_{L/2} = diag(\lambda_{L/2+1},...,\lambda_L)$. For L=2, we obtain the following Hermitian matrix:

$$A_2 = \begin{bmatrix} \cos^2\eta.\lambda_1 + \sin^2\eta.\lambda_2 & -\cos\eta.\sin\eta.e^{j(\theta_1+\theta_2)}.(\lambda_1 - \lambda_2) \\ -\cos\eta.\sin\eta.e^{-j(\theta_1+\theta_2)}.(\lambda_1 - \lambda_2) & \sin^2\eta.\lambda_1 + \cos^2\eta.\lambda_2 \end{bmatrix} \tag{10}$$

Therefore, one can see that for A_L the diagonal elements are equal to:

$$A_{ii} = \frac{2}{L}\sum_{k=0}^{L/2-1}\left(\cos^2\eta.\lambda_{(2k+1)} + \sin^2\eta.\lambda_{(2k+2)}\right) \quad \forall i \in [1...L] \tag{11}$$

and that some non-diagonal elements are similar to:

$$A_{ij} = -\frac{2}{L}\cos\eta.\sin\eta.e^{-j(\theta_1+\theta_2)}\sum_{k=0}^{L/2-1}(\lambda_{(2k+1)} - \lambda_{(2k+2)}) \tag{12}$$

Owing to (8) form, the other terms of interference are also sum of difference between eigenvalues. By simulation, the optimal results were found for pure real or pure imaginay interference. For $L = 2$, $\eta = \pi/4$, $\theta_2 = \theta_1 - \pi/2$, $\theta_1 = 5\pi/4$, we get:

$$A_2 = \frac{1}{2}\begin{bmatrix} \lambda_1 + \lambda_2 & \lambda_2 - \lambda_1 \\ \lambda_2 - \lambda_1 & \lambda_1 + \lambda_2 \end{bmatrix} \tag{13}$$

One may use Fourier matrices instead of these matrices based on Hadamard construction, but we will see in the following section that they lead to worse performance for low values of L when the linear precoding is combined with STBC.

4. LINEAR PRECODING WITH STBC

In this paper, we combine the OSTBC with linear precoding by concatenation according to the equation (5) where Λ_L represents the OSTBC coding and decoding without noise, and Θ_L stands for the linear precoding. It is possible to use more transmit antennas by applying the OSTBC using subgroups of the available transmit antennas [11]. For instance, if we use the Alamouti code with four antennas, the first subgroup including antenna 1 and 2 will code the symbols according to Alamouti while antenna 3 and 4 are switched off. The total transmit power should remain *P*, therefore antenna 1 and 2 will transmit symbol at a power of *P/2*. Then, antennas 3 and 4 will transmit the coded symbols according to Alamouti while antennas 1 and 2 are switched off. All OSTBC can be applied to this scheme as those described in [2][3][6] thus many scenarios can be drawn from this example depending on antenna configurations. As presented in Figure 1, at transmission side, input bits are first mapped into symbol vector $X = [x_1 \quad ... \quad x_L]$ where L is the number of transmitted symbols. Linear precoding is then performed by applying the Θ_L matrix to the X vector. The next step consists in applying an OSTBC to the symbol-rotated vector.

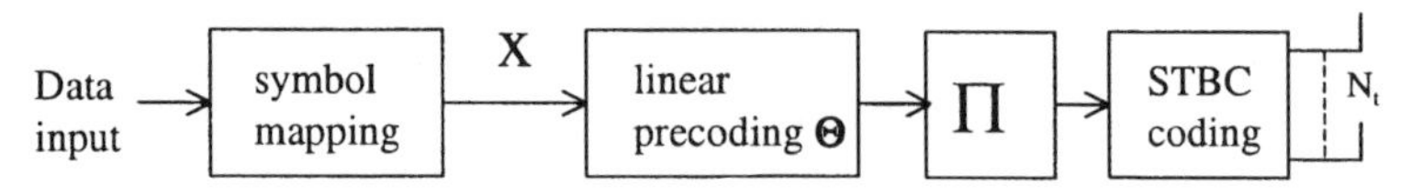

Fig.1 Combination of Linear Precoding with OSTBC: Transmitter scheme

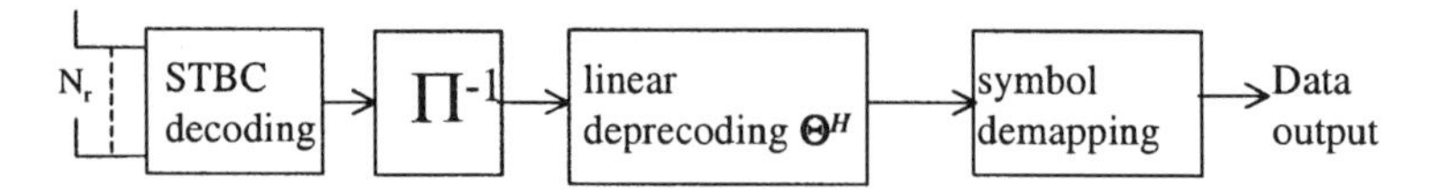

Fig.2 Combination of Linear Precoding with OSTBC: Receiver scheme

The receiver part is described by Figure 2. Without interleaving, the channel representation of the OSTBC codes leads to the equivalent channel coding and decoding matrix with the same diagonal elements. Owing to the Alamouti scheme (1) and the linear precoding described in (6) and (7) with optimal values given in Section 3, for $L=4$ without interleaving we get the following Hermitian matrix:

$$A_4 = \Theta_4 \Lambda_4 \Theta_4^H \tag{14}$$

with $\Lambda_4 = diag(\lambda_1, \lambda_1, \lambda_2, \lambda_2)$ and $\lambda_i = \sum_{j=1}^{2} \left| h_j^i \right|^2 \quad i \in [1...2]$ (15)

where i and j are index used to distinguish different channels related by the OSTBC decoding.

The resulting matrix is then:

$$A_4 = \frac{1}{2} \begin{bmatrix} \lambda_1 + \lambda_2 & 0 & \lambda_1 - \lambda_2 & 0 \\ 0 & \lambda_1 + \lambda_2 & 0 & \lambda_1 - \lambda_2 \\ \lambda_1 - \lambda_2 & 0 & \lambda_1 + \lambda_2 & 0 \\ 0 & \lambda_1 - \lambda_2 & 0 & \lambda_1 + \lambda_2 \end{bmatrix} \tag{16}$$

When applying an interleaving process, the λ's within each block are affected by different channels:

$$\lambda_i = \sum_{j=1}^{N_r} \left| h_j^i \right|^2 \quad i \in [1...L] \tag{17}$$

With interleaving, the Λ_4 matrix and the resulting in formula (14) becomes:

$$\Lambda_4 = diag(\lambda_1, \lambda_2, \lambda_3, \lambda_4) \text{ with } \lambda_i = \sum_{j=1}^{2} \left| h_j^i \right|^2 \quad i \in [1...4] \tag{18}$$

Thus, the resulting matrix A_4 becomes:

$$A_4 = \frac{1}{4}\begin{bmatrix} \lambda_1+\lambda_2+\lambda_3+\lambda_4 & \lambda_1-\lambda_2+\lambda_3-\lambda_4 & \lambda_1+\lambda_2-\lambda_3-\lambda_4 & \lambda_1-\lambda_2-\lambda_3+\lambda_4 \\ -\lambda_1+\lambda_2-\lambda_3+\lambda_4 & \lambda_1+\lambda_2+\lambda_3+\lambda_4 & -\lambda_1+\lambda_2+\lambda_3-\lambda_4 & \lambda_1+\lambda_2-\lambda_3-\lambda_4 \\ \lambda_1+\lambda_2-\lambda_3-\lambda_4 & \lambda_1-\lambda_2-\lambda_3+\lambda_4 & \lambda_1+\lambda_2+\lambda_3+\lambda_4 & \lambda_1-\lambda_2+\lambda_3-\lambda_4 \\ -\lambda_1+\lambda_2+\lambda_3-\lambda_4 & \lambda_1+\lambda_2-\lambda_3-\lambda_4 & -\lambda_1+\lambda_2-\lambda_3+\lambda_4 & \lambda_1+\lambda_2+\lambda_3+\lambda_4 \end{bmatrix} \quad (19)$$

One can notice that, with or without interleaver, at the receiver part a linear decoding can be performed by simply applying the transpose conjugate of the linear precoder. Therefore, we obtain a matrix of the form described in (14). The interleaving has the effect of mixing eigenvalues between different blocks, thus the components of the resulting matrix are different from each others. After OSTBC decoding and linear deprecoding, we can merely detect the signals. When L increases, OSTBC is performed on more subgroups and the resulting matrix corresponds to (5). With OSTBC, the diagonal elements follow a chi square law with $2N_t$ degrees of freedom. When using linear precoding, the diagonal elements reach a chi square law with $N_t.L$ degrees of freedom. With interleaving, the elements of the diagonal matrix reach a chi square law with $2N_t.L$ degrees of freedom, providing more diversity.

When using a Fourier matrix of size $L=4$ instead of the complex Hadamard matrix based on SU(2), we obtain a circulant matrix of the form:

$$A_4 = \frac{1}{4}\begin{bmatrix} a & b & c & d \\ d & a & b & c \\ c & d & a & b \\ b & c & d & a \end{bmatrix} \quad (20)$$

with $a = \lambda_1+\lambda_2+\lambda_3+\lambda_4$, $b = \lambda_1-\lambda_3-j(\lambda_2-\lambda_4)$, $c = \lambda_1-\lambda_2+\lambda_3-\lambda_4$ and $d = \lambda_1-\lambda_3+j(\lambda_2-\lambda_4)$. Hence, the interference terms are different from those obtained with the Hadamard construction. As L increases the interference terms will tend slower towards the gaussian law than the interference terms of the Hadamard construction presented before. Indeed, the interference terms of the Fourier construction follow a chi-square law with twice as less degrees of freedom as the Hadamard one per dimension.

5. LINEAR PRECODING WITH STBC AND MULTI-CARRIER SYSTEMS

In the precedent part, linear precoding is done in the time domain, but it can also be performed in the frequency and/or time domains for OFDM and MC-CDMA systems that provide full frequency diversity owing to the orthogonality between subcarriers of the OFDM modulation. OSTBC will be performed for multicarrier systems as described in [14] for different OSTBC codes.

5.1. Linear Precoding with OFDM

In order to apply linear precoding in the frequency domain, one may use a linear precoder of size $L \leq N_c$, where L always the size of the precoding matrix and N_c is the number of subcarriers. This corresponds to an OFDM linear precoded scheme in the frequency domain. Again, in order to apply linear precoding in the time and frequency domains, one may use a linear precoder of size $L \geq N_c$. This corresponds to an OFDM linear precoded scheme in the time and frequency domains.

5.2. Linear Precoding with MC-CDMA

MC-CDMA combines OFDM modulation and CDMA access technique taking benefits from both the high spectral efficiency and the robustness against multipath channels of OFDM and access flexibility of CDMA [12][13]. In order to linearly precode a MC-CDMA scheme, we simply allocate codes for a specific user of length $L \geq N_c$. This means that allocation of spreading codes is carried out in the time and frequency domain. An interesting analogy can be made when MC-CDMA is applied with a linear precoding. In this case, MC-CDMA is equivalent to an OFDM scheme where our linear precoder is applied in both time and frequency domains. This can be adapted either to MIMO transmissions. We propose linear precoded OFDM scheme with OSTBC or linear precoded MC-CDMA scheme with OSTBC where temporal, spatial and frequency diversities are exploited.

6. RESULTS AND CONCLUSION

We carried out simulations in order to check the behaviour of the proposed system regarding the efficient exploitation of the diversities with multi-carrier systems, and to compare the performance with Hadamard-based and Fourier based precoders.
Figure 3 shows the performance of OSTBC with linear precoding for $L=32$ and spectral efficiency η of 1 bps/Hz for a flat Rayleigh channel. To obtain $\eta=1$, a BPSK is applied to the Alamouti code, whereas QPSK is applied to Tarokh codes. For this spectral efficiency, the Alamouti performs worse than Tarokh codes, but this is not true for higher spectral efficiencies [3][14]. The results confirm that the performance improves with linear precoding for all tested OSTBC code providing a 2 dB gain at BER=10^{-3} with G_4 with a very simple linear receiver. These results have been obtained with a number of transmit antennas N_t corresponding to their respective OSTBC. Since the channel coefficients are uncorrelated, we find the same results by applying the OSTBC using subgroups depending on Nt. For instance, we have the same results with the Alamouti code using $Nt=2,4,8,16$ or 32 transmit antennas for $L=32$ and $N_t/2$ subgroups if $N_t>2$.
Figure 4 shows the performance of the OSTBC Alamouti code ($N_t=2$ and $N_r=1$) with linear precoded OFDM for $L=4$ or $L=4096$ and with $N_c=64$ over uncorrelated Rayleigh channels ($\eta=1$). We see that the rotated Hadamard and FFT precoded OFDM give the same results when N is large ($N=4096$) but this specific Hadamard linear precoder performs better than FFT one when L is small ($L=4$). Moreover, these results are similar to the results of MC-CDMA with OSTBC over Rayleigh channels, adding the benefits of the spreading in time and frequency dimensions.

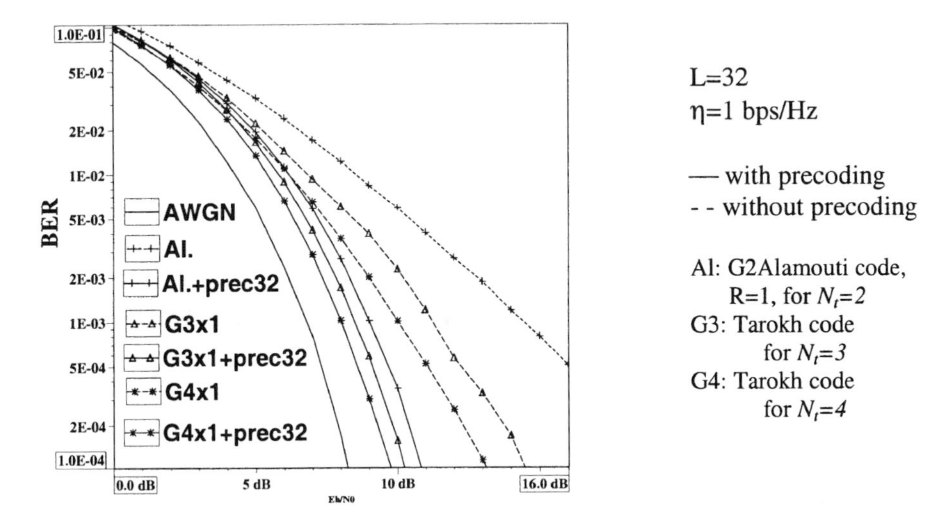

Fig.3 Performance of STBC with linear precoding and different OSTBC

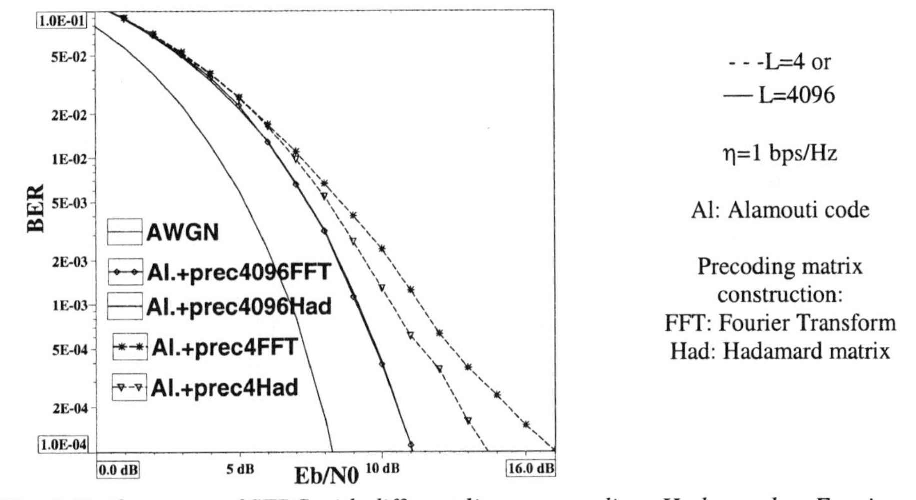

Fig. 4. Performance of STBC with different linear precoding: Hadamard or Fourier

In this paper, we propose to linearly precode and decode OSTBC systems using a particular unitary matrix based on Hadamard or FFT construction. Our scheme has a low complexity, which only grows linearly with the size of the unitary matrix and not exponentially when more complex detectors are used. Simulation results with the specific linear precoders using OSTBC are given for flat independent Rayleigh fadings or OFDM systems. These precoders can be applied to various MIMO transmissions in order to exploit spatial, temporal and frequency diversities. We saw thanks to simulation results that the precoding method is very efficient for multi-carrier modulations. We gave an interesting analogy between linear precoded OSTBC for flat independent Rayleigh fadings, linear precoded OFDM with OSTBC and linear precoded MC-CDMA with OSTBC. One can apply this linear precoding with any OSTBC, keeping the linearity of the transmission chain even at the receiver part. It is also possible to apply this linear precoding with QOSTBC, but at the expense of more complex receiver. Moreover, the proposed scheme suits to several multi-antenna configurations and thus can be adapted to channel characteristics.

7. REFERENCES

[1] Foschini G.J., *Layered space-time architecture for wireless communication in a fading environment when using multi-element antennas,* Bell Labs Tech. Journal, Vol. 1, N° 2, pp. 41-59, 1996.

[2] Alamouti S.M, *A simple Transmit Diversity Technique for Wireless Communications*, IEEE Journal on Selected Areas in Communications, Vol. 16, No. 8, October 1998, pp. 1451-1458, 1998.

[3] Tarokh V., Jafarkhani H., and Calderbank A. R., *Space-Time Block Codes from Orthogonal Designs*, IEEE Transactions on Information Theory, Vol. 45, No. 5, pp. 1459-1467, July 1999.

[4] Jafarkhani H., *A quasi-orthogonal space-time block code*, IEEE Trans. Comm, 49, (1), pp. 1-4, 2001.

[5] Tirkkonen O., Boariu A., Hottinen A., *Minimal orthogonality space-time block code for 3+ Tx antennas, Proc. IEEE Int. Symp. Spr. Spectr. Techn. Appl.* (ISSSTA), New Jersey, USA, September 2000.

[6] Su W., Xia X., *Two Generalized Complex Orthogonal Space-Time Block Codes of Rates 7/11 and 3/5 for 5 and 6 Transmit Antennas*, to appear in IEEE Trans. on Inf. Theory, Jan. 2003

[7] Boutros J., Viterbo E., *Signal Space Diversity: A Power and Bandwidth Efficient Diversity Technique for the Rayleigh Fading Channel,* IEEE Trans. on Information Theory, Vol. 44, No.4, pp. 1453-1467, Jul. 1998.

[8] Damen M.O., K. Abed-Meraim, Belfiore J.C., *Diagonal Algebraic space-time block codes*, IEEE Trans. Inf. Theory, 48, (3), pp.628-636, 2002.

[9] Xin Y., Wang Z., and Giannakis G. B., *Space-Time Diversity Systems Based on Unitary Constellation-Rotating Precoders*, IEEE Conference on Acoustics, Speech, Systems and Signal Processing, Lake Louise, Alberta, Canada, pp. 396-401, Oct. 2000.

[10] Da Silva M. M., Correia A., *Space-Time Block Coding for 4 antennas with Coding rate 1*, IEEE 7th Symp. On Spread Spectrum Technologies and Applications, Prague, Czech Republic, Sept. 2002.

[11] Le Nir V., Hélard M., *Reduced-Complexity Space-Time Block Coding and Decoding schemes with block linear precoding*, IEE Electronic letters, Vol. 39 No.14, 10th July 2003.

[12] Yee N., Linnartz J.P, Fettweis G., *Multicarrier CDMA in Indoor Wireless Radio Networks*, IEEE PIMRC'93, pp. 109-113, Yokohama, Japan, 1993.

[13] Hélard M. , Le Gouable R., Hélard J.F, Baudais J.Y., *Multicarrier techniques for future wideband wireless network,* Annales des Télécom Numéro spécial UMTS, Vol. 56, N°5-6, pp. 260-274, 2000.

[14] Le Nir V., Hélard M., Le Gouable R., *Space-Time Block Coding Applied to Turbo Coded Multicarrier CDMA*, Vehicular Technology Conference, Jeju, South Korea, 22-25 April 2003.

[15] IST MATRICE project, web site http://www.ist-matrice.org.

8. AFFILIATIONS

The authors work at France Télécom R&D in DMR/DDH lab. For more detailed informations, you can contact them at the following email addresses:
{vincent.lenir;maryline.helard;rodolphe.legouable}@francetelecom.com

Part of this work has been carried out in the IST-MATRICE project [15].

PERFORMANCE OF MMSE STBC MC-CDMA OVER RAYLEIGH AND MIMO METRA CHANNELS

J-M. AUFFRAY, J-Y. BAUDAIS, J-F. HELARD

IETR / INSA – 20 avenue des Buttes de Coësmes, 35043 RENNES – FRANCE

Abstract. The performance of MMSE Single-user Detection (SD) and Multi-user Detection (MD) STBC MC-CDMA systems are analysed and compared in the case of two transmit antennas and one or two receive antennas over Rayleigh fading channels and then over the stochastic MIMO METRA channel model. With two transmit and one receive antennas, MD achieves a gain of roughly 1 dB for non-full load systems while the same performance are obtained with MD and SD for full load systems. Besides, with two receive antennas, we present a sub-optimal and an optimal MMSE SD MIMO MC-CDMA schemes, this last one offering a very good performance/complexity trade-off. Finally, the very good behaviour of MMSE STBC MC-CDMA systems is confirmed over the realistic METRA MIMO channel.

1. INTRODUCTION

Nowadays, Multi-Carrier Code Division Multiple Access (MC-CDMA) is the most promising candidate for the air interface downlink of the 4th Generation mobile radio systems. MC-CDMA combines the robustness of Orthogonal Frequency Division Multiplex (OFDM) modulation with the flexibility of CDMA [1]. On the other hand, Multiple Input Multiple Output (MIMO) communication systems, by using several antennas at the transmitter and at the receiver, inherit space diversity to mitigate fading effects. When the channel is not known at the transmitter, taking benefit of the transmit diversity requires methods such as space-time coding which uses coding across antennas and time [2]. For example, Space-Time Block Coding (STBC), as proposed by Alamouti in [3] and Tarokh in [4], provides full spatial diversity gains, no intersymbol interference and low complexity ML receiver if transmission matrix is orthogonal. Moreover with STBC, only one receive antenna can be used, leading in that case to MISO (Multiple Input Single Output) systems.

In [5], it has been shown that unity-rate Alamouti's STBC QPSK MC-CDMA outperforms half-rate Tarokh's STBC 16-QAM MC-CDMA, while offering the same effective throughput of 2 bit/s/Hz without channel coding. Indeed, in order to maintain the same effective throughput, half-rate STBC codes have to be employed in conjunction with higher modulation schemes as 16-QAM, which are more prone to errors and hence degrade the performance of the system. Moreover, unity-rate

K. Fazel and S. Kaiser (eds.), Multi-Carrier Spread-Spectrum, 353–362.

STBC code combined with channel-coded schemes as turbo-codes provides substantial performance improvement over the non-unity-rate STBC. Hence for the same effective throughput, the reduction in coding rate is best invested in turbo-codes, rather than STBC.

In this paper, we compare in the downlink case and without channel coding the performance of Alamouti's STBC MC-CDMA systems combined with Multi-user Detection (MD) or Single-user Detection (SD) schemes. For this comparison, the considered detection schemes are based on Mean Square Error (MSE) criterion, since MMSE detection is known as the most efficient SD technique [6]. In order to obtain asymptotic performance, the algorithms are evaluated over Rayleigh fading channels in the first part. Then, some further results are given over the more realistic stochastic MIMO channel model developped within the European IST METRA (Multi Element Transmit Receive Antennas) project.

2. SYSTEM DESCRIPTION

Figure 1 shows the considered MIMO MC-CDMA system for the jth user based on Alamouti's STBC with $N_t = 2$ transmit antennas and $N_r = 2$ receive antennas [3]. Each user j transmits simultanously from the two antennas the symbol x_j^0 and x_j^1 at time t, and the symbols $-x_j^{1*}$ and x_j^{0*} at time $t+T_x$ where T_x is the OFDM symbol duration. At the output of the space-time encoder, the data symbols $\mathbf{x}^0=[x_1^0 \ldots x_j^0 \ldots x_{Nu}^0]^T$ of the N_u users are multiplied by their specific orthogonal Walsh-Hadamard (W-H) spreading code $\mathbf{c}_j = [c_{j,1} \ldots c_{j,k} \ldots c_{j,Lc}]^T$ where $c_{j,k}$ is the kth chip, and $[.]^T$ denotes matrix transposition (the same goes for symbol 1). $\mathbf{c}_j$ is the jth column vector of the L_c x N_u spreading code matrix $\mathbf{C}$. In this paper, the length L_c of the spreading sequences is equal to the number N_c of subcarriers and to the maximum number N_u of simultaneous active users in the full-load case. Each data symbol x_j is then transmitted in parallel on N_c QPSK modulated subcarriers. The vector obtained at the rth receive antenna after the OFDM demodulation and deinterleaving, at time t and $t+T_x$, is given by:

$$\boldsymbol{R}_r = \boldsymbol{H}_r \boldsymbol{C} \boldsymbol{X} + \boldsymbol{N}_r \quad \textit{with} \quad \boldsymbol{H}_r = \begin{bmatrix} \mathbf{H}_{1r} & \mathbf{H}_{2r} \\ \mathbf{H}_{2r}^* & -\mathbf{H}_{1r}^* \end{bmatrix} \tag{1}$$

where $\boldsymbol{R}_r = [\mathbf{r}_r^T(t)\ \mathbf{r}_r^H(t+T_x)]^T$ with $\mathbf{r}_r(t) = [r_{r,1}(t) \ldots r_{r,k}(t) \ldots r_{r,Nc}(t)]^T$ the vector of the N_c received signals at time t and $[.]^H$ denotes the Hermitian transpose (or conjugate transpose), $\mathbf{H}_{tr} = \mathrm{diag}\{h_{tr,1}, \ldots, h_{tr,Nc}\}$ $(t,r$ ™ $\{1,2\})$ is a N_c x N_c diagonal matrix with $h_{tr,k}$ the complex channel frequency response, for the subcarrier k between the transmit antenna t to the receive antenna r. Time invariance during two MC-CDMA symbols are assumed to permit the recombination of symbols when STBC is used. $\boldsymbol{C} = \mathrm{diag}\{\mathbf{C},\mathbf{C}\}$ with $\mathbf{C}=[\mathbf{c}_1 \ldots \mathbf{c}_j \ldots \mathbf{c}_{Nu}]$ is the L_c x N_u matrix of user's spreading codes with the column vector $\mathbf{c}_j$ equal to the spreading code of user j, $\boldsymbol{X}= [\mathbf{x}^{0T}\ \mathbf{x}^{1T}]^T$. $\boldsymbol{N}_r= [\mathbf{n}_r^T(t)\ \mathbf{n}_r^H(t+T_x)]^T$ with $\mathbf{n}_r(t)=[n_{r,1}(t) \ldots n_{r,k}(t) \ldots n_{r,Nc}(t)]^T$ is the Additive White Gaussian Noise (AWGN) vector with $n_{r,k}(t)$ representing the noise term at subcarrier k, for the rth receive antenna at time t with variance given by $\sigma_k^2=\mathrm{E}\{|n_k|^2\}=N_0\ \forall k$.

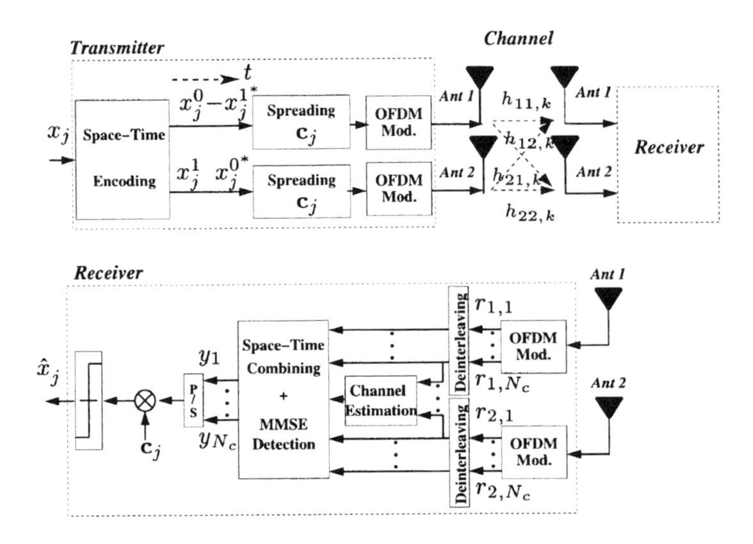

Figure 1. MC-CDMA transmitter and receiver for user j with transmit and receive diversity.

3. MULTI-USER DETECTION VERSUS SINGLE-USER DETECTION IN THE MIMO AND MISO CASES

In the receiver, in order to detect the two transmitted symbols x_j^0 and x_j^1 for the desired user j, SD or MD detection schemes based on the MSE criterion are applied to the received signals in conjunction with STBC decoding. In the SISO case, it has been shown in [6] that MMSE SD is the most efficient SD scheme, while MMSE MD, also called Global-MMSE, is optimal according to the MSE criterion for any number of active users and any power distribution [7]. Here we compare in the MISO case MMSE SD with a new MMSE MD algorithm. Besides, in the MIMO case, an optimal and a sub-optimal MMSE SD algorithms are presented and compared.

3.1 MMSE Single-user detection in the MISO and MIMO cases

After equalisation, for each receive antenna r, the two successive received signals are combined. The resulting signals from the N_r receive antennas are then added to detect the two symbols x_j^0 and x_j^1. After despreading and threshold detection, the detected data symbols $\hat{x}_j^0$ and $\hat{x}_j^1$ for user j are:

$$\begin{bmatrix}\hat{x}_j^0 & \hat{x}_j^1\end{bmatrix}^T = \begin{bmatrix}\mathbf{c}_j^T & \mathbf{c}_j^T\end{bmatrix} \boldsymbol{Y} = \begin{bmatrix}\mathbf{c}_j^T & \mathbf{c}_j^T\end{bmatrix} \sum_{r=1}^{Nr} \boldsymbol{G}_r \boldsymbol{R}_r \text{ with } \boldsymbol{G}_r = \begin{bmatrix}\mathbf{G}_{1r} & \mathbf{G}_{2r}^* \\ \mathbf{G}_{2r} & -\mathbf{G}_{1r}^*\end{bmatrix} \quad (2)$$

where $\mathbf{G}_{tr}$ is a diagonal matrix containing the equalization coefficients for the channel between the transmit antenna t and the receive antenna r. To detect for example x_i^0, the MMSE SD coefficients $g_{tr,k}$ minimises the mean square value of the error ε_k^0 between the signal $\sum_{i=1}^{N_u} c_{i,k} x_i^0$ transmitted on subcarrier k and the assigned output y_k^0 of the equalizer. Besides, no knowledge of the spreading codes $\mathbf{c}_i$ $(i \neq j)$ of the interfering users is required to derive the MMSE SD coefficients.

Table 1. MMSE SD equalization coefficients $g_{tr,k}$ and resulting equalized channel terms $h_{eq,k}^0$ and $h_{eq,k}^1$ to detect the symbol x_j^0 for the sub-optimal MMSE(1) SD and optimal MMSE(2) SD schemes.

	MMSE(1) SD	MMSE(2) SD
$g_{tr,k}$	$h_{tr,k}^* / \left[\sum_{t=1}^{N_t} \lvert h_{tr,k} \rvert^2 + \frac{1}{\gamma_{r,k}} \right]$	$h_{tr,k}^* / \left[\sum_{t=1}^{N_t} \sum_{r=1}^{N_r} \lvert h_{tr,k} \rvert^2 + \frac{1}{\gamma_{r,k}} \right]$
$h_{eq,k}^0$	$\dfrac{\sum_{t=1}^{N_t} \sum_{r=1}^{N_r} \lvert h_{tr,k} \rvert^2}{\sum_{t=1}^{N_t} \lvert h_{tr,k} \rvert^2 + \frac{1}{\gamma_{r,k}}}$	$\dfrac{\sum_{t=1}^{N_t} \sum_{r=1}^{N_r} \lvert h_{tr,k} \rvert^2}{\sum_{t=1}^{N_t} \sum_{r=1}^{N_r} \lvert h_{tr,k} \rvert^2 + \frac{1}{\gamma_{r,k}}}$
$h_{eq,k}^1$	0	0

Table 1 gives the MMSE SD equalization coefficients $g_{tr,k}$ and the resulting equalized channel coefficients $h_{eq,k}^0$ and $h_{eq,k}^1$ to detect x_i^0 and x_i^1 respectively. For the optimal MMSE(2) SD algorithm, $N_t{*}N_r$ channel coefficients $h_{tr,k}$ are taken into account, while only N_t are considered for MMSE(1) SD algorithm. Thus, an excessive noise amplification for low subcarrier signal to noise ratio $\gamma_{r,k}$ is more unlikely with this new MMSE(2) SD algorithm than with MMSE(1) SD algorithm. In both cases, to detect for example x_i^0, the interference terms generated by x_i^1 are cancelled, *i.e.*, $h_{eq,k}^1 = 0$. On the other hand, for large SNR, MMSE SD restores the orthogonality among users, *i.e.*, $h_{eq,k}^0$ tends to one when $\gamma_{r,k}$ increases.

3.2 MMSE Multi-user detection in the MISO case

Contrary to MMSE SD, MD is carried out by exploiting the knowledge of the spreading codes $\mathbf{c}_i$ $(i \neq j)$ of the interfering users. As the optimum Maximum Likelihood (ML) detector is too complex, we consider here sub-optimal linear MMSE MD which is optimal according to the MSE criterion, and applied here for the first time to space-time block coded signals. The MMSE MD technique aims to minimize the mean square error at the input of the threshold detector between the transmitted symbol x_j and the estimated one $\hat{x}_j$. The two detected data symbol $\hat{x}_j^0$ and $\hat{x}_j^1$ for user j are:

$$\left[\hat{x}_{j,opt}^0 \ \hat{x}_{j,opt}^1\right]^T = \mathbf{W}_{j,opt}^H \boldsymbol{R} = \left[\mathbf{c}_j^T \ \mathbf{c}_j^T\right] \boldsymbol{G}_r^H \boldsymbol{R} = \begin{bmatrix} \mathbf{w}_{j,opt}^{0H} \\ \mathbf{w}_{j,opt}^{1H} \end{bmatrix} \begin{bmatrix} \mathbf{r}_r(t) \\ \mathbf{r}_r^*(t+T_x) \end{bmatrix} \tag{3}$$

where $\mathbf{W}_{j,opt}{}^H$ is the optimal $2 \times 2L_c$ weighting matrix and $\boldsymbol{G}_r{}^H$ is the equalization coefficient matrix of the MISO channel at the antenna r. According to the Wiener filtering, the optimal weighting matrix is the matrix which minimises the mean square error $\mathrm{E}|\mathbf{W}_j^H\boldsymbol{R} - [x_j^0\ x_j^1]^T|^2$. The $2L_c$ weighting vectors $\mathbf{w}_{j,opt}{}^0$ and $\mathbf{w}_{j,opt}{}^1$ to detect x_j^0 and x_j^1 respectively are equal to:

$$\mathbf{w}^0_{j,opt} = \Gamma^{-1}_{\boldsymbol{R},\boldsymbol{R}}\Gamma_{\boldsymbol{R},x_j^0} \qquad \mathbf{w}^1_{j,opt} = \Gamma^{-1}_{\boldsymbol{R},\boldsymbol{R}}\Gamma_{\boldsymbol{R},x_j^1} \tag{4}$$

where $\Gamma_{\boldsymbol{R},\boldsymbol{R}}$ is the autocorrelation matrix of the received vector $\boldsymbol{R}$ and $\Gamma_{\boldsymbol{R},x_j^0}$ and $\Gamma_{\boldsymbol{R},x_j^1}$ are the cross-correlation vector between the received signal vector $\boldsymbol{R}$ and the desired symbol x_j^0 and x_j^1 respectively. Hence the optimal weighting matrix is:

$$\mathbf{W}^H_{j,opt} = E_s\left[\mathbf{c}_j^T\ \mathbf{c}_j^T\right]\boldsymbol{H}^H\left(\boldsymbol{H}\boldsymbol{C}\boldsymbol{\Gamma}_{X,X}\boldsymbol{C}^T\boldsymbol{H}^H + \Gamma_{N,N}\right)^{-1} \tag{5}$$

where $\boldsymbol{E_S} = \mathrm{E}|x_j|^2$, $\Gamma_{X,X}$ is the autocorrelation matrix of the transmitted symbols vectors $\boldsymbol{X}$, $\Gamma_{N,N}$ is the autocorrelation matrix of the noise vector $\boldsymbol{N}$.

In the full load case (N_u=L_c) and only in that case, the four equalization coefficients matrix $\mathbf{G}_{tr}$ included in $\boldsymbol{G}_r$ are diagonal matrix with the kth subcarrier equalization coefficient equal to the coefficient obtained for MMSE SD given in table 1 for $N_r=1$. In the non-full load case (N_u<L_c), the matrix $\mathbf{G}_{tr}$ are no more diagonal. In that case, the MMSE MD algorithm outperforms the MMSE SD algorithm, since the MMSE MD scheme minimises the decision error taking into account the despreading process instead of minimizing the error independently on each subcarrier.

4. SIMULATION RESULTS OVER RAYLEIGH FADING CHANNEL

For these results, frequency non-selective Rayleigh fading per subcarrier and time invariance during two MC-CDMA symbols are assumed to permit the recombination of symbols when STBC is used. Based on these assumptions and considering ideal time and frequency interleaving, the complex channel fading coefficients perfectly estimated are considered uncorrelated for each subcarrier k and mutually independent from each transmit antenna to each receive antenna. With these uncorrelation and independence hypothesis, the asymptotic performance of the studied algorithms can be obtained.

Simulation results are carried out without channel coding to evaluate MMSE MD and MMSE SD performance in the MIMO and MISO cases. The length of the spreading codes (L_c=16) is equal to the number of subcarriers N_c. Results are first compared in terms of BER performance versus E_b/N_0. The different subcarriers are supposed to be multiplied by independent non-selective Rayleigh fading perfectly estimated. It is assumed that all the users' signals are received with the same mean power. We do not take into account the power gain provided by the use of multiple antennas, *i.e.* only the spatial diversity gain is taken into account. The Matched

Filter (MF) bound is given as reference as well as the performance of the MC-CDMA system combined with MMSE SD or GMMSE in the SISO case.

4.1 Full load system

Figure 2 shows the performance of MMSE MD and MMSE SD with and without transmit and receive diversity, with full-load systems for $L_c=N_c=N_u=16$. As expected, the performance of MC-CDMA is highly improved when combined with STBC in order to exploit the transmit diversity, and when multiple receive antennas are used in order to take benefit of receive diversity. Moreover, in the MISO case (N_tN_r=2-1), the performance of 21 MMSE MD and 21 MMSE SD are the same because the equalization coefficient matrix $\mathbf{G}_{tr}$ is a diagonal matrix with the kth subcarrier equalization coefficient equal to the coefficient obtained for MMSE SD. Besides, 22 MMSE SD system (N_tN_r=2-2) is the best scheme and offers a very good performance/complexity trade-off.

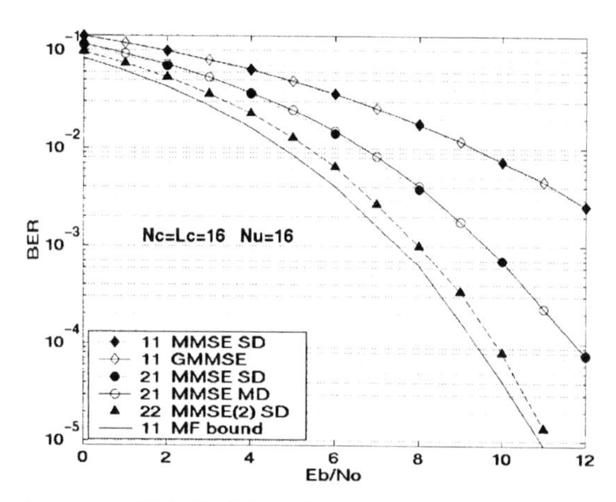

Figure 2. Performance of MMSE MD and SD over Rayleigh fading channels for SISO (N_tN_r=11), MISO (N_tN_r=21) and MIMO (N_tN_r=22) systems with N_c=L_c=16. N_u=16 (full load).

4.2 Half load system

The performance of MMSE MD and MMSE SD with and without transmit and receive diversity, with half-load systems for $L_c=N_c=16$ and $N_u=8$ are presented figure 3. With two transmit antennas and one receive antenna, 21 MMSE MD achieves a gain of roughly 1 dB compared to 21 MMSE SD at high SNR. Again, 22 MMSE SD system is the best scheme even if the gain compared to other systems is lower than in the full load case.

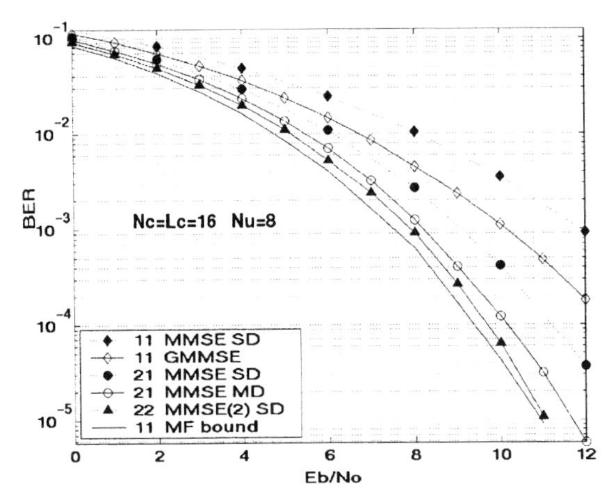

Figure 3. Performance of MMSE MD and SD over Rayleigh fading channels for SISO (N_tN_r=11), MISO (N_tN_r=21) and MIMO (N_tN_r=22) systems with $N_c=L_c=16$. $N_u=8$ (half load).

4.3 Performance versus system load

Finally, in figure 4, the performance of sub-optimal MMSE(1) and optimal MMSE(2) SD MIMO MC-CDMA are compared to MMSE MD MISO MC-CDMA and MMSE SD SISO MC-CDMA. The maximum number N_u of active users versus the required E_b/N_0 to achieve a BER=10^{-3} is given for $L_c=N_c=16$ and equal mean power signals.

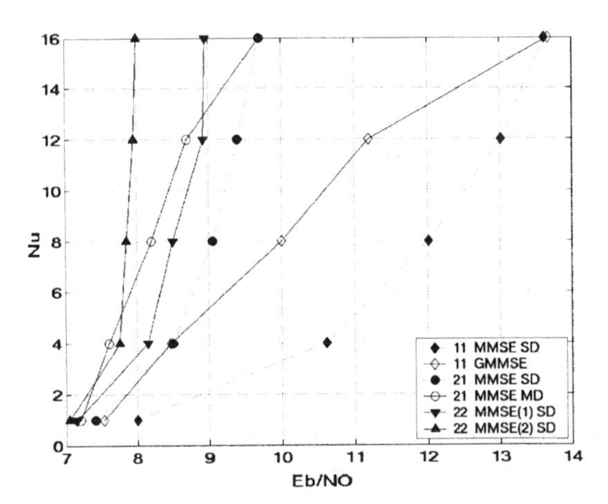

Figure 4. Number N_u of active users versus the required E_b/N_0 to achieve a BER= 10^{-3} with MMSE MD and SD over Rayleigh fading channels for SISO (N_tN_r=11), MISO (N_tN_r=21) and MIMO (N_tN_r=22) systems; $N_c=L_c=16$.

For non full load cases, the gain of MMSE MD compared to MMSE SD, which is roughly equal to 2 dB for SISO systems, decreases to less than 1 dB for MISO systems. Furthermore, the most important result is that, for any load, the low complex and new 22 MMSE(2) SD MIMO scheme outperforms all studied MD and SD MISO and SISO systems. Furthermore, the spatial diversity gain is all the more significant when the number of active users N_u is high.

5. SIMULATION RESULTS OVER METRA CHANNEL

A major characteristic of the stochastic MIMO channel model developed within the European Union IST research METRA project is that, contrary to other directional models, it does not rely on a geometrical description of the environment under study [8]. It is a complex Single-Input Single-Output (SISO) Finite Impulse Response (FIR) filter whose taps are computed so as to simulate time dispersion, fading and spatial correlation. To simulate MIMO radio channels, it has to be inserted between a parallel-to-serial and serial-to-parallel converters. Besides, the correlation properties in the spatial domain of the MIMO radio channel are obtained by the Kronecker product of two independent correlation matrices defining the correlation properties at the Base Station (BS) and Mobile Station (MS).

Table 2. Main system and MIMO channel parameters

OFDM symbol duration (μs)	3.2	Number N_c of subcarriers	64
Guard interval (μs)	0.5	Length L_c of the spreading codes	64
Center frequency (GHz)	5.2	Signal Bandwidth (MHz)	20
Channel Profile	BRAN A	Velocity (km/h)	3.6
Doppler Spectrum	Jakes	Measured coherence bandwidth (MHz)	5.8
Pattern	Omni.	Doppler oversampling factor	2
DoA azimuth	0	Elevation angle (deg)	90

Table 2 summarizes the main system and channel parameters. In the correlated MISO and MIMO cases, we consider in the BS and MS an array of 2 uniformly spaced antennas with an inter-element separation fixed to 1.5 λ and 0.4 λ respectively. Then, the envelope correlation coefficients between antennas are:

$$R_{BS} = \begin{bmatrix} 1 & 0.265 \\ 0.265 & 1 \end{bmatrix} \qquad R_{MS} = \begin{bmatrix} 1 & 0.294 \\ 0.294 & 1 \end{bmatrix}$$

which have been derived from 4x4 correlation matrices obtained through experimental measurements in real indoor scenario [8].

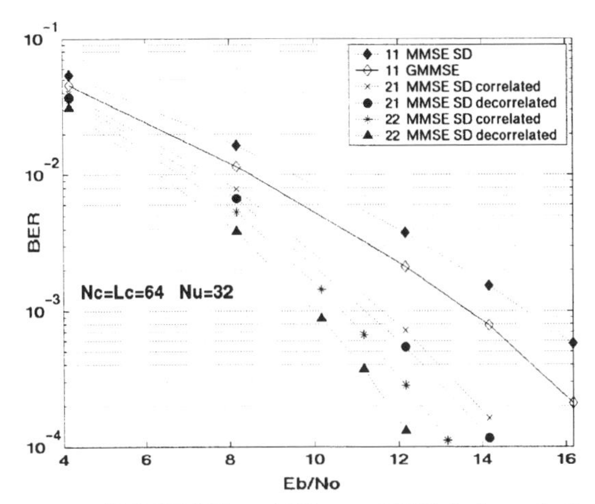

Figure 5. Performance of MMSE MD and SD over METRA channels for SISO (N_tN_r=11), MISO (N_tN_r=21) and MIMO (N_tN_r=22) systems with N_c=L_c=64; N_u=32.

Figure 5 represents for half load systems (L_c=N_c=*64* and N_u=*32*) and indoor environment the performance of MMSE MD and SD with and without transmit and receive diversity. In the SISO case (N_tN_r=1-1), MMSE MD offers, as over Rayleigh channel, a gain of nearly 2 dB (for a BER=10^{-3}) compared to MMSE SD. In the MISO (N_tN_r=2-1) and MIMO (N_tN_r=2-2) cases, the good performance of the 21 MMSE SD and especially 22 MMSE SD measured over the Rayleigh channel are confirmed when the two channels are perfectly decorrelated. Moreover, in the realistic case corresponding to correlated channels with a 1.5 λ and 0.4 λ separation between the two transmit and receive antennas respectively, the performance loss compared to the perfectly decorrelated case is less than 0.5 dB.

6. CONCLUSION

The performance of MMSE Single-user Detection (SD) and Multi-user Detection (MD) MIMO MC-CDMA systems are analysed and first compared over Rayleigh fading channels in the case of two transmit antennas and one or two receive antennas. With two transmit antennas and one receive antenna, MD outperforms SD for non-full load systems while the same performance are obtained for full load systems. Besides, with two receive antennas, 22 MMSE SD STBC MC-CDMA offers a very good performance/complexity trade-off. Finally, over the realistic METRA MIMO channel, the very good behaviour of 22 MMSE STBC MC-CDMA systems is confirmed even in the case of correlated channels.

ACKNOWLEDGEMENTS

The authors would like to express their thanks to FTR&D/DMR/REN which supports this study (contract n° 424 62 728 FT R&D).

7. REFERENCES

[1] S. Hara, R. Prasad, "Overview of multicarrier CDMA", IEEE Communications Magazine, vol.35, N°.12, pp. 126-133, December 1997.

[2] V. Tarokh, N. Seshadri, A.R. Calderbank, "Space-time codes for high data rate wireless communication : Performance criterion and code construction", IEEE Transactions on Information Theory, vol. 44, N°.2, pp. 744-765, March 1998.

[3] S.M.Alamouti, "A simple transmit diversity technique for wireless communications", IEEE Journal on Selected Areas in Commununications, vol.16, pp. 1451-1458, October 1998.

[4] V. Tarokh, H. Jafarkhani, and A.R. Calderbank, "Space-time block coding for wireless communications: Performance results", IEEE Journal on Selected Areas in Commununications, vol.17, pp. 451-460, March1999.

[5] V. Le Nir, J.M. Auffray, M. Hélard, J.F.Hélard, R. Le Gouable, "Combination of Space-Time Block Coding with MC-CDMA Technique for MIMO systems with two, three and four transmit antennas", IST mobile & Wireless Communication Summit, Aveiro, Portugal, June 2003.

[6] S. Kaiser, "Multi-carrier CDMA Radio systems - Analysis and Optimization of Detection, Decoding, and Channel Estimation", PhD. Thesis, VDI-Verlag, Fortschrittberichte VDI, Series 10, N.531,1998.

[7] J-Y. Baudais, J-F. Hélard and J. Citerne, "An improved linear MMSE detection technique for Multi- Carrier CDMA systems : comparaison and combination with interference cancellation", European Transactions on Communications, vol. 11, n°7, pp. 547-554, November-December 2000.

[8] K.I. Pedersen, J.B. Andersen, J.P. Kermoal, and P.E. Mogensen, "A stochastic Multiple-Input-Multiple-Output radio channel model for evluation of space-time coding algorithms", Proceedings of VTC 2000 Fall, pp. 893-897, Boston, United States, September 2000.

MARIAN CODREANU, MATTI LATVA-AHO

COMPARISON BETWEEN SPACE-TIME BLOCK CODING AND EIGEN-BEAMFORMING IN TDD MIMO-OFDM DOWNLINK WITH PARTIAL CSI KNOWLEDGE AT THE TX SIDE

Abstract. In this paper the standard space time block coding, the selection diversity space time block coding and eigen-beamforming techniques are compared in a block fading scenario taking account the channel estimation errors at the transmitter side. The comparison criterion is the signal to noise ratio obtained to the receiver at the output of the antenna combiner. The channel is assumed to be perfectly known at the receiver side and partially known at the transmitter side. A downlink scenario based on OFDM modulation is assumed. We are focusing on a time division duplex link, where the channel at the transmitter side can be estimated based on incoming signal. In order to compare these strategies a simple and robust list square channel estimator was introduced at transmitter side and the estimated channel was used to compute the beamformer vector and for antenna selection. The simulation results show that the eigen-beamforming strategy provide the best performance even under high channel estimation errors and even in a case of independent antenna fading.
Keywords Multiple-input multiple-output (MIMO), diversity, orthogonal frequency division multiplexing (OFDM), space time block coding (STBC), eigen-beamforming.

1. INTRODUCTION

Multiple transmit and receive antennas concepts are one of the most promising techniques for achieving high data rates communication. The high capacity of this technique has been proven in the case of independent Rayleigh flat fading channel in [1,2]. An extension of this work to the case of correlated fading is presented in [3], and the MIMO channel capacity in the case of frequency selective channel is analyzed in [4] where perfect channel state information (CSI) at the transmitter (TX) side was assumed.

By combining the MIMO techniques with OFDM modulation the frequency selective MIMO channel is turned into a set of frequency flat MIMO fading channels which can be individually processed [5]. The channel matrix can be decomposed for each subcarrier using singular value decomposition (SVD). As a result a set of orthogonal subchannels is obtained in space domain. We will call these elementary subchannels eigenmodes.

In this paper the standard STBC, the selection diversity space time block coding (SD-STBC) and eigen-beamforming techniques are compared in a block fading scenario taking account the channel estimation errors at the TX side. The comparison criterion is the signal to noise ratio (SNR) obtained to the receiver at the

K. Fazel and S. Kaiser (eds.), Multi-Carrier Spread-Spectrum, 363–370.

output of the antenna combiner. The channel is assumed to be perfectly known at the receiver side and partially known at the transmitter side.

The paper will be organized as follows: In the section II the system model for the considered signalling scheme is presented and the SNR obtained to the receiver at the output of the antenna combiner is theoretically computed. The numerical results and the conclusions are presented in section III.

Notations used in this paper are as follows. Upper and lower case boldfaces denote matrices and vectors, respectively. Superscripts $(\cdot)^{\mathrm{T}}$, $(\cdot)^{\mathrm{H}}$ stand for transpose, and Hermitian transpose respectively. Matrix $\mathbf{I}_N$ signifies identity matrix of size $N \times N$, $\|\cdot\|_2$ and $\|\cdot\|_{\mathrm{F}}$ denotes the Euclidian and Frobenius norm respectively. $\mathbf{A}(:,c)$ denotes the cth column of the matrix $\mathbf{A}$ and $\mathrm{E}\{\cdot\}$ denotes statistical expectation.

2. SYSTEM MODEL

Consider a MIMO-OFDM system with C subcarriers, T transmit and R receive antennas, denoted by (C,T,R) in the following. At each time instant l, $l = 1,\ldots,L$, the preprocessor generates a $T \times C$ dimensional matrix $\mathbf{X}(l)$. The tth, $t = 1,\ldots,T$, row of matrix $\mathbf{X}(l)$ is associated to the OFDM modulator at the transmit antenna t.

The OFDM modulator comprises IFFT transformer, cyclic prefix (CP) insertion and parallel to serial conversion. The CP is chosen to be longer than maximum excess delay of the channel to avoid the inter symbol interference. At the receiver side the CP is removed, then the resulted signals are serial to parallel converted and passed through the FFT transformers. For each time instant l, the OFDM demodulator generates an $R \times C$ dimensional matrix $\mathbf{Y}(l)$, each row in the matrix corresponding to a receiving antenna.

Assuming perfect frequency and sample clock synchronization between TX and RX, the input-output relation for OFDM modulator/demodulator chain can be written in the form:

$$\mathbf{y}_c(l) = \mathbf{H}_c(l)\mathbf{x}_c(l) + \mathbf{n}_c(l) \tag{1}$$

where $c = 1,\ldots,C$ represents the subcarrier index, $\mathbf{x}_c(l)$ and $\mathbf{y}_c(l)$ denotes the cth columns in the matrices $\mathbf{X}(l)$ and $\mathbf{Y}(l)$ respectively, $\mathbf{n}_c(l)$ represents the noise vector having the covariance matrix $\mathrm{E}\{\mathbf{n}_c(l)\mathbf{n}_c^{\mathrm{H}}(l)\} = N_0\mathbf{I}_R$ and $\mathbf{H}_c(l)$ represents the subchannel matrix at time instant l. The entry $(\mathbf{H}_c(l))_{r,t}$ represents the complex channel gain between TX antenna t and RX antenna r at subcarrier c. We will assume that the elements of $\mathbf{H}_c(l)$ are normalized to the unitary variance $\sigma_H^2 = E\left\{\left|\left(\mathbf{H}_c(l)\right)_{r,t}\right|^2\right\} = 1$. The frame length, L, is chosen to be smaller than the coherence time of the channel and in order to simplify the notation the time index, l, will be skipped in following.

In this paper, we focus our attention to downlink where we have to take into account the difficulty to use a high number of antennas at the handset. More specifically, we have considered a system with $R = 2$ antennas at receiver side and $T = 2 \div 8$ antennas transmitter side.

2.1 *Space Time Block Coding*

Space time block codes (STBC) [6,7] is a technique to map K input data symbols to a codeword spanning N consecutive symbol times. The extension to OFDM was made by encoding independently each subcarrier. In [8] it was shown that STBC decouples the MIMO channel into an equivalent SISO channel, as seen by each transmitted signal. After the receiver combiner, each of the K transmitted data symbol is individual detected based on decision variable:

$$d_{k,c} = \sqrt{\frac{P_c}{T}} \left\|\mathbf{H}_c\right\|_{\mathrm{F}}^2 s_{k,c} + w_{k,c} \tag{2}$$

where $s_{k,c}$, $k = 1,\ldots,K$ represents the complex data symbols allocated to the cth subcarrier, P_c represents the total transmitted power at c^{th} subcarrier for each time instant, l, and $w_{c,k}$ represents the effective noise at the combiner output having the variance $N_0 \left\|\mathbf{H}_c\right\|_{\mathrm{F}}^2$. The input data symbols are normalized to unitary variance. Based on (2), the instantaneous receiver SNR is given by:

$$\mathrm{SNR}_{\mathrm{STBC}} = \frac{P_c}{N_0 T} \left\|\mathbf{H}_c\right\|_{\mathrm{F}}^2 = \frac{P_c}{N_0 T} \sum_{i=1}^{\min(T,R)} \lambda_{i,c} \tag{3}$$

where $\lambda_{i,c}$ denotes the eigenvalues of $\mathbf{H}_c \mathbf{H}_c^{\mathrm{H}}$.

2.2 *Eigen-Beamforming*

If the channel is perfectly known at the transmitter side, the receive SNR can be maximized by using eigen-beamforming signalling, where only the strongest eigenmode at each subcarrier is excited. For each time instant, l, the vector $\mathbf{x}_c$ in (1) is given by:

$$\mathbf{x}_c = s_c \sqrt{P_c}\, \mathbf{v}_c \tag{4}$$

where $\mathbf{v}_c$ is the principal right singular vector of matrix $\mathbf{H}_c$. The optimum ML receiver consists in a linear combiner followed by a detector. The combiner generates the decision variables:

$$d_c = \mathbf{u}_c^{\mathrm{H}} \mathbf{y}_c = \sqrt{P_c}\, s_c \mathbf{u}_c^{\mathrm{H}} \mathbf{H}_c \mathbf{v}_c + \mathbf{u}_c^{\mathrm{H}} \mathbf{n}_c \tag{5}$$

where $\mathbf{u}_c$ is the principal left singular vector of matrix $\mathbf{H}_c$. From (5) the instantaneous receiver SNR can be expressed as:

$$\mathrm{SNR}_{\mathrm{EBMF}} = \frac{P_c}{N_0} \lambda_{M,c} \tag{6}$$

where $\lambda_{M,c}$ represents the largest eigenvalue of $\mathbf{H}_c \mathbf{H}_c^{\mathrm{H}}$.

In this paper we consider the case when the transmitter doesn't know CSI perfectly. At the transmitter side the channel is estimated based on the pilot symbols appended to the reverse frame and, inherently, the estimated CSI will have some errors which are strongly dependent on the SNR at the transmitter side, the channel estimation algorithm, number of pilots, etc. For a more realistic model, the singular vector $\mathbf{v}_c$ from the relation (4) will be replaced by its noisy estimate, $\hat{\mathbf{v}}_c$, obtained by eigen-decomposition of the estimated channel matrix $\hat{\mathbf{H}}_c$. Assuming the channel perfectly known at the receiver side, based on (5) the effective receive SNR in presence of channel estimation errors at the transmitter side is given by:

$$\mathrm{SNR}_{\mathrm{EBMF,E}} = \frac{P_c}{N_0} \lambda_{M,c} \left| \mathbf{v}_c^{\mathrm{H}} \hat{\mathbf{v}}_c \right|^2 = \frac{P_c}{N_0} \lambda_{M,c} \left| \alpha_c \right|^2 \tag{7}$$

where the term $\left|\alpha_c\right|^2$ expresses the decreasing in SNR due to imperfect knowledge of the channel at TX side.

By comparing the relations (3) and (7), the SNR gain, introduced by eigen-beamforming signalling in presence of channel estimation errors at the transmitter side, can be express as:

$$\Gamma_{\mathrm{EBMF,E}} = \frac{\mathrm{E}\left\{\mathrm{SNR}_{\mathrm{EBMF,E}}\right\}}{\mathrm{E}\left\{\mathrm{SNR}_{\mathrm{STBC}}\right\}} = \mathrm{E}\left\{ \frac{\lambda_{M,c}}{\frac{1}{T} \sum_{i=1}^{\min(T,R)} \lambda_{i,c}} \right\} \mathrm{E}\left\{\left|\alpha_c\right|^2\right\} = \Gamma_{\mathrm{EBMF}} \mathrm{E}\left\{\left|\alpha_c\right|^2\right\} \tag{7}$$

where Γ_{EBMF} represents the SNR gain with perfect CSI at the TX side and it depends on the channel eigenvalues distribution and the number of transmit antennas. The second term in (7), $\mathrm{E}\left\{|\alpha_c|^2\right\}$, represents the average amount of power which is directed in the strongest eigenmode when the noisy estimated value of beamforming vector is used instead of the optimum value. It depends on the accuracy of channel estimation at the TX side.

2.3 Selection Diversity Space Time Block Coding

In [7] was proved that a full rate orthogonal STBC exists only for two transmit antennas. Therefore, in order to obtain full rate code for $T > 2$, we have considered a signalling scheme in which an orthogonal Alamouti space time code is used in each subcarrier and the two dimensional encoded vectors are associated to the transmit antennas which provide the highest SNR at the receiver side. We will denote such signalling by selection diversity space time block coding (SD-STBC) in the following. For each subcarrier, the indexes of the selected transmit antennas are given by:

$$(i_c, j_c) = \arg\max_{\substack{i,j=1,\dots,T \\ i \neq j}} \left\{ \left\| \hat{\mathbf{H}}_c(:,i) \right\|_2^2 + \left\| \hat{\mathbf{H}}_c(:,j) \right\|_2^2 \right\} \tag{8}$$

and the SNR at the output of antenna combiner can be express as:

$$\text{SNR}_{\text{SD-STBC}} = \frac{P_c}{N_0 2} \left[\left\| \mathbf{H}_c(:,i_c) \right\|_2^2 + \left\| \mathbf{H}_c(:,j_c) \right\|_2^2 \right] \tag{9}$$

Like in eigen-beamforming signalling case, we define the receive SNR gain introduced by selective diversity as:

$$\Gamma_{\text{SD-STBC}} = \frac{\mathrm{E}\left\{\text{SNR}_{\text{SD-STBC}}\right\}}{\mathrm{E}\left\{\text{SNR}_{\text{STBC}}\right\}} \tag{10}$$

3. NUMERICAL RESULTS AND CONCLUSIONS

The numerical results presented in this section were found by Monte Carlo simulations using the stochastic MIMO channel model presented in [9] and averaging over 10 000 independent channel realizations for each simulation point. The two channel types presented in [9] were simulated: a low correlated channel (LCC) corresponding to an indoor radio link and a high correlated channel (HCC) which corresponds to an outdoor link with the base station antenna located above surrounding scatterers. There was simulated also an independent antennas channel (IAC) which represents the most disadvantageous case for eigen-beamforming transmission. The number of subcarriers in OFDM system was $C = 64$ and the cyclic prefix length was $N_g = 16$ samples. In order to allow direct comparison in terms of number of antennas and channel spatial correlation, the power delay profile was the same for all simulated channels types. It was taken from an exponential JTC model [10] with the normalized path powers $P_\tau = [-1.02 \quad -2.62 \quad -5.72 \quad -11.12 \quad -18.12 \quad -22.7]$ (dB) at delay instances $\tau = [0 \quad 1 \quad 3 \quad 7 \quad 11 \quad 14]$ normalized by sampling period.

The channel is estimated at the TX side using a low complex LS channel estimator based on an orthogonal pilot symbol appended at the end of the uplink frame. It contains N_g pilot tones equidistantly distributed in OFDM symbol which are encoded using Alamouti space time orthogonal code. The channel is first estimated at the subcarriers corresponding to pilot tones and then the estimated channel is interpolated in frequency domain.

Figure 1 shows the dependence of $\Gamma_{\text{EBMF,E}}$ versus SNR at TX side for $R = 2$ and $T = \{2, 4, 8\}$ antennas.

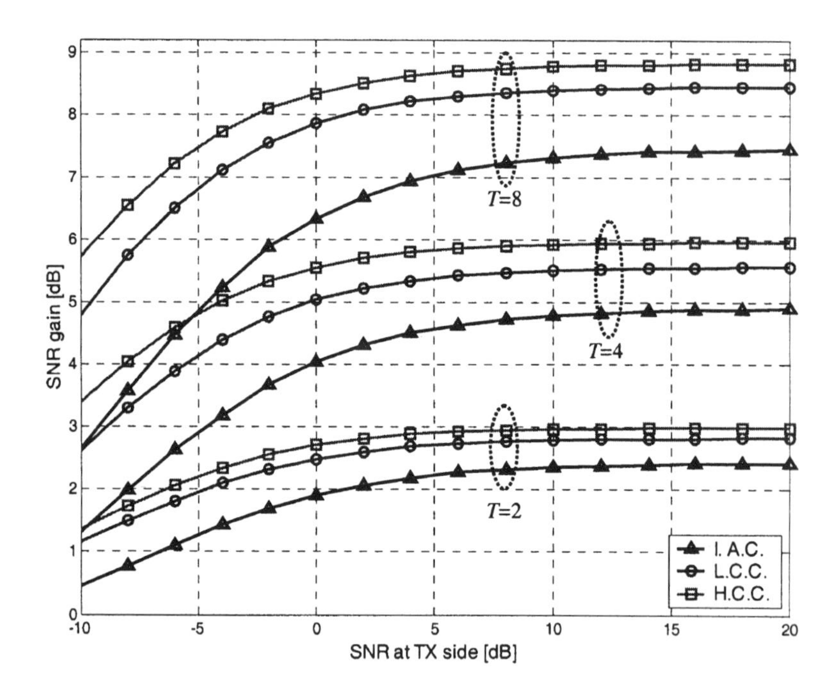

Figure 1. The SNR gain for eigen-beamforming using $R = 2$ and $T = \{2, 4, 8\}$ antennas

As shown in the figure, considerable SNR gain is obtain for all the simulated channel types and for all SNR values which present practical interest. With increasing in SNR the channel estimation become more accuracy and $\Gamma_{\text{EBMF,E}}$ converges asymptotically to the value obtained in case of perfect knowledge of CSI, i.e., Γ_{EBMF}. For values of SNR larger than 10 dB almost all the power is directed in the strongest eigenmode ($E\{|\alpha_c|^2\} \to 1$) and the loss in SNR due channel estimation errors can be neglected.

The maximum SNR gain was obtained for HCC where, due to high correlation between transmit antennas, the channel matrix has only one non-negligible singular value [9] and for such a case $\Gamma_{\text{EBMF}} \to T$.

The Figure 2 shows SNR gain obtained by using SD-STBC with imperfect CSI at the TX side. As we can see from figure, just small SNR gain is provided comparing with eigen-beamforming signalling scheme. The worst performance is obtained when the transmit antennas are highly correlated and consequently the choosing of two most efficient antennas dos not help too much.

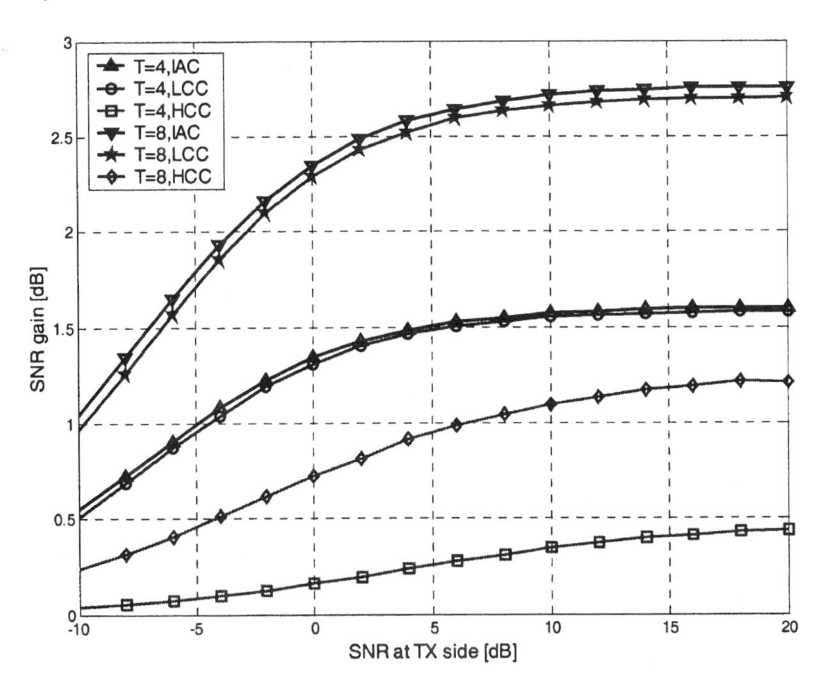

Figure 2. The SNR gain for SD-STBC with $R = 2$ and $T = \{4, 8\}$ antennas

The key feature of SD-STBC is that this method preserves the extremely low encoder/decoder complexity and it provides also some SNR gain in case of low correlated channels.

The most computational complex operation for eigen-beamforming transmission is SVD of the channel matrix. However, in case of $R = 2$, an analytical expression for singular eigenvectors can be easy obtained and it can be used for SVD at the base station. The SVD at the handset can be avoided by directly estimation of the eigenvectors $\mathbf{u}_c$ using the eigenvectors $\hat{\mathbf{v}}_c$ as pilots symbols.

Based on the simulation results we conclude that the eigen-beamforming technique provide better performance than SD-STBC and standard STBC, even under high channel estimation errors at TX side and even in the case of independent antenna fading. By appending only one pilot symbol for channel estimation to the reverse frame, this strategy can be successfully used for increase the data rate and the quality of link, in a TDD transmission.

4. A CKNOWLEDGEMENTS

This research was supported by the National Technology Agency of Finland, Nokia, the Finnish Defence Forces, Elektrobit and Instrumentointi.

5. AFILIATIONS

Centre for Wireless Communications (CWC), Tutkijantie 2 E, P.O. Box 4500 FIN-90014 University of Oulu, Finland, Tel. +358 8 553 1011, Fax. +358 8 553 2845

6. REFERENCES

[1] E. Telatar, "Capacity of multi-antenna Gaussian channels" European Transactions on Telecommunications, vol. 10, no. 6, pp. 585–595, Nov.–Dec. 1999.

[2] Gerard Foschini and Gans, "On limits of wireless communications in a fading environment when using multiple antennas", Wireless Personal Communications, vol. 6, no.3, 1998.

[3] Chen-Nee Chuah, David N. C. Tse, Joseph M. Kahn and Reinaldo A. Valenzuela, "Capacity Scaling in MIMO Wireless Systems Under Correlated Fading" IEEE Trans. on Information Theory, vol. 48, no. 3, pp. 637–650, March 2002.

[4] Gregory G. Raleigh and John M. Cioffi, "Spatio-temporal coding for wireless communication", IEEE Transactions on Communications, vol. 46, no. 3, pp.357-366, March 1998.

[5] Helmut Bolcskei, David Gesbert, Arogyaswami J. Paulraj, "On the Capacity of OFDM-Based Spatial Multiplexing Systems, IEEE Trans on Communications, vol. 50, no. 2, pp. 225, Feb 2002.

[6] Siavash M. Alamouti, "A Simple Transmitter Diversity Technique for Wireless Communications", IEEE Journal on Selected Areas in Communications, vol. 16, no. 8, pp. 1451–1458, Oct. 1998.

[7] Vahid Tarokh, Hamid Jafarkhani and A. R. Calderbank, "Space–Time Block Codes from Orthogonal Designs" , IEEE Trans on Information Theory, vol. 45, no. 5, pp. 1456, July 1999.

[8] S. Sandhu and A. Paulraj, "Space-Time Block Codes: A Capacity Perspective", IEEE Communications Letters, vol. 4, no. 12, pp.384–386, Dec. 2000.

[9] J. P. Kermoal, L. Schumacher; K. I. Pedersen, P.E. Mogensen and F. Frederiksen, " A stochastic MIMO radio channel model with experimental validation ", IEEE Journal on Selected Areas in Communications, vol. 20, no. 6, pp. 1211–1226, August 2002.

[10] K. Pahlavan, Allan Levesque and Allen Levesque, "Wireless Information Networks", Wiley-Interscience, New York, 1995.

Space-Time Block Coding for OFDM-MIMO Systems for Fourth Generation: Performance Results

M. Jankiraman, Ramjee Prasad *Senior Members IEEE*
CPK, Aalborg University, A5-207, 9000 Aalborg, Denmark,
janki, prasad@cpk.auc.dk

Abstract

This paper presents the encoding and decoding algorithms for the various codes and provides simulation results demonstrating their performance in an OFDM-MIMO environment. The system performance is examined in the presence of typical OFDM problems like timing errors, frequency errors and oscillator phase noise. It is shown that using multiple transmit/receive antennas and space-time block coding along with OFDM provides remarkable performance at the expense of almost no extra processing making possible the realization of high bit rates, desirable in a fourth generation system.

I INTRODUCTION

A significant challenging aspect of modern and future wireless communication is the provision of high-capacity, low cost and reliable multimedia communication to carry bursty packet traffic as well as voice and delay constrained streaming traffic. Towards this end, in terms of technology, the most challenging aspects are high bit rate, of the order of 100 Mbps or higher, and high packet repetition rate. The latter is especially necessary because it can give rise to poor reception even though our system can handle high bit rates. Furthermore, the wireless channel suffers attenuation due to destructive addition of multipaths in the propagation channel and due to interference from other users. The channel statistic is often Rayleigh, which makes it difficult for the receiver to reliably determine the transmitted signal unless some less attenuated replica of the signal is provided to the receiver. This technique is called diversity, which can be provided using temporal, frequency, polarization and spatial resources. However, it is not often that wireless channels are significantly time variant or highly frequency selective. This forces system engineers to consider the possibility of spatial diversity by deploying multiple antennas at both the transmitter and receiver.

Fourth generation frequencies have yet to be defined. However the 5 GHz ISM band is one of the possible candidates. Hence, this paper was built around simulations conducted with IEEE 802.11a system. In order to realise the high bit rates expected (100 Mbps or higher, though even this has not yet been defined), the following scheme is proposed as shown in fig 1.

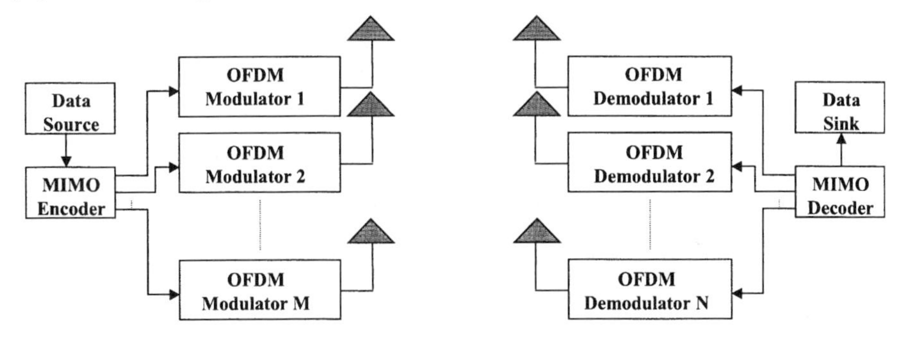

Figure 1: OFDM-MIMO System

K. Fazel and S. Kaiser (eds.), Multi-Carrier Spread-Spectrum, 371–378.

The OFDM systems should be optimized for 54 Mbps mode or in the worst-case 24 Mbps depending upon channel conditions. Normally, a 2x2 or 2x3 system should suffice in order to achieve 100 Mbps throughput. We will need to use a link adaptive bit loading since fourth generation systems are all about OFDM, MIMO and adaptive modulation and coding that allows different data rates to be assigned to different users depending upon their channel conditions.

This design is motivated by the growing demand for broadband Internet access. The challenge lies in achieving a comparable quality to that of wireline technologies. The target frequency band is 2-5 GHz due to the favourable propagation characteristics and low RF equipment cost. The broadband channel is typically non-LOS channel. Only recently has transmit diversity come into focus as a method of combating detrimental effects in wireless fading channels because of its relative simplicity of implementation and feasibility of having multiple antennas at the base station. The key objectives of such a high performance system are reliable transmission, high data rates (>100 Mbps) and high spectrum efficiency (>4 b/s/Hz). These system requirements can be met by the combination of two powerful technologies in the physical layer design: multi-input and multi-output (MIMO) antennas and orthogonal frequency division multiplexing (OFDM) modulation. Henceforth such a system will be referred to as OFDM-MIMO.

We document the performance of space-time block codes[1], which provide a new alternative over Rayleigh fading channels using OFDM-MIMO systems. Data from an IEEE 802.11a WLAN system is encoded using a space-time block code and the encoded data is split into m streams, which are simultaneously transmitted using m transmit antennas. The received signal at each of the n receive antennas is a linear superposition of the m transmitted signals perturbed by noise. Each of these n outputs is fed into an IEEE 802.11a receiver and then given to a space-time decoder. Maximum likelihood decoding is achieved in a simple way through decoupling of the signals transmitted from different antennas rather than joint detection. This uses the orthogonal structure of the space-time block code and gives a maximum likelihood decoding algorithm which is based only on linear processing at the receiver.

Multiple antennas at the transmitter and receiver provide diversity in a fading environment. By employing multiple antennas, multiple spatial channels are created, and it is unlikely all the channels will fade simultaneously. There are many techniques to implement this [1]-[3],[5]-[8]. Alamouti's approach to this problem [1] is a popular one and is expanded by Tarokh et al [2] to cover different types of space-time block codes. The purpose of this paper is to evaluate the performance of these codes in an OFDM-MIMO fixed wireless LAN environment based on IEEE 802.11a systems, which are likely to be workhorses of the future. The simulation is conducted for a Rayleigh channel with a maximum delay spread of 75 ns. The outline of this paper is as follows. In Section II we introduce the reader to the simulation scenario. In Section III we provide simulation results demonstrating the performance of these codes in this environment. Finally, Section IV presents our conclusions and final comments.

II OFDM WLAN Overview

The key resource management problems in packet-based multimedia systems are related to data rates and delay constraints exhibiting very large peak to average capacity demands. Because of the peculiarity of the packet systems, where the delay is constrained in the statistical sense, the time to exchange the resources is very important.

The main advantages of OFDM include simplified equalization, and simple implementation of modulation/multiplexing. OFDM in particular is an efficient high data rate transmission technique for digital transmission on fading channels. It has a huge potential, including flexibility and adaptivity, to improve on spectral efficiency, frequency diversity, and provide for robustness against frequency-selective fading. By using OFDM in the physical layer in the context of an IP wireless architecture and division of the traffic in various levels, transmission bit rates can be increased and bandwidth utilization improved in accordance with the demands of emerging multimedia applications. Use of OFDM in the physical layer and assignment of different levels of protection will also give possibilities to improve on the packet error rate.

Currently there are three approved world WLAN standards that utilize OFDM for their physical layer specifications, namely, IEEE 802.11a (Europe and North America), HiperLAN/2 (Europe and North America) and Multimedia Mobile Access Communication System (MMAC), Japan. Each standard offers data rates ranging from 6 Mbps to 54 Mbps in the 5-GHz band. The major difference between the standards is the medium access control (MAC) used by each. IEEE 802.11a uses a distributed MAC based on CSMA/CA. adopted for the system discussed here. The physical layers of all the standards are very similar and are based on an OFDM baseband modulation. The key parameters of the IEEE 802.11a standard and applied to our simulations can be found in [4]. In this paper we assume an OFDM WLAN IEEE 802.11a system. This is the basic system. We then consider the implications of such a system interfaced with multiple antennas and using space-time block coding. The overall system analysis is done via simulation. An existing OFDM WLAN simulator was used [4]. Effort was made to seamlessly integrate the space-time block codes to the simulator without losing the flavor or the accuracy of the basic simulator. This was done with a view to analyzing the performance of an OFDM-MIMO system using space-time block coding.

The system block diagram for simulating such a system interfaced to a space-time encoder/decoder is shown in fig(2):

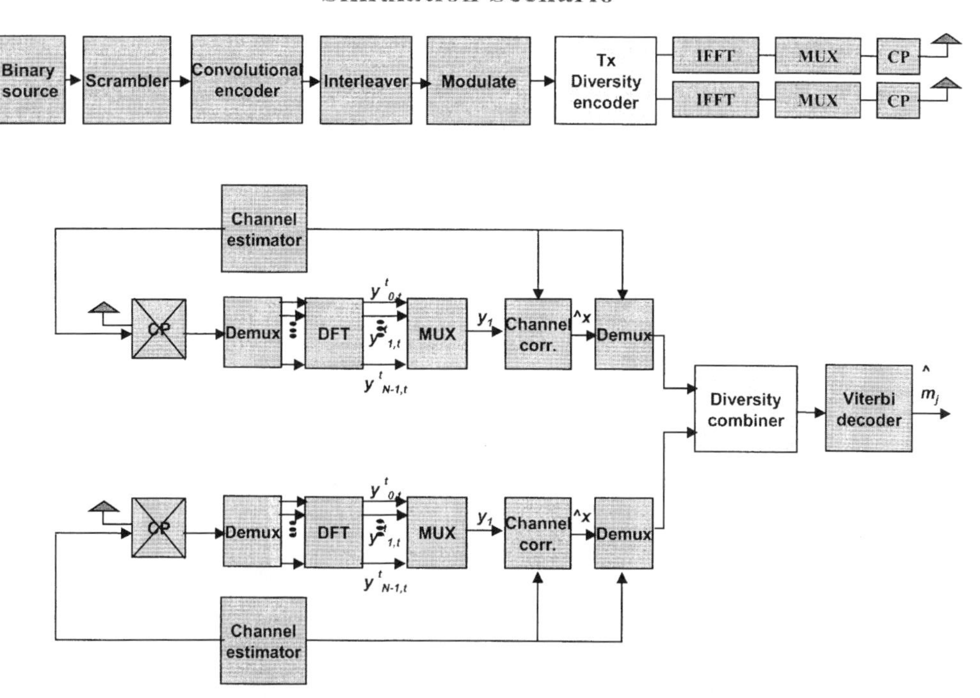

Figure 2: Simulation of OFDM IEEE 802.11a system with Space-Time Block Coding.

The input information is portrayed as a binary source. This is then scrambled with a 127 length PN sequence which is employed to prevent a continuous stream of 1's or 0's as such streams will make packet detection problems error prone. This is then fed into a convolution encoder. In our simulations we have employed a ½ rate coder with constrain length of 7. This is then interleaved up to one symbol depth and then modulated. The modulator is actually a mapper, which maps the bitstream into the chosen type of modulation, which in our case is 16 QAM. We then feed this information stream to a transmission diversity encoder. This is a space-time encoder. The figure shows the structure for a 2 x 2 encoder. The receiver constitutes the reverse process.

III SIMULATIONS

The system uses 16 QAM with ½ rate convolution code with interleaving that can account for as much as 5-dB gain and supports 24 Mbps. In our simulations the receiver was also offset in timing.

From fig.3 we note that for an SNR of 5 dB the BER for a Rayleigh channel without employing synchronization is 0.35 and 0.01 for an OFDM WLAN and OFDM-MIMO, respectively. Use of 2x2 diversity gives us an SNR improvement of 7 dB at a BER of 0.01 confirming the viability of this mode as being superior to OFDM WLAN alone. Employing synchronization algorithms and channel estimation in a Rayleigh channel for the OFDM WLAN system gives a degradation of nearly 2.5 dB for most of the BERs. A wider bandwidth and longer symbol duration will give sufficient improvement or we need to resort to OFDM-MIMO. Employing synchronization algorithms and channel estimation while using diversity gives a degradation compared to the perfect case of 4.5 dB at a BER of 0.01 but still it improves the SNR by 6 dB compared to the OFDM WLAN case.

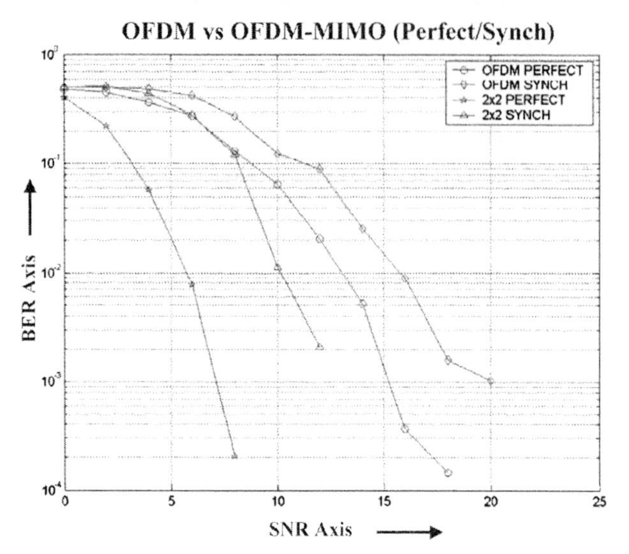

Figure 3: BER for Rayleigh channels with and without synchronization algorithms (OFDM WLAN and OFDM-MIMO).

We now examine the performance of such a system in the presence of phase noise with a corner frequency of 10 Hz and with various combinations of space-time coding. In doing so, we shall evaluate bit error rates.

Phase noise basically introduces a random phase variation that is common to all subcarriers. This causes ICI because the subcarrier spacing undergoes a shift in the frequency domain. If we examine fig(4) in the presence of phase noise we get some very interesting results. The estimated result assuming no phase noise and the result in the presence of phase noise are identical in all graphs. This is not surprising since it is channel estimate that is used to correct the phase drift due to phase noise and moreover, the phase noise introduced in these simulations with a corner frequency of 10 Hz is low. In order to estimate the channel we use the information from the training symbols for each packet. We note that the gap between the perfect estimate and the BER in the presence of phase noise varies depending upon the M x N configuration where M is the transmit diversity and N the receive diversity. This degradation is not attributable to phase noise but rather to the poor condition of the channel transfer matrix caused due to poor channel estimation. This leads to estimation errors causing a sharp fall in performance as compared to the case of perfect estimation. For example, the condition number is around 3.3 on an average for a 2x2 system as compared to 6.6 for a 4x4 system in these simulations. Hence, the gap is more for a 4x4 system as compared to a 2x2 system. Generally if the condition number is around 2 or less, this gap is not much. A 4x1system in fig 4(b) shows a gap of only typically 1.8 dB (condition number: 2.2) as compared to the 4x4 system fig 4(c) which has a gap of 6.3 dB (condition number: 6.6). However, the reader must be cautioned that this does not mean that because of the lower condition number, a 4x1 system has a better capacity than a 4x4 system. Obviously 4x4 is superior as is evident from the curves in fig 4(b) and (c). The condition numbers merely give us a qualitative idea as to how a chosen system performs with respect to *its* perfect estimation curve.

MIMO-OFDM in fact helps to better condition the channel transfer matrix because it takes advantage of the prevailing multipaths, which give rise to frequency selectivity. This leads to better rank distribution across the tones causing the channel matrix to be better conditioned [9]. Furthermore, OFDM systems exploit frequency selectivity caused due to multipaths by utilizing frequency diversity across frequency tones and thereby reducing fading margins. A well-conditioned channel transfer matrix leads to higher capacity or to put it in another way, better BER for the same SNR. Depending upon the chosen transmit/receive diversity we are constrained to live with a sub-optimal performance. In the plots in fig (4), the estimation curves define the limiting performance of the system. If the phase noise is excessive, there will be a gap between the phase noise curve and the estimation curve. There is not much we can do to improve the conditioning of the channel transfer matrix except for going in for better channel estimation techniques than the one used presently in IEEE 802.11a systems. It is pointed out that in these simulations, we have assumed that all the channels are independent. Even otherwise the system will tolerate correlation levels of 0.5 easily [9]. The channel estimation in these simulations is based on an algorithm given in [4], wherein the estimator estimates the channel based on only the two training symbols available with each packet. This makes for speed, but is not so accurate. However, it is sufficient for a SISO system. We can also investigate concatenated coding based on Turbo Codes to make up for the performance shortfall. If we apply synchronization algorithms (based on short and long training sequences) and packet detection algorithms (based on the preamble structure), we note that the OFDM-MIMO performance does not get any worse. This implies that in reality this will be our working curve for a 4x4 system (see fig 4(c)). There is also another interesting observation in fig 4(c). We note that the curves tend to flatten out for high SNRs. These curves have a high diversity order (4 x 4 = 16). Such a value makes the system "sensitive" to minor channel variations i.e. the "flat fading" assumption is no longer valid and the curve flattening occurs due to multipaths between the sub-carriers. This is caused due to poor sub-carrier resolution. If we increase the number of sub-carriers, this aspect will get remedied. Hence, we conclude that increasing of diversity order goes in step with sub-carrier resolution i.e. beyond a point we cannot simply increase the diversity order of a OFDM-MIMO system without increasing the number of sub-carriers.

This implies that a 2x2 or 2x3 system will always be sub-optimal as compared to its "perfect estimation" curve. These curves show that regardless of how low the phase noise is (even if there is no phase noise) the limiting problem is the channel transfer matrix and channel estimation. This is *irrespective of the quality of phase noise in the carrier oscillators*. Poor phase noise will only worsen the situation due to ICI.

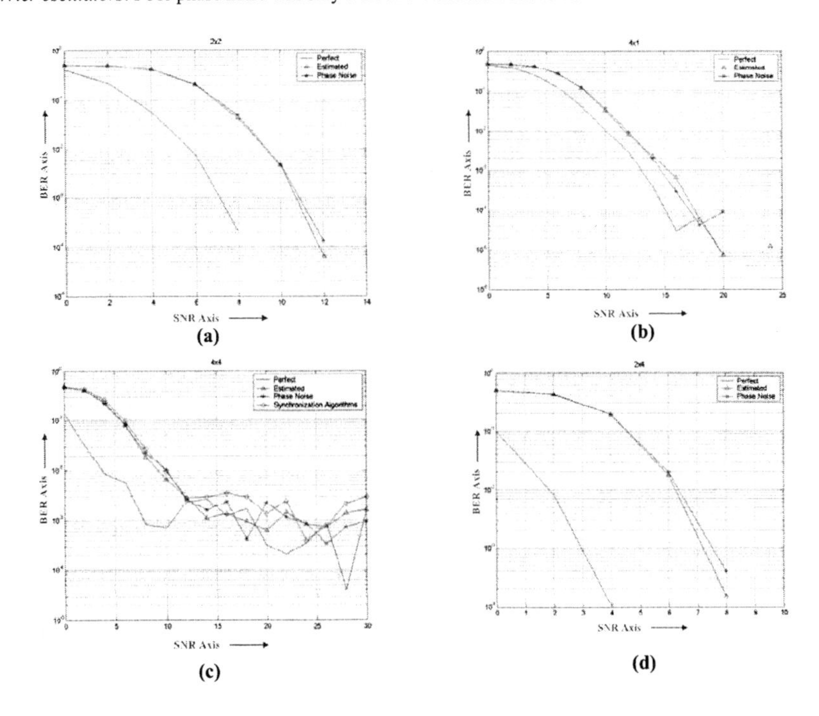

Figure 4: Simulation results of 2x2, 4x1, 2x4 and 4x4 Space-Time Coded Systems

Symbol timing is the task of finding the precise moment of when an individual OFDM symbol starts and ends. A WLAN receiver cannot afford to spend any time beyond the preamble to find the symbol timing. With respect to symbol timing offset OFDM is more robust and symbol timing may vary over an interval equal to the guard time without causing ICI or ISI. In practice, there is always some variation in the timing estimate, although a maximum multipath tolerance is achieved when the symbol timing is fixed to the first sample of an OFDM symbol. Fig. 5 gives the simulation results for different values of the symbol timing offset in terms of PER. Again the OFDM-MIMO is superior to an OFDM-only system by improving the SNR by almost 7 dB. However, these results do not hold in the case of a timing error of 20 samples (Rx= - 20) because of the strong ISI occurring at this value, which is much bigger compared to the guard interval (16 samples).

The fixed wireless channel Doppler spectrum differs from the mobile channel Doppler spectrum [9]. For fixed wireless channels, it was found that the Doppler is in the 0.1 – 2 Hz range and has close to exponential or rounded shape. The effects of such Doppler did not have any adverse effect on our simulations.

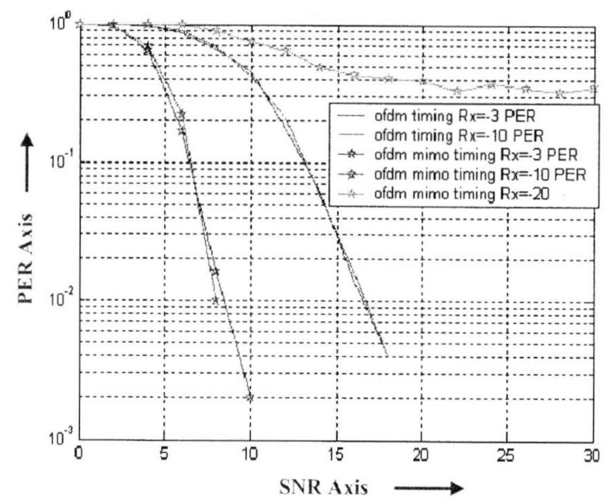

Figure 5: PER comparison for Rayleigh channels in the presence of symbol timing shift (OFDM-WLAN versus OFDM-MIMO).

IV CONCLUSIONS

In this paper we described the simulation strategy for a fourth generation fixed WLAN system. We discussed the need for OFDM–MIMO in order to achieve high throughputs and the implications of such a design when the OFDM system is IEEE 802.11a. We also examined the effects of carrier phase noise and brought out the need to develop algorithms yielding better channel estimates based only on the training symbols carried by each packet, since this is a packet transmission system. The simulation results show that despite our best efforts OFDM-MIMO systems will always operate sub-optimally as compared to the case when they operate using perfect channel estimates *irrespective of the quality of the carrier oscillator phase noise.* We further demonstrated a method for qualitative analysis of performance of a chosen OFDM-MIMO system, so that we can clearly see the potential for further improvement, by comparing the estimated curve with the perfect estimate curve for that system. The important point to note here is that we can realize the fourth generation goal utilizing CDMA-OFDM-MIMO using the existing IEEE 802.11a system and *within the same bandwidth* of 20 MHz.

V ACKNOWLEDGEMENT

Many thanks to John Terry and Juha Heiskala of Nokia Research Center, Dallas, Texas for their invaluable help with the simulations.

REFERENCE

[1] S.M. Alamouti, "A Simple Transmitter Diversity Scheme For Wireless Communications", *IEEE J. Select. Areas Commun.*, vol 16,pp. 1451-1458, Oct 1998.

[2]. Vahid Tarokh et al, "Space-Time Block Coding For Wireless Communications: Performance Results", *IEEE Journal of Selected areas in Communications*, Vol 17,No.3, March 1999

[3] N. Seshadri and J.H. Winters, " Two signalling schemes for improving the error performance of frequency-division-duplex (FDD) transmission systems using transmitter antenna diversity", *Int. J. Wireless Inform. Networks*, vol. 1, no. 1, 1994.

[4] OFDM Wireless LANs: A Theoretical and Practical Guide, Juha Heiskala and John Terry, Ph.D., SAMS 2002.

[5] G.J.Foschini, Jr., "Layered space-time architecture for wireless communication in a fading environment when using multi-element antennas", *Bell Labs Tech. J.*, pp 41-59, Autumn 1996.

[6] V.Tarokh, N.Seshadri and A.R. Calderbank, "Space-time codes for high data rate wireless communication: Performance Analysis and code construction", *IEEE Trans. Inform.Theory*, vol. 44,no. 2, pp 744-765, Mar. 1998.

[7] V.Tarokh, H.Jafarkhani and A.R. Calderbank, "Space-Time Block Codes from Orthogonal Designs", *IEEE Trans. Inform. Theory*, vol. 45, No. 5, July 1999.

[8] D. Agrawal, V.Tarokh, A. Naguib and N.Seshadri, "Space-Time Coded OFDM for High Data-Rate Wireless Communication over Wideband Channels", *Proc. IEEE VTC,* pp 2232-2236, 1998.

[9] H.Sampath, Shilpa Talwar, Jose Tellado , Vinko Erceg and A.J. Paulraj, " A Fourth-Generation MIMO-OFDM Broadband Wireless System: Design, Performance and Field Trial Results", *IEEE Communications Magazine*, pp 143-149, September 2002.

M. VEHKAPERÄ, D. TUJKOVIC, Z. LI & M. JUNTTI

SPACE-FREQUENCY CODING AND SIGNAL PROCESSING FOR DOWNLINK MIMO MC-CDMA

Abstract. Future wireless communication systems will require high data-rates in multiuser environments. Multicarrier code-division multiple-access (MC-CDMA) has emerged as an attractive technique for broadband communications, offering efficient utilization of bandwidth and robust performance in multipath channels. In this paper, coded downlink MC-CDMA is combined with the concept of LST architectures. Suboptimal receiver with soft co-antenna interference (CAI) cancellation is utilized and the performance of the system is evaluated in frequency-flat and frequency-selective fading channels. The results demonstrate that by using multiple-input multiple-output (MIMO) antenna system with coded LST transmission and iterative detection and decoding (IDD) at the receiver, the throughput and error rate performance can be significantly improved over the conventional single-input single-output (SISO) system.

1. INTRODUCTION

The efficient combination of orthogonal frequency division multiplexing (OFDM) with code-division multiple-access (CDMA), known as multicarrier CDMA (MC-CDMA), has gained significant attention as a promising technique for broadband wireless systems [1]. Among other attractive features, MC-CDMA offers low equalization complexity and robust performance in frequency-selective fading channels, given that accurate subcarrier synchronization is guaranteed and cyclic prefix (CP) is longer than the maximum expected channel delay spread.

Assuming a rich scattering environment and uncorrelated transmit antennas, the results in information theory demonstrate a potentially dramatic increase in link capacity of wireless multiple-input multiple-output (MIMO) channels [2, 3]. When the transmitter has no channel state information (CSI), space-time coding (STC) [4] is commonly seen as an optimal signalling strategy, designed to jointly encorrelate symbols across spatial and temporal domains. However, the maximum likelihood (ML) decoding complexity of STC for large antenna arrays is extensive and specific codes have to be designed for different transmit antenna setups. The coded layered space-time (LST) architectures [5, 6], on the other hand, offer pragmatic methods to increase the data-rates by using multi-element antenna arrays (MEAs) and off-the-shelf encoding and decoding blocks. Due to complexity restrictions, sub-optimal receiver based on spatial filtering is commonly utilized in the detection of layered transmission. The cost of this approach is severely degraded error rate performance compared to the optimal ML detection and decoding, if no further means of signal processing are applied at the receiver to guarantee higher orders of receive diversity.

K. Fazel and S. Kaiser (eds.), Multi-Carrier Spread-Spectrum, 379–386.

In this paper, we extend the work of [7] and investigate the downlink performance of a MIMO MC-CDMA system that employs parallel concatenated convolutional codes (PCCCs) [8] or recently proposed space-time turbo coded modulation (STTuCM) [9], applied in space-frequency domain, with spatial multiplexing at the transmitter and optional iterative detection and decoding (IDD) at the receiver. Performance of the underlying system is evaluated in both frequency-flat and frequency-selective fading channels by using Monte Carlo computer simulations.

2. SYSTEM MODEL

A single-cell downlink MC-CDMA system with N_c subcarriers and K users, all having the same spreading factor G is considered. We assume N transmit antennas at the base station, M receive antennas at the mobile terminal, $M \geq N$, and that the CP is longer than the maximum expected delay spread of the channel. Thus, the system has an equivalent frequency domain presentation, as discussed in [10].

Two channel coding strategies for conventional LST architectures can be identified: horizontal and vertical [11]. Regardless of the channel coding method, we assume that each transmit antenna $n = 1, 2, \ldots, N$ is assigned with P coded modulated symbols x from modulation alphabet $\mathcal{M}$ so that the structure of the STC is preserved. A pseudo-random interleaver is used to shuffle the positions of the coded symbols in frequency domain. Before the transmission, for all users $k = 1, 2, \ldots, K$ the coded symbols are multiplied with user specific signature sequences to form a chip-level space-frequency transmit matrix $\boldsymbol{Z}_k \in \mathbb{C}^{NG \times P}$, i.e.,

$$\begin{aligned} \boldsymbol{X}_k &= [\boldsymbol{x}_{k,1}\ \boldsymbol{x}_{k,2}\ \cdots\ \boldsymbol{x}_{k,P}] = [\boldsymbol{x}_{1,k}^{\mathrm{T}}\ \boldsymbol{x}_{2,k}^{\mathrm{T}}\ \cdots\ \boldsymbol{x}_{N,k}^{\mathrm{T}}]^{\mathrm{T}} \in \mathcal{M}^{N \times P} \\ \boldsymbol{Z}_k &= [\boldsymbol{Z}_{k,1}\ \boldsymbol{Z}_{k,2}\ \cdots\ \boldsymbol{Z}_{k,P}], \quad \boldsymbol{Z}_{k,p} = \boldsymbol{x}_{k,p} \otimes \boldsymbol{s}_k \in \mathbb{C}^{NG}, \end{aligned} \tag{1}$$

where $\boldsymbol{X}_k$ represents a space-frequency symbol matrix, $\otimes$ denotes the Kronecker product, $\boldsymbol{s}_k = [s_k^1, s_k^2, \ldots, s_k^G]^{\mathrm{T}} \in \mathcal{S}^G$ is the signature sequence of user k and $\mathcal{S}$ denotes the chip alphabet. We assume that orthogonal spreading codes are used, and, therefore, assign the same set of spreading sequences for each transmit antenna.

Assume that the whole coded frame, that is, $N_f = GP/N_c$ consecutive OFDM symbols, is received. We can now express the frequency domain received signal in terms of code symbol intervals $p = 1, 2, \ldots, P$

$$\boldsymbol{r}_p = \boldsymbol{C}_p \boldsymbol{x}_p + \boldsymbol{\eta}_p, \tag{2}$$

where, after omitting p for notational simplicity, received signal vector, transmit symbol vector and noise vector are defined, respectively, as

$$\begin{aligned} \boldsymbol{r} &= [r_1^1, \ldots, r_1^G, \ldots, r_M^1, \ldots, r_M^G]^{\mathrm{T}} \in \mathbb{C}^{MG} \\ \boldsymbol{x} &= [x_{1,1}, \ldots, x_{1,N}, \ldots, x_{K,1}, \ldots, x_{K,N}]^{\mathrm{T}} \in \mathcal{M}^{NK} \\ \boldsymbol{\eta} &= [\eta_1^1, \ldots, \eta_1^G, \ldots, \eta_M^1, \ldots, \eta_M^G]^{\mathrm{T}} \in \mathbb{C}^{MG}. \end{aligned} \tag{3}$$

The elements of $\boldsymbol{\eta}$ are independent and complex Gaussian with equal power real and imaginary parts, i.e., $\boldsymbol{\eta} \sim \mathcal{CN}(\boldsymbol{0}, N_0 \boldsymbol{I}_{MG})$. We define $\mathrm{E}\{\boldsymbol{x}_{k,p}\boldsymbol{x}_{k,p}^{\mathrm{H}}\} = E_s \boldsymbol{I}_N, \forall k =$

$1, 2, \ldots, K, p = 1, 2, \ldots, P$, since the considered front-end detectors do not have information about the STC structure. The total radiated energy from the base station is held constant regardless of the number of transmit antennas, and, thus, the signal-to-noise ratio (SNR) per receive antenna is given by $\gamma = NE_s/N_0$. Finally, the combined channel-spreading matrix $\boldsymbol{C}$ can be presented as

$$\begin{aligned} \boldsymbol{C} &= [\boldsymbol{c}_{1,1} \cdots \boldsymbol{c}_{1,N} \cdots \boldsymbol{c}_{K,1} \cdots \boldsymbol{c}_{K,N}] \in \mathbb{C}^{MG \times NK} \\ \boldsymbol{c}_{k,n} &= [\boldsymbol{c}_{1,n,k}^{\mathrm{T}}, \boldsymbol{c}_{2,n,k}^{\mathrm{T}}, \ldots, \boldsymbol{c}_{M,n,k}^{\mathrm{T}}]^{\mathrm{T}} \in \mathbb{C}^{MG} \\ \boldsymbol{c}_{m,n,k} &= [H_{m,n}^{\mathfrak{c}(p+1)} s_k^1, H_{m,n}^{\mathfrak{c}(p+2)} s_k^2, \ldots, H_{m,n}^{\mathfrak{c}(p+G)} s_k^G]^{\mathrm{T}} \in \mathbb{C}^G. \end{aligned} \tag{4}$$

The channel is assumed to be constant during the transmission of one coded frame so that $\mathfrak{c}(p+g) \equiv (p-1)G + g \pmod{N_{\mathfrak{c}}}$ represents the subcarrier order. The frequency domain channel coefficients are complex Gaussian with PDF $\mathcal{CN}(0,1)$, uncorrelated between all TX-RX pairs and derived from the time domain tapped delay line presentation of the channel via Fourier transform as discussed in [10].

3. RECEIVER DESIGN

The receiver structures proposed for SISO MC-CDMA systems cannot be immediately used in the considered MIMO system since the non-orthogonal transmissions from different antennas cannot be directly separated before combining. In this section, the receiver design for MIMO MC-CDMA utilizing LST architectures is addressed.

3.1 *Symbol-Level Joint Space-Frequency MMSE Detector*

As shown in [7], a symbol level space-frequency minimum mean squared error (SF-MMSE) detector for a MIMO MC-CDMA system can be written as

$$\boldsymbol{W} = \left(\boldsymbol{C}\boldsymbol{C}^{\mathrm{H}} + N_0\boldsymbol{I}\right)^{-1}\boldsymbol{C} \in \mathbb{C}^{MG \times NK}, \tag{5}$$

where $\boldsymbol{W}$ is a matrix filter that jointly estimates the coded symbols from all transmit antennas. The derivation of (5) assumes perfect CSI at the receiver, symbol energy of unity and uncorrelated noise with the transmitted signals and fading processes.

3.2 *Iterative Symbol-Level MMSE Based Receiver*

Already the simplified LST scheme by Foschini *et al.* [6] proposed successive spatial filtering and serial interference cancellation (SIC) for the reception of LST transmission. In [6], the problem was, however, treated only in the single-user narrowband case. Combining now the same ideas with the SF-MMSE detector, an iterative symbol-level receiver for MIMO MC-CDMA is described in this section.

Assuming that no SIC iterations are done, we can decompose (2) as

$$\boldsymbol{r} = \underbrace{\boldsymbol{C}_{k,j}\,\boldsymbol{x}_{k,j}}_{\text{desired}} + \underbrace{\widetilde{\boldsymbol{C}}_{k,j}\,\tilde{\boldsymbol{x}}_{k,j}}_{\text{self-CAI}} + \underbrace{\widetilde{\boldsymbol{C}}\,\tilde{\boldsymbol{x}}}_{\text{MAI}} + \underbrace{\boldsymbol{\eta}}_{\text{noise}}, \tag{6}$$

where $\boldsymbol{r}_{k,j} \triangleq \boldsymbol{C}_{k,j}\boldsymbol{x}_{k,j}$ represents the desired received signal of user k from layer j, the second right hand side (RHS) denotes the self co-antenna interference (CAI) from layers $j' \neq j$ and the impact of the rest of the users is included in the third RHS term. As we consider synchronous downlink transmission with orthogonal spreading codes, the multiple-access interference (MAI) term is negligible in a frequency-flat fading channel. In case of frequency-selective fading, the MAI term imposes degradation in the system performance, although by using a MMSE-based detector the MAI can be efficiently mitigated in most scenarios. The self-CAI, on the other hand, has severe impact on the system error rate performance in all environments, unless it is mitigated by spatial filtering and/or interference cancellation. It is to be noted that as the SIC process advances, an additional interference term emerges into (6) from the wrong decisions made for the past layers.

In [6], it was proved that in order to minimize the burst error probability, the receiver has to select at each stage of detection the layer that provides the best post-detection SNR. In case of LST MC-CDMA, we have to maximize the post-detection SNR at the output of the SF-MMSE detector, and, thus, over all chips and J_0 sub-layers (antennas within the layer). For the vertically coded case, selection at each stage $j = 1, 2, \ldots, J$ is done on a symbol-by-symbol basis, whereas for horizontal coding the best layer is selected on average over all coded symbols $p = 1, 2, \ldots, P$. Omitting explicit user indexing and defining group-wise index sets

$$\mathcal{J} = \bigcup_{j=1}^{J} \mathcal{J}_j, \qquad \mathcal{Q}(j) = \bigcup_{j'=1}^{j} \mathcal{Q}_{j'} \tag{7}$$

where

$$\begin{aligned} \mathcal{J}_j &= \{(j-1)J_0 + 1, (j-1)J_0 + 2, \ldots, jJ_0\}, & \forall j = 1, 2, \ldots, J \\ \mathcal{Q}_{j'} &= \{q_{(j'-1)J_0+1}, q_{(j'-1)J_0+2}, \ldots, q_{j'J_0}\}, & \forall j' = 1, 2, \ldots, J, \end{aligned} \tag{8}$$

so that $\mathcal{Q}(j)$ is a group-wise permuted subset of $\mathcal{J}$ with requirement $\mathcal{Q}(j) = \emptyset$, $\forall j \leq 0$, we can write the set of undetected layers as $\widetilde{\mathcal{Q}}(j) = \mathcal{J} \setminus \mathcal{Q}(j)$. For horizontally coded case, at each detection step $j = 1, 2, \ldots, J$ the receiver selects the sub-layers

$$\mathcal{Q}_j = \{q_{(j-1)J_0+1}, \ldots, q_{jJ_0}\} = \arg \min_{\{q'_1, \ldots, q'_{J_0}\} \in \widetilde{\mathcal{Q}}(j-1)} \sum_{j'=1}^{J_0} \sum_{p=1}^{P} \| \, [\boldsymbol{W}]_{q'_{j'}} \, \|^2, \tag{9}$$

for detection, where $[\boldsymbol{W}]_j$ denotes the row $(k-1)J_0 + j$ of matrix $\boldsymbol{W}$, corresponding to jth layer of the desired user k. Since the MAI term is constant for all layers, without a performance loss, we can concentrate on the part of $\boldsymbol{W}$ that corresponds to the user of interest. After detection, the soft estimates $\hat{\mathbf{x}}_k = \boldsymbol{W}_k^{\mathrm{H}} \boldsymbol{r}$ are fed to decoder, that uses maximum *a posteriori* (MAP) algorithm to calculate bit log-likelihoods, as discussed in [12]. As in [7], the "Euclidean metric" for the edge transition of interest is taken to be

$$\log \{p\,[\hat{\mathbf{x}}_{k,j} | \, \boldsymbol{b}_{k,j}]\} = \frac{1}{N_0} \left\| \mathbf{x}_{k,j} - \hat{\mathbf{x}}_{k,j} \right\|^2. \tag{10}$$

After decoding, the interference induced by the detected layer is recreated and its effect from $\boldsymbol{r}$ is removed. Prior to proceeding to the detection step $j+1$ and calculating (5) again, columns corresponding to the sub-layers $\mathcal{Q}_j$ are removed from the channel-spreading matrix $\boldsymbol{C}$, $\forall p = 1, 2, \ldots, P$.

3.3 PIC with Iterative Detection and Decoding

In the conventional SIC receiver, the first detected layer usually determines the error rate performance of the system [11]. In an ideal case, when IDD with parallel interference cancellation (PIC) is used, maximum receive diversity order is achieved after the first detection round. However, strong channel coding is required to provide reliable CAI cancellation and increased receive diversity. Also, if hard IC is used, the error propagation may destroy most of the benefits of the IDD receiver. We consider now the IDD with the non-linear SF-MMSE detector (6), assume normalized symbol energy $E_s = 1$ and omit the explicit indexing of p for simplicity of notation.

When the receiver enters the IDD phase, the MMSE detector has to be updated so that for the desired user k the mean squared error minimization is based on [13]

$$(\boldsymbol{W}_{k,j}, \boldsymbol{\Psi}_{k,j}) = \arg \min_{(\boldsymbol{W},\boldsymbol{\Psi})} \mathrm{E}\left\{ \left\| \boldsymbol{x}_{k,j} - \boldsymbol{W}^{\mathrm{H}}\boldsymbol{r} - \boldsymbol{\Psi} \right\|^2 \right\}, \tag{11}$$

where $\boldsymbol{W}_{k,j} \in \mathbb{C}^{MG \times J_0}$ and $\boldsymbol{\Psi}_{k,j} \in \mathbb{C}^{J_0}$ represent the matrix filter and self-CAI cancellation term for the user of interest, respectively. Following [13], by differentiating with respect to $\boldsymbol{\Psi}$ the solution for the self-CAI cancellation term is found to be

$$\boldsymbol{\Psi}_{k,j} = \boldsymbol{W}_{k,j}^{\mathrm{H}} \widetilde{\boldsymbol{C}}_{k,j} \mathrm{E}\{\tilde{\boldsymbol{x}}_{k,j}\}, \tag{12}$$

where $\widetilde{\boldsymbol{C}}_{k,j}$ corresponds to the channel-spreading matrix of the second RHS term in (6). Symbol estimates of the jth layer at the output of SF-MMSE detector are then

$$\hat{\boldsymbol{x}}_{k,j} = \boldsymbol{W}_{k,j}^{\mathrm{H}} \left[\boldsymbol{r} - \widetilde{\boldsymbol{C}}_{k,j}\, \mathrm{E}\{\tilde{\boldsymbol{x}}_{k,j}\}\right], \tag{13}$$

in which the filter coefficients, taking into account MAI are given by

$$\boldsymbol{W}_{k,j} = (\boldsymbol{B} + \boldsymbol{V} + \boldsymbol{M} + N_0\boldsymbol{I})^{-1}\, \boldsymbol{C}_{k,j}, \tag{14}$$

where

$$\begin{aligned} \boldsymbol{B} &= \boldsymbol{C}_{k,j}\boldsymbol{C}_{k,j}^{\mathrm{H}} \\ \boldsymbol{V} &= \widetilde{\boldsymbol{C}}_{k,j}\left[\boldsymbol{I}_{N-J_0} - \mathrm{diag}(\mathrm{E}\{\tilde{\boldsymbol{x}}_{k,j}\}\mathrm{E}\{\tilde{\boldsymbol{x}}_{k,j}\}^{\mathrm{H}})\right] \widetilde{\boldsymbol{C}}_{k,j}^{\mathrm{H}} \\ \boldsymbol{M} &= \widetilde{\boldsymbol{C}}\widetilde{\boldsymbol{C}}^{\mathrm{H}}. \end{aligned} \tag{15}$$

The derivation of (14) assumed that transmitted symbols for different users are independent and $\mathrm{E}\{\tilde{\boldsymbol{x}}_{k,j}\tilde{\boldsymbol{x}}_{k,j}^{\mathrm{H}}\} = \boldsymbol{I}_{J_0}$. The term $\boldsymbol{V}$ represents "co-variance" of the symbol expectations $\mathrm{E}\{\tilde{\boldsymbol{x}}_{k,j}\} \in \mathbb{C}^{J_0 \times J_0}$ so that when interference estimates are correct with high probability $\boldsymbol{V} \to \boldsymbol{0}$. The symbol expectations of the encoder outputs can be found as discussed in [14], i.e.,

$$\begin{aligned} \mathrm{E}\{x\} &= \sum_{x'_l \in \mathcal{M}} x'_l \cdot \mathrm{P}\{x = x'_l\} \\ &= \left(\frac{1}{2}\right)^{\kappa} \sum_{x'_l \in \mathcal{M}} x'_l \cdot \prod_{i=1}^{\kappa} \left(1 + \tilde{b}_{i,l} \tanh\left(\log \mathrm{P}\{c_i\}/2\right)\right), \end{aligned} \tag{16}$$

where $\tilde{b}_{i,l} = 2b_{i,l} - 1$, bits $\{b_{i,l}\}_{i=1}^{\kappa}$ form the constellation point x'_l, κ is the number of encoder output bits per antenna at symbol instant $p = 1, 2, \ldots, P$, and $\log \mathrm{P}\{c_i\}$ are log-likelihoods of the coded bits corresponding to x, calculated by soft-input soft-output (SfISfO) Log-APP decoder as proposed in [12].

4. SIMULATION RESULTS

A coherent downlink communication system with $N_{\mathrm{c}} = P = 512$, frame size of $N_{\mathrm{F}} = GPN$ symbols and $G = K \in \{1, 8\}$ is considered. Maximum path delay of the 24-path NTT DoCoMo rectangular channel [15] is assumed to be shorter than the CP so that ISI is negligible. Flat fading channel at each subcarrier and mutually uncorrelated transmit-receive antenna pairs are assumed. The SNR per receive antenna is $\gamma = NE_{\mathrm{s}}/N_0$, as discussed in Section 2. Rate-$1/2$ punctured PCCCs or STTuCM is used so that with QPSK modulation, a spectral efficiency of N bps/Hz is achieved.

Fig. 1 presents the frame error rate (FER) performance of a MC-CDMA system with turbo codes or STTuCM in a frequency-selective fading channel [7] for reference.

FER performance of a MC-CDMA system employing different LST schemes and receiver with soft PIC iterations is shown in Figs. 2–4. For flat fading case only single user cases are shown for clarity, but due to negligible residual MAI term, $G = K = 8$ cases exhibit practically the same performance. The results show a significant improvement in the FER performance of the system when soft PIC iterations are used at the receiver. Most dramatic performance enhancement is shown by the flat fading case where the increased receive diversity appears as the steeper slopes of the IDD curves. The multiuser cases in frequency-selective channel show less drastic improvement due to limiting MAI term, but a gain of roughly 1 dB is nevertheless achieved from one or two PIC iterations. The horizontally coded case exhibits more reliable CAI cancellation at the first detection round, whereas for vertical coding, the interleaving gain of the turbo codes is higher due to longer block lengths. As a result, the performance differences between horizontally and vertically coded cases are small.

The simulation results for layered STTuCM showed very close the same performance as turbo coded cases and are therefore omitted. The reason for bad performance is that in (10) the output of the MMSE based receiver was not properly modeled, causing almost total loss of transmit diversity in the STC cases. By taking into account the residual channel and MAI-plus-noise after SF-MMSE detector, the transmit diversity can be efficiently regained. This issue is discussed in detail in [16].

5. CONCLUSIONS

In this paper, a downlink MIMO MC-CDMA system utilizing single- or multi-antenna channel coding, LST architectures and iterative receiver with serial and parallel interference cancellation was considered. The throughput and error rate performance of proposed 4×4 MIMO MC-CDMA system was shown to be significant when compared to single-antenna [15], as well as space-frequency coded 2×2 MC-CDMA [7]. Further improvements can be achieved by using proper modeling of the MMSE-detector output, as is addressed in detail in [16].

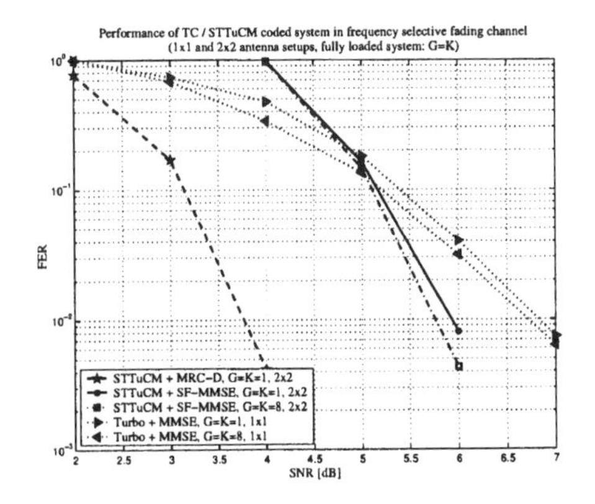

Figure 1. MC-CDMA with turbo codes or STTuCM in a frequency-selective fading channel.

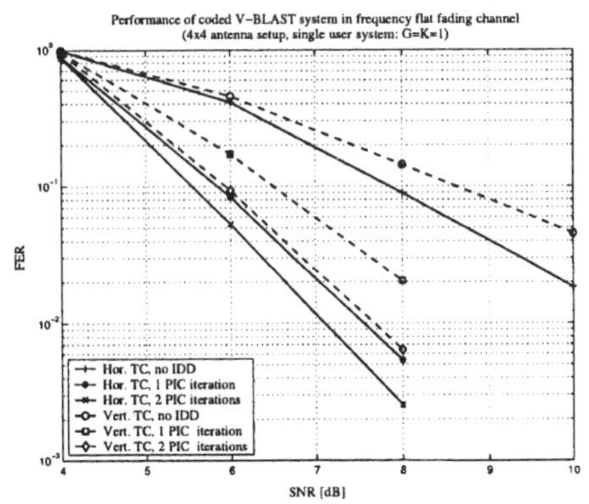

Figure 2. MIMO MC-CDMA with turbo coded LST transmission in a flat fading 4×4 channel.

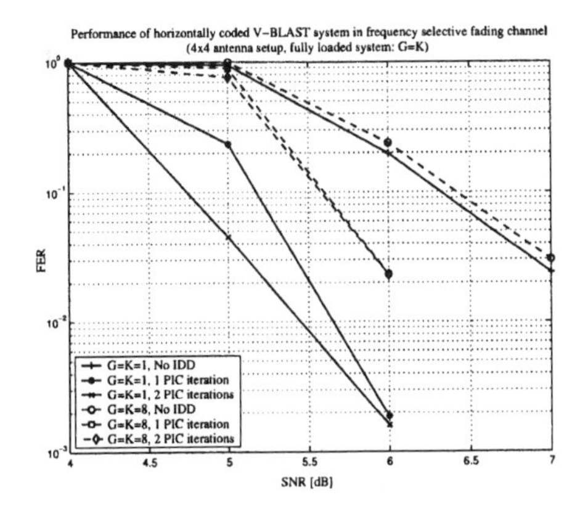

Figure 3. MIMO MC-CDMA with horizontally turbo coded LST transmission in a frequency-selective fading 4×4 channel.

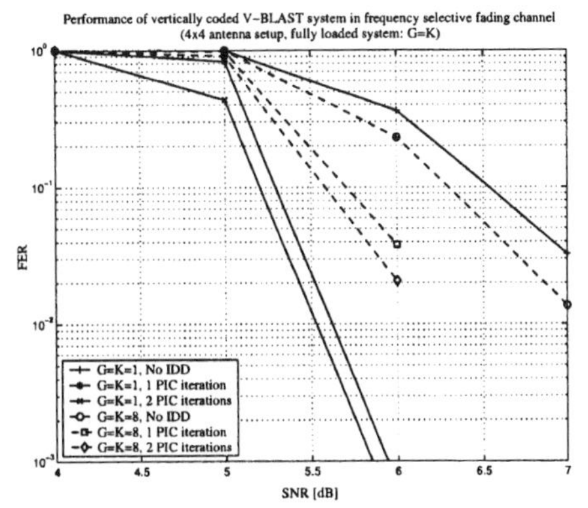

Figure 4. MIMO MC-CDMA with vertically turbo coded LST transmission in a frequency-selective fading 4×4 channel.

6. ACKNOWLEDGEMENTS

This research was supported by the National Technology Agency of Finland, Nokia, the Finnish Defence Forces, Elektrobit and Instrumentointi.

7. AFFILIATIONS

Mikko Vehkaperä, Djordje Tujkovic, Zexian Li and Markku Juntti are with
Centre for Wireless Communications (CWC)
Tutkijantie 2E, P.O. Box 4500, FIN-90014
University of Oulu, Finland

8. REFERENCES

[1] S. Hara and R. Prasad, "Overview of multicarrier CDMA," *IEEE Commun. Mag.*, vol. 35, no. 12, pp. 126–133, Dec. 1997.

[2] I. E. Telatar, "Capacity of multi-antenna Gaussian channels," *European Transactions on Telecommunications*, vol. 10, no. 6, November/December 1999.

[3] G. J. Foschini and M. J. Gans, "On limits of wireless communications in a fading environment when using multiple antennas," *Wireless Personal Communications*, pp. 311–335, 1998.

[4] A. F. Naguib and R. Calderbank, "Space-time coding and signal processing for high data rate wireless communications," *Wireless Communications and Mobile Computing*, vol. 1, pp. 13–34, 2001.

[5] D. Gesbert, M. Shafi, D. Shiu, P. J. Smith, and A. Naguib, "From theory to practice: An overview of MIMO space-time coded wireless systems," *IEEE J. Select. Areas Commun.*, vol. 21, no. 3, pp. 281–302, 2003.

[6] G. J. Foschini, G. D. Golden, R. A. Valenzuela, and P. W. Wolniansky, "Simplified processing for high spectral efficiency wireless communication employing multi-element arrays," *IEEE J. Select. Areas Commun.*, vol. 17, no. 11, pp. 1841–1852, Nov. 1999.

[7] Z. Li, M. Vehkaperä, D. Tujkovic, M. Juntti, M. Latva-aho, and S. Hara, "Performance evaluation of space-frequency coded MIMO MC-CDMA system," in *Proc. Information Society Technologies (IST) Summit*, Aveiro, Portugal, June 15–18 2003.

[8] C. Berrou and A. Glavieux, "Near optimum error correcting coding and decoding: Turbo-codes," *IEEE Trans. Commun.*, vol. 44, no. 10, pp. 1261–1271, Oct. 1996.

[9] D. Tujkovic, "Space-time turbo coded modulation for wireless communication systems," Ph.D. dissertation, volume C184 of Acta Universitatis Ouluensis. University of Oulu Press, Oulu, Finland, 2003.

[10] Z. Wang and G. B. Giannakis, "Wireless multicarrier communications: Where Fourier meets Shannon," *IEEE Signal Processing Mag.*, vol. 17, no. 3, pp. 29–48, May 2000.

[11] X. Li, H. Huang, G. J. Foschini, and R. A. Valenzuela, "Effects of iterative detection and decoding on the performance of BLAST," in *Proc. GLOBECOM*, vol. 2, 2000, pp. 1061–1066.

[12] S. Benedetto, D. Divsalar, G. Montorsi, and F. Pollara, "A soft-input soft-output APP module for iterative decoding of concatenated codes," *IEEE Commun. Lett.*, vol. 1, no. 1, pp. 22–24, Jan. 1997.

[13] M. Sellathurai and S. Haykin, "TURBO-BLAST for wireless communications: Theory and experiments," *IEEE Trans. Signal Processing*, vol. 50, no. 10, pp. 2538–2546, Oct. 2002.

[14] M. Tüchler, A. C. Singer, and R. Koetter, "Minimum mean squared error equalization using *A Priori* information," *IEEE Trans. Signal Processing*, vol. 50, no. 3, pp. 673–683, Mar. 2002.

[15] H. Atarashi and M. Sawahashi, "Variable spreading factor orthogonal frequency and code division multiplexing (VSF-OFCDM)," in *Multi-Carrier Spread-Spectrum & Related Topics*, K. Fazel and S. Kaiser, Eds. Kluwer, 2002, pp. 113–122.

[16] M. Vehkaperä, D. Tujkovic, Z. Li, and M. Juntti, "Receiver design for spatially layered downlink MC-CDMA system," to be submitted to EURASIP Journal on Wireless Communications and Networking, Oct. 2003.

ADAPTIVE V-BLAST BASED BROADBAND MIMO SYSTEMS IN SPATIALLY CORRELATED CHANNELS

Augusto del Cacho, Santiago Zazo, Ivana Raos

Department of Signals, Systems and Radiocommunications
Technical University of Madrid, Spain
E-Mail: augusto@gaps.ssr.upm.es

Keywords: V-BLAST, OFDM-MIMO, adaptive loading

Abstract: In this paper an extension of the V-BLAST (Vertical Bell Labs Layered Space-Time) algorithm with adaptive rate and power allocation for broadband multiple-input multiple-output (MIMO) systems is considered in order to overcome the poor behavior of conventional V-BLAST in spatially correlated channels.

1. INTRODUCTION

Systems beyond 3G will need to provide high spectral efficiencies in order to cope with heavy multimedia traffic. Using multiple antennas at both the transmitter and the receiver is a promising technique to enhance wireless link capacity without requiring extra bandwidth allocation. Multiple-input multiple-output algorithms require flat fading channels so it seems natural to combine these systems with multi-carrier transmission dividing the frequency-selective fading channel into multiple flat fading sub-channels using the Orthogonal Frequency Division Multiplexing (OFDM).

K. Fazel and S. Kaiser (eds.), Multi-Carrier Spread-Spectrum, 387–394.

The optimum transmission scheme in broadband OFDM-MIMO systems when full knowledge of the channel is available at the transmitter is the use of the "water-pouring" solution with adaptive loading across the spatial and frequency eigenmodes obtained by the singular value decomposition (SVD) [1] of the matrix channel at each sub-carrier. This well-known solution maximizes the channel capacity.

In order to transmit in the eigenmodes of the channel, the right singular vectors at each sub-carrier and the power and bit allocation of each mode must be known by the transmitter. This may result in a very heavy overload for the feedback channel which makes this solution impractical in most cases. In Time Division Duplex (TDD) systems where reciprocity between the uplink and downlink may be applied this feedback channel may be avoided only if the number of receiving and transmitting antennas is the same.

To overcome this limitations an alternative scheme based in an adaptive extension of the V-BLAST algorithm may be used with less CSI needed at the transmitter. As it is stated in [2], using adaptive rate and power allocation in the V-BLAST approaches the eigenmode solution. From computer simulation its behavior has been analyzed and compared with the optimum solution in spatially correlated channels.

2. SYSTEM MODEL

The block structure of the proposed broadband adaptive V-BLAST system is shown in figure 1. Multi-carrier transmission allows to extend narrowband MIMO techniques since each sub-carrier behaves as a flat fading channel using a L-IDFT which may be efficiently implemented by the fast Fourier transform and the cyclic prefix. In the proposed system, frequency division multiplex is applied in the sub-carrier level asigning K sub-carriers to each user so different users do not suffer multiple access interference. Furthermore, the different users sub-carriers shall be interleaved to efficiently exploit frequency diversity.

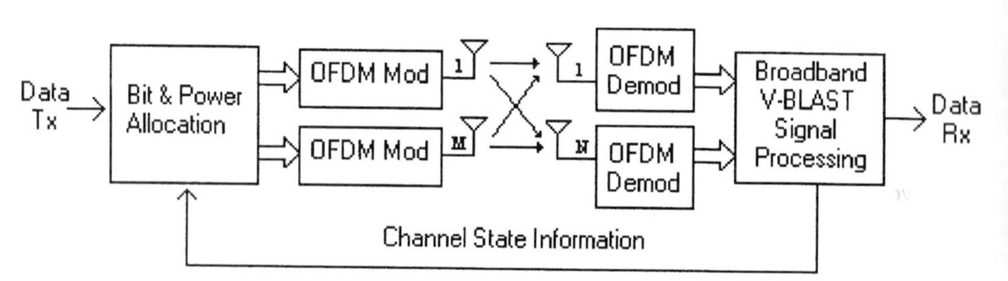

Figure 1. Broadband adaptive V-BLAST block structure.

3. LOADING IN THE V-BLAST

The V-BLAST is a spatial multiplex architecture developed by the Bell Laboratories [3] to improve wireless link capacity on narrowband rich scattering scenarios. On the transmitter side the fast data rate is demultiplexed into slower sub-channels which are transmitted by M antennas. Since the signals received by the N antennas ($N \geq M$) suffer multi-antenna interference, an iterative detector based on nulling and cancelling is performed by the receiver. If the nulling vector ($\boldsymbol{w}$) is designed following the zero forcing criterion, the different sub-channels do not suffer multi-antenna interference when the previous symbols have been correctly detected.

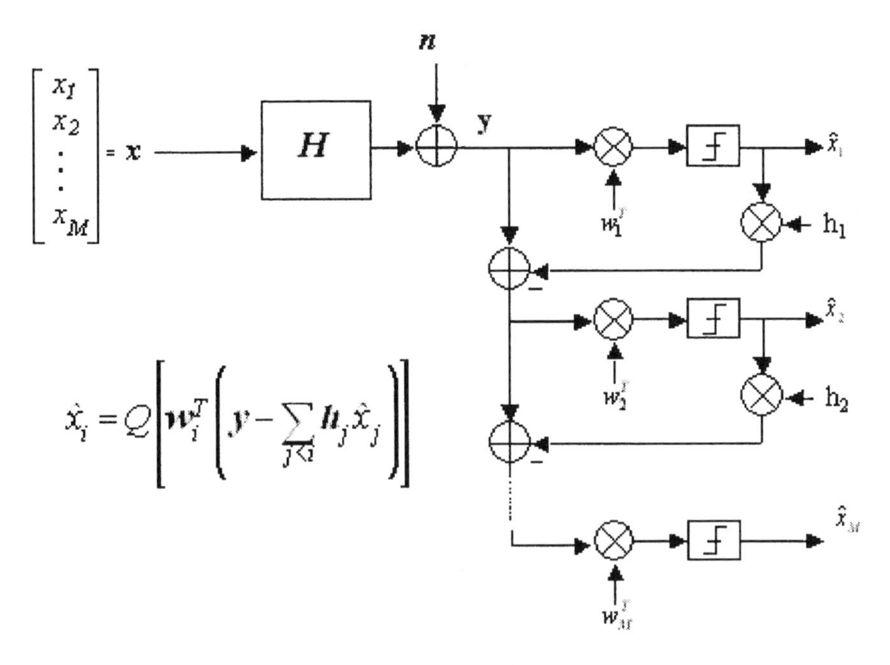

Figure 2. V-BLAST nulling and cancelling

In order to maximize the overall performance of the different sub-channels, these must be detected following a specific order to maximize the signal to noise ratio of the worst sub-channel. On the conventional algorithm all the transmitting antennas use the same modulation and power allocation.

The performance of the V-BLAST may be improved when the antennas on the transmitter side use different loading and power allocation adapting themselves to the channel state. The transmitter needs to know the loading strategy and power allocation, but in contrast to the eigen-beamforming performed by the singular value decomposition, no precoding vectors are needed so the feedback information is reduced.

The conventional ordering used when all antennas have the same modulation is not optimum for the adaptive bit loading case, unsorted detection or the one proposed in [4] obtain better results.

To simplify the loading strategy only M-QAM of 0, 2, 4 and 6 bits are considered. Once the gains of the different sub-channels are computed with the zero forcing V-BLAST algorithm at each frequency, the bits are assigned iteratively in couples to the sub-channel which needs less energy to increase its data rate until the target data rate is achieved.

4. SPATIAL CHANNEL MODEL

The V-BLAST has been designed for rich scattering scenarios where the coefficients of the channel matrix can be modelled with an uncorrelated Rayleigh distribution so independent fading among the antennas is assumed. When the channel is not "rich enough" the V-BLAST performance degrades because correlated fading takes place. Since the goal of this paper is not to develop a spatial channel model, a simple way of introducing spatial channel correlation is introducing a pair of correlation matrixes at both ends of the channel link so this will be the model used to study the adaptive extension of the algorithm.

Let $\boldsymbol{G}$ be an N x $\boldsymbol{M}$ matrix obtained from the uncorrelated Rayleigh distribution. The correlated channel matrix $\boldsymbol{H}$ can be written as

$$\boldsymbol{H} = \boldsymbol{C}_{rx}^{1/2} \boldsymbol{G} \boldsymbol{C}_{tx}^{1/2} \tag{1}$$

where $\boldsymbol{C}_{\text{rx}}$ and $\boldsymbol{C}_{\text{tx}}$ are the correlation matrixes at both ends. The correlation matrixes depend of the antenna separation and radiation pattern, and the angle spread of the received signals. The correlation matrixes of this paper will be characterised only by a correlation coefficient ($1 \geq \rho \geq 0$) following an exponential decay.

$$\langle C \rangle_{ij} = \delta^{|i-j|} \tag{2}$$

For instance, for N=3 the correlation matrix at the receiver would be

$$\boldsymbol{C}_{rx} = \begin{bmatrix} 1 & \delta & \delta^2 \\ \delta & 1 & \delta \\ \delta^2 & \delta & 1 \end{bmatrix} \tag{3}$$

More realistic correlation matrixes are being obtained from measurement campaigns by different research groups as the ones presented in [5]. In the proposed system frequency interleaving of the sub-carriers is used so uncorrelation between sub-carriers will be assumed in the model.

5. SIMULATION RESULTS

A set of computer simulations have been performed to study the behaviour of the proposed system in spatially correlated channels. The frame length is assumed to be smaller than the time coherence interval of the channel so no variation takes place inside each snapshot. No channel coding has been used so the results in the curves are uncoded bit error rates. The proposed architecture is a 2 x 2 MIMO with one sub-carrier assigned to each user with a spectral efficiency of 8 bps/Hz, which corresponds to a 16-QAM per antenna for the conventional V-BLAST. First of all figure 3 shows how the adaptive V-BLAST proposal achieves almost the same performance than the eigenmode solution in an uncorrelated channel which clearly outperforms the conventional V-BLAST with no channel knowledge at the transmitter.

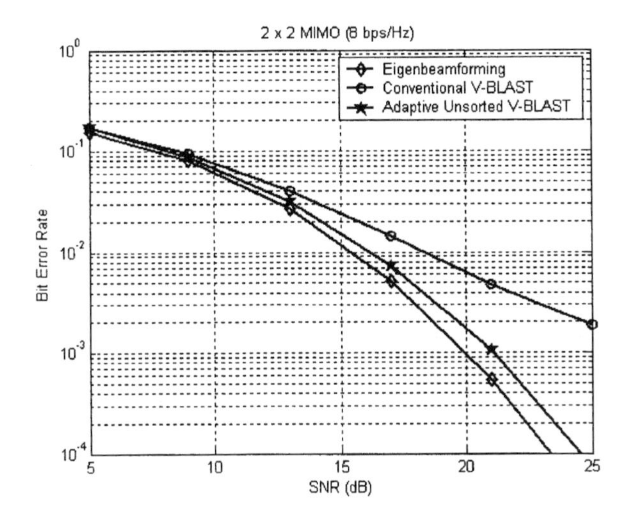

Figure 3. Performance comparison in an uncorrelated channel

At a target bit error rate of 10^{-3} the gap between the optimum solution and the adaptive V-BLAST is only of 1 dB. This gap can be reduced using the detection order proposed in [4] instead of an unsorted detection, but the cost is a higher complexity since for the unsorted detection only one QR decomposition must be computed while for the other case a set of pseudo inverse operations are needed.

Figures 4 and 5 show the spatial correlation influence at the receiver for a low and high correlated transmitter respectively. The adaptive V-BLAST seems more robust since has a lower degradation because of the different slope of the curves. As expected, the high correlated transmitter obtains a worse performance.

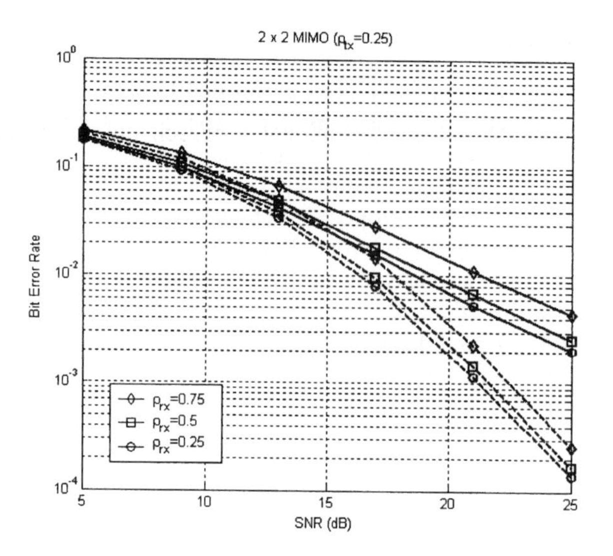

Figure 4. Spatial correlation influence with ρ_{tx}=0.25

(Solid: Conventional, Dashed: Adaptive)

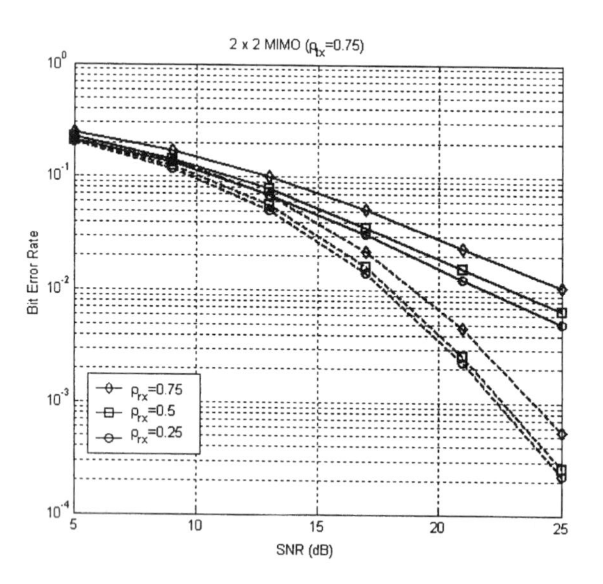

Figure 5. Spatial correlation influence with ρ_{tx}=0.75

(Solid: Conventional, Dashed: Adaptive)

When more than one sub-carrier with independent fading are assigned to each user the performance increases due to the extra degrees of freedom in the adaptive extension of the algorithm which allows exploiting frequency diversity. Adding new sub-carriers does not benefit the conventional approach unless channel coding is used. Since using more sub-carriers per user reduces the number of simultaneous users in a fixed bandwidth increasing the data rate per user, a time division multiplex can be applied to keep the same data rate per user and number of users in the system.

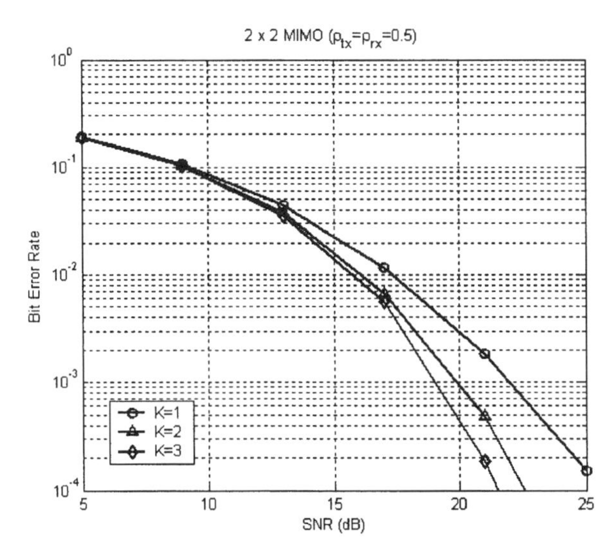

Figure 6. Adaptive V-BLAST performance with more sub-carriers per user

Further improvement can be achieved increasing the number of antennas at either the transmitter or receiver or at both of them since a higher diversity can be exploited but it has not been considered in the paper.

6. CONCLUSIONS

In this paper the adaptive extension of the V-BLAST has been considered. It has been proved how this approach which needs less feedback than the eigenmode solution reaches nearly the same performance. The robustness of the V-BLAST and its adaptive extension in spatially correlated channels has been also analyzed. The adaptive solution results less sensitive than the conventional approach because it has a steeper slope.

Finally it has been shown how adding extra degrees of freedom through assigning more sub-carriers per user achieves better performance maintaining the capacity of the system.

The influence of the time variation due to the Doppler spread must be studied in future research in order to implement the proposed system on mobile environments.

Acknowledegements

The work presented in this contribution was supported by the Matrice project (www.ist-matrice.org) and TIC200-1395-C02-02 and 07T/0032/2000.

REFERENCES

[1] B. Noble, J. W. Daniel, "Applied linear algebra", Prentice-Hall, Englewood Cliffs, New Jersey 1988

[2] S. T. Chung, A. Lozano, H. C. Huang, "Approaching Eigenmode BLAST Channel Capacity Using V-BLAST with Rate and Power Feedback", Vehicular Technology Conference, 2001 (VTC 2001 Fall), Vol 2 , pp. 915-919

[3] P. W. Wolniansky, G. J. Foschini, G. D. Golden, R. A. Valenzuela, "V-BLAST: An Architecture for Realizing Very High Data Rates Over the Rich Scattering Wireless Channels", Signal, Systems and Electronics, 1998, ISSSE 98

[4] T. Vencel, C. Windpassinger, R. F. H. Fischer, "Sorting in the V-BLAST Algorithm and Loading", Proceedings of Communication Systems and Networks (CSN2002), September 2002 pp. 304-309

[5] IST 1999-11729 Metra Project, (www.ist-metra.org)

JIANFENG LIU, ANDRE BOURDOUX,
HUGO DE MAN, MARC MOONEN

DESIGN OF THE LOW COMPLEXITY TURBO MIMO RECEIVER FOR WLAN

Abstract: In this paper, the techniques for turbo detection of single user spatial division multiplexing data streams were discussed. Simulation results and complexity study shows the tradeoff between performance and complexity. It is shown that MMSE initialised PIC receiver achieves good performance complexity tradeoff when the number of transmit antennas and constellation size is large.

1. INTRODUCTION

Results from information theory has shown a huge capacity increase when information is transmitted through a Multiple Input Multiple Output (MIMO) channel, which provides a promising solution to meet the need for high data rate transmission in the future wireless broadband communication systems. On the other hand, to tackle the multipath indoor channel, a popular low complexity modulation is Orthogonal Frequency Division Multiplexing (OFDM) [1], a multi-carrier transmission technique, which divides the available spectrum into many flat-fading subchannels. OFDM modulation scheme has been adopted in WLAN standards such as IEEE 802.11a and ETSI Hiperlan2.

The MIMO technology has been successfully combined with OFDM technology to further enhance the Quality of Service (QoS) of WLAN [2,3]. The high spectral efficiency of MIMO SDM OFDM system comes along with multiple stream interference (MSI) on each sub-carrier, which severely limits the performance of the conventional MIMO OFDM receivers. When combined with bit-interleaved coded modulation, the complexity of the direct implementation of joint detection and decoding simply is simply impossible. The turbo principle [4], i.e. iterative feedback processing, can be applied in coded MIMO SDM OFDM system to jointly detect and decode the data streams, which achieves the optimal performance in a sub-optimal way but with reduced complexity.

In this paper, the algorithm design and the complexity study of the low complexity turbo MIMO SDM OFDM receiver for WLAN is investigated. The MMSE initialized hard-input Parallel Interference Cancellation technique [5] is generalized from the OFDM-SDMA scenario to the coded MIMO SDM OFDM system with soft input. By combining both the good convergence property of the MMSE detector [6] and the low complexity of PIC receiver [7], the MMSE initialized soft PIC achieves the good performance and complexity tradeoff in the mild bit error rate region.

The organization of the paper is as follows. In Section 2, the MIMO SDM OFDM system is defined and the general turbo MIMO OFDM receiver is described.

K. Fazel and S. Kaiser (eds.), Multi-Carrier Spread-Spectrum, 395–402.

In Section 3, several different turbo detection algorithms for MIMO SDM OFDM system are described, and MMSE-PIC algorithm is proposed. Their complexity study also comes along with the algorithm description. Section 4 provides the performance comparison between the different detection algorithms. In Section 5, we summarize our results and come to the conclusions.

2. SYSTEM MODEL

Consider a multiple antenna SDM OFDM wireless communication system with M transmit and N receive antennas. Bit interleaved coded modulation scheme is adopted to extract frequency diversity and provides a good coding gain for a wide range of different fading channels. The block diagram in Fig.1 describes the data flow model in MIMO SDM OFDM transmitter and turbo receiver.

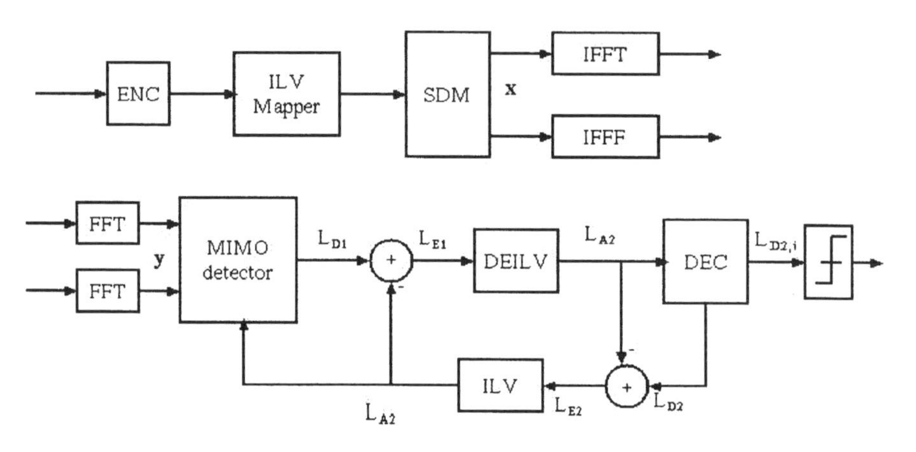

Fig.1 A general transmitter and turbo receiver diagram for MIMO SDM OFDM system

At the transmitter side, information bits from a source generator are encoded by a channel encoder, which could be either a convolutional encoder or turbo encoder. The coded stream is multiplexed into M parallel data-streams in the spatial domain. The M parallel streams are mapped to a signal constellation $\{s_1, s_2, \cdots, s_{2^{M_c}}\}$ with $s_j \sim \lfloor s_{j,1}, s_{j,2}, \cdots, s_{j,M_c} \rfloor$ M_c bit per constellation point. $s_{j,q} \in \{0,1\}$ $q = 1,2,\cdots, M_c$. Each modulated stream from the mapper goes through the conventional OFDM with K sub-carriers and cyclic prefix of length P. Each antenna m groups "its" K frequency-domain symbols $x_m[1]$ to $x_m[k]$ into one OFDM symbol, applies an inverse Fourier transform, adds the cyclic prefix of length P, filters and up-converts the signal to RF, and transmits the time-domain symbols over the channel to the receiver. The N antennas are perfectly synchronized i.e., the M OFDM symbols are perfectly "time-aligned". The MIMO receiver collects the M transmitted signals with its N antennas; each antenna sees the sum of the M convolutions added with AWGN noise. In each antenna branch, the MIMO receiver then down-converts and filters the signals, removes the cyclic prefix and performs direct Fourier transform, which yields the frequency domain received signals $y_n[k]$. The outputs of the K DFT operators are then grouped per sub-carrier; the MIMO detector takes N frequency-domain samples and outputs M frequency-domain symbols, per sub-carrier.

If the cyclic prefix is sufficiently large and with proper symbol synchronisation, the MIMO receiver observes the linear channel convolutions as cyclic. In the frequency domain, these convolutions are equivalent to a scalar multiplication on each sub-carrier k with the corresponding coefficient of the discrete Fourier transformed channel, $h_{nm}[k]$ which stands for the frequency domain channel coefficient between the transmit antenna m and receive antenna n, at the sub-carrier. Mathematically, the input-output relationship of MIMO-OFDM system on the k-th sub-carrier is:

$$\underbrace{\begin{bmatrix} y_1[k] \\ \vdots \\ y_N[k] \end{bmatrix}}_{\mathbf{y}[k]} = \underbrace{\begin{bmatrix} h_{11}[k] & \cdots & h_{1M}[k] \\ \vdots & & \vdots \\ h_{N1}[k] & \cdots & h_{NM}[k] \end{bmatrix}}_{\mathbf{H}[k]} \bullet \underbrace{\begin{bmatrix} x_1[k] \\ \vdots \\ x_M[k] \end{bmatrix}}_{\mathbf{x}[k]} + \underbrace{\begin{bmatrix} n_1[k] \\ \vdots \\ n_N[k] \end{bmatrix}}_{\mathbf{n}[k]} \tag{1}$$

where:

- **x**[k] is the column vector of the M frequency domain symbols at sub-carrier k transmitted by the MIMO transmitter, right before the Inverse DFT operation
- **y**[k]is the column vector of the N signals received the MIMO receiver antennas, right after the DFT operations
- **H**[k] is the frequency-domain channel
- **n**[k] is the additive white gaussian noise (AWGN)

The turbo MIMO OFDM receiver operates in the following way: the MIMO OFDM detector take as input the channel input x[k] at the k-th sub-carrier and the corresponding a priori information $L_{A1}[k]$ from the decoder, and outputs the refined posteriori probability $L_{D1}[k]$. After substracting the contributions of $L_{A1}[k]$, the extrinsic information $L_{E1}[k]$ is obtained. Together with the extrinsic information from the other sub-carriers and OFDM symbols, L_{E1} is de-interleaved and further refined into the posteriori probability L_{D2} for all the coded bits and $L_{D2,i}$ for the information bits. The extrinsic information from the decoder L_{E2} is obtained by substracting L_{E1} from L_{D2}. The extrinsic information L_{E2} is interleaved and provides the a priori information L_{A1} of the MIMO detectors. The above process can be executed for several times, which is called turbo processing. Finally, the posteriori probability of the information bits $L_{D2,i}$ are sliced and formed the decisions.

3. THE MIMO DETECTION ALGORITHMS AND COMPLEXITY

As all the following MIMO detector algorithms are based on per-carrier processing, the index k of the sub-carrier is omitted without confusion.

3.1 Maximum likelihood receiver

The MIMO detector performs the maximum a posteriori probability (MAP) bit detection on **y** and L_{A1} [8]. It is noticed that the number of codeword searched in the optimal MAP detection is prohibitive if the product M*M_c is large. Therefore sub-optimal algorithms such as list sphere detection algorithm try to generate a subset L of the whole codeword set, and apply MAP criterion on the subset G_{SD}. Due to the complexity of the sphere detection algorithm can only be computed in the statistical sense [5], no exact complexity analysis is conducted here, however, it is sure that its complexity is much more higher than the MMSE receiver described below.

3.2. The Minimum Mean Square Error Receiver

The MMSE receiver with a priori input is shown in Fig. 2. There are four processing steps in MMSE receiver with a piori input:

1) the computation of the 1st order and 2nd order statistics of the a priori input

$$\overline{x_i} = E\{x_i\} = \sum_{j=0}^{2^{M_c}-1}\left(s_j \cdot \prod_{q=1}^{M_c} P(s_{j,q})\right) = \sum_{j=0}^{2^{M_c}-1}\left(s_j \cdot \frac{\exp\left(\sum_{q=1}^{M_c} \tilde{s}_{j,q} L_{A1}((i-1)\cdot M_c + j)\right)}{\prod_{q=1}^{M_c}\left(1+\exp\left(L_{A1}((i-1)\cdot M_c + j)\right)\right)}\right) \quad (2)$$

with $\tilde{s}_{j,q} = 2 \cdot s_{j,q} - 1$

$$\sigma_i^2 = E\{x^2\} - \bar{x}_i^2 = \sum_{j=0}^{2^{M_c}-1}\left(s_j^2 \cdot \prod_{q=1}^{M_c} P(s_{j,q})\right) - \bar{x}_i^2 = \sum_{j=0}^{2^{M_c}-1}\left(s_j^2 \cdot \frac{\exp\left(\sum_{q=1}^{M_c} \tilde{s}_{j,q} L_{A1}((i-1)\cdot M_c + j)\right)}{\prod_{q=1}^{M_c}\left(1+\exp\left(L_{A1}((i-1)\cdot M_c + j)\right)\right)}\right) - \bar{x}_i^2 \quad (3)$$

2) the computation of the soft symbol output

$$\hat{x}_i = \mathbf{H}(:,i)^H \cdot \left(\mathbf{H}\mathbf{R}_i\mathbf{H}^H + \sigma^2 \mathbf{I}_{NxN}\right)^{-1} \cdot \left(y - \mathbf{H}\tilde{\mathbf{x}}_i\right) \quad (4)$$

3) The parameters μ_i and $\sigma_{\eta_i}^2$ in the AWGN modelling $\hat{x}_i = \mu_i x_i + \eta_i$

$$\mu_i = \mathbf{w}_i^H \cdot \mathbf{H}(:,i) \quad (5)$$

$$\sigma_{\eta_i}^2 = \left(\mu_i - \mu_i^2\right) \cdot \sigma_x^2 \quad (6)$$

4) the computation of the bit LLR output

$$L_{E1}\left((i-1)\cdot M_c + j\right) = \ln \frac{\sum_{s_p:s_{p,j}=1} \exp\left[-\frac{\left|\hat{x}_i - \mu_i s_p\right|^2}{2\sigma_{\eta_i}^2} + \sum_{\substack{q:q=1,\dots,M_c,\\ s_{p,q=1},q\neq j}} L_{A1}\left((i-1)\cdot M_c + q\right)\right]}{\sum_{s_p:s_{p,j}=0} \exp\left[-\frac{\left|\hat{x}_i - \mu_i s_p\right|^2}{2\sigma_{\eta_i}^2} + \sum_{\substack{q:q=1,\dots,M_c,\\ s_{p,q=0},q\neq j}} L_{A1}\left((i-1)\cdot M_c + q\right)\right]} \quad (7)$$

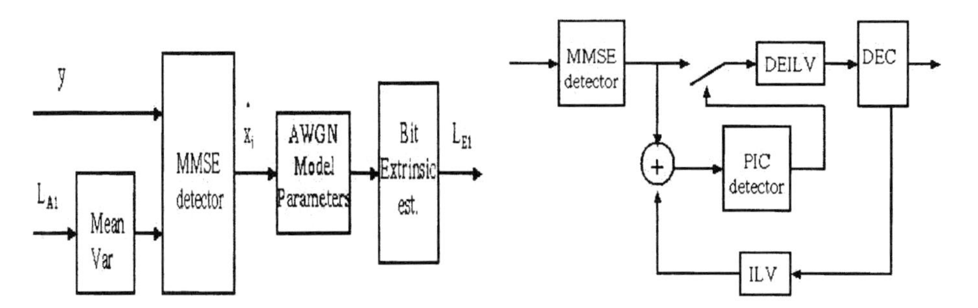

Fig.2 MIMO MMSE detector

Fig.3 MIMO MMSE-PIC detector

3.3 The MMSE-PIC Receiver

As shown above, the computation of MMSE filter coefficients has a matrix inversion as it's major component, and it has to be updated for every symbol, which makes the mmse filter a quite expensive method for implementation. However, to the best of our knowledge, pure turbo PIC receiver has problems in convergence when the constellation size large. To achieve the performance and complexity tradeoff, we propose to generalize the hard decision based MMSE initialised PIC to deal with soft inputs, and essentially we use full LLR information as the inputs to the PIC receiver instead of only a priori information. Although this will make the extrinsic information becoming more correlated, its performance approaches the turbo MMSE receiver.

Fig.3 shows the block diagram for MMSE-PIC receiver. For the 1st iteration, as there is no a priori information, the MIMO detector is a conventional MMSE receiver. For following iterations, the extrinsic information from the 1st iteration MMSE receiver and the extrinsic information from the decoder are first combined, and then are processed by a PIC receiver.

Similar to turbo MMSE algorithm, the PIC algorithm consists of 3 major steps:

1) The symbol mean calculation from equation (2), note that the a priori input for calculation of the mean is $L_{A1}+L_{MMSE}$
2) The soft symbol estimate is

$$\hat{x}_i = \mathbf{H}(:,i)^H (y - \mathbf{H}\tilde{\mathbf{x}}_i) / (\mathbf{H}(:,i)^H \mathbf{H}(:,i)) \quad (8)$$

3) The LLR extrinsic information can be obtained from the equation (7) with $\mu = 1$ and $\sigma_{\eta_i}^2 = \sigma_n^2 / (\mathbf{H}(:,i)^H \mathbf{H}(:,i))$.

As the filter coefficients in equation (8) depends only on $\mathbf{H}(:,i)^H \mathbf{H}(:,i)$, the filter coefficients calculation is only a simple scalar division, and further the it only depends on channel's i-th column, which doesn't need to be updated as long as the channel remains unchanged.

3.4 complexity study of MMSE and MMSE-PIC algorithm

Due to space limitations, only the complexity of the step 2 of MMSE and PIC algorithm are studied. Assume the channel remains unchanged for N_s OFDM symbols. The complexity is studied by counting the number of complex multiplications (MUL), additions (ADD) and data transfers (DT).

Table 1. The peration counts per coded bit for MMSE and MMSE-PIC

Receiver type	*Operation type*	*Operation counts*	*N=M=2 M_c=2*	*N=M=4 M_c=4*
MMSE	MUL	$\frac{N^3}{3}+\frac{3}{2}N^2M+2NM-\frac{1}{3}N$	18	148
	ADD	$\frac{N^3}{3}+\frac{3}{2}N^2M-\frac{1}{2}N^2+\frac{5}{3}N$	12	116
	DT	$2N^3+7N^2M+3N^2$ $+10NM+4N-3M-3$	117	772
PIC	MUL	$NM+(N+1)/N_S$	4	4
	ADD	$2N-1+(N-1)/N_S$	3	7
	DT	$3NM+6N-3+6N/N_S$	21	69

The above table shows that the calculation of filter coefficients of the PIC receiver is almost 1 order less complex than MMSE receiver for larger number of transmit antennas and constellation size.

4. SIMULATION RESULTS

Assume that a MIMO communication system has equal number of transmit and receive antennas. As defined above, it's assumed that power delay profile is defined as in ETSI channel A, and channel is quasi-static, i.e. the channel doesn't change during one transmission burst, changes independently from burst to burst. Assume that every pair of channel is perfectly uncorrelated with the others. As in the conventional communication system, we use the gray mapper. The parameter Eb, i.e., Energy per bit, used in our simulations is defined as the total received energy at all receive antennas in one receive vector symbol divided by the number of information bits transmitted in one transmit vector symbol.

Fig.4 shows that, for a 2x2 QPSK system with constraint length of 5, rate ½ convolutional codes, the different methods, ML, MMSE and MMSE-PIC achieves about same performance, there is little performance penalty when using simpler methods like MMSE or MMSE initialized PIC instead of ML method. Further, all these methods approaching the Single Input Multiple Output (SIMO) Maximum Ratio Combining bound. Finally, it's shown that the ML approach combined with turbo coding had better performance.

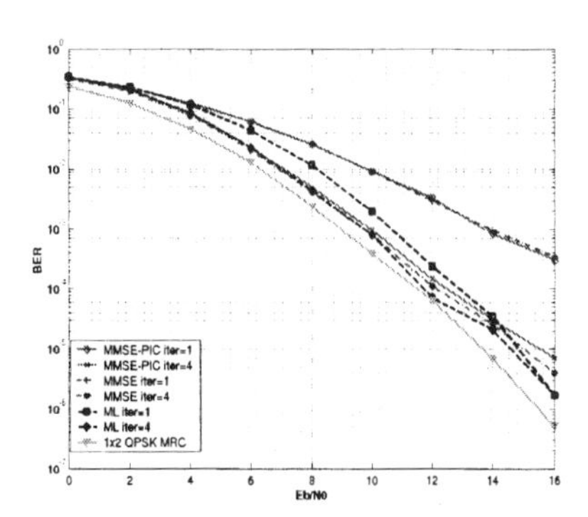

Fig.4 2x2 QPSK turbo MIMO receiver

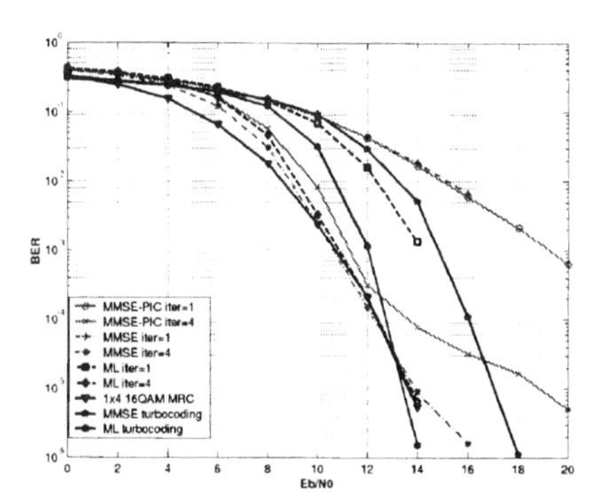

Fig.5 4x4 16-QAM turbo MIMO receive

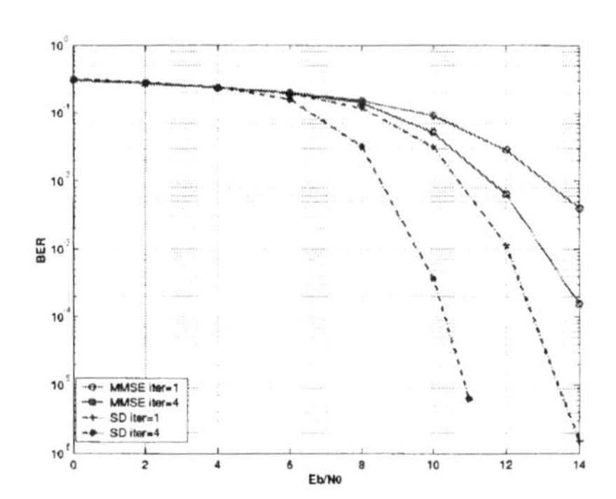

Fig.6 4x4 16-QAM turbo MIMO receiver with turbo decoding rate 1/2

In Fig.5, a simulation was made for a MIMO system with 4 transmit 4 receive antennas using Gray-mapped 16-QAM modulation with constraint length of 5, coding rate ½ convolutional codes. It's shown that turbo MMSE processing with convolutional codes has much better performance than a conventional MMSE receiver with turbo coding. It's also shown that turbo MMSE-PIC receiver is much has a little performance penalty at moderate BER until 1e-4. Surprisingly, the turbo MMSE even outperform the ML+turbo coding at moderate BER region.

To further enhance the performance of turbo MIMO receiver, turbo encoder is used as the channel coding scheme. We adopt the turbo convolutional codes which was proposed 3gpp standard. Coding rate ½ is used in the simulations, and for each turbo MIMO processing loop, turbo decoder runs 4 iterations. The performance of the turbo coded turbo MIMO processing of the 4x4 16-QAM modulation is shown in Fig.6. Another 2 dB is gained by turbo processing when using complicated ML detection method. However, It seems that the turbo decoding characteristics doesn't match well with MMSE detector, and the turbo processing over MMSE doesn't gain much compared with conventional receiver.

5. CONCLUSIONS

The MMSE intialized hard-input Parallel Interference Cancellation technique is generalized from the OFDM-SDMA scenario to the coded MIMO SDM OFDM system with soft input. By combining both the good convergence property of the MMSE detector and the low complexity of PIC receiver, the MMSE initialized soft PIC achieves the good performance and complexity tradeoff for a large number of transmit antennas and constellation size at mild bit error rates.

AFFILIATIONS

Jianfeng Liu, Andre Bourdoux, Hugo De Man are with Inter-university MicroElectronics Center (IMEC), Kapeldreef 75, Leuven 3001, Belgium
Jianfeng Liu is also a PhD student of Katholieke University of Leuven
Hugo De Man is also a professor at Katholieke Univeristy of Leuven
Marc Moonen is a professor with Department of Electrical Engineering (ESAT), Katholieke Univeristy of Leuven, Kasteelpark Arenberg 10, Leuven 3001, Belgium

REFERENCE:

[1] J. Bingham, "Multicarrier Modulation for data transmission, An idea whose time has come", IEEE Com. Mag., pp.5-14, May 1990
[2] Sampath, H., Talwar, S., et al, "A fourth-generation MIMO-OFDM broadband wireless system: design, performance, and field trial results", IEEE Com. Mag., pp143-149, Sep. 2002
[3] A. van Zelst, R. van Nee, and G.A. Awater, "Space division multiplexing (SDM) for OFDM system", Proc. of IEEE VTC, vol. 2, pp.1070-1074, 2000
[4] J.Hagenauer, "The turbo principle: Tutorial introduction and state of the art", Proc. 1st Intern. Symp. Turbo Codes, pp1-12, June 1997
[5] M. Munster, L. Hanzo, "Co-channel interference cancellation techniques for antenna array assisted multiuser OFDM system", Proc. of 3G Inter. Conf. On Mobile commun. Tech., pp. 256-260, 2000
[6] M. Tuchler, A. Singer, and R. Koetter, "Minimum mean squared error equalization using a priori information," IEEE Trans. Signal Proc., vol. 50, pp673-683, March 2002
[7] M. Sellathurai, S Haykin, "Turbo-BLAST for high-speed wireless communications," Proc. WCNC, Sept. 2000, pp315-320
[8] B. M. Hochwald, S. ten Brink, "Achieving near-capacity on a multiple-antenna channel," submitted to IEEE Trans on Commun., July 2001

Section V

MULTIPLEXING, DETECTION & INTERFERENCE CANCELLATION

DISTRIBUTED MULTIPLEXING IN MULTICARRIER WIRELESS NETWORKS

J. THOMAS

Abstract. Multiple-element antenna channels provide increased spectral efficiency via spatial multiplexing. It is therefore of interest to consider their emulation in a distributed fashion in both ad hoc and structured networks where communicating nodes cannot be equipped with multiple-element antennas for practical reasons. The work proposes conceptually simple algorithms towards this end. These distributed algorithms are spectrally efficient in that they require only the same number of degrees of freedom as their multiple-element antenna counterparts while performing almost as well as the latter. However, several practical issues relating to processing complexity remain to be addressed.

1. INTRODUCTION

Signaling with multiple-element antennas at both the transmitter and receiver in a wireless channel that is impaired primarily by fading and interference is known to provide attractive performance gains [1, 2, 3, 4]. In such high signal-to-noise (SNR) ratio environments, the use of an N_T-element transmitter and an N_R-element receiver results in a spectral efficiency that is $\min(N_T, N_R)\log_2\Gamma$ bits/s/Hz, where Γ is the SNR. In other words, the system sees a $\min(N_T, N_R)-$fold increase (called the spatial-multiplexing gain) in its available degrees of freedom, in the high SNR regime provided the propagation environment has a rich fading profile, and the channel response is known at the receiver. As with other results on channel capacity, the above fact assumes an infinite interleaving depth. This assumption cannot be approximated in practice in slowly fading environments. With quasi-static fading conditions, multiple-element antennas can be used with an appropriate signaling scheme to yield diversity gains, i.e. with an (N_T, N_R) transmit-receive system as described above, the bit error probability decays as $\Gamma^{-N_T N_R}$ (again assuming large Γ) rather than as Γ^{-1}, yielding a maximum spatial-diversity gain factor of $N_T N_R$. In general, any given signaling scheme with multiple-element antennas provides both a spatial-multiplexing gain and a spatial-diversity gain; the optimal tradeoff between these gains is described by the fact that achieving a spatial-multiplexing gain r in a (N_T, N_R) configuration implies that the achievable diversity gain is then at most $(N_T - r)(N_R - r)$.

The above progress in the understanding of multiple-antenna communications is remarkable because it strengthens the view that spatial dimensions are an important source for increasing the available degrees of freedom and diversity-order of a configuration that already exploits the temporal and frequency domains. There is, however, at least one practical problem, namely that of mounting multiple-element antennas on communicating nodes in a network. This is especially true for mobile nodes in both ad hoc and infrastructure-based (e.g. cellular or local area) networks. It is also true for access points and base stations in the latter from an economic perspective since

K. Fazel and S. Kaiser (eds.), Multi-Carrier Spread-Spectrum, 405–412.

an increasing number of these is necessary to cope with increasing user demands. The study of relay channels [5] in Shannon theory suggests a means of circumventing the physical need for multiple-element antennas, while effectively emulating their role and thence obtaining their benefits. Whereas in the models assumed in these early studies the relay's role was confined to assisting the source node to transmit its message to the destination node, the present setting has no provision for special nodes dedicated to such roles, and must use the existing nodes in the network to assist their peers while also transmitting their own messages (to their respective destinations) all via mutually orthogonal channels. Such cooperative mechanisms [6, 7, 8, 9] have been recently studied in the context of effectively emulating spatial diversity in the absence of physical multiple-element antennas[1]. These algorithms essentially have both the intended destination nodes and the cooperating node(s) receive independently faded versions of the source node's transmission. The relay(s) then either employ distributed space-time coding or forward amplified (or decoded-and-rencoded) versions of these signals to their transmissions in accordance with specified schedules yielding a family of distributed diversity algorithms.

The present work deals with the notion of distributed multiplexing which is significantly more involved than distributed diversity. A spatial-multiplexing gain is possible only if both the transmitting and receiving nodes have multiple-element antennas, i.e. both $N_T > 1$ and $N_R > 1$, in contrast with the case of spatial-diversity gains. Distributed multiplexing requires the emulation of interleaving an encoded data stream (originating from one source) over both time and distributed nodes, such that multiple mixtures of the resulting component faded streams are available to the destination node. An efficient layered spatial multiplexing scheme will first be presented, and distributed algorithms will be introduced for both infrastructure-based and ad hoc networks. Multicarrier signaling is assumed thereby avoiding the need for delay spread equalization and thus retaining simplicity. Issues of fundamental theoretical significance such as achievable rate regions and outage analysis for the schemes proposed here are discussed in [9].

2. LAYERED SPATIAL MULTIPLEXING

An efficient spatial-multiplexing scheme that avoids wastage of space-time slots at packet boundaries (in contrast with the diagonal-BLAST architecture [1]) is presented below, following [4].

Let $\mathbf{w}^{(c)}$, $c = 0, 1, \ldots$, denote the codeword packets, each interleaved in time, according to some specified pseudorandom function across B bursts, $\mathbf{w}_b^{(c)}$, $0 \leq b \leq B-1$. $\mathbf{w}^{(c)}$ is then expressed as a row vector $(\mathbf{w}_0^{(c)}\ \mathbf{w}_1^{(c)}\ \ldots\ \mathbf{w}_{B-1}^{(c)})$, each burst being comprised of J (coded) symbols. A burst serves as a convenient unit for spatially interleaving a codeword packet, across multiple transmit-elements. A packet is proposed to be transmitted across N_T elements (the total transmission power being fixed, regardless of the value of N_T) per the scheme introduced below, first, for two examples which help illustrate the idea, and then for the general case. With $N_T = 4$ and time being indexed in units corresponding to burst durations, β, the cases $B = 3$ ($< N_T$)

[1]The notion of macrodiversity in cellular networks has existed for decades.

and 5 ($> N_T$) are depicted, as below

$\underline{B=3}$

	$\beta=$ 0	1	2	3	4	...
$q=0$	$\mathbf{w}_0^{(0)}$	$\mathbf{w}_1^{(3)}$	$\mathbf{w}_2^{(2)}$	$\mathbf{w}_0^{(4)}$	$\mathbf{w}_1^{(7)}$	...
1	$\mathbf{w}_0^{(1)}$	$\mathbf{w}_1^{(0)}$	$\mathbf{w}_2^{(3)}$	$\mathbf{w}_0^{(5)}$	$\mathbf{w}_1^{(4)}$	...
2	$\mathbf{w}_0^{(2)}$	$\mathbf{w}_1^{(1)}$	$\mathbf{w}_2^{(0)}$	$\mathbf{w}_0^{(6)}$	$\mathbf{w}_1^{(5)}$	...
3	$\mathbf{w}_0^{(3)}$	$\mathbf{w}_1^{(2)}$	$\mathbf{w}_2^{(1)}$	$\mathbf{w}_0^{(7)}$	$\mathbf{w}_1^{(6)}$	...

$\underline{B=5}$

	$\beta=$ 0	1	2	3	4	5	6	...
$q=0$	$\mathbf{w}_0^{(0)}$	$\mathbf{w}_1^{(3)}$	$\mathbf{w}_2^{(2)}$	$\mathbf{w}_3^{(1)}$	$\mathbf{w}_4^{(0)}$	$\mathbf{w}_0^{(4)}$	$\mathbf{w}_1^{(7)}$	...
1	$\mathbf{w}_0^{(1)}$	$\mathbf{w}_1^{(0)}$	$\mathbf{w}_2^{(3)}$	$\mathbf{w}_3^{(2)}$	$\mathbf{w}_4^{(1)}$	$\mathbf{w}_0^{(5)}$	$\mathbf{w}_1^{(4)}$	...
2	$\mathbf{w}_0^{(2)}$	$\mathbf{w}_1^{(1)}$	$\mathbf{w}_2^{(0)}$	$\mathbf{w}_3^{(3)}$	$\mathbf{w}_4^{(2)}$	$\mathbf{w}_0^{(6)}$	$\mathbf{w}_1^{(5)}$	...
3	$\mathbf{w}_0^{(3)}$	$\mathbf{w}_1^{(2)}$	$\mathbf{w}_2^{(1)}$	$\mathbf{w}_3^{(0)}$	$\mathbf{w}_4^{(3)}$	$\mathbf{w}_0^{(7)}$	$\mathbf{w}_1^{(6)}$	...

In the general case, with N_T transmit-elements, the codeword transmitted in the space-time slot (n, β),

$$\mathbf{s}_{\beta}^{(n)} = \mathbf{w}_{\beta \bmod B}^{(N_T \lfloor \frac{\beta}{B} \rfloor + n - \beta \bmod B)} \tag{1}$$

and burst b of codeword $\mathbf{w}^{(c)}$ is transmitted during the space-time slot defined by

$$n = (c+b) \bmod N_T, \quad \beta = B\lfloor \frac{c}{N_T} \rfloor + b \tag{2}$$

For simplicity, all subsequent discussions will focus on the case $N_T = N_R = 2$, but generalizations to arbitrary numbers are easy. Also in the above scheme, successive columns depicted in the examples will be multiplexed onto different carriers rather than different time-slots such that every burst of a packet is transmitted at a different frequency, implying space-frequency (rather than space-time) signaling.

3. DISTRIBUTED MULTIPLEXING IN STRUCTURED NETWORKS

The adaptive reconfigurability of infrastructure-based networks is very desirable in view of the fact that demands on channel access resources vary with time. A novel arrangement is to have a powerful central base station (or a master access point in the case of a local area network) to which are connected several small stations via very high bandwidth wired links, e.g. fiber. These local stations have limited processing power, are inexpensive, and are readily introduced from an economic perspective. They serve primarily to receive and transmit signals between the central station and the mobile nodes that they serve. An example of such a system is an optical backbone infrastructure where the use of wavelength-division multiplexing allows easy insertation of local stations [10] facilitating the availability of bandwidth on demand. In practice, a mobile node would be served by a set of local stations which mutually exchange information via the central station. This set of local stations assigned to a given mobile node is dynamically updated depending on the quality of the channel between the mobile node and each potentially available local station.

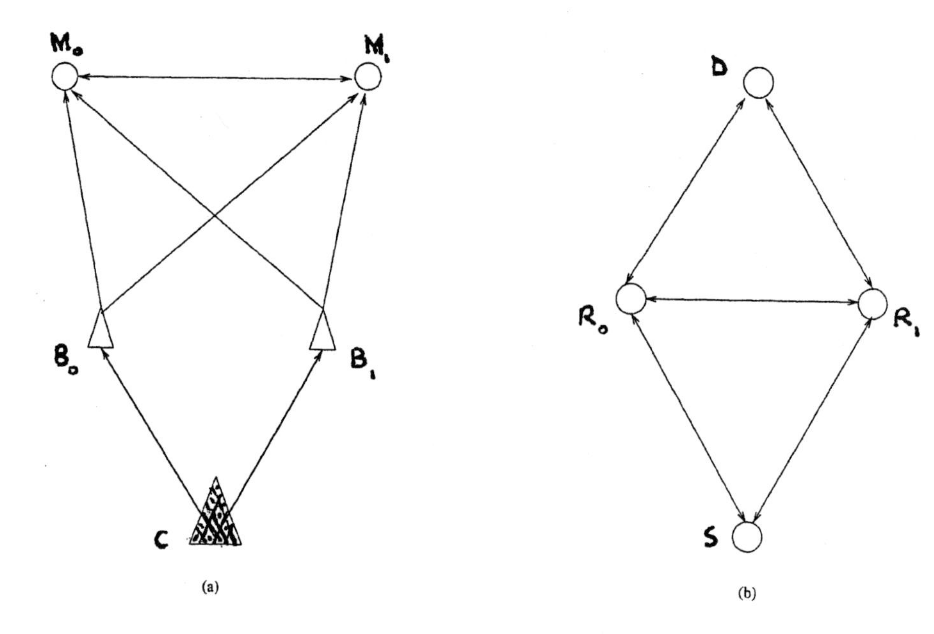

Figure 1: Schematic representations for distributed multiplexing in (a) structured, and (b) ad hoc networks

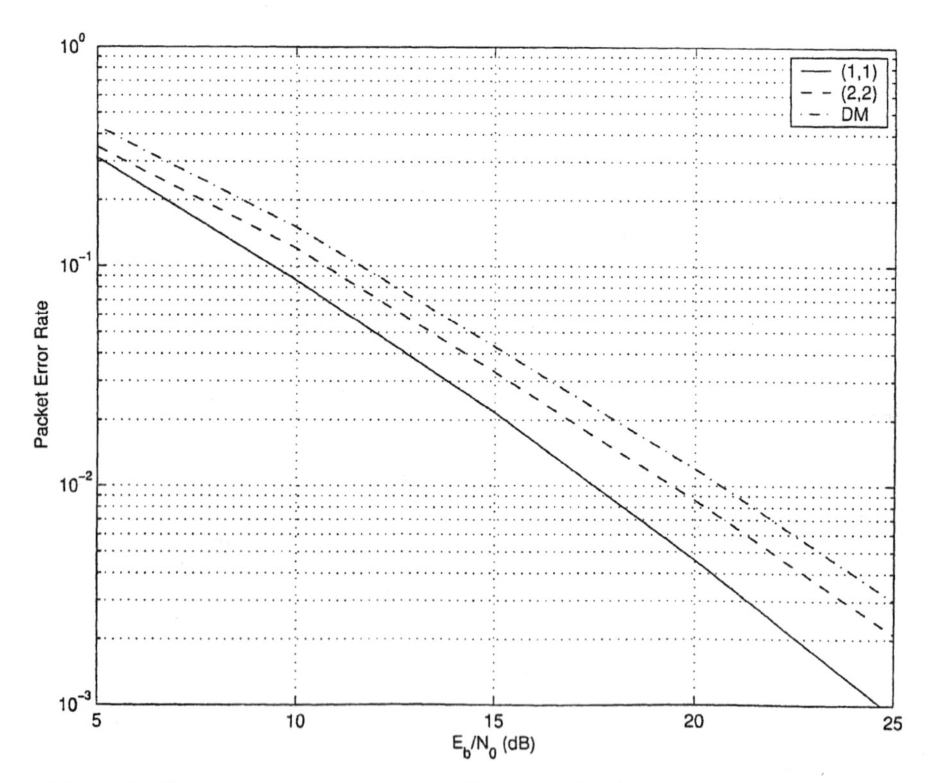

Figure 2: Performance curves for distributed multiplexing in structured networks

The above system is particularly amenable to the implementation of distributed multiplexing from the transmitter's perspective in the downlink, since the local stations play the role of the multiple elements of the central station's virtual antenna[2]. This is a valid emulation of a multiple-element transmitting antenna because the fading gains of the transmissions from different local stations are indeed uncorrelated.

A distributed scheduling algorithm that operates with the assistance of other mobile nodes to emulate a multiple-element receiver is now necessary. For simplicity, the sequel will be restricted to the emulation of the simple $N_T = N_R = 2$ case with the help of just one additional mobile node as in Fig. 1(a) where C represents the central station, L_0 and L_1 represent local stations (connected to C via high-bandwidth wired links, depicted by the dark lines), and M_0 and M_1 represent mobile nodes. In the benchmark scheme ($N_T = N_R = 2$) each of M_0 and M_1 has two-element antennas and receives packets that comprise of two bursts each, per the layering scheme described in Section 2 (now in the context of space-frequency multiplexing, where two carrier frequencies are available). If $\mathbf{w}^{(0)}$ and $\mathbf{w}^{(1)}$ be the first two packets intended for M_0 then the bursts $\mathbf{w}_0^{(0)}$ and $\mathbf{w}_0^{(1)}$ are transmitted from the local stations L_0 and L_1 respectively using the first carrier frequency, while the bursts $\mathbf{w}_1^{(0)}$ and $\mathbf{w}_1^{(1)}$ are concurrently transmitted from L_0 and L_1 respectively using the second carrier frequency. The packets for M_1 are similarly transmitted using some orthogonal multiaccess scheme so that there is no multiaccess interference. Consider now the transmission intended for M_0 in the distributed multiplexing scheme. First, L_0 and L_1 broadcast $\mathbf{w}_0^{(0)}$ and $\mathbf{w}_0^{(1)}$ respectively so that the received faded mixtures at M_0 and M_1 corresponding to the nth bit of these bursts are

$$\begin{aligned}(\mathbf{x}^{(0)})_n &= a_n^{(00)}(\mathbf{w}_0^{(0)})_n + a_n^{(10)}(\mathbf{w}_0^{(1)})_n + v_n^{(0)} \\ (\mathbf{x}^{(1)})_n &= a_n^{(01)}(\mathbf{w}_0^{(0)})_n + a_n^{(11)}(\mathbf{w}_0^{(1)})_n + v_n^{(1)}\end{aligned} \tag{3}$$

where $\mathbf{x}^{(m)}$ is the received burst at mobile node M_m, $m = 0, 1$, and a_n^{ij} is the fading gain seen by the nth bit $(\mathbf{w}_i^{(j)})_n$ of burst $\mathbf{w}_i^{(j)}$. Similar relations may be written for the bursts on the other carrier and for the orthogonal transmissions intended for M_1. Two independent mixture sums $\mathbf{x}^{(0)}$ and $\mathbf{x}^{(1)}$ of independently faded versions of the bursts in question are now available and it only remains to transmit $\mathbf{x}^{(1)}$ to M_0; this is done next on an orthogonal channel so that M_0 may now detect (using e.g. the multichannel minimum mean square error algorithm) the transmitted messages once the entire packet is available for deinterleaving and decoding, thus effectively providing the benefits of spatial multiplexing (and diversity). It must be emphasized that the above distributed multiplexing scheme uses only the same number of degrees of freedom as the benchmark multiple-antenna scheme (instead of two antenna-elements per mobile node it requires an extra orthogonal channel to transmit $\mathbf{x}^{(1)}$ from M_1 to M_0). As expected however, it does require more processing than the latter.

Fig. 2 shows a performance simulation of this distributed multiplexing algorithm. A rate$-(1/2)$ convolutional code with a constraint length of 7 was used following by psuedorandom interleaving across the two bursts (of 1024 bits each) constituting every packet. The two carriers were assumed to be separated by a sufficiently wide

[2]In the uplink of such an infrastructure-based network, distributed diversity rather than distributed multiplexing is a practically conceivable proposition.

guard band. Flat Jakes fading with a normalized Doppler rate of 0.01 on each carrier and the availability of perfect channel state information at the receiver were assumed. The curves labeled '(1,1)' and '(2,2)' respectively refer respectively to the cases of the single-element antenna ($N_T = N_R = 1$) and the two-element antenna ($N_T = N_R = 2$) systems. In this benchmark system, at a packet error rate of 10%, the two-element antenna system doubles the spectral efficiency with a mere 2 dB increase in the bit energy to noise ratio E_b/N_0. The proposed distributed multiplexing ('DM') algorithm achieves the same doubling of spectral efficiency at the cost of an additional dB increase in the E_b/N_0. This additional dB arises due to the amplify-and-forward action on $\mathbf{x}^{(1)}$ by M_1 in the final step, where the additive noise is amplified along with the signal.

4. DISTRIBUTED MULTIPLEXING IN AD HOC NETWORKS

Distributed multiplexing in ad hoc networks is considerably more difficult than in structured environments such as that in Section 3 because the former do not have any convenient emulation for multiple-element transmitters. The distributed algorithm proposed below is therefore more complex than that in Section 3, though in the present case too, the number of degrees of freedom required is the same as in the benchmark $N_T = N_R = 2$ system for peer-to-peer communication.

Figure 1(b) shows the basic elements of the ad hoc distributed multiplexing system: S and D represent the source and destination nodes while R_0 and R_1 represent the intermediate relay nodes which play a vital role in emulating the multiple-element transmitter for the source and the multiple-element receiver for the destination. With essentially the same basic signaling format as in the infrastructure-based network, one requires four orthogonal channels in order to avoid excessive complexity. (The benchmark system also uses four degrees of freedom, i.e. from two-element antennas and two orthogonal channels to allow a mechanism for duplexing transmissions.) In the first of the four orthogonal channels in the present case, the source S transmits $\mathbf{w}_0^{(0)}$ to relay R_0. In the next orthogonal channel, relay R_0 transmits this faded version of $\mathbf{w}_0^{(0)}$ to relay R_1, while S concurrently transmits $\mathbf{w}_0^{(1)}$ to R_1. In the third channel, R_1 broadcasts its faded mixture to R_0 and the destination D, and in the last channel R_0 transmit its faded mixture to D. Thus, at the end of this process D has two independent mixture sums $\mathbf{y}^{(0)}$ and $\mathbf{y}^{(1)}$ of independently faded versions of $\mathbf{w}_0^{(0)}$ and $\mathbf{w}_0^{(1)}$

$$\begin{aligned} (\mathbf{y}^{(0)})_n &= b_n^{(00)}(\mathbf{w}_0^{(0)})_n + b_n^{(10)}(\mathbf{w}_0^{(1)})_n + u_n^{(0)} \\ (\mathbf{y}^{(1)})_n &= b_n^{(01)}(\mathbf{w}_0^{(0)})_n + b_n^{(11)}(\mathbf{w}_0^{(1)})_n + u_n^{(1)} \end{aligned} \tag{4}$$

which has the same form and interpretation as (3).

The results of a simulation of this algorithm under the same conditions as with the infrastructure-based case in Section 3 are shown in Fig. 3. Here, an additional 2 dB is required by the distributed multiplexing scheme over the $N_T = N_R = 2$ benchmark. This is explained by the fact that this scheme involves four amplify-and-forward steps leading to greater amplification of the additive noise. Nevertheless, its performance is promising. An important issue in this set-up is the assumption of the availability of perfect channel state information at the receiver. This could prove difficult even in a moderately fast fading ad hoc environment. On the other hand, spatial- (or distributed-) multiplexing is not very attractive in a slow fading

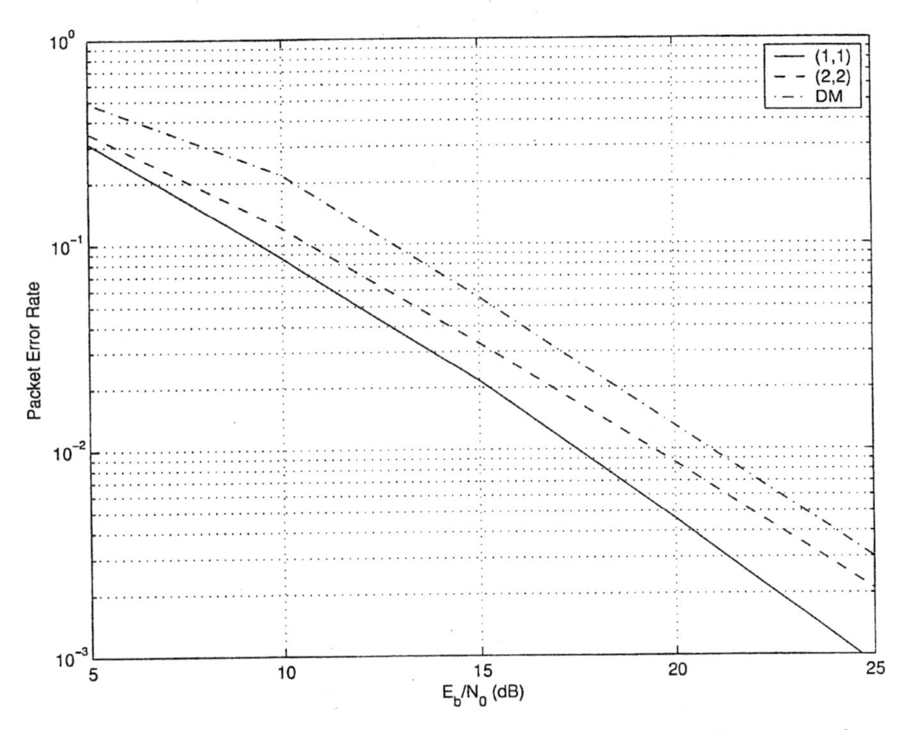

Figure 3: Performance curves for distributed multiplexing in ad hoc networks

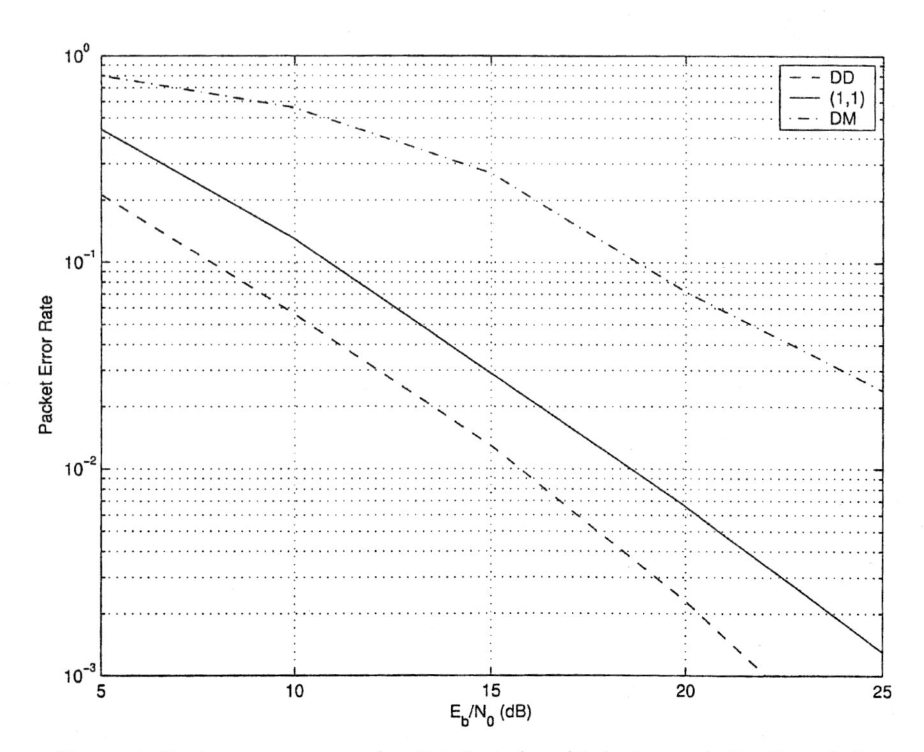

Figure 4: Performance curves for distributed multiplexing and distributed diversity in block fading ad hoc networks

environment since the delays due to the large interleaving depths required in such scenarios is excessive. Distributed diversity ('DD') using a single amplify-and -forward relay is more appropriate here as borne out by the simulation results (in Fig. 4) where the fading remains constant over the length of the burst.

5. CONCLUSIONS

Distributed signaling in wireless networks is a topic of emerging interest. The present work has considered conceptually simple algorithms to effectively emulate spatial multiplexing in a distributed fashion in both structured and ad hoc networks where nodes cannot be equipped with multiple-element antennas. While it was observed that it is possible to obtain spectrally efficient schemes (i.e. ones that require the same number of degrees of freedom as their multiple antenna counterparts) there are several practical issues relating to processing complexity that must be addressed. In particular, the notion of adaptively choosing a distributed signaling strategy based on information about link quality exchanged among the communicating nodes will probably be an important factor in practice.

The author is with the University of Maryland, Baltimore, USA.

References

[1] G. Foschini, "Layered space-time architecture for wireless communication in a fading environment when using multi-element antennas," *Bell Labs Technical Journal,* vol. 1, pp. 41-59, Jun. 1996

[2] E. Telatar, "Capacity of multi-antenna Gaussian channels," *European Transactions on Telecommunications,* vol. 10, pp. 585-595, Nov. - Dec. 1999 (also *Bell Labs Technical Memo,* 1995)

[3] L. Zheng and D. Tse, "Diversity and multiplexing: A fundamental tradeoff in multiple antenna channels," *IEEE Transactions on Information Theory,* vol. 49, pp. 1073-1096, May 2003

[4] J. Thomas, "Efficient turbo receivers for MIMO communications," *Research Report,* University of Maryland, College Park, MD, Jan. 1998 (also *Proc. Int. Soc. Opt. Engng.,* vol. 4529, pp. 104-117)

[5] T. Cover and A. El Gamal, "Capacity theorems for the relay channel," *IEEE Transactions on Information Theory,* vol. 25, p. 572-584, Sep. 1979

[6] A. Sendonaris, E. Erkip, and B. Aazhang, "User cooperation diversity," submitted to *IEEE Transactions on Communications.*

[7] J. Laneman, D. Tse, and G. Wornell, "Cooperative diversity in wireless networks: Efficient protocols and outage behavior," submitted to IEEE Transactions on Information Theory.

[8] P. Anghel and M. Kaveh, "Multiuser space-time coding in cooperative networks," submitted to *ICASSP*, 2003

[9] J. Thomas, "Distributed multiplexing and diversity in wireless networks," *Research Report,* University of Maryland, Jan. 2000; revised Nov. 2001

[10] F. Choa, Private communication, Mar. 2002

SHIGEHIKO TSUMURA, MATTI LATVA-AHO
AND SHINSUKE HARA

AN INTER-CELL INTERFERENCE SUPPRESSION TECHNIQUE USING VIRTUAL SUBCARRIER ASSIGNMENT (VISA) FOR MC-CDMA UPLINK

Abstract. Multi-carrier code division multiple access (MC-CDMA) system, which is a combination of OFDM (Orthogonal Frequency Division Multiplexing) and CDMA, is robust to hostile mobile radio environments because the receiver can effectively combine all the received signal energy scattered in the frequency domain. Moreover, MC-CDMA system can establish a quasi-synchronous transmission due to the cyclic prefix, and easily employ a multi-user detection in an uplink channel. In a multiple-cell environment, the inter-cell interfering deteriorates the performance of the MC-CDMA multi-user detection.

In this paper, utilizing VISA (VIrtual Subcarrier Assignment), we propose an inter-cell interference suppression method for an MC-CDMA uplink with a multi-user detection, where the base station can easily *spatially filter in* only the signals from the own cell and *spatially filter out* the signal from the other cells.

1. INTRODUCTION

Multi-carrier code division multiple access (MC-CDMA) system, which is a combination of OFDM and CDMA, is robust to hostile mobile radio environments because the receiver can effectively combine all the received signal energy scattered in the frequency domain. Moreover, MC-CDMA system can establish a quasi-synchronous transmission due to the guard interval (GI) which is cyclically extended to fast Fourier transform (FFT) output, and easily employ a multi-user detection in an uplink channel. Therefore, MC-CDMA is now considered to be a promising multiple access technique for 4G uplink.

To obtain a good transmission performance in an uplink, an MC-CDMA base station requires a multi-user detection scheme, which generally requires information on all the received signals from user terminals. However, in a multiple cell environment, information on signals from neighboring cells (inter-cell interference) is unknown. For a base station, so even when the base station employs a multi-user detection scheme to mitigate multiple user (intra-cell) interference, the inter-cell interfering deteriorates the performance of the MC-CDMA multi-user detection. Therefore, an inter-cell interference suppression technique is still required for successful MC-CDMA uplink.

In [1], the principle on a novel spatial filtering technique for OFDM-based signals is shown. VISA employs the position of virtual subcarrier (subcarrier that is not used for actual data transmission) as an ID (identifier) to distinguish an individual user. In

K. Fazel and S. Kaiser (eds.), Multi-Carrier Spread-Spectrum, 413–420.

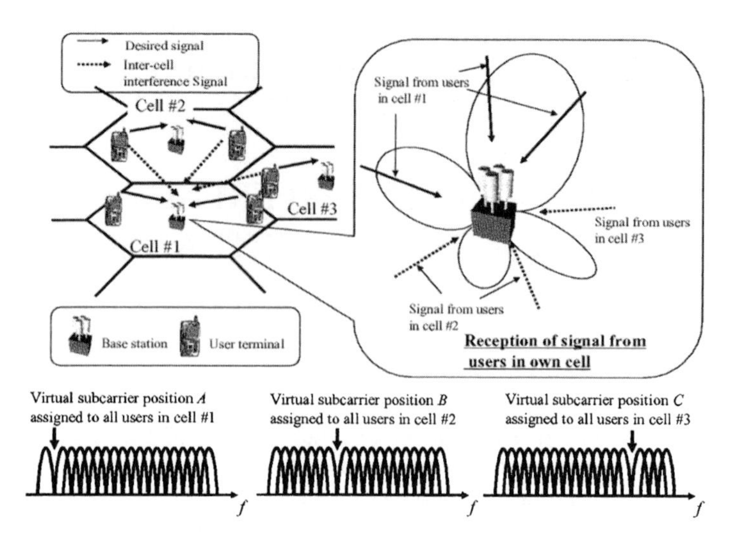

Figure 1. Principle of inter-cell interference suppression with VISA.

this paper, as an application of VISA, we propose an inter-cell interference suppression method for an MC-CDMA uplink with a multi-user detection, where the base station can easily *spatially filter in* only the signals from the own cell and *spatially filter out* the signal from the other cells.

2. PROPOSED METHOD

2.1. Principle

For the proposed method, different base stations assign different virtual subcarrier positions to all the user terminals in their cells. Note that they are all the same for user terminals in their own cell. Each base station is equipped with an adaptive array antenna, and the base station controls the array weights so as to force the output of the virtual subcarrier assigned to the users in the own cell to be zero. As a result, the base station can suppress inter-cell interference, and spatially filter in only the signal of the users in the own cell. For example, we give an outline of the proposed method as shown in Figure 1. Here, the base station assigns the virtual subcarrier position A, B and C to all the active user terminals in cell #1, #2 and #3, respectively. The base station in the cell #1 controls the array weights so as to force the output of the virtual subcarrier A assigned to the all user terminals in the cell #1, and can spatially filter in only the signal of all the user terminals in the cell #1.

2.2. Configuration

Figure 2 shows the configuration of the base station with a pre-FFT type adaptive array antenna with M antenna elements for the proposed system. The virtual subcarrier

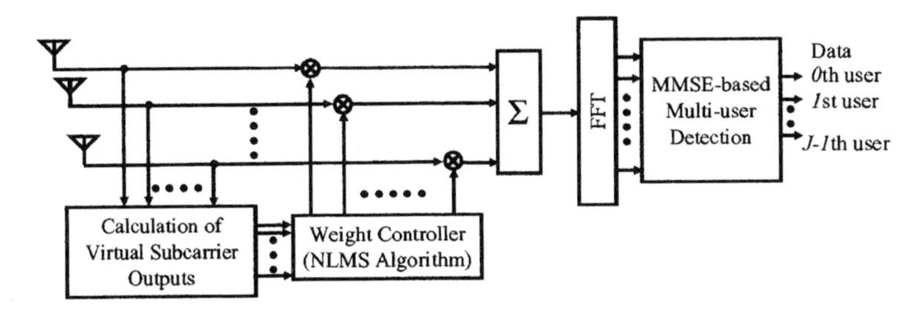

Figure 2. Configuration of proposed method.

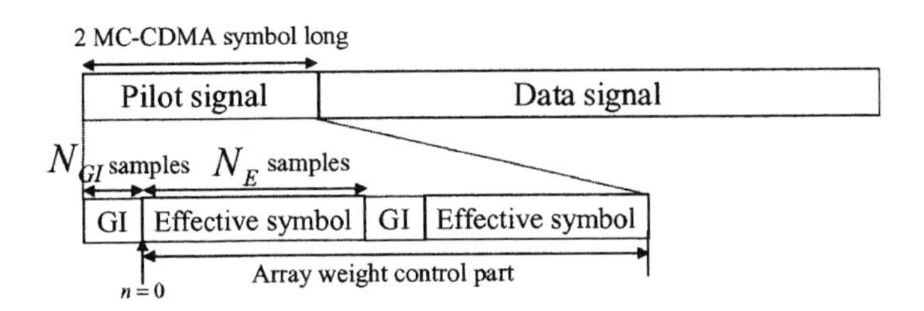

Figure 3. Structure of Signal burst.

component assigned to the own cell is first calculated from the received *mixed desired and interfering* pilot signal which is inserted in the forefront of the burst (see Fig. 3). The data signal is composed of 32 MC-CDMA symbols, where in one MC-CDMA symbol, the GI and the effective symbol are N_E samples and N_{GI} samples, respectively. The pilot signal and the data signal are the same signal format. The pilot signal is 2 MC-CDMA symbol long and the same cell-specific virtual subcarrier is assigned to two pilot MC-CDMA symbols. Then, the weight controller with a normalized least mean square (NLMS) algorithm computes the array weights by forcing the output of the virtual subcarrier to be zero in the array weight control part. Only the signals with the assigned virtual subcarrier position can go through the array antenna, so the array antenna output can contain only the signals from the own cell. This means that the array can suppress the inter-cell interference. Finally, the base station separates each user with the minimum mean square error (MMSE)-based multi-user detection. In the following, details of the proposed system are discussed.

2.2.1 *Proposed Interference Cancellation Method*

Here, the kth virtual subcarrier component at nth iteration $\boldsymbol{v}_k(n) = [v_{0,k}(n), \cdots, v_{M-1,k}(n)]^T$, where superscript T stands for transposition, is given by

$$\boldsymbol{v}_{\boldsymbol{k}}(n) = \frac{1}{t_s} \int_{\frac{t_s}{N}n}^{\frac{t_s}{N}n+t_s} \boldsymbol{u}(t) \exp\left(-\frac{j2\pi f_k t}{t_s}\right), n = 0, 1, \cdots, N_E + N_{GI} \quad (1)$$

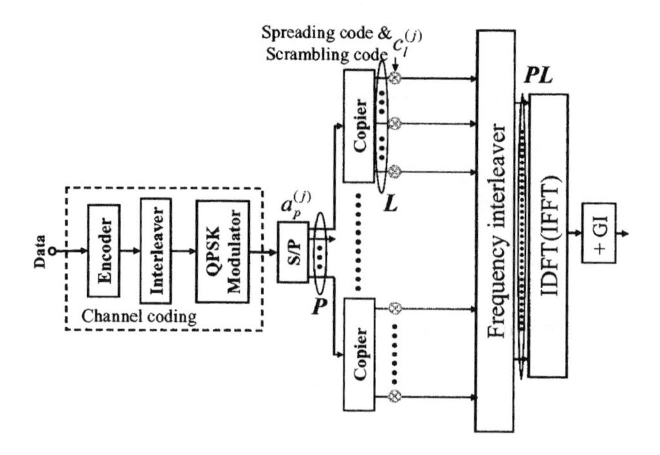

Figure 4. Transmitter for jth user.

where t_s and f_k are the subcarrier duration and the kth subcarrier frequency for one pilot MC-CDMA symbol, respectively, and $\boldsymbol{u}(t) = [u_0(t), \cdots, u_{M-1}(t)]$ shows the received signal before the FFT processing. On the other hand, the weight controller calculates the array weights $\boldsymbol{w}$(n)= $[w_0(n), \cdots, w_{M-1}(n)]$ so as to make the virtual subcarrier component after array combining closer to zero:

$$\min_{\arg \boldsymbol{w}} \sum_k |0 - \boldsymbol{w}^{\boldsymbol{H}}(n)\boldsymbol{v}_{\boldsymbol{k}}(n)|^2, \tag{2}$$

where superscript H stands for Hermitian transpose. The NLMS algorithm is applicable for the minimization problem of Eq. (2). Note that Eq. (2) has a solution of $\boldsymbol{w}(n) = \boldsymbol{0}$. In order to avoid the trivial solution, we give a constraint for the proposed algorithm: $w_0(n) = 1$. Here, the weight update algorithm for the proposed system is as follows:

$$e_k(n) = 0 - \boldsymbol{w}'^{\boldsymbol{H}}(n)\boldsymbol{v}_{\boldsymbol{k}}(n) \tag{3}$$

$$\boldsymbol{w}'(n+1) = \boldsymbol{w}'(n) - \mu \sum_k \frac{\boldsymbol{v}_{\boldsymbol{k}}(n)e_k^*(n)}{a + |\boldsymbol{v}_{\boldsymbol{k}}(n)|^2}, \tag{4}$$

where μ and a denote the step size and the positive constant for the NLMS algorithm, respectively, and $\boldsymbol{w}'(n) = [1, w_1(n), \cdots, w_{M-1}(n)]$.

2.2.2 *Multi-User Detection Method*

The code orthogonality among users is totally distorted by the difference of instantaneous frequency response for any active users. Therefore, in the uplink application, a multiuser detection scheme is required, in order to recovery the code orthogonality distorted for each of the channels of users.

Figure 4 shows the transmitter for the jth user. In the following, we discuss the pth subcarrier block where pth serial-to-parallel(S/P) converted and QPSK-modulated symbol is spread. For analytical simplicity, here, we can omit the order of subcarrier block p without loss of generality. The jth user's transmitted signal $\boldsymbol{s}^{(j)} = [s_0^{(j)}, s_1^{(j)}, \cdots, s_{L-1}^{(j)}]^T, j = 0, 1, \cdots, J-1$ before the frequency interleaver where L and J denote the spreading code length and the number of active users, is given by

$$\boldsymbol{s}^{(j)} = \boldsymbol{c}^{(j)} a^{(j)}, \tag{5}$$

where it is defined that jth user spreading code vector $\boldsymbol{c}^{(j)} = [c_0^{(j)}, c_1^{(j)}, \cdots, c_{L-1}^{(j)}]^T$ which is element as jth user's spreading code multiplied by scrambling code $c_l^{(j)} \in \{+1, -1\}$, and $a^{(j)} \in \{\pm 1 \pm j\}$ is jth user's data symbol which is channel-encoded with random bit-interleaver whose length is one packet data size and QPSK-modulated. Moreover, the received signal $\boldsymbol{r} = [r_0, r_1, \cdots, r_{L-1}]^T$ after the array antenna and FFT at the base station is shown by

$$\boldsymbol{r} = \widetilde{\boldsymbol{C}}\boldsymbol{a} + \widetilde{\boldsymbol{C}_i}\boldsymbol{a_i} + \boldsymbol{n} \tag{6}$$

where we define the noise vector $\boldsymbol{n} = [n_0, n_1, \cdots, n_{L-1}]^T$, the symbol vector $\boldsymbol{a} = [a^{(0)}, a^{(1)}, \cdots, a^{(J-1)}]^T$ and the distorted code matrix $\widetilde{\boldsymbol{C}} = [\ \widetilde{\boldsymbol{c}}^{(0)}, \cdots, \widetilde{\boldsymbol{c}}^{(J-1)}\ ]$ whose column is jth user's distorted code vector $\widetilde{\boldsymbol{c}}^{(j)} = [\widetilde{c}_0^{(j)}, \widetilde{c}_1^{(j)}, \cdots, \widetilde{c}_{K-1}^{(j)}]^T$. In addition, $\widetilde{\boldsymbol{C}_i}\widetilde{a}_i$, where $\widetilde{\boldsymbol{C}_i} = [\widetilde{\boldsymbol{c}}_i^{(0)}, \cdots, \widetilde{\boldsymbol{c}}_i^{(J'-1)}]$ and $\boldsymbol{a_i} = [a_i^{(0)}, a_i^{(1)}, \cdots, a_i^{(J'-1)}]^T$, ($J'$ denotes number of the users in the adjacent cells) shows the residual inter-cell interfering signal component after the adaptive array antenna processing. Here, the jth user's spreading code distorted by channel in the lth subcarrier after the adaptive array antenna $\widetilde{\boldsymbol{c}}_l^{(j)}$ is given by

$$\widetilde{\boldsymbol{c}}_l^{(j)} = \frac{1}{t_s} \int_0^{t_s} c_l^{(j)} \boldsymbol{w}^H \boldsymbol{h}^{(j)}(t) \exp\left(\frac{-j2\pi f_l t}{t_s}\right) dt, \tag{7}$$

where $\boldsymbol{h}^{(j)}(t) = [h_0^{(j)}(t), \cdots, h_{M-1}^{(j)}(t)]$ denotes the channel impulse response in the adaptive array antenna elements for the jth user. Assuming that the inter-cell interfering can be sufficeintly suppressed by the proposed method, i.e., $\widetilde{\boldsymbol{C}_i}\boldsymbol{a_i} \approx \boldsymbol{0}$ in Eq. (6), we can derive the correlation matrix $\boldsymbol{R}$ of received signals as

$$\boldsymbol{R} = \boldsymbol{E}[\boldsymbol{r}\boldsymbol{r}^H] = \widetilde{\boldsymbol{C}}\widetilde{\boldsymbol{C}}^H + \sigma_n^2 \boldsymbol{I_N}, \tag{8}$$

where σ_n^2 is variance of the noise and $\boldsymbol{I_N}(L \times L)$ is the identity matrix. Finally, decision valuable vector $\boldsymbol{D} = [d^{(0)}, d^{(1)}, \cdots, d^{(J-1)}]^T$ is given by

$$\boldsymbol{D} = \widetilde{\boldsymbol{C}}^H \boldsymbol{R}^{-1} \boldsymbol{r} \tag{9}$$

Therefore, Eq. (9) shows MMSE-multiuser detection method for the proposed system.

3. COMPUTER SIMULATION

We evaluated the bit error rate (BER) performance of the proposed method in an uplink by computer simulation. For an antenna configuration, we assumed an 8 element-circular array antenna with element spacing of half wavelength. An MC-CDMA signal was generated with 512 point-FFT, a 50 sample-long cyclic prefix was inserted in each MC-CDMA symbol, and Walsh Hadamard code was employed as the spreading code with length of 32. For modulation/demodulation and channel coding formats, we assumed a coherent QPSK and a half-rate convolutional encoding/Viterbi decoding with constraint length of 9. For a spatial channel model, each path which had a Rayleigh distributed amplitude with the same power arrived in a cluster that was composed of 5 waves with root mean square (RMS) angle spread of 5 deg (see Fig.5). This model is true for a case where the position of the antenna is relatively high, as installed on top of a tall building. For the channel model, the arrival time for all active users and inter-cell interfering signal was uniformly distributed within the GI. Also, the number of virtual subcarriers for the pilot signal was 2 and the 127th subcarrier and the 383rd subcarrier assigned to the virtual subcarriers. The step size and the positive constant for the NLMS in the weight control were 0.1 and 1, respectively, i.e., $\mu = 0.1$ and $a = 1$ in Eq. (4). Moreover, we assume that the reciver knows the channel state information after the array antenna.

Figure 6 shows the BER versus the $\overline{E_b/N_0}$ per antenna for $J = 2$, when the number of desired paths was 6 and the direction of arrival paths was fixed to be 126, 144 and 162 and 198, 216 and 234 [deg] for 0th user and 1st user, respectively, and the number of inter-cell interfering paths was 1 and the DoA was fixed to 306 [deg]. In addition, the average desired signal per user to inter-cell interfering signal power ratio (D/I) was -10, 0 and 10 [dB]. Also, Fig. 7 shows example of an antenna beam pattern of the proposed method in the above channel response when $\overline{E_b/N_0}$=20[dB]. For D/I = -10 and 0 [dB], we can see that the proposed method outperforms the no-array system because the proposed method forms null toward the interfering signal. However, the performance of the proposed system is worse than that of the no-array system for

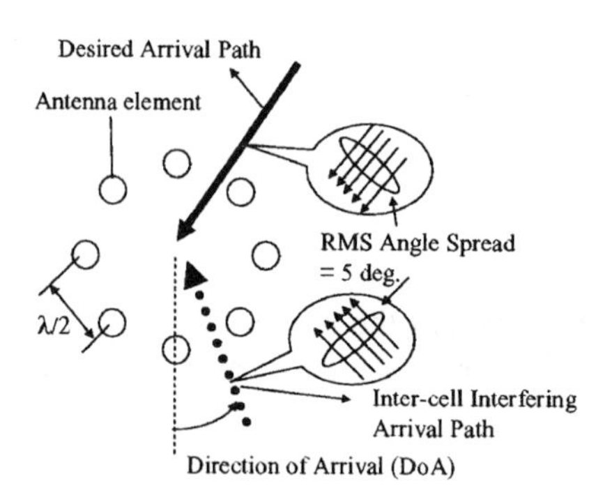

Figure 5. A spatial channel model.

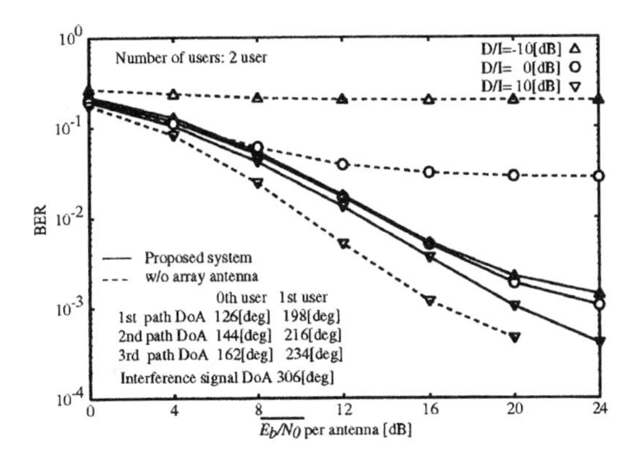

Figure 6. BER versus $\overline{E_b/N_0}$.

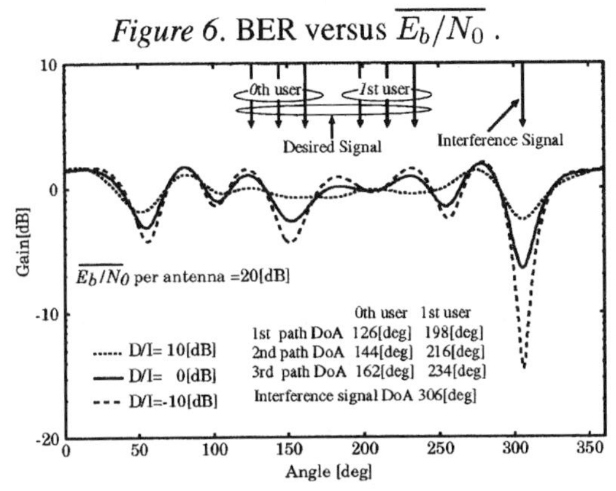

Figure 7. Antenna beam pattern (Channel response of Fig.6).

D/I=10[dB]. This is because the proposed system could not give gain to the desired signals as shown in Fig. 7.

Figure 8 shows the BER versus the $\overline{E_b/N_0}$ per antenna, when the number of desired paths and inter-cell interfering paths was 1 and 1,respectively, D/I was -10, 0 and 10 [dB]. Here, the direction of arrival for desired path and inter-cell interfering path was randomly chosen from [0, 360 deg). There is the BER floor for the no-array antenna system, and the BER of the proposed system is better than that of the no-array antenna system in the higher $\overline{E_b/N_0}$ region.

Figure 9 shows the BER versus the number of active users, when the number of desired paths per user and interfering paths was 3 and 3, respectively and $\overline{E_b/N_0}$ = 15 [dB]. For D/I=0 and -10 [dB], the proposed method is superior to the no-array antenna system and the proposed method can keep a good BER performance. From the figures, however, the BER of the proposed system is worse than that of the no-array antenna system in the lower $\overline{E_b/N_0}$ region or D/I = 10 [dB]. This is because the proposed system has the two causes of the degradation. The first is that, when the

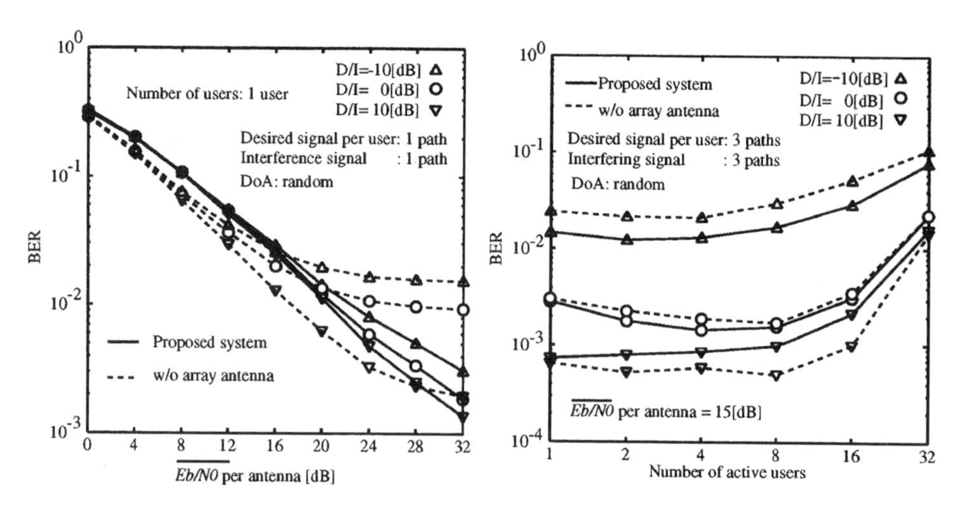

Figure 8. BER versus $\overline{E_b/N_0}$.

Figure 9. BER versus J.

directions of arrival for the desired signal and the interfering signal is close, the proposed system suppress not only the interfering signal but also the desired signal. The second is that, the accuracy of calculated array weights is poor in the lower $\overline{E_b/N_0}$, and in this case, the proposed system cannot sufficiently reject the interfering signal.

4. CONCLUSIONS

In this paper, we have proposed an inter-cell interference suppression method using VISA, which has simple weight control algorithm for an adaptive antenna array making effective use of the virtual subcarrier, for an MC-CDMA uplink with an MMSE-based multi-user detection. Computer simulation results have shown that the proposed system can achieve good BER performance in the presence of inter-cell interfering signals and is effective especially when the ratio of desired signal power to inter-cell interfering signal power is low. From the results, it can be concluded that the proposed system with VISA can reject inter-cell interfering signal when the inter-cell interering signal is much stronger than intra-cell multiple access signal with high $\overline{E_b/N_0}$.

References

[1] S. Hara, "VIrtual Subcarrier Assignment (VISA): Principle and Applications" *Proc. Fourth International Workshop on Multi-Carrier Spread-Spectrum (MC-SS 2003)*, Oberpfaffenhofen, Germany, Sept. 2003.

Shigehiko Tsumura and Shinsuke Hara are with Graduate School of Engineering, Osaka University, Osaka, Japan.
Matti Latva-aho is with Centre for Wireless Communications, University of Oulu, Oulu, Finland.

F. BADER & S. ZAZO

SYNCHRONISM LOSS EFFECT ON THE SIGNAL DETECTION AT THE BASE STATION USING AN OFDM-CDMA SYSTEM

Abstract. This proposal work focus on the analysis of the synchronism loss effect upon the number of active users detected at the base station using an orthogonal frequency division multiplexing (OFDM) modulation with a code division multiple access (CDMA) scheme in the reverse transmission mode.

1. INTRODUCTION

The first combination between the multiple carrier modulation and the CDMA scheme have appeared during the nineties more exactly in 1993 under divers transmission forms [1], up to 1998 big efforts have been made in the optimization detection, decoding, and channel estimation [2]. Although the OFDM-CDMA scheme is an attractive candidate for future wireless communication systems and its application rise in many communication area as in broadband cellular systems (European project MATRICE[1]) or in VHF Communications for Air Traffic Management domain [3]. There's a tremendous amount to do before it can achieves its full potential. One of these challenges is the detector design for reverse transmission mode.

As it's well know the OFDM-CDMA detection complexity in the reverse transmission mode is higher whether it is compared with the forward mode. Exist a lot of factors closely related with the reverse transmission mode that affect the signal detection at the base station and need to be process with more attention than in the forward mode. In the forward case, a same pilot pattern for channel estimation is used for all the active users [2]. However in the reverse mode, each user needs a specific pilot pattern distribution not overlapped by the rest of the user's pilots since the active users experience different channel attenuations in this mode. The different propagation distances between the different mobile systems (MS) of the different users and the base station (BS) in the reverse mode introduce significant time offset. Considering the synchronism loss phenomena means to develop more complex receivers dealing with the asynchronous mode in the multiple user detection (MUD) process at the BS than that used in the forward case. In the MUD detection of several synchronous signals, the interferences are basically originate from the current symbol of the rest of the active users (see figure 2). However, when we are confronted to a synchronism loss in the time domain (asynchronous case) the interferences are not only originate from the current symbol but also from part of the past and further symbols of the rest of the active users (see figure 3). It is clear that

[1] Multicarrier CDMA Transmission Techniques for Integrated Broadband Cellular Systems-*http://www.ist-matrice.org/*

K. Fazel and S. Kaiser (eds.), Multi-Carrier Spread-Spectrum, 421–428.

the performances of the detection process is limited by the presence of the multiple users interferences, but it is possible to design receivers that have the capacity to reduce this effect.

2. TRANSMISSION AND RESEPTION MODELS

In this paper we are dealing with the scenario based on the inter block interference (IBI) free signal transmission and reception scheme for an OFDM-CDMA system [5]. Figure (1) shows the transmission scheme used for transmitting the signal of the *i-th* user expressed by the follow equation

$$\mathbf{y}_i(v) = \mathbf{H}_0{}^i\mathbf{T}\mathbf{F}^{-1}\mathbf{c}_i D_i(v) + \mathbf{H}_1{}^i\mathbf{T}\mathbf{F}^{-1}\mathbf{c}_i D_i(v-1) \tag{1}$$

Where $\mathbf{H}^i_0$ and $\mathbf{H}^i_1$ means the Toeplitz convolutive channel matrices of the *i-th* user, $\mathbf{T}$ is the transmitted redundancy matrix (introduce the cyclic prefix) [5], $\mathbf{F}$ and $\mathbf{F}^{-1}$ are the Fourier and the inverse Fourier matrices respectively, the vector column $\mathbf{c}_i$ means the spreading code word, and D_i is the transmitted data symbol of the *i-th* active user ($i = 1, \ldots, K$) [5, 6].

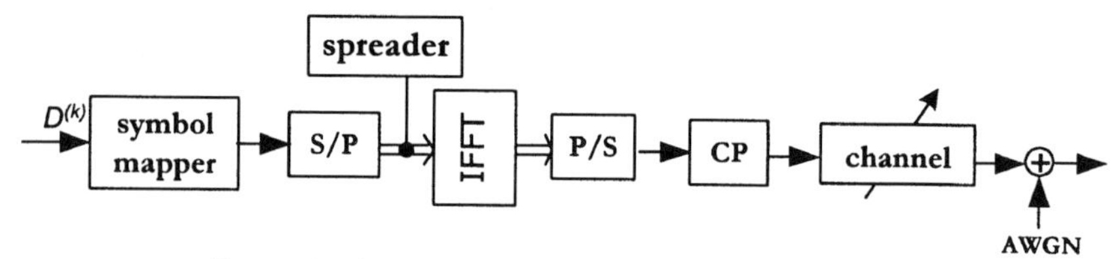

Figure 1: The OFDM-CDMA transmission scheme.

The first treated case is that when all the active user's signals reach the *BS* at the

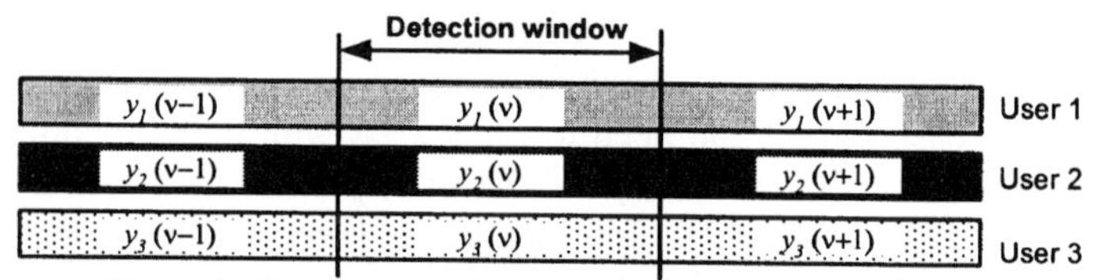

Figure 2: Synchronous detection model at the base station.

same time. In such case the frames structures received does not need major measures of synchronisation [2]. Thus, the signals received at the *BS* can be expressed as in [4] by

$$\mathbf{r}_{syn}(v) = \mathbf{y}_i(v) + \sum_{\substack{k=1\\k\neq i}}^{K} \mathbf{y}_k(v) + \mathbf{n}(v) \tag{2}$$

where the vector $\mathbf{r}_{syn}(\mathrm{v})$ represent in this case the total received signal, and $\mathbf{n}$ (v) represent the additive white Gaussian noise. When the detected signals experience a synchronism loss means that the signals reach the base station with different time offsets. The observed delays are closely linked with the distance between the mobile station (MS) of each user and the base station. Considering this signal users' experience at the BS is a more realistic scenario for the general reverse mode wireless communication [4, 6]. In this case the signals are detected asynchronously (with synchronism loss) at the *BS* (see figure 3) can be expressed as

$$\mathbf{r}^{i}_{asyn}(\nu)=\sum_{k=1,k<i}^{i-i}\underset{k<i,i}{\mathbf{r}}(\nu)+\underset{i,i}{\mathbf{r}}(\nu)+\sum_{k>i}^{K}\underset{k>i,i}{\mathbf{r}}(\nu)+\mathbf{n}(\nu) \tag{3}$$

Where the first term of (3) $\underset{k<i,i}{\mathbf{r}}(\nu)$, represent the multiple access interference (MAI) caused essentially by all the users' signals whose reach the BS before ($\tau_k < \tau_{i,}$) ($k\neq i$, $k=1...K$) the desired signal of the user i, the second sum express the desired detected signal of the users i without any interference, and the third sum encompass the signals of all the interferers whose reach the BS after the user i with a delay $\tau_k > \tau_i$ ($k \neq i$, $k=1...K$).

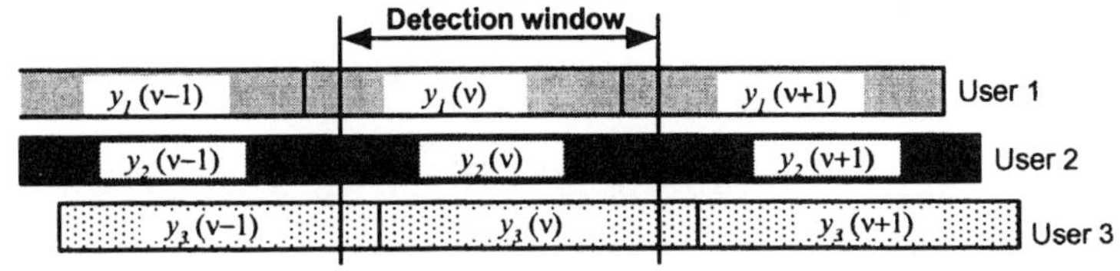

Figure 3: Asynchronous detection scheme at the base station .

The values $\underset{k<i,i}{\mathbf{r}}(\nu)$ and $\underset{k>i,i}{\mathbf{r}}(\nu)$ in (3) are defined as

$$\begin{aligned}\underset{k<i,i}{\mathbf{r}}(\nu)=&\left[\Gamma(k,i)\mathbf{H}_1^k\mathbf{T}\mathbf{F}^{-1}\mathbf{c}_k\right]D^k(\nu-1)+\left[\Gamma(k,i)\mathbf{H}_0^k\mathbf{T}\mathbf{F}^{-1}\mathbf{c}_k\right.\\&\left.+\Psi(k,i)\mathbf{H}_1^k\mathbf{T}\mathbf{F}^{-1}\mathbf{c}_k\right]D^k(\nu)+\left[\Psi(k,i)\mathbf{H}_0^k\mathbf{T}\mathbf{F}^{-1}\mathbf{c}_k\right]D^k(\nu+1)\end{aligned} \tag{4}$$

and

$$\begin{aligned}\underset{k>i,i}{\mathbf{r}}(\nu)=&\left[\Gamma(k,i)\mathbf{H}_1^k\mathbf{T}\mathbf{F}^{-1}\mathbf{c}_k\right]D^k(\nu-2)+\left[\Gamma(k,i)\mathbf{H}_0^k\mathbf{T}\mathbf{F}^{-1}\mathbf{c}_k\right.\\&\left.+\Psi(k,i)\mathbf{H}_1^k\mathbf{T}\mathbf{F}^{-1}\mathbf{c}_k\right]D^k(\nu-1)+\left[\Psi(k,i)\mathbf{H}_0^k\mathbf{T}\mathbf{F}^{-1}\mathbf{c}_k\right]D^k(\nu)\end{aligned} \tag{5}$$

Both matrices $\Gamma(k,i)$ and $\Psi(k,i)$ with size ($P \times P$) represents the synchronism loss degree in the *k-th* interferer user ($k \neq i$ and $k = 1,..., K$) when the signal of the *i-th* users is detected [6, 7].

In this proposal, we have developed the results obtained from the asynchronous detection case in [6] using more complex detectors at the *BS*, as the minimum mean square error (MMSE) detector coupled with a parallel interference canceller (PIC) scheme to reduce the interferences of the unwanted users' signals. Note that the reconstruction of the multiple access interference (MAI) signal within the PIC canceller must take into account the different delays (τ_i , $i=1...K$) (see figure (4)). The PIC performance depends essentially on the quality of the MAI estimated and regenerated with the coefficients of the channel transmission and the corresponding data's of the different active users. By that reason, the PIC performance yields it's closely linked with the behaviour of the initial stage, which is the first estimation process of the transmitted data by the different active users as it is indicated in equation (6), values **a**, **e**, and **b** are defined in [6,7].

$$\hat{D}_{j,i} = Q\left\{\left(\mathbf{c}^i\right)^H \mathbf{G}_i \left(\mathbf{r}^i{}_{asyn} - \sum_{\substack{k=1 \\ k\neq i}}^{K} \mathbf{s}^{k,i}_{MAI,[j-1]}\right)\right\}, \text{ with } j > 0 \tag{6}$$

$$\mathbf{s}^{k,i}_{MAI,[j-1]} = \underbrace{\mathbf{a}^{k,i}\hat{D}^k_{j-1}(\nu-1) + \mathbf{e}^{k,i}\hat{D}^k_{j-1}(\nu) + \mathbf{b}^{k,i}\hat{D}^k_{j-1}(\nu+1)}_{\mathbf{MAI}}$$

The calculus of the MMSE is equivalent to evaluate the trace of the covariance operation in (6) with minimizing the matrix **M**

$$\min_{\mathbf{M}} E\left\{\left\|\mathbf{d}(\nu) - \mathbf{M}\,\underline{\mathbf{r}}_{asyn}\right\|^2\right\} = \min_{\mathbf{M}}\left\{tr\left(\text{cov}\left(\mathbf{d}(\nu) - \mathbf{M}\,\underline{\mathbf{r}}_{asyn}\right)\right)\right\} \tag{7}$$

where *tr* (.) is the trace and *Cov* (.) the covariance. The covariance operation in (7) permits to define the minimizes factor in the *MMSE* detector in such way that

$$Cov\left(\mathbf{d}(\nu) - \mathbf{M}\,\underline{\mathbf{r}}_{asyn}\right) = E\left\{\left(\mathbf{d}(\nu) - \mathbf{M}\,\underline{\mathbf{r}}_{asyn}\right)\left(\mathbf{d}(\nu) - \mathbf{M}\,\underline{\mathbf{r}}_{asyn}\right)^H\right\}$$

$$= \mathbf{I} - \mathbf{M}\mathbf{S}_0 - \mathbf{S}_0^H\mathbf{M}^H - \mathbf{M}\left(\underbrace{\mathbf{S}_0\mathbf{S}_0^H + \mathbf{S}_{-1}\mathbf{S}_{-1}^H + \mathbf{S}_{+1}\mathbf{S}_{+1}^H + \sigma^2\mathbf{I}}_{\mathbf{B}}\right)\mathbf{M}^H \tag{8}$$

$\mathbf{s}_{-1}$ and $\mathbf{s}_{+1}$ are the vectors that affect the past and the further symbols respectively, and $\mathbf{s}_0$ concern the actual symbols. Let to introduce the factor $\overline{\mathbf{M}} = \mathbf{S}_0^*\mathbf{B}^{-1}$ in order to express equation (8) in a compact form

$$\text{cov}\left(\mathbf{d}(\nu) - \mathbf{M}\,\underline{\mathbf{r}}_{asyn}\right) = \left(\mathbf{I} - \mathbf{S}_0^H\mathbf{B}^{-1}\mathbf{S}_0\right) + \left(\mathbf{M} - \overline{\mathbf{M}}\right)\mathbf{B}\left(\mathbf{M} - \overline{\mathbf{M}}\right)^* \tag{9}$$

From (9), for minimizing the trace of the covariance matrix we assume that the matrix **B** is defined positive. If we substitute the $\overline{\mathbf{M}}$ matrix values in the **M** matrix the conditional minimum trace will be

$$\mathbf{M} = \mathbf{S}_0^H\left(\mathbf{S}_{-1}\mathbf{S}_{-1}^H + \mathbf{S}_0\mathbf{S}_0^H + \mathbf{S}_{+1}\mathbf{S}_{+1}^H + \sigma^2\mathbf{I}\right)^{-1} \tag{10}$$

The detection can be finally performed by $\hat{\mathbf{d}} = Q\{\mathbf{M}.\mathbf{r}\}$, where $Q\{.\}$ indicate the quantification process. The minimum error will be based on the values

$$\varepsilon_{\min} = tr\left(\mathbf{I} - \mathbf{S}_0^H\mathbf{B}^{-1}\mathbf{S}_0\right) \tag{11}$$

3. SIMULATION RESULTS

All along simulation tests, we have considered channel parameters knowledge obtained after a channel estimation process (CSI: *Channel State Information*) and also for the arrival time of all the users and their respective code word. We assume that all the transmission delays concerning each users signal reaching the BS have been measured properly during the frame detection process. The system frequency bandwidth W= 3 KHz and Δf= 93.75 Hz. The length of the spreading code vector is N=32. The used code is an orthogonal Gold code in view of the fact of its correlation characteristics with a length L_{sc}=32 [6]. The length of the cyclic prefix (*CP*) is L_{cp}=6, two channel paths are implemented for each user, and the channel power is normalized in the frequency domain. The total length of each OFDM frame is P (N_c +L_{cp}). The different delays τ_i have been generated randomly in such way that the maximum signal delay detection at the BS does not exceed 10 T_c . Neither, the channel encoder nor the interleaver block have been considered in our simulations.

We have used in the synchronous detection mode at the BS a simple OFDM-CDMA decorrelators that detect the transmitted data of the different active users used in [6, 7]. For the asynchronous case we have used at the BS the decorrelator developed in [7], where all the past, current and further symbols that actuate as interferers in the detection of each desired user's signal have been considered.

Figures (5), (6) and (7) shows the bit error rate (BER) performances using different values of signal noise ratios (SNRs) for the two detection modes. Figures (5) and (6) allow us to have a notion of the system damage comparison between the synchronous and asynchronous detection modes. However, figure (7) shows the BER performance of the OFDM-CDMA system with synchronism loss, firstly with the MMSE detector and secondly when it is combined with a parallel interference canceller (MMSE-PIC). The figures (8) and (9) encompass the results obtained in figures (5), (6) and (7) but focused from point of view of the system capacity in terms of the number of active users supported by the base station for specific values of bit error rates.

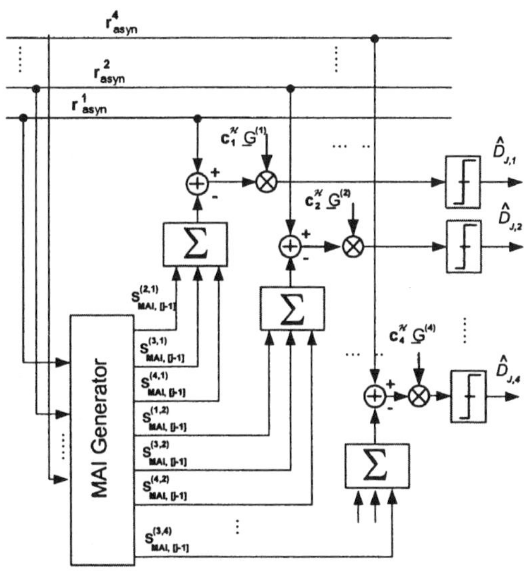

Figure 4: The structure of the parallel interference canceller (PIC) for four active users adapted to the reverse transmission mode with synchronism loss

4. CONLUTIONS

We develop in this proposal the analysis of the synchronism loss effect at the base station of an OFDM-CDMA system using the MMSE and the MMSE-PIC detectors. The analysis has focused firstly on a comparative study between the performance of the synchronous and the asynchronous detection mode at the BS used in [6]. It is clear if we compare the two figures (5) and (6) we can note that the synchronism loss affect negatively the system performance. For four active users at the BS and a signal noise ration of 10 dB, we can achieve a BER near to 10^{-3}. On the other hand, to obtain the same BER value in the asynchronous detection mode it is necessary to work with an SNR margin greater or equal to 20 dB. In few words, we should work in an SNR margin two times that used in the synchronous detection whether we wont to obtain the same performance.

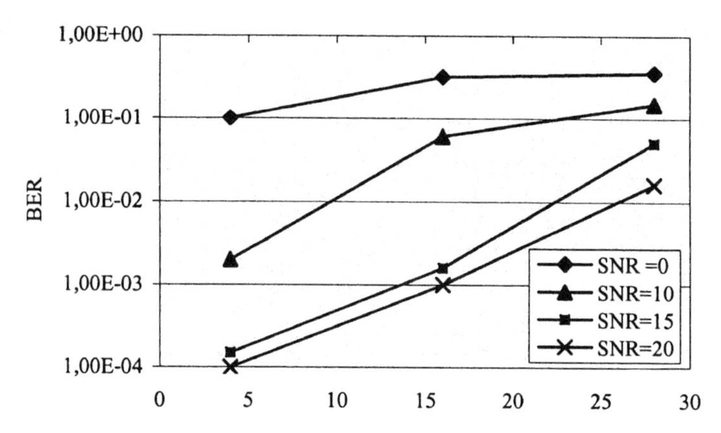

Figure 5 :The BER performance versus the number of active users at the BS with a synchronous detection mode. The SNR values are in [dB].

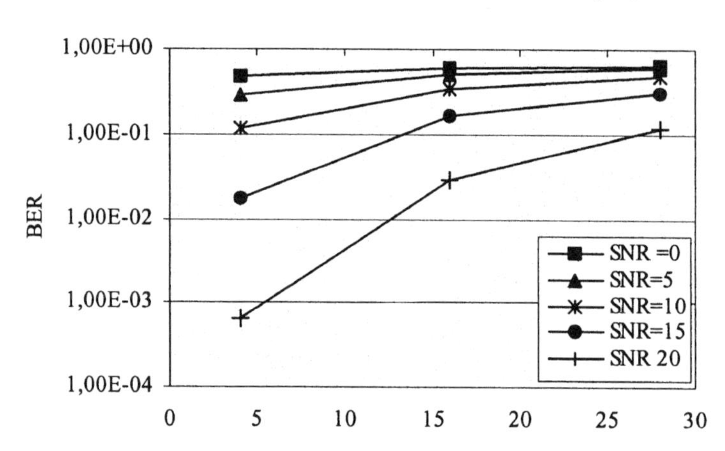

Figure 6: The BER performance versus the number of active users at the BS with the asynchronous detection mode. The SNR values are in [dB].

The performance of the decorrelator simulated in figure (5) can be improved using other detectors as the MMSE detector, also the performance obtained with this last detector can be up performed substantially with a PIC canceller. The system performance with a BER = 10^{-3} is achieved in the case when the asynchronous decorrelator developed in [6] and the MMSE detector worked under a minimum value of SNR equal to 20 dB. Nevertheless the same BER performance is achieved with a minimum value of SNR = 15 whether a MMSE-PIC detection strategy is employed.

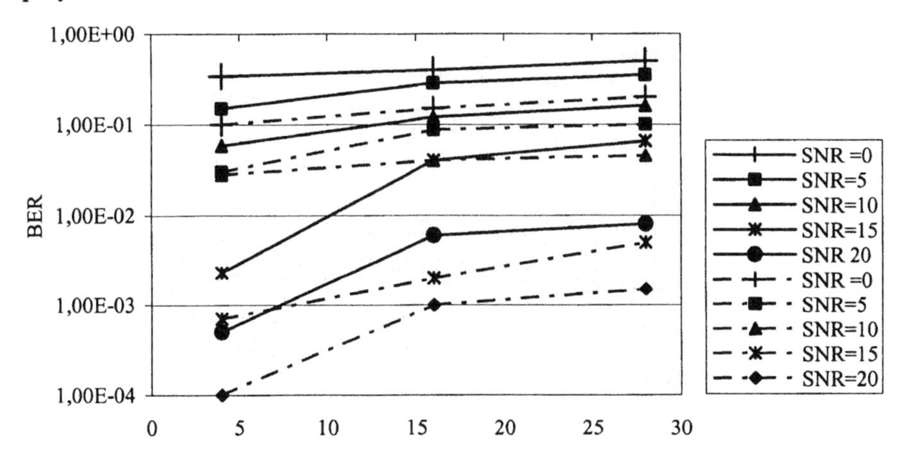

Figure 7: The BER performance versus the number of users for the asynchronous detection using both detectors (solid lines for the MMSE and dashed lines for the MMSE-PIC) at the BS. The SNR values are in [dB].

Concerning the possible active users supported by the BS with these different detectors. It is clear (see figures (8, 9)) that in the case where the system experience a synchronism loss (no major than 10 T_c) the MMSE-PIC strategy achieves a maximum number of possible active users at the BS. The number of active users in figure (9) for the SNRs values (15, 20) are not represented for MMSE nor the MMSE-PIC detector, because working from an SNR=15 with this two systems we will always obtain a BER system performance lower than 10^{-2} with a full loaded system capacity (*means that the number of active users supported by the BS is upper than 30 (K = 32)*).

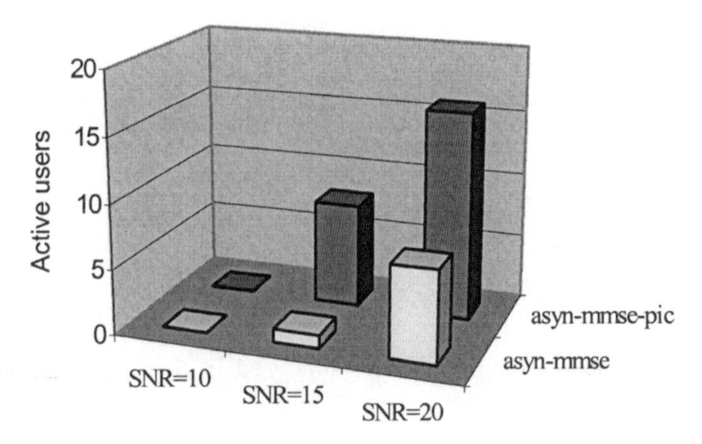

Figure 8: The number of active users supported by the BS for a BER=10^{-3}, using both detectors strategies (MMSE and the MMSE-PIC).

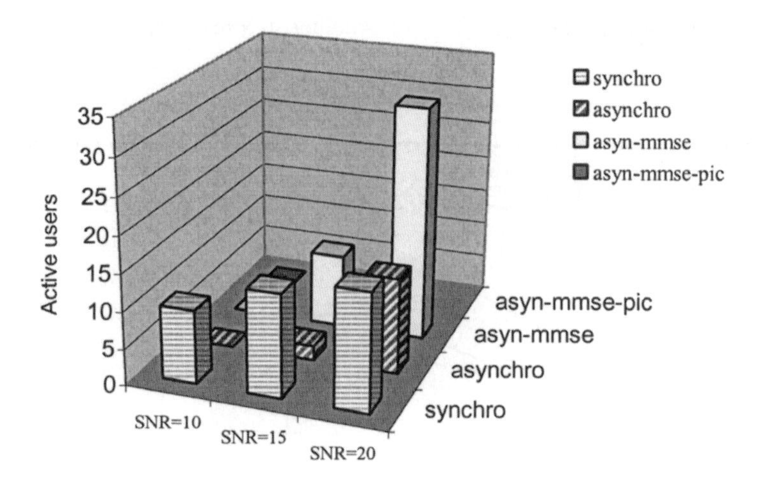

Figure 9: The number of active users supported by the BS for a BER=10^{-2}, using the different detection modes and detectors.

Note that in our simulations we do not take into account of any kind of coding, which in any case will improve the obtained performance.

5. REFERENCES

[1] S. Hara, R. Prasad, "DS-CDMA, MC-CDMA and MT-CDMA," in the Proceedings of the IEEE *Vehicular Technology Conference* 1996 (VTC'96), pp. 1680- 1686. Atlanta, USA.

[2] S. Kaiser, "Multi-Carrier Mobile Radio Systems, Analysis and Optimisation of Detection, Decoding and Channel Estimation," Ph. D Thesis published with, VDI Verlag. Dusseldorf. Germany 1998.

[3] Bernhard Haindl, "Multi-Carrier CDMA for Air Traffic Control Air/Ground Communication", - *Spread Spectrum & Related Topics*. Edited by Khaled Fazel and Stefan Kaiser in Kluwer Academic Publishers (KAP) © 2002. ISBN 0-7923-7653-6. pp. 77-84.

[4] Sergio Verdú,- Multiuser Detection-. Ed. Cambridge University Press ©1998.

[5] Zhengdao Wang, G. B. Giannakis, "Wireless Multicarrier Communications Where Fourier Meets Shannon," in the Proceedings of the IEEE *Signal Processing Magazine*. pp 29 -48. May 2000.

[6] F. Bader, S. Zazo,I. Raos," Improvement on the Multi-User Detection Decorrelator of a MC-CDMA Used in the Reverse Link," in the proceedings of the IEEE of the 13TH International *Symposium on Personal, Indoor and Mobile Radio Communication* (PIMRC'2002). Sept 2002. Lisbon, Portugal.

[7] F. Bader, S. Zazo," MC-CDMA System Evaluation Using the MMSE Decorrelator and the PIC in the Reverse Link Over an Asynchronous Channel Environmen," in the proceedings of the IEE *European Mobile Personal Communications Conference* (EMPCC'2003), Glasgow- Scotland, UK. 22-25 of april 2003.

AFFILIATIONS

Faouzi Bader
Centre Tecnològic de Telecomunicacions de Catalunya- CTTC
C/ Gran Capità, 2-4.
08034- Barcelona, Spain.
Faouzi.bader@cttc.es

Santiago Zazo
Universidad Politécnica de Madrid
ETSI de Telecomunicación
S/n Ciudad Universitaria
28040-Madrid, Spain.
Santiago@gaps.ssr.upm.es

RALF IRMER, WOLFGANG RAVE AND GERHARD FETTWEIS

MINIMUM BER MULTIUSER TRANSMISSION FOR SPREAD-SPECTRUM SYSTEMS IN FREQUENCY-SELECTIVE CHANNELS

Vodafone Chair Mobile Communications Systems
Dresden University of Technology, Germany

Abstract. Spread Spectrum CDMA systems suffer from interference in frequency-selective channels. For the uplink, Multiuser Detection (MUD) mitigates the interference. For the downlink, Multiuser Transmission (MUT) is possible, provided that an estimation of the channel impulse response is available. A novel MUT approach is proposed, which outperforms other MUT methods, like Joint Transmission (JT) and die Transmit Wiener filter. The Bit Error Probability (BER) at the receivers is minimized directly, by calculating appropriate transmit-symbol pre-processing coefficients. For this approach, nonlinear optimization algorithms are used. The focus of this paper is on the comparison of Minimum BER MUT in conjunction with Rake and Pre-RAKE. The complexity of the numerical optimization algorithm is also investigated.

1. INTRODUCTION

Spread-Spectrum CDMA systems are subject to Multiple Access Interference (MAI) in frequency-selective channels, which can be mitigated by Multiuser Detection (MUD). This is in most cases only feasible for the uplink from the mobile to the base station (BS). However, the downlink (DL) usually carries the most traffic for multimedia applications. In the downlink direction, the signal can be pre-processed in the transmitter before transmission to improve the signal quality in the mobile receivers. The transmitter-based approaches are called Multiuser Transmission (MUT). One requirement for MUT is advance knowledge about the wireless channel impulse response (CIR) in the transmitter. For Time Division Duplex (TDD) the CIR is measured in the BS in the uplink and can be used for subsequent downlink pre-processing due to the channel reciprocity.

CDMA systems where MUT and MUD is reasonable include Multi-Carrier Spread-Spectrum (MC-SS) CDMA and Direct-Sequence Spread-Spectrum (DS-SS) CDMA, where in both cases time-division processing is possible [1]. In this paper, 3GPP TDD-CDMA and TD-SCDMA are considered exemplary [2].

Most of the proposed concepts of multiuser transmission (MUT) use a linear transformation, i.e. data-independent matrix multiplication or filtering in the transmitter. The linear Zero-Forcing MUT concepts are known as Transmitter Precoding [6], Joint Transmission [7], Joint Pre-Distortion or Joint Signal Precoding [8]. The linear Transmit Wiener filter [9] or MMSE exploits additionally the noise

K. Fazel and S. Kaiser (eds.), Multi-Carrier Spread-Spectrum, 429–438.

variance knowledge, thus achieving a better performance in noise-dominated scenarios.

In this paper, a novel concept to calculate the pre-processor is proposed which exploits the knowledge about the transmitted data sequence. The performance figure for the TDD-CDMA DL is the BER at the receivers for a limited total transmit power. Therefore, instead of optimizing criteria like Zero-Forcing or MMSE, we propose to minimize the BER directly. In [3], Minimum BER Multiuser Transmission is introduced, but only in conjunction with a RAKE receiver like in [6]. This is extended to multiple antennas in [4], where also the impact of non-ideal channel knowledge is investigated. A similar way is followed in [7]. For the MUD problem, the BER is optimized directly in [10].

Each transmitted symbol has an impact on the received symbols. This is expressed by the symbol-to-symbol system matrix. It includes the influence of the spreading codes, the instantaneous channel impulse response, and the receiver (i.e. code-matched filter or RAKE) and transmitter filter coefficients (i.e. Pre-RAKE. Using the actually transmitted symbols, which are pre-processed by a symbol- and user-specific coefficient, and the system matrix, the received symbols can be predicted. Using the additive noise variances, the BER at the detectors can be predicted already in the transmitter. The BER is then minimized with respect to all pre-processing coefficients, where also a transmit power constraint has to be fulfilled. Unfortunately, the calculation of the pre-processor coefficients is nonlinear and can be solved only numerically. State-of-the-art nonlinear optimization algorithms like Sequential Quadratic Programming (SQP) [12] can be used to find the BER-minimizing pre-processor coefficients. Although a global optimum can not be guaranteed, simulations show significant performance improvements using this scheme compared to conventional transmission without pre-processing and also compared to linear MUT methods. Additionally, using the exact received symbol calculation, the BER can be predicted for any MUT method, including linear Joint Transmission. Thus simulation speed for high SNR values can be increased considerably.

In this paper, Pre-RAKE and RAKE in conjunction with symbol-based Minimum BER Multiuser Transmission and other MUT methods are compared. Furthermore, the convergence behaviour and the computational complexity of the iterative optimization algorithm are investigated.

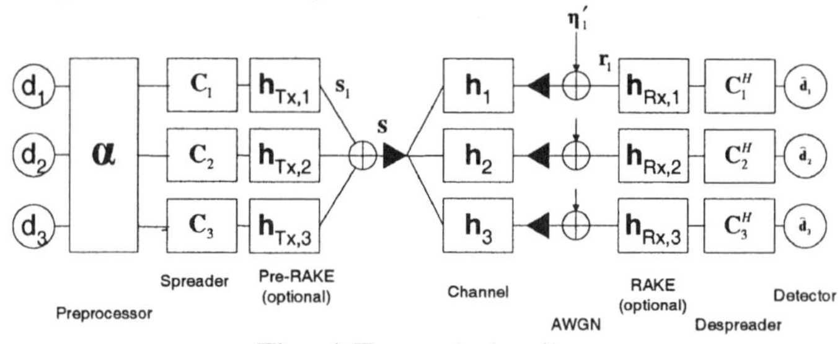

Fig. 1 Transmission line

2. SYSTEM MODEL

A CDMA downlink with U active users is shown in Fig. 1. In this paper, only one Tx antenna in the BS and one receive antenna are considered. Without loss of generality QPSK is assumed for modulation, with the symbols $d_{u,n} \in \sqrt{\frac{1}{2}}\{1+j,-1+j,1-j,-1-j\}$. They are organized in the vector $\mathbf{d}_u$ for user u, and in $\mathbf{d}=\left[\mathbf{d}_1^T,..,\mathbf{d}_U^T\right]^T$. They are then multiplied by the pre-processing matrix $\mathbf{T}$, which has diagonal elements $\alpha_{u,n}$ for the case of Minimum BER Multiuser Transmission or an identity matrix, if conventional transmission is used. The user specific spreading code (here: short code) $\mathbf{c}_u=\left[c_u(1),..,c_u(G)\right]^T$ with spreading gain G is arranged in the spreading matrices $\mathbf{C}_u=\text{blockdiag}\left(\mathbf{c}_u,..,\mathbf{c}_u\right)\in\mathbb{C}^{GN\times N}$ and $\mathbf{C}=\text{blockdiag}\left(\mathbf{C}_1,..,\mathbf{C}_U\right)\in\mathbb{C}^{UGN\times UN}$. The spread sequence $\mathbf{s}'_u=\mathbf{C}_u\mathbf{d}_u\in\mathbb{C}^{GN\times 1}$ can be optionally filtered by a user specific FIR filter with the impulse response $\mathbf{h}_{Tx,u}$. The convolution can be expressed by the multiplication with the Toeplitz matrix $\mathbf{H}_u\in\mathbb{C}^{GN+L_{Tx}-1\times GN}$ if the transmitted signal is allowed to be longer, or with $\mathbf{H}_u\in\mathbb{C}^{GN\times GN}$ if the transmitted signal is required to have length GN. The former is considered in this paper. For all users, the Block-Toeplitz Tx filter matrix is $\mathbf{H}_{Tx}=\left[\mathbf{H}_{Tx,1},..,\mathbf{H}_{Tx,U}\right]\in\mathbb{C}^{GN+L_{Tx}-1\times UGN}$. The total transmit vector is

$$\mathbf{s}=\mathbf{H}_{Tx}\mathbf{C}\mathbf{T}\mathbf{d}\in\mathbb{C}^{GN+L_{Tx}-1\times 1}. \tag{1}$$

The frequency-selective downlink channel for each user is modelled as a block-constant chip-spaced tapped-delay line with maximum length L. The channel impulse response (CIR) of user u is $\mathbf{h}_u=\left[h(0),..,h(L-1)\right]^T$. With the Toeplitz channel convolution matrices $\mathbf{H}_u$ and $\mathbf{H}=\left[\mathbf{H}_1^T,..,\mathbf{H}_U^T\right]^T\in\mathbb{C}^{U(GN+L_{Tx}+L-2)\times GN+L_{Tx}-1}$, the received signal is

$$\mathbf{r}_u=\mathbf{H}_u\mathbf{s}+\boldsymbol{\eta}'_u \text{ and } \mathbf{r}=\mathbf{H}\mathbf{s}+\boldsymbol{\eta}'\in\mathbb{C}^{U(GN+L_{Tx}+L-2)\times 1} \tag{2}$$

for user u and all users, respectively. The proposed MUT scheme can be applied for systems with any linear receiver. In the downlink, the receivers can usually not cooperate, in contrast to multi-layered MIMO transmission.

As one option, a simple code-matched filter could be used in the receivers, making them very cheap. As another option, RAKE receivers can be used to improve the performance further. In most terminals, several RAKE fingers have to be present anyway for cell search, channel estimation etc. In a third option, combined transmitter- and receiver FIR filter optimization is possible to maximize the SNR at the detectors.

The receiver filter with length L_{Rx}, represented by $\tilde{\mathbf{H}}_{Rx,u}\in\mathbb{C}^{GN+L_{all}-1\times GN+L_{Tx}+L-2}$, is followed by the despreader (code-matched filter). It correlates at latency time w

with a typical value of $w=\frac{1}{2}(L_{all}-1)$ with $L_{all}=L_{Tx}+L+L_{Rx}-2$. This can be expressed by the submatrix $\mathbf{H}_{Rx,u}=\tilde{\mathbf{H}}_{Rx,u}(w+1,..,w+1+GN,:)$. The Rx matrix for the whole system is $\mathbf{H}_{Rx}=\text{blockdiag}(\mathbf{H}_{Rx,1},..,\mathbf{H}_{Rx,U})\in\mathbb{C}^{UGN\times U(GN+L_{Tx}+L-2)}$. In the case of a code-matched filter only, the Rx filter is omitted or set formally to the identity matrix. Now, the despreader $\mathbf{C}^H$ is applied, leading to the symbol at the detector

$$\hat{\mathbf{d}}=\overbrace{\mathbf{C}^H\mathbf{H}_{Rx}\mathbf{H}\mathbf{H}_{Tx}\mathbf{C}\mathbf{T}}^{\mathbf{B}}\mathbf{d}+\mathbf{C}^H\mathbf{H}_{Rx}\boldsymbol{\eta}'=\tilde{\mathbf{d}}+\boldsymbol{\eta}. \tag{3}$$

System Matrix

The symbol-symbol system matrix $\mathbf{B}\in\mathbb{C}^{UN\times UN}$ in (3) describes the influence of each transmitted symbol to each received symbols. Following, the calculation of its distinct elements is regarded. $\mathbf{B}$ is sparse and has a banded structure, which will be exploited. It is assumed that one symbol is only influenced by the previous, current and next transmitted symbol. This is valid for $L_{all}<2G+1$ for a typical w, or $L\leq G+1$ for a Pre-RAKE or RAKE. For a well-designed short-code system, this condition is usually fulfilled.

The partial short-code cross-correlation functions between the users v and u are

$$\varphi_{v,u}^{(1)}(m)=\sum_{i=0}^{m-1}c_v(G-m+i)c_u^*(i) \text{ and } \varphi_{v,u}^{(2)}(m)=\sum_{i=0}^{G-m-1}c_v(i)c_u^*(i+m). \tag{4}$$

The effects of transmit, channel and receive filters are summarized by $\mathbf{p}_{v,u}=\mathbf{h}_{Tx,v}*\mathbf{h}_u*\mathbf{h}_{Rx,u}\in\mathbb{C}^{L_{all}\times 1}$, where $*$ denotes convolution. The influence of the previous, current and next symbol of user v on the current symbol of user u is

$$\begin{aligned}
\gamma_{a,v,u}&=\sum_{x=w+1}^{L_{all}-1}\varphi_{v,u}^{(1)}(x-w)\,p_{v,u}(x)\\
\gamma_{b,v,u}&=\sum_{x=0}^{w-1}\varphi_{v,u}^{(1)}(G-x-w)\,p_{v,u}(x)+\sum_{x=w}^{L_{all}-1}\varphi_{v,u}^{(2)}(x-w)\,p_{v,u}(x)\\
\gamma_{c,v,u}&=\sum_{x=0}^{w-1}\varphi_{v,u}^{(2)}(G-x-w)\,p_{v,u}(x),
\end{aligned} \tag{5}$$

respectively. For Min BER MUT, only $3U^2$ elements in (5) have to be calculated and stored for the system matrix $\mathbf{B}$ once per block. An alternative calculation of the system matrix can be found in [5].

MINIMUM BER MULTIUSER TRANSMISSION

In the proposed Minimum BER Transmission, each symbol of each user is multiplied in the transmitter with the complex pre-processing factor $\alpha_{u,n}$, arranged in the vector $\boldsymbol{\alpha}=[\alpha_{1,1},..,\alpha_{1,N},..,\alpha_{U,N}]$. With (5), the n-th noise-free data symbol at the detector of user u (3) is

$$\tilde{d}_{u,n}=\sum_{v=1}^{U}\left(\alpha_{v,n-1}d_{v,n-1}\gamma_{a,v,u}+\alpha_{v,n}d_{v,n}\gamma_{b,v,u}+\alpha_{v,n+1}d_{v,n+1}\gamma_{c,v,u}\right). \tag{6}$$

Note that the pre-distortion $\boldsymbol{\alpha}$ and all interference are included in (6). The AWGN noise variance at the detectors is $\sigma_u^2 = \sigma^2 \mathbf{h}_{Rx,u}^H \mathbf{h}_{Rx,u}$, with $\sigma^2 = \frac{1}{2} N_0$.
For QPSK modulation, the distances of each symbol to the decision thresholds are with (6) :

$$\xi_{I,u,n} = \Re\left(\tilde{d}_{u,n}\right)\operatorname{sgn}\left(\Re\left(d_{u,n}\right)\right) \text{ and } \xi_{Q,u,n} = \Im\left(\tilde{d}_{u,n}\right)\operatorname{sgn}\left(\Im\left(d_{u,n}\right)\right). \tag{7}$$

Note that in contrast to the MMSE, not the distance to the *transmitted symbol*, but the distance to the *decision threshold* is the figure of merit for the BER, which is

$$P_e = \frac{1}{4UN} \sum_{u=1}^{U} \sum_{n=1}^{N} \operatorname{erfc}\left(\frac{\xi_{\mathrm{I,u,n}}}{\sqrt{2}\sigma_u}\right) + \operatorname{erfc}\left(\frac{\xi_{\mathrm{Q,u,n}}}{\sqrt{2}\sigma_u}\right) \tag{8}$$

The BER P_e includes $\boldsymbol{\alpha}$, which should be calculated to minimize(8). The analytic partial derivative of P_e with respect to each pre-processing coefficient $\alpha_{u,n}$ and the analytic Hessian matrix $\mathbf{W}_{Hess}$ of second derivatives can be calculated also [3].
The allowed transmit power of one TDD-burst is E_{Bl}. The transmit power constraint is $g(\boldsymbol{\alpha}) = \mathbf{s}^H\mathbf{s} - E_{Bl} \overset{!}{=} 0$, with $\mathbf{s}^H\mathbf{s} = \boldsymbol{\alpha}^H \mathbf{R}_{Tx} \boldsymbol{\alpha}$ and $\mathbf{R}_{Tx} = \operatorname{diag}(\mathbf{d})\mathbf{C}^H \mathbf{H}_{Tx}^H \mathbf{H}_{Tx} \mathbf{C} \operatorname{diag}(\mathbf{d})$.
Its derivative is $2\mathbf{R}_{Tx}\boldsymbol{\alpha}$. The optimization problem reads now as

$$\boldsymbol{\alpha}_{opt} = \arg\min P_e(\boldsymbol{\alpha}) \quad \text{s.t.} \quad g(\boldsymbol{\alpha}) = 0. \tag{9}$$

4. NUMERICAL OPTIMIZATION

Unfortunately, there is no analytic solution of the nonlinear problem with a quadratic (nonlinear) constraint (9) nor is it convex. Thus there is no guarantee that a local minimum is the global minimum. But state-of-the-art numerical methods show satisfactory results. The algorithm used in this paper is Sequential Quadratic Programming (SQP) [12].
The complex vector $\boldsymbol{\alpha} \in \mathbb{C}^{UN\times 1}$ has to be transformed to the real vector $\boldsymbol{\alpha}' \in \mathbb{R}^{2UN\times 1}$ before optimization, since almost all optimization algorithms require real optimization variables. The result can be mapped back to the complex space.
Starting from an initial vector $\boldsymbol{\alpha}_0$, a quadratic approximation of the objective function $P_e(\boldsymbol{\alpha})$ at the point $\boldsymbol{\alpha}_0$ is calculated and minimized. A line search is performed in the calculated direction of the function minimum. This is repeated iteratively, until a stopping condition is fulfilled. For the quadratic approximation, analytical values for $P_e(\boldsymbol{\alpha})$ at $\boldsymbol{\alpha}_0$ and its derivative and a matrix $\mathbf{W}$ are necessary. The matrix $\mathbf{W}$ should be a positive definite matrix which should mimic the curvature of P_e in $\boldsymbol{\alpha}_0$. On the one hand, the Hessian matrix $\mathbf{W}_{Hess}$ of the second derivatives fulfils the latter requirement, whereas it is not necessarily positive definite. $\mathbf{W}_{Hess}$ can be calculated exactly quite easily. However, the positive definiteness is necessary for a convex solution of the quadratic sub-problem. This is fulfilled for the approximation of the Hessian $\mathbf{W}_{BFGS}$ by the BFGS Algorithm, also

called Quasi-Newton update [12]. The problem is, that $\mathbf{W}_{BFGS}$ is dense, whereas $\mathbf{W}_{Hess}$ is sparse (banded), thus simplifications are difficult.

Choice of Initialization Vector

A good choice of the initialization vector $\boldsymbol{\alpha}_0$ is not only important for fast convergence, but also to find a good local minimum in the non-convex optimization problem:

- An arbitrary initialization vector, like $\boldsymbol{\alpha}_0 = [1,..,1]^T$ or even a random vector offers in most cases already a good performance and a moderate convergence speed.
- Any other MUT method, like the linear methods TxZF and TxWF can deliver an equivalent $\boldsymbol{\alpha}_0$. Thus, the BER performance of that MUT method is already guaranteed and can be increased further by the proposed method. The proposed nonlinear method needs only a very small number of further iterations. For the linear MUT methods, computational efficient implementations or approximations are possible. Using a cascade of linear and nonlinear MUT, the required performance and the necessary complexity can be adjusted according to the specific requirements and abilities.
- Adaptive optimization is proposed as follows to reduce the number of necessary iterations and to achieve better performance. The BER (8) is first optimized for a higher noise variance $\tilde{\sigma}_u^2$ to find a start vector for the actually present σ_u^2. Then the optimum solution $\boldsymbol{\alpha}_{opt}\left(\tilde{\sigma}_u^2\right)$ is used as $\boldsymbol{\alpha}_0$ for σ_u^2. The reduction of $\tilde{\sigma}_u^2$ till σ_u^2 can be repeated iteratively. In the first steps, the termination tolerance can be relaxed considerably. The total number of iterations can be considerably reduced. The reason is, that the error function *erfc* in (8) is almost linear for a large $\tilde{\sigma}_u^2$, thus making the problem almost a linear one. In contrast to that, for a low $\tilde{\sigma}_u^2$, *erfc* in (8) is almost a step function. Thus the problem becomes "more nonlinear" with more local optima. The proposed adaptive optimization first solves an easier substitute problem and uses this solution for the actual problem.

Stopping Conditions

The algorithm iterates until one of the stopping conditions is fulfilled. A local minimum is reached if the value of the objective function derivative is small enough. Furthermore, the maximum number of iterations and a minimum function value are other stopping conditions.

Computational Complexity

	Complexity	Typical Ratio
Function Evaluation	$O(U^2N)$	5 %
Gradient Evaluation	$O(U^2N)$	5 %
BFGS Gradient Update	$O(U^2N^2)$	10 %
Quadratic Sub-Problem	$O(U^3N^3)$	80 %

Computational Complexity Order of Symbol-Based Nonlinear Minimum BER MUT

The computational complexity of the major steps of a single iteration is given in Tab. 1. The major part is the solution of the quadratic sub-problem. It is expected, that the exploitation of the sparsity of $\mathbf{W}_{Hess}$ could further reduce the complexity of the quadratic sub-problem, at least to achieve $O(U^3N)$, i.e. to be independent from the block size N. With an appropriate $\boldsymbol{\alpha}_0$, the number of iterations and the achievable BER performance are also independent from the block size N.

PERFORMANCE EVALUATION

In the simulations, 3GPP-TDD / TD-SCDMA spreading codes (random scrambling codes) with spreading factor $G=16$ are used. The channel has an exponentially decaying power delay profile with four chip-spaced taps.

Fig. 2 shows the performance of the proposed symbol based minimum BER pre-processing scheme for $U=12$ users. Two scenarios are considered, a simple code matched filter receiver (left, with a Pre-RAKE filter in the transmitter) and a RAKE receiver (right). For the later, in the transmitter additionally a filter matched to the effective channel (Pre-RAKE) is used, like in [7],[8],[9]. With that, the RAKE performance is superior to [6] and [3]. For comparison of the proposed algorithm, a conventional channel matched filter (RAKE or Pre-RAKE), a linear zero forcing Joint Transmitter (TxZF) and a Tx Wiener filter (TxWF) are shown. The good agreement of the BER simulation and the proposed analytical prediction (8) is visible in Fig. 2 for all MUT methods. The conventional RAKE/Pre-RAKE saturates at $P_e=10^{-1}$, whereas the TxZF has no saturation, but has a worse performance for the noise-dominated region. The TxWF is the lower bound of the former schemes. The Minimum BER Multiuser Transmission scheme outperforms all other concepts, especially for high SNRs.

In Fig. 3, the necessary SNR to achieve $P_e=10^{-2}$ is plotted for different system loads. The conventional schemes can only support 2-4 users, whereas all MUT methods can be used up to full system load, but with different necessary transmit powers. Interestingly, for the TxZF an additional RAKE receiver makes no sense up to 12 users, in contrast to the TxWF and the proposed nonlinear methods (right. The symbol-based minimum BER MUT approach (right) outperforms the other methods (left). The use of an additional RAKE in the receiver results in a performance gain of about 2 dB.

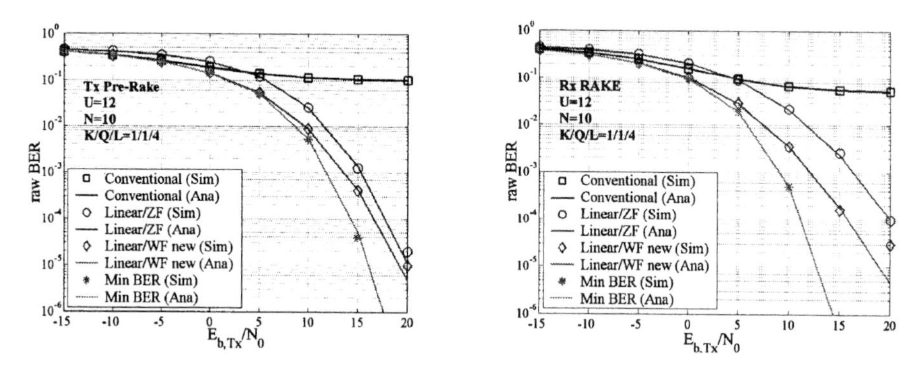

Fig. 2: Uncoded BER performance of different MUT methods, Pre-RAKE configuration (left) and RAKE configuration (right), L=4 channel taps, spreading factor G=16, orthogonal codes (3GPP-TDD / TD-SCDMA)

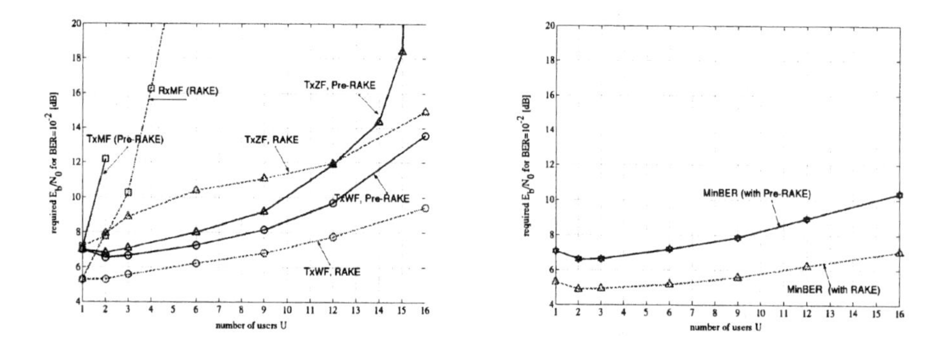

Fig. 3: Required SNR to achieve BER=10^{-2} for linear MUT Methods (left), nonlinear Symbol-Based Minimum BER Multiuser Transmission (right)

CONCLUSIONS

A novel approach to Multiuser Transmission (MUT) for the CDMA downlink in frequency-selective channels was proposed. The key idea is to minimize the BER at the receivers directly. It can be predicted by calculating the signal at the receiver (including interference) exactly and treating only the additive noise at the receiver statistically. Using nonlinear optimization methods like SQP, the BER minimizing coefficients for the symbol pre-processing can be found. Simulations have shown that the proposed approach outperforms other MUT methods. Using a RAKE in the receivers opposed to a simple code-matched filter, performance improvements are possible for the TxWF and the proposed method, but not for the TxZF.

REFERENCES

[1] G. Fettweis, A. Nahler and J. Kühne, "A Time Domain View to Multi-Carrier Spread Spectrum", Proc. IEEE ISSSTA, NY, pp 141-144, Sep. 2000.

[2] R. Esmailzadeh and M. Nakagawa, "TDD-CDMA for Wireless Communications", Artech House, Boston, 2003.

[3] R. Irmer, W. Rave and G. Fettweis, "Minimum BER Transmission for TDD-CDMA in Frequency-Selective Channels", Proc. IEEE PIMRC, Beijing, Sep. 2003.

[4] R. Irmer, R. Habendorf, W. Rave and G. Fettweis, "Nonlinear Multiuser Transmission using Multiple Antennas for TDD-CDMA", Proc. IEEE WPMC, Yokosuka, Oct. 2003.

[5] F. Wathan, R. Irmer, and G. Fettweis, "On Transmitter-Based Interference Mitigation in TDD-Downlink with Frequency-Selective Fading Environment", Proc. Asia-Pacific Conference on Communications (APCC), Bandung, Indonesia, pp 125-128, Sept. 2002.

[6] B. Vojcic and W. Jang, "Transmitter Precoding in Synchronous Multiuser Communications", IEEE Trans. Commun., vol. 46, no. 10, pp 1346-1355, Oct. 1998.

[7] P. W. Baier, M. Meurer, T. Weber, and H. Tröger, "Joint Transmission (JT), an Alternative Rationale for the Downlink of Time Division CDMA using Multi-Element Transmit Antennas", Proc. IEEE ISSSTA, NY , pp. 1-5, Sep. 2000.

[8] A. N. Baretto and G. Fettweis, "Capacity Increase in the Downlink of Spread Spectrum Systems Through Joint Signal Precoding", Proc. IEEE ICC, vol. 4, pp. 1142-1146, 2001.

[9] M. Joham, K. Kusume, M. Gzara, W. Utschick, and J. Nossek, "Transmit Wiener Filter for the Downlink of TDD DS-CDMA Systems", Proc. IEEE ISSSTA, Sept. 2002, vol. 1, pp. 9-13.

[10] S. Chen, A. Samingan, B. Mulgrew, and L. Hanzo, "Adaptive Minimum-BER Linear Multiuser Detection for DS-CDMA Signals in Multipath Channels", IEEE Trans. Signal Processing, vol. 49, no. 6, pp. 1240-1247, 2001.

[11] T.Weber, M. Meurer, and A. Sklavos, "Optimum Nonlinear Joint Transmission". Proc. COST 273 TD(03)008, Barcelona, 2003.

[12] J. Nocedal and S. Wright, "Numerical Optimization", Springer, New York, 1999.

ACKNOWLEDGEMENTS

The authors acknowledge the contributions of Robert Jäschke and Rene Habendorf.

AFFILIATIONS

The authors are with Vodafone Chair Mobile Communications Systems at Dresden University of Technology, Mommsenstrasse 13, D-01062 Dresden, Germany. Email {irmer,rave,fettweis}@ifn.et.tu-dresden.de. Internet www.vodafone-chair.de.

IVAN COSOVIC, MICHAEL SCHNELL AND ANDREAS SPRINGER

COMBINED PRE- AND POST-EQUALIZATION FOR UPLINK TIME DIVISION DUPLEX MC-CDMA IN FADING CHANNELS

Abstract. Equalization at the receiver of multi-carrier code-division multiple-access (MC-CDMA) systems is a well known and well investigated topic in multi-carrier communications. Moreover, pre-equalization techniques for uplink time division duplex (TDD) MC-CDMA systems have been introduced and investigated recently.

In this paper, combined pre- and post-equalization for uplink TDD/MC-CDMA systems in fading channels is proposed. Especially, it is shown that by choosing the corresponding pre- and post-equalization techniques in a preferred way considerable performance improvements are achieved compared to pre- or post-equalization alone. For combined equalization the design criterion for pre-equalization at the transmitter is optimal power assignment with respect to the fading channel rather than suppression of multiple-access interference, whereas multiple-access interference cancellation followed by maximum ratio combining of the almost interference free received signal is the main task at the receiver. As a result, new single-user bounds for MC-CDMA systems transmitting over fading channels and, thus, considerable performance improvements are achieved by applying this novel concept of combined equalization.

1. INTRODUCTION

Uplink time division duplex multi-carrier code-division multiple-access (TDD/MC-CDMA) systems recently attained increasing significance in mobile radio communications [1] [2] [3]. TDD/MC-CDMA operates on the premise, that there is a strong correlation between up- and downlink channel conditions. The main advantage of TDD/MC-CDMA is that the effects of the channel on the uplink signal can be pre-equalized at the mobile station based on the channel estimation from the downlink signal.

Both pre- and post-equalization techniques for TDD/MC-CDMA are well known and well investigated. However, investigations on combined pre- and post-equalization for multi-carrier transmission systems are not yet available. In this paper, it is proposed to combine well-chosen pre-equalization techniques at the transmitter with well-chosen post-equalization techniques at the receiver within an uplink TDD/MC-CDMA system. Moreover, simulation results for TDD/MC-CDMA in independent Rayleigh fading are presented which show that considerable performance improvements are achieved by applying this novel concept of

K. Fazel and S. Kaiser (eds.), Multi-Carrier Spread-Spectrum, 439–450.

combined equalization compared to the case where pre- or post-equalization techniques are applied alone.

The paper is organized as follows. Section 2 describes the considered TDD/MC-CDMA transmission system. In Section 3, the applied pre- and post-equalization techniques are briefly reviewed and the concept of combined equalization is introduced. The performance results achieved with pre-, post-, and combined equalization are presented and discussed in Section 4. Finally, Section 5 summarizes the results.

2. TRANSMISSION SYSTEM

The uplink transmitter of a mobile user $k, k=1,\ldots,K$, within a synchronous TDD/MC-CDMA system is depicted in Fig. 1.

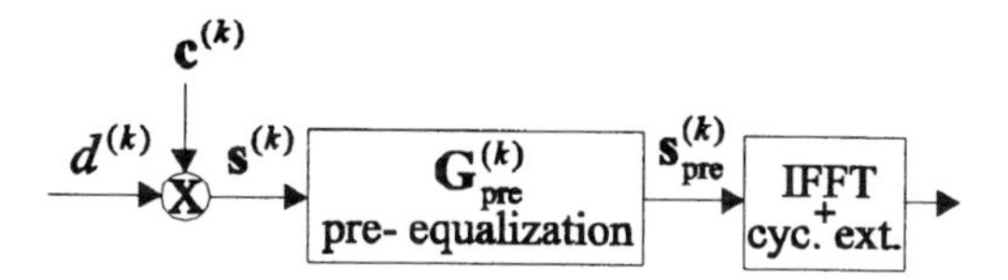

Fig. 1. TDD/MC-CDMA uplink transmitter of mobile user k.

The complex-valued data symbol $d^{(k)}$ is obtained from the symbol mapper and spread by a user-specific spreading code. Throughout this paper, orthogonal spreading is assumed by applying Walsh-Hadamard codes. The length of the spreading code $\mathbf{c}^{(k)} = (c_1^{(k)}, c_2^{(k)}, \ldots, c_L^{(k)})^{\mathrm{T}}$ is L and the duration of the spreading chip is T_c. The data symbol $d^{(k)}$ is of duration $T = LT_c$. The spreading process results in the transmission vector $\mathbf{s}^{(k)}$ given by

$$\mathbf{s}^{(k)} = \mathbf{c}^{(k)} d^{(k)} = (s_1^{(k)}, \ldots, s_L^{(k)})^{\mathrm{T}}, \tag{1}$$

where $(.)^{\mathrm{T}}$ denotes transposition. The transmission vector $\mathbf{s}^{(k)}$ is pre-equalized with an $L \times L$ diagonal pre-equalization matrix $\mathbf{G}_{\text{pre}}^{(k)}$ resulting in

$$\mathbf{s}_{\text{pre}}^{(k)} = \mathbf{G}_{\text{pre}}^{(k)} \mathbf{s}^{(k)} = (s_{\text{pre},1}^{(k)}, \ldots, s_{\text{pre},L}^{(k)})^{\mathrm{T}}. \tag{2}$$

The diagonal elements $G_{\text{pre},l,l}^{(k)}$, $l=1,\ldots,L$, of the pre-equalization matrix $\mathbf{G}_{\text{pre}}^{(k)}$ are calculated from the channel state information derived from the downlink channel estimation. In the following, perfect downlink channel estimation is assumed.

The pre-equalized transmission vector $\mathbf{s}_{\text{pre}}^{(k)}$ is modulated onto the L subcarriers through the inverse fast Fourier transform (IFFT) and finally a guard interval in the form of a cyclic extension is added.

During the propagation through the frequency-selective, time-variant multipath channel, the transmission signal is influenced by the multipath channel itself, by additive white Gaussian noise (AWGN), and by multiple-access interference (MAI).

At the receiver, the received signal after guard interval removal and fast Fourier transform (FFT) is given by

$$\mathbf{r} = \sum_{k=1}^{K} \mathbf{r}^{(k)} + \mathbf{n} = \sum_{k=1}^{K} \mathbf{H}^{(k)} \mathbf{G}_{\text{pre}}^{(k)} \mathbf{s}^{(k)} + \mathbf{n} = (r_1, \ldots, r_L)^{\mathrm{T}}, \tag{3}$$

where the $L \times L$ diagonal matrix $\mathbf{H}^{(k)}$ represents the channel matrix with the diagonal elements $H_{l,l}^{(k)}$. The vector $\mathbf{n} = (n_1, \ldots, n_L)^{\mathrm{T}}$ represents the AWGN with variance σ^2.

In the case of combined equalization, additional post-equalization is applied at the receiver. For equalization at the receiver the well-known techniques [4] [5] can be used. To asses the possible gains of additional post-equalization over pre-equalization alone, it is assumed that the fading channel is perfectly estimated at the receiver. Fig. 2 illustrates the block diagram of the generalized TDD/MC-CDMA uplink receiver as applied for combined equalization. It consists of a minimum mean-square error multi-user detector (MMSE-MUD) and a parallel interference cancellation (PIC) block combined with single-user detection (SUD). For multi-user detection (MUD) and SUD several possible equalization techniques might be used as will be described in the Section 3.2.

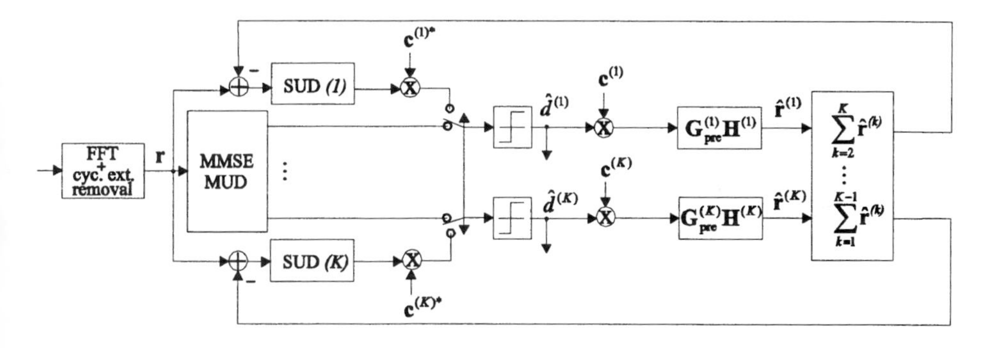

Fig. 2. Generalized TDD/MC-CDMA uplink receiver.

3. PRE-, POST-, AND COMBINED EQUALIZATION TECHNIQUES

3.1 Pre-equalization

In this paper, only constrained pre-equalization is considered where the total transmission power is equal to the transmission power without pre-equalization as introduced in [6]. Note, by applying constrained pre-equalization the transmission power is re-distributed between the used subcarriers and no additional power is invested in the transmission signal.

In the following, the pre-equalization techniques chosen for combined equalization are briefly reviewed. These techniques are maximum ratio combining pre-equalization (pre-eq MRC), quasi minimum mean-square-error pre-equalization (quasi-MMSE), and modified quasi-MMSE pre-equalization (mod quasi-MMSE). For other possible pre-equalization techniques please refer to [3].

Maximum ratio combining pre-equalization weights the transmission signal with a coefficient proportional to the complex conjugate channel fading coefficient, resulting in

$$G_{\mathrm{pre},l,l}^{(k)} = H_{l,l}^{(k)*} \sqrt{\frac{L}{\sum_{l=1}^{L} \left|H_{l,l}^{(k)}\right|^2}}, \tag{4}$$

where the superscript '*' denotes complex conjugation.

The drawback of pre-eq MRC is that it further destroys the orthogonality of the spreading codes and, therefore, increases multiple access interference (MAI). Note, the pre-eq MRC does not take into account information about MAI, but tries to maximize the signal-to-noise ratio (SNR).

Quasi minimum mean-square error pre-equalization does not represent the real uplink MMSE pre-equalization technique. This solution is termed quasi-MMSE due to its similarity with downlink MMSE per-subcarrier equalization (MMSE-PS) [4]. The assigned pre-equalization coefficient is [3] [6]

$$G_{\mathrm{pre},l,l}^{(k)} = \frac{H_{l,l}^{(k)*}}{(K-1)\left|H_{l,l}^{(k)}\right|^2 + \sigma^2 L} \sqrt{\frac{L}{\sum_{l=1}^{L} \frac{\left|H_{l,l}^{(k)}\right|^2}{\left((K-1)\left|H_{l,l}^{(k)}\right|^2 + \sigma^2 L\right)^2}}}. \tag{5}$$

Since MMSE-PS is the best known SUD technique, it is expected that quasi-MMSE pre-equalization performs quite good. Additionally, in [2] it has been shown that quasi-MMSE represents a suboptimal solution of pre-equalization based on maximization of signal-over-interference-plus-noise ratio (SINR). Thus, quasi-MMSE takes into account MAI and tries to minimize its influence.

Quasi-MMSE pre-equalization requires knowledge about both the number of active users K and the noise variance σ^2. This introduces additional computational complexity at the transmitter.

Modified quasi-MMSE) is a suboptimal solution of quasi-MMSE. Mod quasi-MMSE reduces the computational complexity of quasi-MMSE by choosing a fixed value λ that replaces K. The pre-equalization coefficient is given by

$$G_{\mathrm{pre},l,l}^{(k)} = \frac{H_{l,l}^{(k)^*}}{(\lambda-1)\left|H_{l,l}^{(k)}\right|^2+\sigma^2 L}\sqrt{\frac{L}{\sum_{l=1}^{L}\frac{\left|H_{l,l}^{(k)}\right|^2}{\left((\lambda-1)\left|H_{l,l}^{(k)}\right|^2+\sigma^2 L\right)^2}}}. \tag{6}$$

In addition, by setting a fictive number of users equal to λ a certain trade-off between the pre-eq MRC and quasi-MMSE is made. Note, by choosing $\lambda=1$ mod quasi-MMSE reduces to pre-eq MRC. The selection of λ has a great impact on the system performance as will be shown in Section 4.

3.2 Post-equalization

There is a general division of post-equalization techniques into SUD and MUD techniques. The main difference between these two general concepts is that MUD techniques use information about other users in order to reduce MAI, while SUD techniques do not. Investigations of different post-equalization techniques are reported in, e.g. [4]. In the following, only post-equalization techniques relevant to the proposed combined equalization are briefly reviewed.

Single-User Detection techniques are, in principle, much simpler to implement than MUD techniques. A simple and important SUD technique is maximum ratio combining post-equalization (post-eq MRC) which yields equalization coefficient

$$G_{\mathrm{post},l,l}^{(k)} = H_{l,l}^{(k)^*}. \tag{7}$$

Multi-User Detection techniques can be divided into linear and non-linear techniques. An interesting linear technique is MMSE-MUD which is described by the $K \times L$ post-equalization matrix

$$\mathbf{G}_{\mathrm{post}} = \left(\left(\mathbf{B}\mathbf{B}^{\mathrm{H}}+\sigma^2\mathbf{I}_L\right)^{-1}\mathbf{B}\right)^{\mathrm{H}}, \tag{8}$$

where $(.)^{\mathrm{H}}$ denotes the Hermitian transposition and $\mathbf{I}_L$ is an $L \times L$ identity matrix. $\mathbf{B}$ is an $L \times K$ matrix with coefficients

$$b_{l,k} = H_{l,l}^{(k)} c_l^{(k)}, l=1,\ldots,L; k=1,\ldots,K. \tag{9}$$

A promising non-linear MUD technique applied to combined equalization is parallel interference cancellation (PIC). The initial detection stage of PIC is given by

$$\hat{d}^{(k)}[0] = Q\{(\mathbf{G}_{\text{post}}\mathbf{r})_k\}, \tag{10}$$

where a linear MUD technique described by the post-equalization matrix $\mathbf{G}_{\text{post}}$ is used. Note, $Q\{.\}$ denotes the data detection and $(.)_k$ the kth element of a vector.

The following detection stages $(t > 0)$ involve a SUD technique with post-equalization matrix $\mathbf{G}_{\text{post}}^{(k)[t]}$ to produce estimates $\hat{d}^{(k)}[t]$ of the transmitted data symbols. The tth $(t > 0)$ detection stage is described by

$$\hat{d}^{(k)}[t] = Q\left\{\left(\mathbf{c}^{(k)*}\right)^{\mathrm{T}} \mathbf{G}_{\text{post}}^{(k)[t]}\left(\mathbf{r} - \sum_{p=1,p\neq k}^{K} \mathbf{H}^{(p)}\mathbf{c}^{(p)}\hat{d}^{(p)}[t-1]\right)\right\}. \tag{11}$$

As already illustrated in Fig. 2, post-equalization can combine several SUD and MUD techniques together in order to obtain a better performance [4] [5]. When using post-equalization alone, promising results are obtained by applying MMSE-MUD in the initial stage $(t = 0)$ of PIC, followed by several iterations of PIC in combination with a certain SUD, e.g. post-eq MRC.

3.3 Combined pre- and post-equalization

Combining pre- and post-equalization enables an additional degree of freedom within the system design of a TDD/MC-CDMA system. Using pre-equalization alone leads to a design criterion where two contrary goals, namely optimal power distribution with respect to the fading channel and avoidance of MAI, have to be taken into account solely at the transmitter. Considering post-equalization as the only equalization technique no channel-adjusted power distribution can be applied at all. Thus, applying combined equalization allows performing channel-adjusted power distribution at the transmitter and MAI cancellation at the receiver. Moreover, combined equalization applies MRC as final step, since after MAI cancellation the optimal receiver strategy is MRC. This novel approach to equalization within a TDD/MC-CDMA system is capable of improving the performance significantly.

To explain combined equalization and its advantages over pre- or post-equalization alone, the single-user case $(K = 1)$ is considered. Since in this case no MAI is present both transmitter and receiver apply MRC – the transmitter in order to achieve the optimal power distribution with respect to the fading channel, the receiver, because MRC is the optimal combining technique if no MAI is present. The post-equalization coefficients are calculated from the residual fading coefficients which remain after pre-equalization

$$G_{\text{post},l,l}^{(1)} = \left(G_{\text{pre},l,l}^{(1)} H_{l,l}^{(1)}\right)^{*} = \left|H_{l,l}^{(1)}\right|^{2} \sqrt{\frac{L}{\sum_{l=1}^{L}\left|H_{l,l}^{(1)}\right|^{2}}}. \tag{12}$$

The ratio of useful signal power to noise power after post-equalization and despreading at the receiver can be represented as

$$\frac{\mathrm{P}_{\text{signal}}}{\mathrm{P}_{\text{noise}}} = \frac{\left(\sum_{l=1}^{L}\left|H_{l,l}^{(1)}\right|^{4}\right)^{2} \dfrac{L^{2}}{\left(\sum_{i=1}^{L}\left|H_{i,i}^{(1)}\right|^{2}\right)^{2}} \dfrac{\mathrm{P}_{\text{tx}}}{L}}{\sum_{l=1}^{L}\left|H_{l,l}^{(1)}\right|^{4} \dfrac{L}{\sum_{i=1}^{L}\left|H_{i,i}^{(1)}\right|^{2}} \sigma^{2}} = \frac{\sum_{l=1}^{L}\left|H_{l,l}^{(1)}\right|^{4}}{\sum_{l=1}^{L}\left|H_{l,l}^{(1)}\right|^{2}} \frac{\mathrm{P}_{\text{tx}}}{\sigma^{2}}, \tag{13}$$

where $\mathrm{P}_{\text{signal}}$ represents the useful signal power after despreading at the receiver, $\mathrm{P}_{\text{noise}}$ the noise power after despreading at the receiver, and P_{tx} the transmission power.

From Equation (13) it can be concluded that by applying combined channel equalization the SNR can be improved and, thus, the MC-CDMA transmission single-user bound, also known as matched filter (MF) bound, can be significantly enhanced. Note, for $L \rightarrow \infty$ the factor $\left(\sum_{l=1}^{L}\left|H_{l,l}^{(1)}\right|^{4}\right) \Big/ \left(\sum_{l=1}^{L}\left|H_{l,l}^{(1)}\right|^{2}\right)$ converges to the value 2 which corresponds to a gain of 3 dB. In the following section, these theoretical considerations are proved by simulation results.

4. SIMULATION RESULTS

In this section, several simulation results are given that illustrate the performance improvements obtained by the usage of combined channel equalization over using pre- and post- equalization alone. In order to asses the potential of combined equalization, additionally the single-user case is considered. It is shown, that considerably improved single-user bounds are valid for combined equalization.

The underlying mobile radio channel for the simulations is the independent Rayleigh fading channel. QPSK symbol mapping is used, while channel coding is not applied. Spreading length, number of users and applied pre- and post-equalization techniques are varied among the simulations.

Different power constrained pre-equalization techniques for a fully-loaded uplink TDD/MC-CDMA system with parameters $L = 16$ and $K = 16$ are compared in Fig. 3. From Fig. 3 it can be seen that controlled equalization (CE) and quasi-

MMSE outperform other pre-equalization techniques, like pre-eq MRC, equal gain combining (EGC), or zero-forcing (ZF) equalization.

Fig. 4 illustrates the performance of different post-equalization techniques. The focus is on post-equalization techniques which combine MMSE-MUD, PIC and post-eq MRC. It can be seen that the combination of MMSE-MUD with 2 iterations of PIC and post-eq MRC outperforms the performance of MMSE-MUD alone. In this case the spreading length is set to $L=16$ and the number of active users is $K=4,8,12,\text{or } 16$.

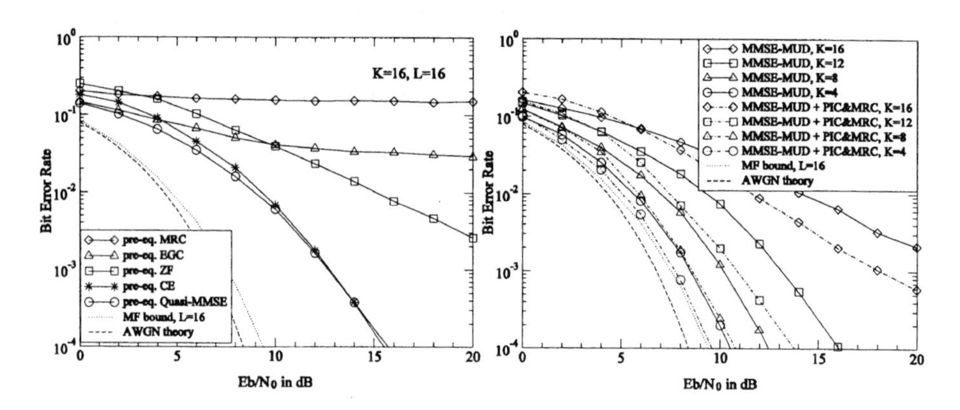

Fig. 3. Performance of basic TDD/MC-CDMA pre-equalization techniques.

Fig. 4. Performance of investigated post-equalization techniques.

Fig. 5 shows the benefits of combined equalization in the single-user case $(K=1)$. The spreading length L is kept variable. Single-user performances for pre- and post-equalization are equal to the MF bound. For combined equalization two different bounds are observed, the MRC-MRC bound and the extreme bound. For the MRC-MRC bound pre-eq MRC and post-eq MRC are applied, while for the extreme bound, all available transmit power is concentrated on only one subcarrier, the subcarrier with the highest SNR. From Fig. 5 it can be seen that in the single-user case the MRC-MRC bound is around 2-3 dB better than the MF bound. Results for different spreading lengths are given for comparison. Moreover, it can be noticed that combined equalization outperforms even the AWGN curve. As already noticed in Section 3.3 for $L \to \infty$ the MRC-MRC bound lies 3 dB below the AWGN bound, while the extreme bound for $L \to \infty$ can achieve the error-free transmission.

In the following, several combined equalization concepts are investigated. The spreading length is set to $L=16$ and the number of active users K is kept variable. For post-equalization, MMSE-MUD followed by 2 iterations of PIC combined with post-eq MRC is applied, while the pre-equalization technique varies for the different concepts. For reference the AWGN performance, MF bound, MRC-MRC bound, the best pre-equalization and the best post-equalization performance are given.

Fig. 6 illustrates the performance gains achieved with combined channel equalization in the case of quasi-MMSE as pre-equalization technique. The performance gains achieved with combined equalization are around 2 dB. In this concept, named 'Concept I', performance gains cannot be larger since the MRC-MRC bound cannot be achieved when quasi-MMSE pre-equalization is applied at the transmitter, except for the single-user case where pre-eq MRC and quasi-MMSE are the same.

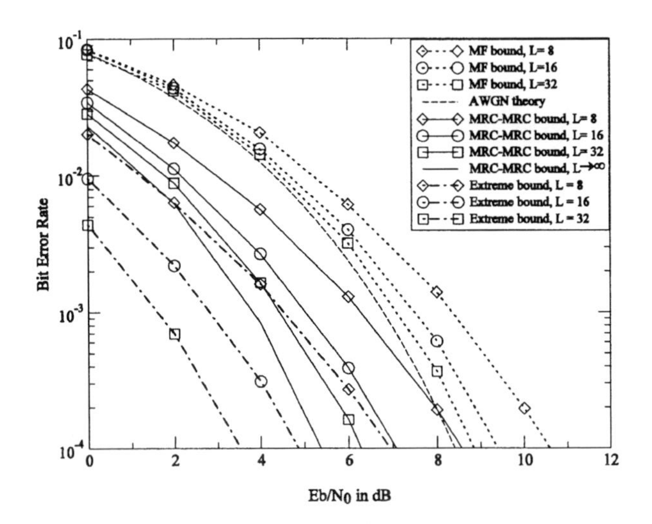

Fig. 5. Single-user bounds of post-, pre-, and combined equalization for different spreading lengths.

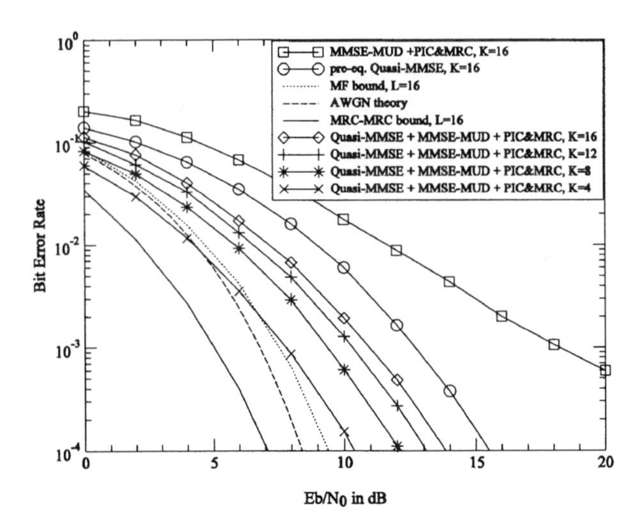

Fig. 6. Combined equalization, Concept I.

Performance gains achieved with 'Concept II' are shown in Fig. 7. 'Concept II' performs pre-eq MRC at the transmitter. In this case the performance of the fully-loaded system is quite bad since MAI is not taken into account at the transmitter, while for lower system loads performance is very good. It can be noticed that already for a half-loaded system the performance is better than the AWGN curve.

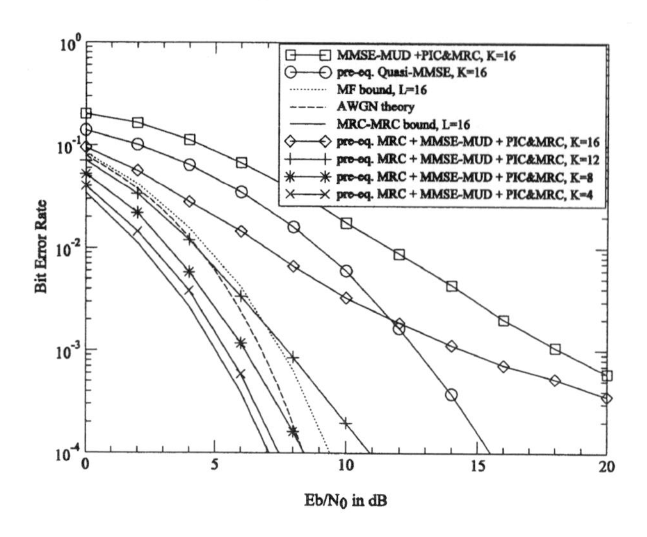

Fig. 7. Combined equalization, Concept II.

Fig. 8 shows the performance achieved with 'Concept III'. 'Concept III' represents a trade-off between investing more power on good subcarriers and partly suppressing MAI at the transmitter. The pre-equalization method applied is mod quasi-MMSE in which the number of active users is set to 2 independent of the actual number of users. This is done by setting $\lambda = 2$ in Equation (6). As it can be seen, the performance for the fully-loaded and the 75 % loaded systems are very good, and nearly achieve the MF bound, while performances for the lower loads are not as good as in the case of 'Concept II'.

Finally, the performance obtained with 'Concept II+III' is shown in Fig. 9. 'Concept II+III' applies 'Concept II' in the case when the system load is lower than 50%, and 'Concept III' in the case when the system load is higher than 50%. Thus, 'Concept II+III' combines the advantages of 'Concept II' and 'Concept III' in a pragmatic approach.

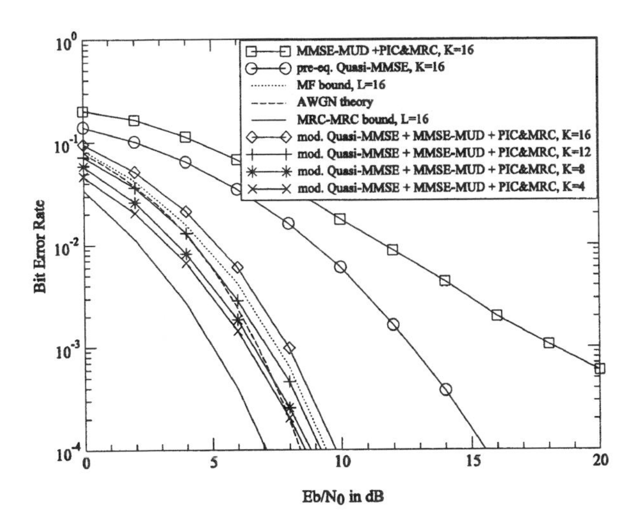

Fig. 8. Combined equalization, Concept III.

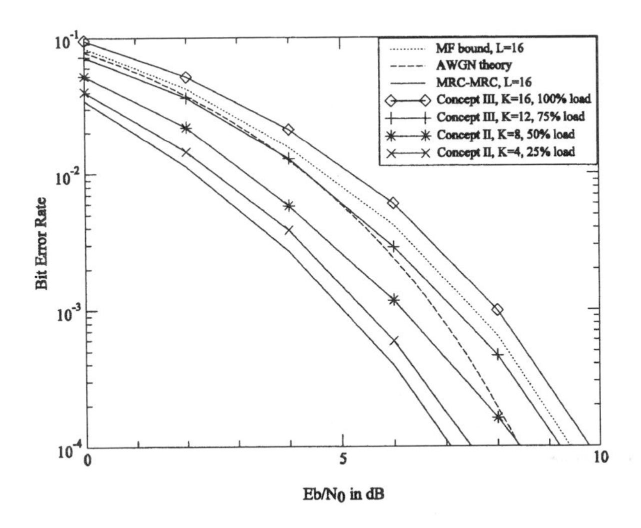

Fig. 9. Combined equalization, Concept II+III.

5. CONCLUSIONS

In this paper the novel concept of combined equalization is proposed which combines pre- and post-equalization. Although the concept is proposed for uplink TDD/MC-CDMA, the same concept can be applied to downlink MC-CDMA and several similar multi-carrier and even single-carrier transmission concepts.

The results show that even for uncoded uplink TDD/MC-CDMA system considerable improvements with combined equalization are achievable. Combined equalization clearly outperforms both pre- and post-equalization applied alone.

The new single-user bounds of combined equalization are presented and it is shown that these bounds indicate considerable performance improvements compared to the well-known MF bounds for MC-CDMA transmission. Due to the improved promising single-user bounds obtained when combining pre-eq MRC and post-eq MRC it can be expected that in systems with channel coding very promising results can be achieved when applying pre-eq MRC in combination with different post-equalization techniques.

6. REFERENCES

[1] D.G. Jeong and M.J. Kim, "Effects of channel estimation error in MC-CDMA/TDD systems," *in Proceedings IEEE Vehicular Technology Conference (VTC'00, Spring)*, pp. 1773-1777, May 2000.

[2] D. Mottier and D. Castelain, "SINR-based channel pre-equalization of fading channels for uplink multi-carrier CDMA systems," *in Proceedings IEEE International Symposium on Personal, Indoor and Mobile Radio Communications (PIMRC'02)*, pp. 1488-1492, September 2002.

[3] I. Cosovic, M. Schnell and A. Springer, "On the performance of different channel pre-compensation techniques for uplink time division duplex MC-CDMA," Accepted for Publication in the Proceedings of the IEEE Vehicular Technology Conference (VTC'03, Fall), October 2003.

[4] S. Kaiser, *Multi-Carrier CDMA Mobile Radio Systems - Analysis and Optimization of Detection, Decoding and Channel Estimation*, Fortschrittberichte VDI, Düsseldorf: VDI Verlag, Reihe 10, Nr.531, 1998.

[5] V. Kühn, "Combined MMSE-PIC in coded OFDM-CDMA systems," *in Proceedings IEEE Global Telecommunications Conference (GLOBECOM'2001)*, pp. 231-235, November 2001.

[6] M. Schnell, *Systeminhärente Störungen bei "Spread-Spectrum" – Vielfachzugriffsverfahren für die Mobilfunkübertragung*, Fortschrittberichte VDI, Düsseldorf: VDI Verlag, Reihe 10, Nr. 505, 1997.

7. AFFILIATIONS

Ivan Cosovic and Michael Schnell are with the German Aerospace Center (DLR), Institute of Communications and Navigation, Oberpfaffenhofen, D-82234 Wessling, Germany. Andreas Springer holds the position of an Assistant Professor at the Institute for Communications and Information Engineering at the University of Linz, A-4040 Linz, Austria.

E-mails: ivan.cosovic@dlr.de, michael.schnell@dlr.de, and a.springer@icie.jku.at.

T. ZEMEN, C.F. MECKLENBRÄUKER AND R.R. MÜLLER

TIME VARIANT CHANNEL EQUALIZATION FOR MC-CDMA VIA FOURIER BASIS FUNCTIONS

Abstract. When users move at vehicular speed the radio channel is time variant. Sayeed et al. proposed a Fourier basis expansion model for time variant channels. We apply this concept to MC-CDMA for channel equalization and give simulation results for the forward link. To gain further insights we give a detailed discussion of the benefits and weaknesses of the Fourier basis expansion channel model.

1. INTRODUCTION

When users move at vehicular speed the radio channel is time variant. For channel equalization an accurate channel model is required otherwise the time variation is a significant cause of symbol errors.

The Doppler effect in multipath propagation is the starting point for the channel description given by Höher in [4]. This description depends in nonlinear manner on the Doppler frequencies for each individual path which is a major obstacle when used for channel equalization.

The representation of the time variant radio channel by means of the scattering function was used by Bello [2]. The continuous time equation that links the impulse response $h(t, \tau)$ to the scattering function $S_{\boldsymbol{H}}(\omega, \tau)$ is subsequently discretized in both time and frequency. The discretization is implemented by replacing the Fourier integral with a finite sum [8, 7]. Therefore the time variant impulse response is expanded linearly in terms of Fourier basis functions.

In this paper our contribution is to analyze the performance of a multi carrier (MC) code division multiple access (CDMA) forward link when the time variant channel is modeled by the Fourier Basis Expansion.

The paper is organized as follows: We introduce the signal model for MC-CDMA in Sec. 2 and describe the Fourier basis expansion model in Sec. 3. We present simulation results and analyze the approximation properties of the Fourier basis expansion model in Sec. 4. We conclude with some remarks in Sec. 5.

K. Fazel and S. Kaiser (eds.), Multi-Carrier Spread-Spectrum, 451–458.

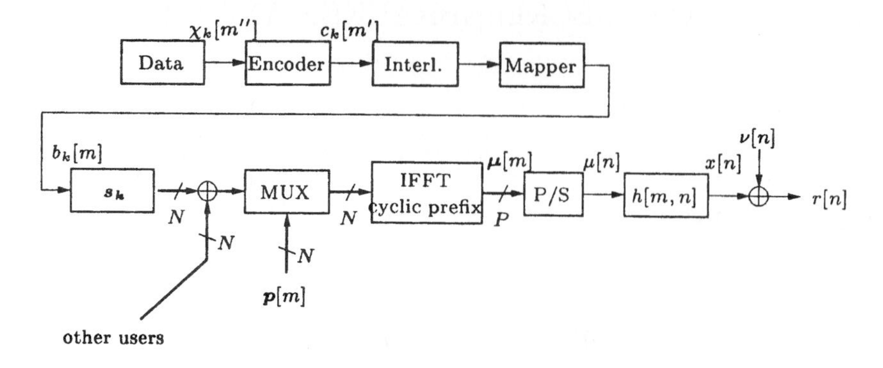

Figure 1. Model for the MC-CDMA transmitter in the forward link.

2. SIGNAL MODEL

In [11] we presented a MC-CDMA multi user receiver for *block fading* channels. MC-CDMA is based on orthogonal frequency division multiplexing (OFDM). Spreading codes are applied in the frequency domain to distinguish each user. Therefore each chip is transmitted over a different subcarrier.

In this paper we will deal with channels which vary significantly over the duration of a data block. To focus the presentation we will limit ourselves to the forward link. The base station transmits quaternary phase shift keying (QPSK) modulated symbols $b_k[m]$ drawn from the alphabet $\frac{1}{\sqrt{2}}\{\pm 1 \pm \mathrm{j}\}$ in blocks of length M. Discrete time is denoted by m. There are K users in the system, the user index is denoted by k. Each symbol is spread by a unique Walsh-Hadamard sequence $\boldsymbol{s}_k$ of length N with elements $\frac{1}{\sqrt{N}}\{\pm 1\}$, see Fig. 1.

We will use the following notation: All vectors are defined as columns vectors and denoted with bold lower case letters. Matrices are given in bold upper case, $(\cdot)^T$ denotes transpose, $(\cdot)^*$ denotes complex conjugate, $(\cdot)^H$ denotes Hermitian (i.e. complex conjugate) transpose and $\boldsymbol{I}_N$ denotes the $N \times N$ identity matrix. The $M \times N$ upper left part of matrix $\boldsymbol{A}$ is referenced as $\boldsymbol{A}_{M\times N}$ and the element on ith row and ℓth column of matrix $\boldsymbol{A}$ is referenced by $[\boldsymbol{A}]_{i,\ell}$. The result of diag$(\boldsymbol{a})$ is a diagonal matrix with the elements of $\boldsymbol{a}$ on its main diagonal.

In each data block there are M data symbols. They result from the binary information sequence $\chi_k[m'']$ of length $2MR_C$ by convolutional encoding, random interleaving and QPSK modulation with Gray labeling. The code rate is denoted by R_C. The spread signals of all users are added together and multiplexed with optional pilot symbols. Then, an N point inverse discrete Fourier transform (IDFT) is performed and a CP (cyclic prefix) with length G is inserted. The resulting signal is transmitted over a time variant multipath fading channel $h(t,\tau)$ with a delay spread of L chips, $T_D = LT_C$. The chip duration is denoted by T_C. We denote the sampled time variant impulse response by $h[m,n] = h(mPT_C, nT_C)$, $\boldsymbol{h}[m] = [h(mPT_C, 0), \cdots, h(mPT_C, (L-1)T_C)]^T$ in vector notation, where $P = N + G$ gives the length of the OFDM symbol in chips.

The receiver removes the CP and performs a discrete Fourier transform (DFT). The received signal vector

$$\boldsymbol{y}[m] = \mathrm{diag}\,(\boldsymbol{g}[m])(\boldsymbol{S}\boldsymbol{b}[m] + \boldsymbol{p}[m]) + \boldsymbol{\nu}[m]$$

where $\boldsymbol{g}[m] = \sqrt{N}\boldsymbol{F}_{N\times L}\boldsymbol{h}[m]$, and $\boldsymbol{\nu}[m]$ denotes complex additive white Gaussian noise with zero mean and covariance $\sigma_\nu^2 \boldsymbol{I}_N$. The unitary DFT matrix $\boldsymbol{F}$ has elements $[\boldsymbol{F}]_{i,\ell} = \frac{1}{\sqrt{N}} \mathrm{e}^{\frac{-\mathrm{j}2\pi i \ell}{N}}$, $i, \ell = 0, \cdots, N-1$. Matrix $\boldsymbol{S}$ is defined as $\boldsymbol{S} = [\boldsymbol{s}_1, \cdots, \boldsymbol{s}_K]$, and $\boldsymbol{b}[m] = [b_1[m], \cdots, b_K[m]]^T$ contains the stacked data symbols for K users.

We define the effective spreading vector for user k as

$$\tilde{\boldsymbol{s}}_k[m] = \mathrm{diag}\,(\boldsymbol{g}[m])\boldsymbol{s}_k,$$

collect all users in matrix $\tilde{\boldsymbol{S}}[m] = [\tilde{\boldsymbol{s}}_1[m], \cdots, \tilde{\boldsymbol{s}}_K[m]]$ and finally get $\boldsymbol{y}[m] = \tilde{\boldsymbol{S}}[m]\boldsymbol{b}[m] + \mathrm{diag}\,(\boldsymbol{g}[m])\boldsymbol{p}[m] + \boldsymbol{\nu}[m]$.

This signal model is valid for time variant channels when the intercarrier interference (ICI) is small. This is true when the (one sided) Doppler bandwidth $B_D = v/\lambda$ is smaller than $\varepsilon = 1\%$ of the subcarrier bandwidth $\Delta f = 1/(NT_C)$ [5]: $B_D < \varepsilon \Delta f$. The speed of the mobile station is denoted by v and the wave length by λ.

The single user receiver is a time variant matched filter $z_k[m] = \frac{\tilde{\boldsymbol{s}}_k[m]^H}{\|\tilde{\boldsymbol{s}}_k[m]\|^2}\boldsymbol{y}[m]$. To calculate the effective spreading sequence we need a model to describe the time variant channel. A possible solution is a Fourier basis expansion model which we describe in the next section.

3. FOURIER BASIS EXPANSION MODEL

The sampled time variant channel impulse response $h[m,n]$ is represented by the scattering function in the Doppler delay domain

$$S_{\boldsymbol{H}}(f,n) = \sum_{m=-\infty}^{\infty} h[m,n]\mathrm{e}^{-\mathrm{j}2\pi m f}.$$

For a wireless system the Doppler bandwidth B_D is generally known, $S_{\boldsymbol{H}}(f,n)$ is therefore band limited and vanishes for $|f| > W$ $h[m,n] = \int_{-W}^{W} S_{\boldsymbol{H}}(f,n)\mathrm{e}^{\mathrm{j}2\pi m f}\mathrm{d}f$, where $W = B_D P T_C$ and $0 < W < 1/2$.

By limiting the time interval to $[0, M-1]$ the scattering function is discretized in the frequency domain, $S_{\boldsymbol{H}}[d,n] = \sum_{m=0}^{M-1} h[m,n]\mathrm{e}^{-\mathrm{j}2\pi md/M}$, which is equivalent to the discrete Fourier transform (DFT). The rectangular windowing results in spectral leakage (see [6] Sec. 5.4). Therefore $S_{\boldsymbol{H}}[d,n]$ does not vanish for $|d| > \lceil WM \rceil$. But $S_{\boldsymbol{H}}[d,n]$ will decay with increasing $|d|$.

The Fourier basis expansion model for the time variant channel is given by $\hat{h}[m,n] = \sum_{d=0}^{D-1} \gamma_d[n] u_d[m]$ with $D = 2\kappa\lceil WM \rceil + 1$ and $u_d[m] = \frac{1}{\sqrt{M}} \mathrm{e}^{\frac{\mathrm{j}2\pi m(d-(D-1)/2)}{M}}$. The parameters $\gamma_d[n]$ are calculated by $\gamma_d[n] = \sum_{m=0}^{M-1} h[m,n] u_d^*[m]$. The mean square approximation error $e' = \sum_{m=0}^{M-1} \sum_{n=0}^{L-1} \left| h[m,n] - \hat{h}[m,n] \right|^2$ is controlled by κ.

To express the basis expansion model in vector matrix notation we introduce the delay time channel matrix with dimension $L \times M$ $\boldsymbol{\mathcal{H}} = \left[\boldsymbol{h}[0], \boldsymbol{h}[1], \ldots, \boldsymbol{h}[M-1]\right]$ where each column in $\boldsymbol{\mathcal{H}}$ represents the time variant impulse response at discrete time m. The frequency time channel matrix $\boldsymbol{\mathcal{G}} = \left[\boldsymbol{g}[0], \boldsymbol{g}[1], \ldots, \boldsymbol{g}[M-1]\right]$ is related to the delay time matrix via $\boldsymbol{\mathcal{G}} = \sqrt{N} \boldsymbol{F}_{N\times L} \boldsymbol{\mathcal{H}}$.

We define the $D \times M$ synthesis matrix $\boldsymbol{U} = \left[\boldsymbol{u}[0], \boldsymbol{u}[1], \cdots, \boldsymbol{u}[M-1]\right]$, and the $D \times M$ analysis matrix $\boldsymbol{V} = \boldsymbol{U}^*$.

The Fourier basis expansion model in matrix vector notation is given by $\hat{\boldsymbol{h}}[m] = \boldsymbol{\Gamma}^T \boldsymbol{u}[m]$. To obtain the delay basis expansion parameter matrix $\boldsymbol{\Gamma} = \boldsymbol{V}\boldsymbol{\mathcal{H}}^T$, we apply the analysis matrix $\boldsymbol{V}$ on $\boldsymbol{\mathcal{H}}$. The elements of the $D \times L$ matrix $\boldsymbol{\Gamma}$ are defined as $[\boldsymbol{\Gamma}]_{d,n} = \gamma_d[n]$.

In the context of MC-CDMA the channel is naturally represented in the frequency time domain. We introduce the frequency basis expansion parameter matrix $\boldsymbol{\Phi} = \boldsymbol{V}\boldsymbol{\mathcal{G}}^T$, and express the time variant frequency characteristic $\hat{\boldsymbol{g}}[m] = \boldsymbol{\Phi}^T \boldsymbol{u}[m]$ through the Fourier basis expansion model.

This gives the time variant effective spreading sequence as $\tilde{\boldsymbol{s}}_k[m] \approx \operatorname{diag}\left(\boldsymbol{\Phi}^T \boldsymbol{u}[m]\right) \boldsymbol{s}_k$, and the received data signal

$$\boldsymbol{y}[m] = \operatorname{diag}\left(\boldsymbol{\Phi}^T \boldsymbol{u}[m]\right) \left(\boldsymbol{S}\boldsymbol{b}[m] + \boldsymbol{p}[m]\right) + \boldsymbol{\nu}[m].$$

4. PERFORMANCE ANALYSIS

To generate the time variant channel realization for $\mathcal{H}$ we use the model

$$\boldsymbol{h}[m] = \operatorname{diag}(\boldsymbol{h}) \left[\mathrm{e}^{\frac{\mathrm{j}2\pi f_0 m}{M}}, \cdots, \mathrm{e}^{\frac{\mathrm{j}2\pi f_{(L-1)} m}{M}} \right]^T e^{\frac{\mathrm{j}2\pi \Delta f_C m}{M}}.$$

Vector $\boldsymbol{h}$ models the Rayleigh fading large scale channel statistics and is normalized so that $\mathrm{E}\{|\boldsymbol{h}|^2\} = 1$. We use the exponentially decaying typical urban (TU) power-delay-profile (PDP) from COST 259 [3], the chip rate $1/T_C = 3.84$Mcps as in UMTS and the delay spread $L = 15$ corresponding to $T_D = 3.9\mu$s.

The time variant characteristic is modeled as random Doppler component f_ℓ for every channel tap. To achieve a Jakes spectrum [4] $f_\ell = \sin(2\pi\xi_\ell) B_D M P T_C$, $\ell = 0, \cdots, L-1$, where ξ_ℓ is a random variable, uniformly distributed in the interval $(0, 1)$. The normalized carrier frequency offset at the receiver Δf_C is incorporated in the time variant channel model as common additional Doppler component for all channel taps. For every channel realization Δf_C is randomly sampled from a uniform distribution on the interval $(-\Delta f_{C,max}, +\Delta f_{C,max})$.

The system is operated at carrier frequency $f_C = 2\,\mathrm{GHz}$, the users move with velocity $v = 50\,\mathrm{km/h}$, this gives $B'_D = 90\,\mathrm{Hz}$. The complete OFDM symbol with cyclic prefix has length of $P = G + N = 79$. The number of subcarriers $N = 64$ and the data block length $M = 256$. The maximum carrier frequency offset was chosen to be $\Delta f_{C,max} = 90\,\mathrm{Hz} M P T_C = 0.47$. For the Fourier basis expansion model this results in $\lceil WM \rceil = 1$ with $B_D = B'_D + \Delta f_{C,max}/(M P T_C)$.

The Fourier basis expansion model approximates for every channel tap a complex exponential with a rational frequency f in the range $|f| < WM$. We analyze the MSE of the Fourier basis expansion model for a single path channel with normalized Doppler frequency f varied over the interval $(0, 1)$, $h[m, 1] = \mathrm{e}^{\frac{\mathrm{j}2\pi m f}{M}}$, and define $\boldsymbol{h}_1 = \left[h[0, 1], \cdots, h[M-1, 1] \right]^T$.

We evaluate the MSE according to $\hat{\boldsymbol{h}}_1 = \boldsymbol{U}^T \boldsymbol{V} \boldsymbol{h}_1$, $e = \left| \boldsymbol{h}_1 - \hat{\boldsymbol{h}}_1 \right|^2$. The MSE for the Fourier basis expansion is given in Fig. 2. With increasing number of basis functions, controlled by κ the MSE is reduced. At $f = 0.5$ the error is maximum.

To get more insight in the detailed approximation behavior we plot $h[m, 1]$ and $\hat{h}[m, 1]$ for $f = 0.5$ and $\kappa = \{1, 4\}$ in the complex plane in Fig. 3. The amplitude and phase error at the beginning and end of the data block is significant for $\kappa = 1$, and decreases with increasing κ.

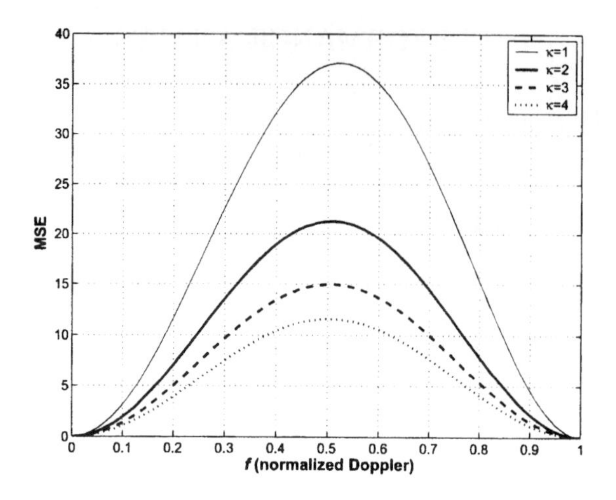

Figure 2. Mean squared error (MSE) of the Fourier basis expansion model for $\lceil WM \rceil = 1$ and $D = 2\kappa\lceil WM \rceil + 1$ basis functions, with $\kappa = \{1, 2, 3, 4\}$. The normalized doppler frequency f was varied over the interval $(0, 1)$.

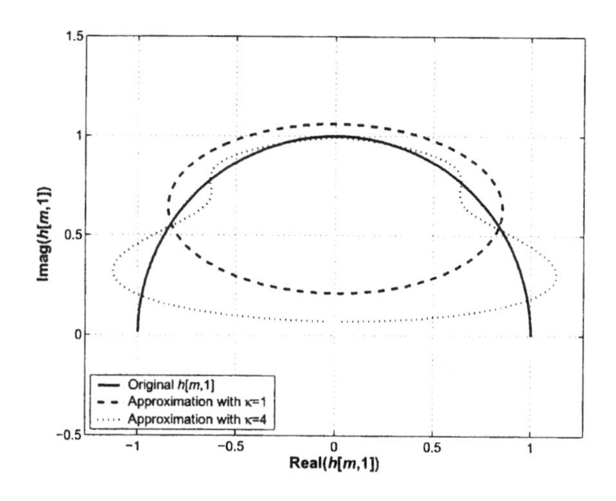

Figure 3. Trajectory of $h[m, 1]$ and its approximation by the Fourier basis expansion for $\kappa = \{1, 4\}$.

For data detection we use a time variant matched filter. The resulting code symbol estimates are demapped, deinterleaved and decoded by a BCJR decoder [1] to obtain estimates for the transmitted data bits $\hat{\chi}_k[m'']$.

We evaluate the receiver performance subject to parameter κ of the channel approximation model. The additional simulation parameters for the MC-CDMA system are chosen as follows: The spreading sequence has length $N = 64$ equal to the number of subcarriers. For data transmission the convolutional code used, is a non-systematic,

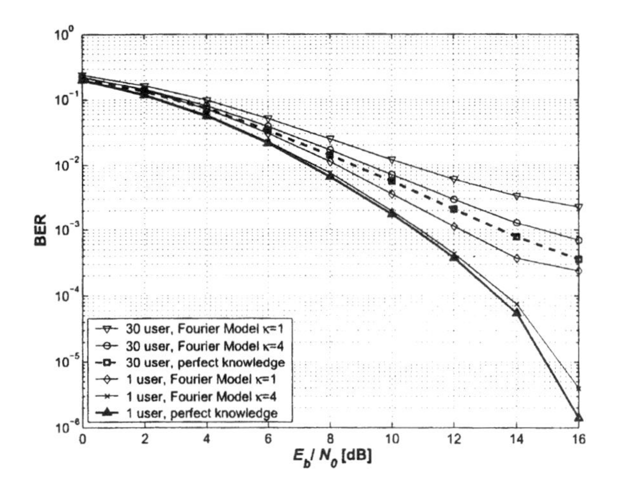

Figure 4. We compare the MC-CDMA receiver performance for the Fourier basis expansion channel model with $\kappa = \{1, 4\}$ and the perfectly known channel. The performance in the forward link is given in terms of BER versus SNR for $K = 1$ and $K = 30$ users. The spreading sequence length $N = 64$ is equal to the number of subcarriers. The data block length $M = 256$. The carrier frequency $f_C = 2\,\text{GHz}$, the Doppler bandwidth $B_D = 90\,\text{Hz}$ and the maximum carrier frequency offset is 90 Hz.

non-recursive, 4 state, rate $R_C = 1/2$ code with generator polynomial $(5, 7)_8$. We average all simulations over 2000 independent channel realizations. The QPSK symbol energy is normalized to 1, the E_b/N_0 is therefore defined as $\frac{E_b}{N_0} = \frac{1}{2R\sigma_\nu^2}\frac{P}{N}$.

We analyze the performance for perfectly known *model parameters* of the channel. This means the model parameters $\mathbf{\Gamma}$ are calculated from a perfectly known channel $\boldsymbol{\mathcal{H}}$. Applying the Fourier basis expansion model, we give the MC-CDMA receiver performance for the forward link in Fig. 4 for $\kappa = \{1, 4\}$ and for $K = 1$ and $K = 30$ users. We additionally give the performance with the perfectly known channel itself.

5. CONCLUSION

The Fourier basis expansion model allows to describe a time variant channels. Its inherent rectangular windowing results in spectral leakage. Therefore the number of Fourier basis functions given by the time bandwidth product $2\lceil WM \rceil + 1$, usually reported in literature, is not sufficient. Increasing the number of Fourier basis functions $D = 2\kappa \lceil WM \rceil + 1$ by a factor κ we can decrease the MSE of the channel approximation and the BER in the MC-CDMA forward link. The principal error floor behavior can not be removed for practical values $\kappa \leq 4$. Therefore a new basis expansion model is needed [9, 10].

6. REFERENCES

1. Bahl, L. R., J. Cocke, F. Jelinek, and J. Raviv: 1974, 'Optimal decoding of linear codes for minimizing symbol error rate'. *IEEE Transactions on Information Theory* **20**, 284–287.
2. Bello, P.: 1963, 'Characterization of randomly time-variant linear channels'. *IEEE Trans. Comm. Syst.* **CS-11**, 360–393.
3. Correia, L. M.: 2001, *Wireless Flexible Personalised Communications*. Wiley.
4. Hoeher, P.: 1992, 'A Statistical Discrete-Time Model for the WSSUS Multipath Channel'. *IEEE Transactions on Vehicular Technology* **41**(4), 461–468.
5. Li, Y. G. and L. J. Cimini: 2001, 'Bounds on the Interchannel Interference of OFDM in Time-Varying Impairments'. *IEEE Transactions on Communications* **49**(3), 401–404.
6. Proakis, J. G. and D. G. Manolakis: 1996, *Digital Signal Processing*. Prentice-Hall, 3rd edition.
7. Sayeed, A. M. and B. Aazhang: 1999, 'Joint Multipath-Doppler Diversity in Mobile Wireless Communications'. *IEEE Transactions on Communications* **47**(1), 123–132.
8. Sayeed, A. M., A. Sendonaris, and B. Aazhang: 1998, 'Multiuser Detection in Fast-Fading Multipath Environment'. *IEEE Journal on Selected Areas in Communications* **16**(9), 1691–1701.
9. Zemen, T. and C. F. Mecklenbräuker: 2003a, 'Time Variant Channel Equalization for MC-CDMA via Prolate Spheroidal Sequences'. to be presented as invited paper at *Asilomar Conference on Signals, Systems and Computers, Pacific Grove (CA), USA*.
10. Zemen, T. and C. F. Mecklenbräuker: 2003b, 'Time Variant Channel Estimation via Generalized Prolate Spheroidal Sequences'. submitted to *IEEE Transactions on Signal Processing*.
11. Zemen, T., J. Wehinger, C. Mecklenbräuker, and R. Müller: 2003, 'Iterative Detection and Channel Estimation for MC-CDMA'. In: *IEEE International Conference on Communications, ICC, Anchorage (Alaska), USA*.

7. AFFILIATION

T. Zemen is with Siemens AG, PSE PRO RCD, Erdbergerlände 26, A-1031 Vienna, Austria, E-mail: `thomas.zemen@siemens.com`.

C. Mecklenbräuker and R. Müller are with Telecommunication Research Center Vienna (ftw.), Tech Gate Vienna, Donau-City Str. 1/III, A-1220 Vienna, Austria, Email: {`cfm`, `mueller`}`@ftw.at`.

The work is funded by the Radio Communication Devices department (RCD), part of the Siemens AG Austria, Program and System Engineering (PSE) and the Telecommunications Research Center Vienna (ftw.) in the C0 project.

THOMAS P. KRAUSS & KEVIN L. BAUM

IMPACT OF CHANNEL VARIATION ON A CODE MULTIPLEXED PILOT IN MULTICARRIER SYSTEMS

Abstract. In spread multi-carrier systems with a code-multiplexed pilot channel, variations of the channel within the spreading block impact channel estimation and data demodulation through the loss of orthogonality of the spreading codes. In this paper the mean-squared error (MSE) between the true channel and a channel estimate obtained through simple despreading of the pilot code is derived. The closed-form MSE expression is a function of the channel fading statistics, the currently employed spreading codes, and the powers of the different users. By evaluating the MSE for a time-spread orthogonal frequency-division multiplexed (OFDM) system, it is observed that while spreading in time will have little impact in a pico-cell system scenario, a wide-area cell system experiences significant degradation in channel estimation performance at moderate Doppler frequencies. The impact on data demodulation is evaluated through simulations with the data users employing a turbo code.

1. INTRODUCTION

Multicarrier CDMA was suggested in [1] as a way to combine the beneficial characteristics of both CDMA and OFDM systems. In [1], spreading was introduced across subcarriers (i.e., in the frequency dimension) of an OFDM system. Moreover, spreading in the time dimension, which is the conventional approach for single carrier systems, can also be applied to multicarrier systems [2], [3]. Finally, the spreading can be performed in both the time dimension and the frequency dimension, as noted in [4].

For the downlink of a cellular system based on CDMA principles, users' signals can be code-multiplexed based on orthogonal codes. A multicarrier system that includes spreading (in time, frequency, or both) together with orthogonal code multiplexing can be referred to as OFCDM (orthogonal frequency and code division multiplexing), which is a term used in [5]. Regardless of the type of spreading that is used in an OFCDM system, accurate channel estimation plays a critical role in the detection process, enabling MMSE chip combining to reduce the effects of channel-induced inter-code interference.

Pilot symbol insertion in an OFCDM system can be performed like a conventional OFDM system (time and/or frequency multiplexed pilot), or like a conventional CDMA system (code multiplexed pilot). In this paper, we focus on the case where the pilot is code multiplexed and analyze the impact of channel variations on the quality of the despread pilot signal. Channel variation will occur in the frequency dimension due to multipath delay spread, and in the time dimension due to Doppler spread. While the analysis is applicable to both frequency spreading and time spreading, we will focus on time spreading to illustrate the impact of high mobility on the code multiplexed pilot performance.

This paper quantifies the degradation the time variation incurs in terms of mean squared error (MSE). An expression for the MSE is provided in terms of the

K. Fazel and S. Kaiser (eds.), Multi-Carrier Spread-Spectrum, 459–470.

channel covariance matrix, the spreading factor and spreading codes currently employed, and their relative powers. Moreover, the MSE is investigated through simulations of a 12-ray "Typical Urban" channel with a flat Doppler spectrum for each ray.

2. SYSTEM MODEL

Transmitter: Consider an OFCDM system with spreading factor of *SF* in the time dimension. Up to *SF* symbols can be code multiplexed onto the same set of resources. At least one of the Walsh codes is used as a pilot channel. The composite signal to be transmitted at a particular location in the time-frequency grid is described as

$$x(b,n,k) = c(b,n,k)\left(\underbrace{A_p(b,k)d_p(b,k)W_p(n,k)}_{\text{pilot channel}} + \underbrace{\sum_{\substack{i\neq p\\ i=1:SF}} A_i(b,k)d_i(b,k)W_i(n,k)}_{\text{data channels}}\right) \quad (1)$$

where: b is the spreading block index (b increases by one every *SF* OFDM symbol periods); n is the chip index within the bth spreading block interval, $1 \le n \le SF$; k is the subcarrier index, $1 \le k \le K$; c denotes the scrambling code; i is the Walsh code index, $1 \le i \le SF$; p denotes the Walsh code index that is used for the pilot channel; W_i denotes the ith Walsh code; A_i denotes the (real) gain applied to the ith Walsh code channel (e.g., based on power control settings, if any); $_i$ denotes the data symbol that modulates the ith Walsh code; d_p denotes the pilot symbol that modulates the pth Walsh code channel (i.e., the pilot channel).

Receiver: Assuming a cyclic prefix that is longer than the channel impulse response and proper receiver synchronization, the received signal (after the FFT) is given by

$$r(b,n,k) = h(b,n,k)x(b,n,k) + \eta(b,n,k) \quad (2)$$

where $h(b,n,k)$ is the frequency domain channel, and $\eta(b,n,k)$ is thermal noise at the bth block, n^{th} OFDM symbol, k^{th} subcarrier.

3. RECEIVE PROCESSING: CHANNEL ESTIMATION

In this section, closed-form expressions are derived for determining the impact of channel variations on the pilot channel used for coherent channel estimation. A channel estimate is obtained for each (*b*,*k*) by despreading the pilot channel as follows:

$$\hat{h}(b,k) = \frac{1}{A_p(b,k)} \cdot \frac{1}{d_p(b,k)} \cdot \frac{1}{SF} \sum_{n=1}^{SF} c^*(b,n,k) W_p^*(b,n,k) r(b,n,k) \tag{3}$$

How well does the channel estimate $\hat{h}(b,k)$ approximate the true channel $h(b,n,k)$? We consider the mean square error (MSE) between them. The block and subcarrier indices (b,k) are omitted for clarity. First an expression for the MSE at a given chip will be evaluated conditioned on the channel,

$$MSE(n) = E\left\{ \left| \hat{h} - h(n) \right|^2 \right\}, n = 1,\ldots,SF \tag{4}$$

Define h_c as the "mean" or DC component of the channel during the block: $h_c \equiv \frac{1}{SF} \sum_{n=1}^{SF} h(n)$ and $h_t(n)$ as the "time-varying" AC part of $h(n)$: $h(n) = h_t(n) + h_c$. Note that if the channel is constant over the block, then $h_c = h(n)$ for all chips $n = 1,\ldots,SF$. With this partitioning, the received signal can be written as the sum of the DC and AC channel contributions:

$$r(n) = h_c x(n) + h_t(n) x(n) + \eta(n) \tag{5}$$

Further partitioning of the transmit signal into pilot and data channels results in a four term expansion for the received signal (not including noise):

$$r(n) = \underbrace{h_c A_p d_p W_p(n)}_{\substack{\text{pilot channel} \\ \text{DC - term}}} + \underbrace{h_t(n) A_p d_p W_p(n)}_{\substack{\text{pilot channel} \\ \text{AC - term}}} + \underbrace{h_c \sum_{\substack{i \neq p \\ i=1:SF}} A_i d_i W_i(n)}_{\substack{\text{data channels} \\ \text{DC - term}}} + \underbrace{h_t(n) \sum_{\substack{i \neq p \\ i=1:SF}} A_i d_i W_i(n)}_{\substack{\text{data channels} \\ \text{AC - term}}} \tag{6}$$

Despreading by the pilot channel's Walsh code as in Equation (3), the channel estimate $\hat{h}$ reduces to (including a despread noise term)

$$\underbrace{h_c}_{\substack{\text{pilot channel} \\ \text{DC - term}}} + \underbrace{\frac{1}{SF} \sum_{n=1}^{SF} h_t(n)}_{\substack{\text{pilot channel} \\ \text{AC - term}}} + \underbrace{0}_{\substack{\text{data channels} \\ \text{DC - term}}} + \underbrace{\frac{1}{A_p} \cdot \frac{1}{d_p} \cdot \sum_{\substack{i \neq p \\ i=1:SF}} A_i d_i \frac{1}{SF} \sum_{n=1}^{SF} h_t(n) W_p^*(n) W_i(n)}_{\substack{\text{data channels} \\ \text{AC - term}}} + \eta' \tag{7}$$

The term due to the data channel codes and DC part of the channel is 0 due to the orthogonality of the spreading codes over a constant channel. Since by definition the time-varying portion of the channel is zero-mean, $\sum_{n=1}^{SF} h_t(n) = 0$, it is seen that the channel estimate $\hat{h}$ is actually an estimate of the mean h_c, degraded by inter-code interference caused by the time-varying portion of the channel, as well as the despread noise term:

$$\hat{h} = h_c + \frac{1}{A_p} \cdot \frac{1}{d_p} \cdot \sum_{\substack{i \neq p \\ i=1:SF}} A_i d_i \frac{1}{SF} \sum_{n=1}^{SF} h_t(n) W_p^*(n) W_i(n) + \eta' \tag{8}$$

Returning to the error expression, we obtain

$$\begin{aligned} e(n) = h(n) - \hat{h} &= h(n) - h_c - \frac{1}{A_p} \cdot \frac{1}{d_p} \cdot \sum_{\substack{i \neq p \\ i=1:SF}} A_i d_i \frac{1}{SF} \sum_{n_1=1}^{SF} h_t(n_1) W_p^*(n_1) W_i(n_1) - \eta' \\ &= h_t(n) - \gamma - \eta' \end{aligned} \tag{9}$$

Due to independence of the channel, data symbols, and noise, the per-chip $MSE(n)$ is the sum of three terms, only the first of which depends on n:

$$MSE(n) = E\{|h_t(n)|^2\} + E\{|\gamma|^2\} + E\{|\eta'|^2\} \tag{10}$$

Next, we will find expressions for each of these terms. The interference term is the most involved. It has variance

$$E\{|\gamma|^2\} = E\left\{ \frac{1}{A_p^2} \cdot \frac{1}{SF^2} \sum_{\substack{i1 \neq p \\ i1=1:SF}} A_{i1} d_{i1} \sum_{n1=1}^{SF} h_t(n1) W_p^*(n1) W_{i1}(n1) \sum_{\substack{i2 \neq p \\ i2=1:SF}} A_{i2} d_{i2}^* \sum_{n2=1}^{SF} h_t^*(n2) W_p(n2) W_{i2}^*(n2) \right\} \tag{11}$$

where this assumes the pilot data symbol d_p is unit modulus, and the A_i are all real. Taking the expectation over the random data symbols of the users, where $E\{|d_i|^2\} = 1$,

$$E\{|\gamma|^2\} = E\left\{ \frac{1}{A_p^2} \cdot \frac{1}{SF^2} \sum_{n1=1}^{SF} \sum_{n2=1}^{SF} h_t(n1) h_t^*(n2) \sum_{\substack{i \neq p \\ i=1:SF}} A_i^2 W_p^*(n1) W_i(n1) \left(W_p^*(n2) W_i(n2) \right)^* \right\} \tag{12}$$

Now define vector quantities

$$\mathbf{h}_t = \begin{bmatrix} h_t(1) \\ h_t(2) \\ \vdots \\ h_t(SF) \end{bmatrix}, \; \mathbf{h} = \begin{bmatrix} h(1) \\ h(2) \\ \vdots \\ h(SF) \end{bmatrix}, \; \mathbf{w}_i = \begin{bmatrix} W_i(1) \\ W_i(2) \\ \vdots \\ W_i(SF) \end{bmatrix}, \; \tilde{\mathbf{w}}_i = \frac{1}{A_p} \begin{bmatrix} W_p^*(1)W_i(1) \\ W_p^*(2)W_i(2) \\ \vdots \\ W_p^*(SF)W_i(SF) \end{bmatrix},$$

$$\begin{aligned} \tilde{\mathbf{W}}_{\bar{p}} &= \left[\tilde{\mathbf{w}}_1 \;\cdots\; \tilde{\mathbf{w}}_{p-1} \;\; \tilde{\mathbf{w}}_{p+1} \;\cdots\; \tilde{\mathbf{w}}_{SF}\right], \\ \mathbf{A}_{\bar{p}} &= \text{diag}([A_1 \;\cdots\; A_{p-1} \;\; A_{p+1} \;\cdots\; A_{SF}]) \end{aligned} \tag{13}$$

With these definitions, the power expression becomes

$$E\{|\gamma|^2\} = E\left\{\frac{1}{SF^2}\mathbf{h}_t^T \tilde{\mathbf{W}}_{\bar{p}} \mathbf{A}_{\bar{p}}^2 \tilde{\mathbf{W}}_{\bar{p}}^H \mathbf{h}_t^*\right\} \tag{14}$$

Now define the unit-norm constant vector $\mathbf{c}$ of length SF (not to be confused with the scrambling code $c(n)$), $\mathbf{c} = \frac{1}{\sqrt{SF}}[1 \;\; 1 \;\; \cdots \;\; 1]^T$.

Using the fact that a vector may be written as the product of its diagonal matrix times a vector of 1's, $\mathbf{b} = \sqrt{SF}\,\text{diag}(\mathbf{b})\mathbf{c}$ for arbitrary vector $\mathbf{b}$, the expression may be written in terms of diagonal matrices for the $\mathbf{h}_t$,

$$\begin{aligned} E\{|\gamma|^2\} &= E\{\frac{1}{SF}\mathbf{c}^T \,\text{diag}(\mathbf{h}_t)\tilde{\mathbf{W}}_{\bar{p}}\mathbf{A}_{\bar{p}}^2\tilde{\mathbf{W}}_{\bar{p}}^H \,\text{diag}(\mathbf{h}_t)^H \mathbf{c}\} \\ &= \frac{1}{SF}\mathbf{c}^T E\{\text{diag}(\mathbf{h}_t)\tilde{\mathbf{W}}_{\bar{p}}\mathbf{A}_{\bar{p}}^2\tilde{\mathbf{W}}_{\bar{p}}^H \,\text{diag}(\mathbf{h}_t)^H\}\mathbf{c} \end{aligned} \tag{15}$$

Now using the fact that $\text{diag}(\mathbf{b})\mathbf{A}\,\text{diag}(\mathbf{b})^H = \left(\mathbf{b}\mathbf{b}^H\right)\circ\mathbf{A}$ where $\mathbf{A}\circ\mathbf{B}$ is the element-wise product of matrices $\mathbf{A}$ and $\mathbf{B}$, we can write the term inside the expectation as

$$E\{|\gamma|^2\} = \frac{1}{SF}\mathbf{c}^T E\{\left(\mathbf{h}_t\mathbf{h}_t^H\right)\circ\left(\tilde{\mathbf{W}}_{\bar{p}}\mathbf{A}_{\bar{p}}^2\tilde{\mathbf{W}}_{\bar{p}}^H\right)\}\mathbf{c} \tag{16}$$

The only random part remaining inside the expectation is the term with the channel,

$$E\{|\gamma|^2\} = \frac{1}{SF}\mathbf{c}^T\left(E\{\mathbf{h}_t\mathbf{h}_t^H\}\circ\left(\tilde{\mathbf{W}}_{\bar{p}}\mathbf{A}_{\bar{p}}^2\tilde{\mathbf{W}}_{\bar{p}}^H\right)\right)\mathbf{c} \tag{17}$$

Now express the vector "AC-part" of the channel in terms of the actual channel vector with the projection onto the constant subspace subtracted off:

$$\mathbf{h}_t = \mathbf{h} - h_c = \mathbf{h} - \mathbf{c}\mathbf{c}^T\mathbf{h} = (\mathbf{I} - \mathbf{c}\mathbf{c}^T)\mathbf{h} \tag{18}$$

$$E\{\mathbf{h}_t\mathbf{h}_t^H\} = E\{(\mathbf{I} - \mathbf{c}\mathbf{c}^T)\mathbf{h}\mathbf{h}^H(\mathbf{I} - \mathbf{c}\mathbf{c}^T)\} = (\mathbf{I} - \mathbf{c}\mathbf{c}^T)\mathbf{R}_{hh}(\mathbf{I} - \mathbf{c}\mathbf{c}^T) \tag{19}$$

where $\mathbf{R}_{hh}$ is the covariance of the channel vector, $\mathbf{R}_{hh} = E\{\mathbf{h}\mathbf{h}^H\}$. The final expression for the interference power is hence

$$E\{|\gamma|^2\} = \frac{1}{SF}\mathbf{c}^T\left(\left((\mathbf{I} - \mathbf{c}\mathbf{c}^T)\mathbf{R}_{hh}(\mathbf{I} - \mathbf{c}\mathbf{c}^T)\right)\circ\left(\tilde{\mathbf{W}}_{\bar{p}}\mathbf{A}_{\bar{p}}^2\tilde{\mathbf{W}}_{\bar{p}}^H\right)\right)\mathbf{c} \tag{20}$$

This may be described as the mean of all the elements of a matrix, where the matrix is given by the element-wise product of two matrices: 1) the covariance matrix of the time-varying portion of the channel, and 2) the outer-product of a pilot-code-modified spreading code matrix of the interfering users.

The noise power is

$$E\{|\eta'|^2\} = \frac{1}{A_p^2}\cdot\frac{1}{SF^2}\sum_{n1=1}^{SF}c^*(n1)W_p^*(n1)\sum_{n2=1}^{SF}c(n2)W_p(n2)E\{\eta(n1)\eta^*(n2)\} \tag{21}$$

$$= \frac{1}{A_p^2}\cdot\frac{1}{SF^2}\sum_{n=1}^{SF}\sigma^2 = \frac{1}{A_p^2}\cdot\frac{\sigma^2}{SF} \tag{22}$$

where the noise samples are assumed i.i.d. with per-chip noise power σ^2, that is $E\{\eta(n1)\eta^*(n2)\} = \sigma^2\delta(n1 - n2)$.

The per-chip power in the time varying portion of the channel is

$$E\{|h_t(n)|^2\} = \boldsymbol{\delta}_n^T(\mathbf{I} - \mathbf{c}\mathbf{c}^T)\mathbf{R}_{hh}(\mathbf{I} - \mathbf{c}\mathbf{c}^T)\boldsymbol{\delta}_n \tag{23}$$

where $\boldsymbol{\delta}_n$ is the length-SF column vector with a "1" in the nth position and zeros elsewhere. Combining terms, we obtain

$$E\{|e(n)|^2\} = E\{|h_t(n)|^2\} + E\{|\gamma|^2\} + E\{|\eta'|^2\} = \boldsymbol{\delta}_n^T(\mathbf{I} - \mathbf{c}\mathbf{c}^T)\mathbf{R}_{hh}(\mathbf{I} - \mathbf{c}\mathbf{c}^T)\boldsymbol{\delta}_n + \frac{1}{SF}\mathbf{c}^T\left(\left((\mathbf{I} - \mathbf{c}\mathbf{c}^T)\mathbf{R}_{hh}(\mathbf{I} - \mathbf{c}\mathbf{c}^T)\right)\circ\left(\tilde{\mathbf{W}}_{\bar{p}}\mathbf{A}_{\bar{p}}^2\tilde{\mathbf{W}}_{\bar{p}}^H\right)\right)\mathbf{c} + \frac{1}{A_p^2}\cdot\frac{\sigma^2}{SF} \tag{24}$$

The average over all chips is

$$MSE=\frac{1}{SF}\sum_{n=1}^{SF}E\{|e(n)|^2\}=\frac{1}{SF}\operatorname{trace}\left((\mathbf{I}-\mathbf{cc}^T)\mathbf{R}_{hh}(\mathbf{I}-\mathbf{cc}^T)\right)+ \\ \frac{1}{SF}\mathbf{c}^T\left(\left((\mathbf{I}-\mathbf{cc}^T)\mathbf{R}_{hh}(\mathbf{I}-\mathbf{cc}^T)\right)\circ\left(\tilde{\mathbf{W}}_{\bar{p}}\mathbf{A}_{\bar{p}}^2\tilde{\mathbf{W}}_{\bar{p}}^H\right)\right)\mathbf{c}+\frac{1}{A_p^2}\cdot\frac{\sigma^2}{SF} \tag{25}$$

This expression is the average mean squared error between the true channel $h(n)$ and the constant estimate $\hat{h}$ obtained by simple despreading of the pilot channel. An interesting aspect of these results is that they do not depend on the frequency selective characteristics of the channel (qualified by the assumption that the cyclic prefix is longer than the channel delay spread). Conversely, if the analysis is applied to spreading in the frequency dimension, the results will not depend on the Doppler spectrum (qualified by the assumption that the Doppler-induced OFDM inter-subcarrier interference is negligible).

4. SIMULATION RESULTS

In this section, the analysis is applied to some typical system scenarios to provide further insights. First, the impact of the *shape* of the Doppler spectrum is examined by both simulation and analysis. The simulation and analysis results are compared to confirm their accuracy. Next, example results are given for two different sets of OFDM system parameters that correspond to a pico-cell system and a wide-area system. Finally, results of turbo-coded link simulations are presented for the wide-area system scenario.

Since the MSE expressions depend on the covariance matrix of the channel, different results can be expected for different Doppler spectra. Here, we examine three cases for which closed form expressions are readily available [9]. For a flat rectangular (brick-wall) Doppler spectrum, the channel covariance is:

$$E\{h(n_1)h^*(n_2)\}=r_u((n_2-n_1)T_s)=\operatorname{sinc}(2f_d(n_2-n_1)T_s) \tag{26}$$

where the sinc function is defined $\operatorname{sinc}(x)\equiv\sin(\pi x)/\pi x$, f_d is the Doppler frequency, and T_s is the OFDM symbol time.

For Jakes fading, the channel covariance is

$$E\{h(n_1)h^*(n_2)\}=r_J((n_2-n_1)T_s)=J_0(2\pi f_d(n_2-n_1)T_s) \tag{27}$$

where $J_0(x)$ is the zeroth-order Bessel function of the first kind.

For two-ray, sinusoidal fading, the channel covariance is

$$E\{h(n_1)h^*(n_2)\}=r_t((n_2-n_1)T_s)=\cos(2\pi f_d(n_2-n_1)T_s)\,. \tag{28}$$

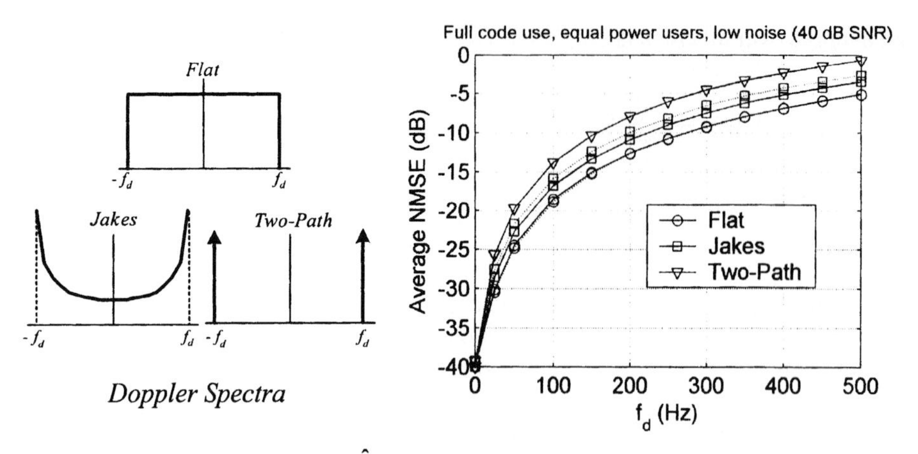

Figure 1. Comparison of NMSE of $\hat{h}$ for different fading statistics; solid lines are theoretical results, dashed lines are from simulations (the solid lines usually obscure the dashed lines).

Figure 1 compares the normalized MSE (denoted as NMSE, which is the MSE divided by the average power of $h(n)$) of $\hat{h}$ vs. f_d for channels with these three Doppler spectra, for a system with $SF = 16$ and an OFDM symbol duration of 50 μs. Since the Jakes and two-path models have more energy concentrated near the maximum Doppler frequency, they produce more rapid channel variations and hence a larger NMSE than the flat Doppler spectrum (an interesting topic for further investigation is whether results can be predicted based on a simpler statistic such as RMS Doppler spread). In addition, this figure provides a comparison between the theoretical NMSE and the simulated NMSE. The curves are in excellent agreement, although the simulated Jakes channel actually has more time-variation than the analytical model due to imperfect filtering in generating the fading coefficients for the simulation (the spectrum is not perfectly limited to +/- f_d). Regardless of the Doppler spectrum assumption, the NMSE increases dramatically (by 35 to 40 dB) as the Doppler frequency increases from 0 to 500 Hz.

Next, the relationship between different system scenarios and the resulting NMSE characteristics is examined. The first scenario is a system designed for reasonably wide-area cellular coverage (e.g., 2 km cell radius) [6][7][8]. This system can handle up to 10 μs delay spread and has a total OFDM symbol duration of 50 μs. The second scenario is a system designed for pico-cell coverage with parameters similar to an IEEE802.11a wireless LAN, namely a 0.8 μs cyclic prefix and a total OFDM symbol duration of 4 μs.

The NMSE of the despread pilot $\hat{h}$ is evaluated in Figure 2 for spreading factors from 4 to 128 and Doppler rates up to 500 Hz based on simulations using the flat Doppler spectrum. The channels are independent realizations of a 12-path "TU" channel model from the COST-207 specification; this channel has maximum delay spread of 5 μs (however, for the pico-cell scenario, the channel is scaled down to fit all 12 rays within the cyclic prefix).

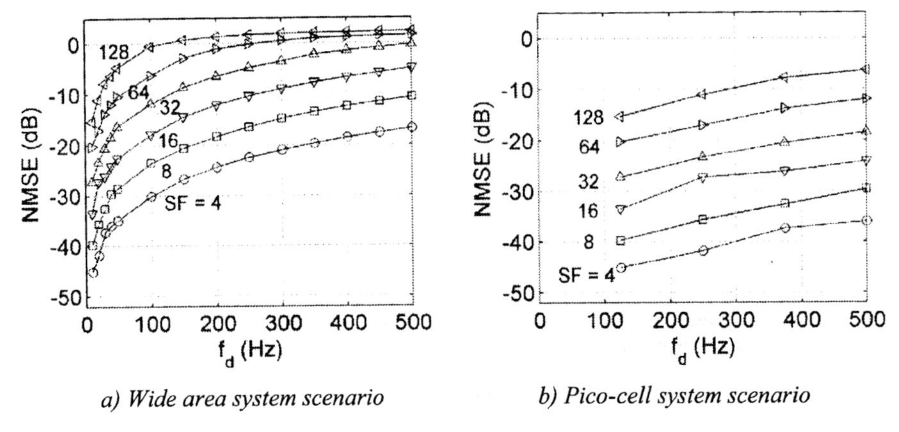

Figure 2. Average NMSE, no noise added, random QPSK data symbols, full code usage.

For the wide-area system scenario, high mobility leads to high Doppler rates, and it can be seen that the spreading factor must be kept fairly small to limit distortion (due to variation over the spreading block and the breakdown of code orthogonality) on the despread pilot channel. For the pico-cell system scenario, the shorter spreading block duration (due to shorter OFDM symbols) provides much more robustness to channel variation for the same Doppler rate and spreading factor. Moreover, the typical Doppler rates for the pico-cell scenario will likely be much lower than the wide-area scenario.

Since the wide-area system scenario is more challenging and relevant to 4G cellular, additional coded link simulations were performed using the wide-area system parameters. The simulation employs multicode transmission with 15 equal-power 16-QAM Walsh channels and a single pilot ($SF = 16$). Each data channel is rate ½ turbo-coded and interleaved over 751 OFDM subcarriers (1496 information bits per code word). Max-log-map decoding is used with 8 decoding iterations. Three cases were simulated: first with the channel estimate provided by simple despreading of the pilot channel ($\hat{h}$); next using the "true mean" of the channel over each spreading block (h_c), providing a lower bound for detection using one channel estimate per spreading block; and finally a per-chip MMSEC (minimum mean-squared-error combining) using ideal channel knowledge on a chip-by-chip basis ($h(n)$), providing a lower bound for MMSE detection based on a per-chip channel estimator.

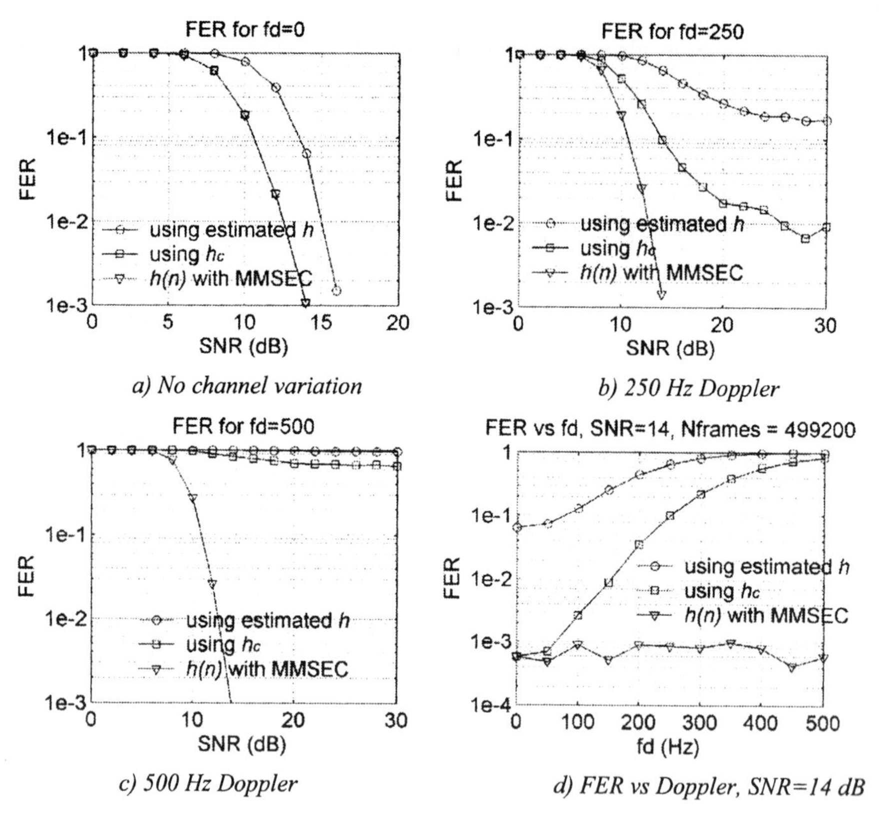

a) No channel variation *b) 250 Hz Doppler*

c) 500 Hz Doppler *d) FER vs Doppler, SNR=14 dB*

Figure 3. FER curves for a turbo coded system at 0 Hz, 250 Hz, and 500 Hz Doppler frequencies (a, b and c respectively), and versus Doppler frequency at an SNR of 14 dB.

Figure 3 presents frame error rate (FER) curves versus thermal noise power (SNR) in subfigures a), b) and c) at 0, 250, and 500 Hz Doppler frequencies, respectively, and FER versus Doppler at a fixed SNR of 14 dB in subfigure d). In subfigure a), there is no channel variation so the only impairment on the simple estimator $\hat{h}$ is the thermal noise. Since the noise power is the same on the despread pilot and the despread data, the observed degradation of ~3 dB matches expectations (note that for this case the channel estimates could be improved by filtering $\hat{h}$ over subcarriers). For subfigures b), c) and d), the impact of Doppler induced channel estimation error becomes clear. Even if the thermal noise and inter-code interference could be completely removed from $\hat{h}$ by additional filtering, the performance with any estimator that generates one channel estimate per spreading block is lower-bounded by the curves using h_c. On the other hand, it can be seen that with ideal per-chip channel knowledge $h(n)$ and MMSE detection, the system is robust to high Dopplers. In order to move towards the ideal performance curve, time interpolation must be added to the simple channel estimator. However, note that this would require multiple spreading blocks to be received and buffered (which may not be applicable to a burst-mode transmission) since there is only one despread pilot per spreading block. Moreover, the tracking capability of time interpolation is fundamentally limited by the ratio of the spreading block rate to the maximum Doppler frequency (e.g., for the wide area system scenario with a desired oversampling factor of two, the maximum allowable Doppler frequency is 312 Hz). Further investigation of these issues and comparisons with conventional time/frequency multiplexed pilot symbols are interesting topics for future work.

5. CONCLUSIONS

This paper analyzed the impact of channel variation on a code-multiplexed pilot channel in time-spread multi-carrier systems. A closed-form MSE expression was derived as a function of the channel fading statistics, the currently employed spreading codes, and the powers of the different users. Simulation results were provided to confirm the analysis and show the impact of the assumed OFDM symbol duration, spreading factor, and maximum Doppler frequency through representative examples.

The authors are with Motorola Labs, Schaumburg, IL, USA. E-mail: krauss@labs.mot.com, baum@labs.mot.com

7. REFERENCES

[1] N. Yee, J.-P. Linnartz, & G. P. Fettweis, "Multi-Carrier CDMA in Indoor Wireless Radio Networks," in *Proc. Of the IEEE PIMRC*, Yokohama, pp. 1670-1674, 1994.

[2] R. Prasad and S. Hara, "Overview of multicarrier CDMA," IEEE Commun. Mag., pp. 126-133, Dec. 1997.

[3] S. Kondo & L. B. Milstein, "Performance of Multicarrier DS CDMA Systems," IEEE Transactions on Communications, Vol. 44, No. 2, February 1996, pp. 238-246.

[4] J. Egle, M. Reinhardt, & J. Lindner, "Equalization and Coding for Extended MC-CDMA over Time and Frequency Selective Channels," in *Multi-Carrier Spread Spectrum*, ed. K. Fazel and G. P. Fettweis, Kluwer Academic Publishers, pp. 127-134, 1997.

[5] H. Atarashi & M. Sawahashi, "Variable Spreading Factor Orthogonal Frequency and Code Division Multiplexing VSF-OFCDM)," in *Multi-Carrier Spread Spectrum and Related Topics*, ed. K. Fazel and S. Kaiser, Kluwer Academic Publishers, 2002.

[6] M. D. Batariere, T. K. Blankenship, J. F. Kepler, T. P. Krauss, I. Lisica, S. Mukthavaram, J. W. Porter, T. A. Thomas, F. W. Vook, "Wideband MIMO Mobile Impulse Response Measurements at 3.7 GHz," *IEEE VTC-2002/Spring*, Birmingham, AL, May 4-9, 2002.

[7] J. F. Kepler, T. P. Krauss, & S. Mukthavaram, "Delay Spread Measurements on a Wideband MIMO Channel at 3.7 GHz," *IEEE VTC-2002/Fall*, Vancouver CA, September 24-29, 2002.

[8] T. P. Krauss, T. A. Thomas, & F. W. Vook, "Direction of Arrival and Capacity Characteristics of an Experimental Broadband Mobile MIMO-OFDM System," *IEEE VTC-2003/Spring*, Jeju Island, Korea, April 2003.

[9] Y. Li & L. J. Cimini, Jr., "Bounds on the Interchannel Interference of OFDM in Time-Varying Impairments," IEEE Transactions on Communications, Vol. 49, No. 3, March 2001, pp. 401-404.

Section VI

REALIZATION AND IMPLEMENTATION

JAN TUBBAX, BORIS CÔME, LIESBET VAN DER PERRE,
STÉPHANE DONNAY, MARC ENGELS

JOINT COMPENSATION OF IQ IMBALANCE, FREQUENCY OFFSET AND PHASE NOISE.

Abstract

Zero-IF receivers are gaining interest because of their potential to enable low-cost OFDM terminals. However, Zero-IF receivers introduce IQ imbalance which may have a huge impact on the performance. Rather than increasing component cost to decrease the IQ imbalance, an alternative is to tolerate the IQ imbalance and compensate it digitally. Current solutions require extra analog hardware or converge too slowly for bursty communication. Moreover, the impact of a frequency offset and phase noise on the IQ estimation/compensation problem is not considered. In this paper, we analyze the joint IQ imbalance-frequency offset-phase noise estimation and propose a low-cost, highly effective, all-digital compensation scheme. For large IQ imbalance ($\epsilon = 10\%$, $\Delta\phi = 10^o$), large frequency offsets and in the presence of phase noise, our solution results in an average implementation loss below 1 dB. It therefore enables the design of low-cost, low-complexity OFDM receivers.

1 INTRODUCTION

OFDM [1] is a widely recognized and standardized modulation technique for broadband wireless systems: because of its ability to elegantly cope with a multi-path environment, it is used for WLAN [2][3], DVB [4], Fixed Wireless Access [5], ...

Recently, a lot of effort is spent in developing cost efficient, more power efficient, more integrated OFDM receivers. The Zero-IF or Direct-Conversion receiver [6] is an attractive candidate, since it avoids costly IF filters and allows for easier integration than the super-heterodyne structure. On other hand, due to the absence of a digital IF, IQ imbalance is introduced. Moreover, OFDM is very sensitive to carrier frequency offsets (CFO) [7]. Finally, Zero-IF architectures do not solve the phase noise problem.

Unfortunately, OFDM is sensitive to non-idealities in the receiver front-end [8][9]. This leads either to stringent front-end specifications and thus an expensive device, or large performance degradations. IQ imbalance, frequency offset and phase noise have been identified as key non-idealities for OFDM receivers.

Therefore, we investigate the joint IQ imbalance, frequency offset and phase noise estimation and compensation and introduce a low-complexity fast convergence compensation scheme. To the best of our knowledge, no such solution has been proposed so far. A number of IQ compensation schemes exist, but they rely on slow convergence estimation schemes [10], [11] or use a calibration solution [12], [13] which requires extra, carefully designed, analog hardware. In [14] we proposed an IQ compensation with a fast convergence, but it only handles small frequency offsets.

In section 2, we first tackle the IQ imbalance problem. Section 3 explains the effect of phase noise on OFDM. In section 4, we analyze the effect of phase noise on

K. Fazel and S. Kaiser (eds.), Multi-Carrier Spread-Spectrum, 473–480.

the IQ compensation scheme and show how we can do a joint IQ and phase noise compensation. Section 5 summarizes the conclusions.

2 IQ COMPENSATION IN THE PRESENCE OF CFO

2.1 IQ imbalance model

IQ imbalance can be characterized by 2 parameters: the amplitude imbalance ϵ between the I and Q branch, and the phase orthogonality mismatch $\Delta\phi$. The complex baseband equation for the IQ imbalance effect on the ideal time domain signal $\mathbf{r}$ is given by [15] as

$$\mathbf{r_{iq}} = \alpha \cdot \mathbf{r} \cdot e^{\jmath 2\pi CFO \cdot \mathbf{t}} + \beta \cdot \mathbf{r}^* \cdot e^{-\jmath 2\pi CFO \cdot \mathbf{t}} \tag{1}$$

with $\mathbf{r_{iq}}$ the time domain signal with IQ imbalance, $\Re()$ denotes the real part, $\Im()$ the imaginary part and $()^*$ the complex conjugate and

$$\alpha = \cos\Delta\phi + \jmath\epsilon \sin\Delta\phi \tag{2}$$

$$\beta = \epsilon \cos\Delta\phi - \jmath \sin\Delta\phi \tag{3}$$

Frequency domain signals are underscored, while time domain signals are not. Signals are indicated in bold and scalar parameters in normal font.

2.2 Estimation/Compensation

[16] shows the receive time domain signal can be restored as

$$\hat{\mathbf{r}} = \frac{\alpha^* \cdot \mathbf{r_1} - \beta \cdot \mathbf{r_2^*}}{|\alpha|^2 - |\beta|^2} \tag{4}$$

from

$$\mathbf{r_1} = \mathbf{r_{iq}} \cdot e^{-\jmath 2\pi CFO \cdot \mathbf{t}} = \alpha \cdot \mathbf{r} + \beta \cdot \mathbf{r}^* \cdot e^{-2\jmath 2\pi CFO \cdot \mathbf{t}} \tag{5}$$

$$\mathbf{r_2} = \mathbf{r_{iq}} \cdot e^{\jmath 2\pi CFO \cdot \mathbf{t}} = \alpha \cdot \mathbf{r} \cdot e^{2\jmath 2\pi CFO \cdot \mathbf{t}} + \beta \cdot \mathbf{r}^* \tag{6}$$

Equivalently, we write the equation for the channel correction based on $\underline{\mathbf{h_1}} = fft(\mathbf{r_1}) \cdot \underline{\mathbf{t}}$ and $\underline{\mathbf{h_2}} = fft(\mathbf{r_2^*}) \cdot \underline{\mathbf{t}}$ (as indicated in figure 1) as follows

$$\underline{\hat{\mathbf{c}}} = \frac{\alpha^* \cdot \underline{\mathbf{h_1}} - \beta \cdot \underline{\mathbf{h_2}}}{|\alpha|^2 - |\beta|^2} \tag{7}$$

(7) provides the corrected channel response if we know (α,β). The estimation of α and β is based on the information that the corrected channel response should have a smooth channel characteristic: since the coherence bandwidth of the channel is (a lot) larger than the inter-carrier-spacing in a WLAN system, the channel response does not change substantially between successive frequency taps (the x-line in figure 2). With IQ imbalance, sharp transitions occur in the measured channel response $\underline{\mathbf{h}}$ due to the β degradation term (the o-line in figure 2). Thus, correcting the IQ imbalance

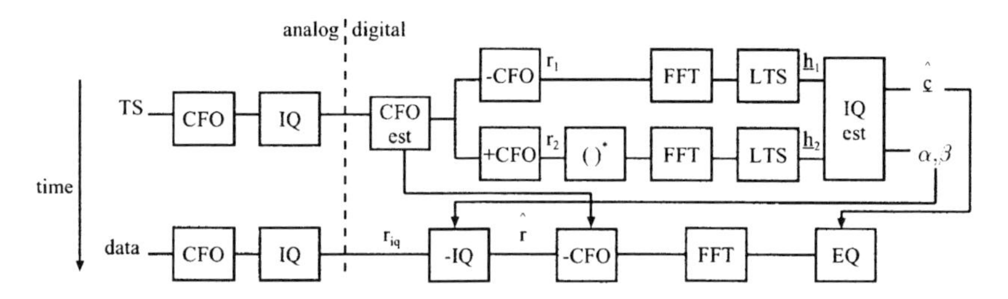

Fig. 1. Overview of the joint IQ-CFO estimation/compensation.

means making the channel response 'smooth' again. Therefore, we select the set of IQ parameters (α, β) which renders the corrected channel response $\hat{c}$ as smooth as possible ; in other words we minimize the Mean Square Error (MSE) between consecutive channel coefficients

$$MSE = \sum_l |\hat{\underline{c}}_{l+1} - \hat{\underline{c}}_l|^2 \tag{8}$$

[16] shows that (α,β) can be estimated as

$$\hat{\beta} = \frac{\sum_l (\underline{\mathbf{h}}_{1,l+1} - \underline{\mathbf{h}}_{1,l})(\underline{\mathbf{h}}^*_{2,l+1} - \underline{\mathbf{h}}^*_{2,l})}{\sum_l |\underline{\mathbf{h}}_{2,l+1} - \underline{\mathbf{h}}_{2,l}|^2} \tag{9}$$

$$\hat{\alpha} = \sqrt{1 - \Im^2\{\hat{\beta}\}} - \jmath\frac{\Re\{\hat{\beta}\}\Im\{\hat{\beta}\}}{\sqrt{1 - \Im^2\{\hat{\beta}\}}} \tag{10}$$

Once $(\hat{\alpha},\hat{\beta})$ are known, (4) can be used to correct the IQ imbalance on the received data signal, followed by a frequency offset correction.

The resulting block diagram is depicted in figure 1.

2.3 *Performance*

Figure 2 shows that we can correct the influence of the IQ imbalance on the channel estimate extremely well: the corrected channel response (the Δ-line) coincides (almost) perfectly with the exact channel response (the x-line). Note that carriers 28 to 38 are zero carriers, which means no channel estimate is needed on those carriers.

We illustrate the performance of this joint IQ-CFO compensation scheme for a WLAN case study with coded ($R = 3/4$ from the IEEE or HIPERLAN standard) 64QAM transmission.

The results in [16] illustrates the IQ compensation works for large frequency offsets as well: the degradation doesn't exceed 0.5 dB at a BER of 10^{-5} coded 64QAM even for frequency offsets exceeding 100 kHz.

The IQ estimation/compensation algorithm requires no extra analog hardware and a very small additional digital complexity. The IQ imbalance may occur anywhere in the receiver, because the RF, LO and baseband contributions are jointly estimated and compensated.

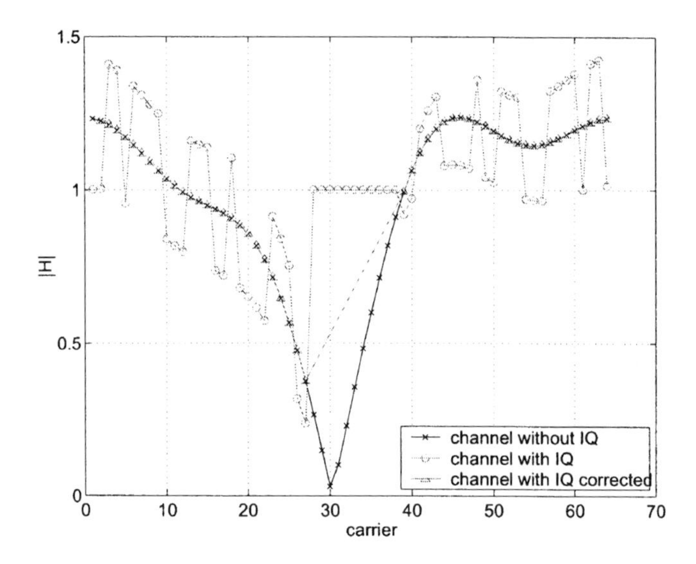

Fig. 2. The effect of IQ imbalance and correction on channel estimation.

3 PHASE NOISE

The origin of phase noise is located in the local oscillator (LO). Ideally, the LO should produce all its output power at the required frequency. In reality, the power appears not only at the desired frequency, but also in a band around it.

After down-conversion, we find the same spectrum at baseband around DC, as shown in figure 3. We assume a piece-wise linear PSD model, which is widely used in literature [10], [17], [18], [19], [20] and corresponds quite well with measurements [17]. We apply the following parameters:

n_1	-57.5 dBc/Hz	the noise floor of the PLL
n_2	-97.5 dBc/Hz	the overall noise floor
f_1	10 kHz	the loop bandwidth of the PLL
s	-20dB/decade	the sloop of the transition

An important parameter in phase noise models is the integrated phase noise power, which is defined as the power of the phase noise relative to the power at the carrier frequency. For this set of parameters (and a signal bandwidth of 20 MHz according to the standards [2], [3]) the total integrated phase noise is -32 dBc as in [21].

From [17] we learn that phase noise consists of 2 components. First, an Own Noise Contribution (ONC), which causes a Common Phase Error (CPE), identical on all carriers, and slowly varying from one symbol to the next. Second, phase noise has a Foreign Noise Contribution (FNC), which causes Inter-Carrier-Interference (ICI), different for all carriers and stochastic by nature. This difference between CPE and ICI readily inspires a compensation scheme: estimate the common rotation (the CPE)

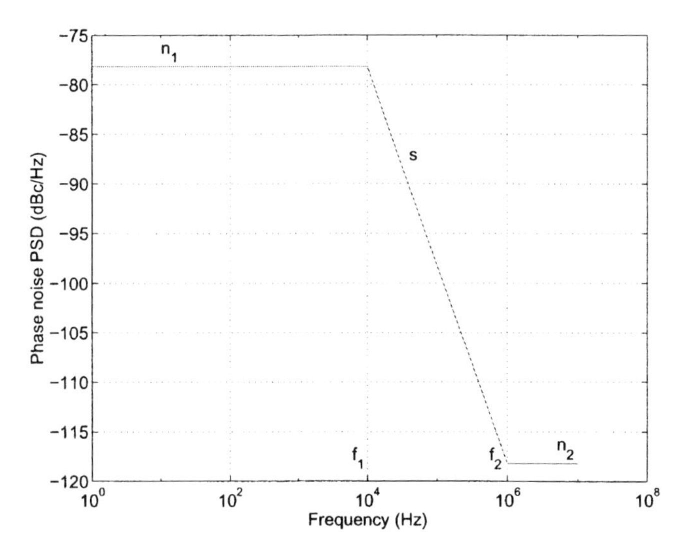

Fig. 3. The PSD of the phase noise model.

on all carriers and compensate it. This has been proposed by e.g. [18], [19], [22]. The estimation of the CPE can be based on the known pilot carriers and the decisions on the data carriers.

4 JOINT IQ IMBALANCE PHASE NOISE COMPENSATION

We studied IQ imbalance (section 2) and phase noise (section 3) separately, and presented a solution for both problems: the IQ imbalance is estimated and corrected during the Channel Estimation, based on a training symbol, while the phase noise is compensated by a decision-directed approach, based on the data. In this section, we analyze the joint estimation/compensation of IQ imbalance and phase noise.

4.1 Model

To study the combined effect of IQ imbalance and phase noise, we concatenate the models for both effects as they are described in the corresponding sections.

Analysis of the signals at RF shows that IQ and phase noise can be modelled by first applying the phase noise (θ) and frequency offset (CFO) on the baseband signal and then all IQ imbalance contributions (RF-LO-BB), combined in (α,β). This means

$$\mathbf{r}_{\mathbf{iq}}(\theta) \quad = \quad \alpha\, \mathbf{r} \cdot e^{\jmath\theta} \cdot e^{\jmath 2\pi CFO\mathbf{t}} + \beta\, \mathbf{r}^* \cdot e^{-\jmath\theta} \cdot e^{-\jmath 2\pi CFO\mathbf{t}} \qquad (11)$$

with $\mathbf{r}$ the incoming baseband signal (no IQ imbalance and phase noise), $\mathbf{r}_{\mathbf{iq}}(\theta)$ the received baseband signal after the front-end (with IQ imbalance, frequency offset and phase noise).

4.2 *Analysis*

To analyze the effect of phase noise on the IQ correction scheme, we first verify the effect of a CPE. Later, simulations incorporate the ICI contribution of the phase noise.

Analysis shows that the estimates of α and β are insensitive to a CPE rotation ; for the correction of the channel estimation and of the time domain data signal (12) and (13) hold

$$\hat{\mathbf{r}}(\theta_{CPE}) = \hat{\mathbf{r}}(0)e^{\jmath\theta_{CPE}} \tag{12}$$

$$\hat{\underline{\mathbf{c}}}(\theta_{CPE}) = \hat{\underline{\mathbf{c}}}(0)e^{\jmath\theta_{CPE}} \tag{13}$$

Equations (12) and (13) show that the phase rotation doesn't affect the IQ compensation: for both the channel estimation as the time domain correction, the IQ compensation still eliminates the IQ imbalance (as if there were no phase rotation) and the rotation angle θ_{CPE} appears intact at the output. This means that the channel estimate is rotated by the initial CPE, which is thus removed from the data through equalization. Therefore we correct the time-variant CPE rotation afterwards by the decision-directed approach. To summarize, we have separated the IQ-phase-noise compensation in 2 steps: first, we compensate the IQ imbalance based on the training symbol and second, the phase rotation based on the data.

However, phase noise not only consists out of the CPE, but also of a ICI contribution. This means we still need to test the above joint IQ-phase-noise compensation scheme in a real phase noise scenario.

4.3 *Performance*

Figure 4 shows the performance of our scheme for a multi-path channel with coded (R=3/4) 64QAM for an IQ imbalance $\epsilon = 10\%$ and $\Delta\phi = 10^o$, a frequency offset of 80 kHz and a total integrated phase noise of $K = -32dBc$ (with the parameters from section 3). The figure shows that without compensating anything the system is not usable with an error floor at 10^{-1} ; compensating the IQ imbalance (and frequency offset) gets the implementation loss down to 4.1 dB and the additional (decision-directed) phase compensation gets the residual degradation down to 0.6 dB at 10^{-5}.

Both front-end non-idealities combined consume 0.6 dB of the budget for the front-end implementation loss, which is perfectly acceptable for most designs. By allowing 0.6 dB of implementation loss, an IQ imbalance of $\epsilon = 10\%$ and $\Delta\phi = 10^o$ can be tolerated, even in the presence of phase noise and frequency offsets. This greatly simplifies the design process. Moreover, the IQ imbalance can occur anywhere in the receiver because all IQ imbalance contribution (from RF, LO and baseband) are jointly estimated and compensated. Therefore, our IQ estimation/compensation technique enables Zero-IF receivers.

5 CONCLUSIONS

IQ imbalance, frequency offset and phase noise can each cause large degradations in an OFDM receiver. We have introduced an estimation/compensation method to jointly

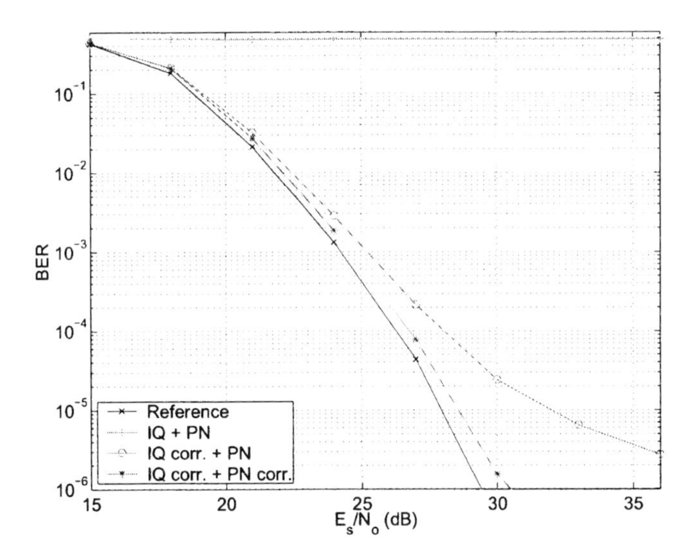

Fig. 4. IQ Imbalance and Phase noise Compensation for coded 64QAM.

combat all 3 effects. An OFDM training symbol provides sufficient information to accurately estimate the channel and the IQ imbalance. The phase noise is estimated and corrected by a decision-directed approach on the data. This approach leads to tremendous performance enhancements. Allowing IQ imbalance up to $\epsilon = 10\%, \Delta\phi = 10^o$, frequency offsets up to 240 kHz and phase noise up to -32dBc without compensation, renders the system unusable (an error floor at 10^{-1}). Our estimation/compensation scheme brings the remaining degradation down to 0.6 dB. Thus this scheme allows to greatly relax the front-end mismatch specs. Since all IQ imbalance contributions are jointly estimated/compensated, the mismatches are allowed anywhere in the receiver up to a combined IQ imbalance of $\epsilon = 10\%, \Delta\phi = 10^o$. Therefore, our method enables zero-IF receivers employing cheap analog components. The IQ estimation is based on 1 OFDM training symbol and thus ensures immediate convergence.

AFFILIATIONS

Jan Tubbax[1], Boris Côme, Liesbet Van der Perre, Stéphane Donnay, Marc Engels[2]
IMEC, Kapeldreef 75, 3001 Leuven, Belgium

REFERENCES

[1] J. Bingham. Multicarrier modulation for data transmission: An idea whose time has come. *IEEE Comm. Mag.*, pages 5–14, May 1990.

[2] HIPERLAN type 2 standard - functional specification data link control (DLC) layer. October 1999.

[3] Draft supplement to standard for LAN/MAN part 11: MAC and PHY specifications: High speed physical layer in the 5GHz band. Technical report, IEEE P802.11a/D7.0, July 1999.

[1] Jan Tubbax is also a Ph.D Student at the KULeuven.

[2] Marc Engels is also a professor at the KULeuven.

[4] U. Reimers. DVB-T: the COFDM-based system for terrestrial television. *IEE Electronics and Communication Engineering Journal*, 9(1):28–32, February 1997.

[5] I. Koffman and V. Roman. Broadband wireless access solutions based on OFDM access in IEEE 802.16. *IEEE Communications Magazine*, 40(4):96–103, April 2002.

[6] Asad A. Abidi. Direct-conversion radio tranceivers for digital communications. *IEEE Journal of Solid-State Circuits*, 30(12):1399–1410, Dec. 1995.

[7] T. Pollet, M. Van Bladel, and M. Moeneclaey. BER sensitivity of OFDM systems to carrier frequency offset and wiener phase noise. *IEEE Transactions on Communications*, 43(2/3/4):191–193, Feb./Mar./Apr. 1995.

[8] J. Tubbax, B. Côme, L. Van der Perre, L. Deneire, and M. Engels. OFDM vs. single-carrier with cyclic prefix: a system-based comparison. In *IEEE VTC Fall*, volume 2, pages 1115–1119, 2001.

[9] B. Côme, R. Ness, S. Donnay, L. Van der Perre, W. Eberle, P. Wambacq, M. Engels, and I. Bolsens. Impact of front-end non-idealities on bit error rate performance of WLAN-OFDM transceivers. *Microwave Journal*, 44(2):126–140, February 2001.

[10] R. Hasholzner, C. Drewes, and J.S. Hammerschmidt. The effects of phase noise on 26 Mb/s OFDMA broadband radio in the local loop systems. *Proc. ACTS Mobile Communications Summit '97*, pages 105–112, October 1997.

[11] M. Valkama, M. Renfors, and V. Koivunen. Advanced methods for I/Q imbalance compensation in communication receivers. *IEEE Transactions on Signal Processing*, 49(10):2335–2344, Oct. 2001.

[12] S. Simoens, M. de Courville, F. Bourzeix, and P. de Champs. New I/Q imbalance modeling and compensation in OFDM systems with frequency offset. *Personal, Indoor and Mobile Radio Communications*, 2:561–566, 2002.

[13] I. Bouras, S. Bouras, T. Georgantas, N. Haralabidis, G. Kamoulakos, and C. Kapnistis et al. A digitally calibrated 5.15-5.825 GHz transceiver for 802.11a wireless LANs in 0.18μm CMOS. *IEEE Internat. Solid-State Circuits Conf.*, pages 352–353, 2003.

[14] J. Tubbax, B. Côme, L. Van der Perre, S. Donnay, and M. Engels. IQ imbalance compensation for OFDM systems. *IEEE International Conference on Communications ICC*, 5:3403–3407, May 2003.

[15] Behzad Razavi. *RF Microelectronics*. Prentice Hall, 1998.

[16] J. Tubbax, B. Côme, L. Van der Perre, S. Donnay, H. De Man, and M. Moonen. Joint iq imbalance and frequency offset compensation for OFDM systems. *IEEE Radio and Wireless Conference*, 2003.

[17] Claus Muschallik. Influence of RF oscillators on an OFDM signal. *IEEE Trans. on Consumer Electronics*, 41(3):592–603, Aug. 1995.

[18] P. Robertson and S. Kaiser. Analysis of the effects of phase-noise in orthogonal frequency division multiplex (OFDM) systems. *IEEE International Conference on Communications,ICC'95*, 3:1652–1657, 1995.

[19] Ana Garcia Armada and Miguel Calvo. Phase noise and sub-carrier spacing effects on the performance of an OFDM communication system. *IEEE Communications Letters*, 2(1):11–13, Jan. 1998.

[20] Nevio Benvenuto. Achievable bit rates of DMT and FMT systems in the presence of phase noise and multipath. *IEEE VTC-Spring Tokyo*, 3:2108 –2112, May 2000.

[21] B. Côme, S. Donnay, and L. Van der Perre et al. Impact of front-end non-idealities on bit error rate performances of WLAN-OFDM transceivers. *IEEE Radio and Wireless Conference*, pages 91–94, September 2000.

[22] M.S. El-Tanany and Yiyan Wu. Impact of phase noise on the performance of OFDM systems over frequency selective channels. *IEEE Vehicular Technology Conference*, 49(10):1802–1806, May 1997.

ZHANYUN DUAN, TOBIAS HIDALGO STITZ, MIKKO VALKAMA, AND MARKKU RENFORS

PRACTICAL ISSUES OF PIC IN MC-CDMA SYSTEMS

Abstract. Multicarrier techniques are of high interest in modern communication systems due to their spectral efficiency and simplicity of channel equalization even in the case of heavily frequency selective channels. Further advantages of MC techniques include, e.g., their insensitivity to timing errors as well as to limited narrowband interference. The combination with CDMA provides an interesting multiple access technique for multiple users. Multiaccess interference in highly loaded systems is a major factor for performance loss and diverse methods exist to mitigate it. This paper focuses on a fully loaded MC-CDMA system with linear MMSE and nonlinear PIC receivers in a Rayleigh fading channel. We study the system performance with relatively low number of subchannels and compare it with the theoretically predicted performance when the number of subchannels tends towards infinity. QPSK and higher order QAM modulations are studied, and a scaling issue in the PIC receiver is identified for the latter. We come to the general conclusion that the higher the modulation order, the more subchannels we need to achieve performance close to that predicted by theory. We also conclude that in the (quasi-) synchronous uplink case, PIC performs similarly to downlink, offering a good improvement compared to the linear MMSE receiver.

1. INTRODUCTION

In general, multicarrier modulation seems to be a key ingredient in future beyond 3G and 4G communication system developments. MC-CDMA in particular tries to combine the advantages of OFDM and CDMA, allowing several users or code channels to be transmitted over the same set of subchannels [1]. Each user signal is spread using a unique spreading code (assumed mutually orthogonal) and the spread signal is transmitted using OFDM, on a one chip per subcarrier basis. Interestingly, sending the same information over several subcarriers provides natural frequency diversity.

We distinguish between the downlink, where all the code channels are subject to identical channel distortion, and the uplink, where the channel responses for different users are different. One fundamental difference here is that in the (single-cell) downlink case, the different user signals are perfectly synchronized, whereas in the uplink accurate synchronization is not possible. We consider here the quasi-synchronous [2] uplink case, in which the timing differences are small compared to the guard interval, and thus the useful parts of the consecutive symbols of different users are not overlapping.

In [3] it is shown that an MC-CDMA system with number of subchannels tending to infinity and MMSE equalization transforms Rayleigh flat-fading ergodic subchannels of OFDM into AWGN subchannels with modified noise variance, performing thus better than OFDM. Among others, parallel interference cancellation (PIC) techniques can be used to improve the system performance by reducing the multiaccess interference (MAI).

This research was supported by Nokia and the Graduate School in Electronics, Telecommunications, and Automation (GETA).

K. Fazel and S. Kaiser (eds.), Multi-Carrier Spread-Spectrum, 481–488.

To increase flexibility and reduce detection complexity, an MC-CDMA multiplex could use only a relatively small subset of the subcarriers of the underlying OFDM system. It should also be noted that to increase frequency diversity, it is beneficial to allocate to each MC-CDMA multiplex subcarriers that are not very close to each other.

In this paper, we are targeting to show how many/few subchannels should/can be used in combination with PIC in order to obtain the predicted results of [3]. Thus we base our experiments on that work, expanding it further to higher order QAM modulations and the uplink case. In general, if the needed number is too high regarding the total number of subchannels of the underlying OFDM system, the separation between the subcarriers of the MC-CDMA multiplex could become smaller than the coherence bandwidth, compromising thus the available frequency diversity.

2. SYSTEM MODELS

We consider a fully loaded MC-CDMA system with N subcarriers and $K=N$ users. The size of the symbol alphabet is M. Considering the signal detection, perfect channel state knowledge is assumed. Given that the guard interval (GI) is longer than the channel delay spread, intersymbol interference is avoided, and due to the cyclic prefix (CP) the transmission channel is effectively flat within each subchannel [4], [5]. (Quasi-)Synchronicity is also assumed.

With the previous assumptions, only one symbol from each user is contributing to the observed data within one detection interval, and a direct frequency domain signal model can be used. This being the case, the signal within one detection interval after GI removal and FFT appears as (time index omitted)

$$\mathbf{r} = \begin{bmatrix} H_{1,1}c_{1,1} & H_{2,1}c_{2,1} & \cdots & H_{K,1}c_{K,1} \\ H_{1,2}c_{1,2} & H_{2,2}c_{2,2} & \cdots & H_{K,2}c_{K,2} \\ \vdots & \vdots & \ddots & \vdots \\ H_{1,N}c_{1,N} & H_{2,N}c_{2,N} & \cdots & H_{K,N}c_{K,N} \end{bmatrix} \begin{bmatrix} A_1 \\ A_2 \\ \vdots \\ A_K \end{bmatrix} + \begin{bmatrix} n_1 \\ n_2 \\ \vdots \\ n_N \end{bmatrix} = \mathbf{Sa} + \mathbf{n} \qquad (1)$$

where $\mathbf{r} = [r(1), r(2), \ldots, r(N)]^T$ with $r(i)$ being the i-th subchannel observation, A_k is the transmitted symbol for user k and $\mathbf{n}$ denotes the noise vector. In general, the code chips of user k are denoted as $c_{k,1}, c_{k,2}, \ldots, c_{k,N}$ and $H_{k,i}$ denotes the i-th subchannel response for user k. In downlink, $H_{1,i} = H_{2,i} = \ldots = H_{K,i}$ for all i, i.e., all the users or code channels experience the same channel response. For simultaneous detection of all the data symbols, we use the nonlinear PIC technique.

2.1. Parallel interference cancellation (PIC) detection

As a nonlinear multiuser detector, the PIC approach is based on a very intuitive idea: if we can estimate the data of an interfering user, this information can be used to reproduce the interfering signal of this user and that interference can then be removed from the received signal to obtain a signal with less interference. Logically, the performance of this principle will depend on how well we can estimate the interfering data. Assuming that the signal on which the decision is made is improved, the PIC principle can be used in an iterated manner, as in the iterated-decision receivers of [6]. These receivers use optimised iterative algorithms to successively cancel interference from a block of received data and generate symbol decisions so that the slicer input signal-to-interference and noise ratio (SINR) is maximized. The iterated-decision multiuser detector is depicted in Figure 1.

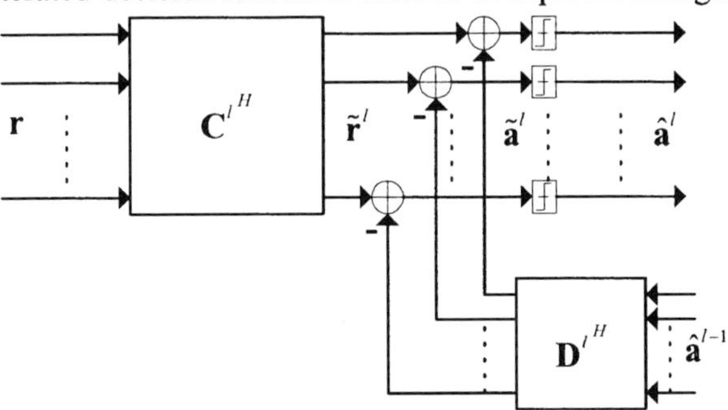

Figure 1. Iterated-decision multiuser PIC detector.

The algorithm is formulated in terms of two matrices evolving as the iteration proceeds. A forward $K \times N$ matrix $\mathbf{C}^l$ is used for linear equalization of the received signal vector $\mathbf{r}$ and a backward $K \times K$ matrix $\mathbf{D}^l$ for feeding back the interfering symbols. Optimal $\mathbf{C}^l$ and $\mathbf{D}^l$ are obtained by maximizing the SINR at the input of the slicers. Thus, at each iteration, the received signal $\mathbf{r}$ is multiplied by $(\mathbf{C}^l)^H$, producing the equalized $K \times 1$ vector $\tilde{\mathbf{r}}^l = (\mathbf{C}^l)^H \mathbf{r}$. Then, based on the symbol estimates of the previous iteration, an estimate of the MAI is subtracted from $\tilde{\mathbf{r}}^l$ to produce $\tilde{\mathbf{a}}^l$, i.e., $\tilde{\mathbf{a}}^l = \tilde{\mathbf{r}}^l - (\mathbf{D}^l)^H \hat{\mathbf{a}}^{l-1}$. In order to only deal with interference, the diagonal elements of $\mathbf{D}^l$ are forced to zero. A bank of slicers applied to $\tilde{\mathbf{a}}^l$ generates the estimate $\hat{\mathbf{a}}^l$ of the transmitted symbols. Given perfect interference estimates and thus perfect MAI cancellation, the performance of single user transmission would be obtained. In general, a major benefit of this scheme is that iterations can be stopped when the desired bit error rate is reached, thus giving control over the overall system complexity.

In [3], optimal values for $\mathbf{C}^l$ and $\mathbf{D}^l$ were derived for QPSK modulation with unit signal energy. Here, we extend the method to general QAM modulations. As $\mathbf{a}$ and $\hat{\mathbf{a}}^l$ can be ideally considered as vectors of zero-mean uncorrelated symbols with energy ε (equal energy symbols assumed), their correlation is of the form

$$E[\mathbf{a}(\hat{\mathbf{a}}^l)^H] \approx \epsilon \boldsymbol{\rho}^l = \varepsilon \operatorname{diag}[\rho_1^l, \rho_2^l, \cdots \rho_K^l] \tag{2}$$

The output SINR is now maximized for the iterated-decision algorithm with the following choice of matrices:

$$\mathbf{D}^l = (\boldsymbol{\rho}^{l-1})^H[(\mathbf{C}^l)^H\mathbf{S} - \mathrm{diag}\{((\mathbf{C}^l)^H\mathbf{S})_{1,1}, \cdots, ((\mathbf{C}^l)^H\mathbf{S})_{K,K}\}]^H \tag{3}$$

and

$$\mathbf{C}^l \propto \varepsilon[\sigma^2\mathbf{I} + \varepsilon\mathbf{S}(\mathbf{I} - \boldsymbol{\rho}^{l-1}(\boldsymbol{\rho}^{l-1})^H)\mathbf{S}^H]^{-1}\mathbf{S} \tag{4}$$

where σ^2 denotes the additive noise variance. The confidence coefficient ρ_i^{l-1} measures the closeness of the decision $\hat{A}_i^{l-1}$ to A_i at iteration l. If $\hat{A}_i^{l-1}$ is a poor estimate of A_i, ρ_i^{l-1} will be small, and the estimated MAI that is subtracted from $\tilde{\mathbf{r}}^l$ will be weighted accordingly low. On the other hand, if $\hat{A}_i^{l-1}$ is a good estimate of A_i, ρ_i^{l-1} will be close to one, and the corresponding MAI estimate is used with high confidence. In particular, if $\hat{A}_i^{l-1} = A_i$ (with probability one), then $\rho_i^{l-1} = 1$ and MAI can be exactly reproduced. In this case, the interference produced by user i to other users can be cancelled completely. Note also, that in the starting situation, $\rho_i^0 = 0$, thus also $\mathbf{D}^1 = \mathbf{0}$ and $\mathbf{C}^1$ coincides with the general linear minimum mean-squared error (MMSE) estimator [7]. Further, in the downlink and in our fully loaded case, the general MMSE estimation principle coincides with the per-carrier MMSE detection [8].

The expression for the ith user SINR β_i^l is of the form

$$\beta_i^l = \frac{\eta_i^l\varepsilon}{1-(1-\left|\rho_i^{l-1}\right|^2\eta_i^l)\varepsilon} \tag{5}$$

where $\eta_i^l = \mathbf{S}[i]^H[\sigma^2\mathbf{I} + \varepsilon\mathbf{S}(\mathbf{I} - \boldsymbol{\rho}^{l-1}(\boldsymbol{\rho}^{l-1})^H)\mathbf{S}^H]^{-1}\mathbf{S}[i]$, and $\mathbf{S}[i]$ denotes the ith column of $\mathbf{S}$.

The desired confidence coefficient ρ_i^l can be expressed in terms of the symbol error probability P_i^l at the output of the corresponding slicer. Next we will derive an expression for ρ_i^l in the general QAM modulation case.

For rectangular M-QAM modulation, the average energy of signal is $\varepsilon = 2(M-1)/3$. Let the error at the output of the associated slicer be $e_i^l = \hat{A}_i^l - A_i$ at iteration l. In order to obtain ρ_i^l, we rewrite $E[A_i(\hat{A}_i^l)^*] = \varepsilon + E[A_i(e_i^l)^*]$ and evaluate $A_i(e_i^l)^*$ for the different points of the constellation.

If the transmitted symbol is one of the four corner points, we have the approximation (assuming that an error yields one of the nearest neighbouring symbols):

$$\overline{A_i(e_i^l)^*} \approx -2P_i^l(\sqrt{M}-1) \tag{6}$$

where $\overline{(.)}$ refers to expectation *within* the constellation points under consideration. When the transmitted signal is one of the $(\sqrt{M}-2)^2$ inside points, $\overline{A_i(e_i^l)^*}=0$, due to the symmetry of errors. For the remaining $4(\sqrt{M}-2)$ "side" points, we obtain

$$\overline{A_i(e_i^l)^*} \approx -\frac{2}{3}P_i^l(\sqrt{M}-1) \tag{7}$$

Combining the above results we can finally write

$$E[A_i(\hat{A}_i^l)^*] = \varepsilon + E[A_i(e_i^l)^*] \approx \varepsilon - \frac{8}{3M}P_i^l(M-1) \tag{8}$$

and for $l \geq 1$,

$$\rho_i^l = \frac{E[A_i(\hat{A}_i^l)^*]}{\varepsilon} = 1 - \frac{4}{M}P_i^l \tag{9}$$

and the symbol error probability is

$$P_i^l = 4\left(1-\frac{1}{\sqrt{M}}\right)Q\left(\sqrt{\frac{3}{M-1}\beta_i^l}\right) - 4\left(1-\frac{1}{\sqrt{M}}\right)^2 Q\left(\sqrt{\frac{3}{M-1}\beta_i^l}\right)^2 \tag{10}$$

Notice that for $M = 4$, $\rho_i^l = 1 - P_i^l$, and the results are consistent with those given in [3] for QPSK modulation.

3. OBTAINED RESULTS

In the following, some simulation results are presented. The previously discussed model is used, and the subchannels are assumed to be independently Rayleigh fading and non-frequency-selective. This holds if a proper cyclic prefix is used and if the subchannel spacing is greater than the channel coherence bandwidth. The spreading is done by Walsh-Hadamard codes of the same length as the number of subchannels. The performance of the studied systems is presented in terms of raw bit error rate BER vs. E_b/N_0 (bit energy to noise spectral density ratio). In addition to simulated BER results, also theoretical BER curves are presented for the studied cases. The theoretical values are obtained by averaging the SINR values β_i^l over different channel realizations and then applying the expression for the *symbol* error probability presented in the previous section. The final *bit* error probability is then obtained by assuming that one symbol error produces exactly one bit error (Gray coding).

In the first case, we study the performance of a fully loaded downlink QPSK MC-CDMA system with different number of subchannels. Figure 2 presents the performance of the theoretical and the simulated results for 32 subchannels (left) and for 64 subchannels (right). The curves represent the linear MMSE receiver and the PIC receiver after 1, 3 and 5 iterations. It can be noticed that the higher the number of subchannels, the closer the theory and the simulations match. 32 subchannels are clearly not enough to obtain the results predicted by the theory. We also see that there is some crossing of the 5-th iteration curve with the 3-rd iteration curve, with the lower iteration performing somewhat better at high E_b/N_0 range. This is probably due to error propagation. Only when having 256 subchannels or more the 5-th PIC iteration is consistently the best as expected.

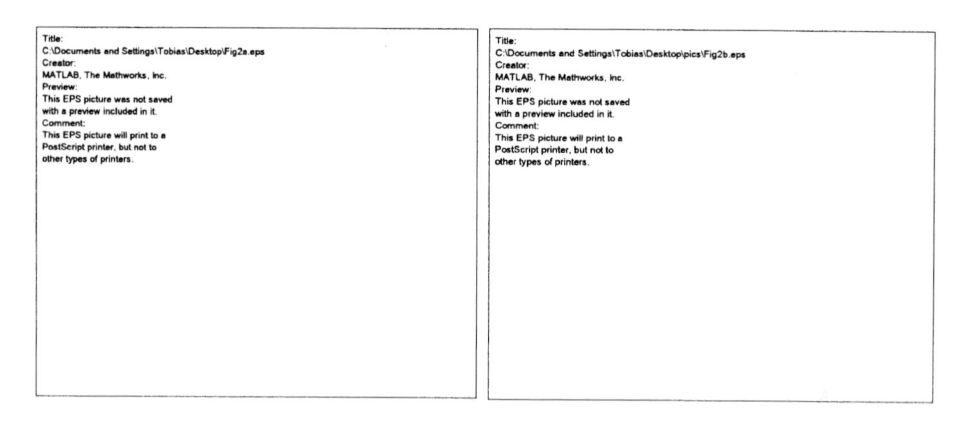

Figure 2. Theoretical and simulated performance of linear MMSE and PIC detectors with 32 and 64 subchannels with QPSK modulation. Downlink.

After that we studied the behaviour for higher order QAM modulations and were confronted with a "scaling problem" in the basic PIC algorithm, as described next.

3.1. PIC scaling for higher order modulations

In the PIC detector, the input of the slicers is

$$\begin{aligned}\tilde{\mathbf{a}}^l &= (\mathbf{C}^l)^H \mathbf{r} - (\mathbf{D}^l)^H \hat{\mathbf{a}}^{l-1} = (\mathbf{C}^l)^H (\mathbf{Sa}+\mathbf{n}) - (\mathbf{D}^l)^H \hat{\mathbf{a}}^{l-1} \\ &= (\mathbf{C}^l)^H \mathbf{Sa} + (\mathbf{C}^l)^H \mathbf{n} - [(\mathbf{C}^l)^H \mathbf{S} - \mathrm{diag}\{((\mathbf{C}^l)^H \mathbf{S})_{1,1}, \cdots, ((\mathbf{C}^l)^H \mathbf{S})_{K,K}\}]\boldsymbol{\rho}^{l-1}\hat{\mathbf{a}}^{l-1}\end{aligned} \tag{11}$$

Assuming correct decisions at iteration l–1 (i.e., $\hat{\mathbf{a}}^{l-1} = \mathbf{a}$) and $\boldsymbol{\rho}^{l-1} = \mathbf{I}$, then

$$\tilde{\mathbf{a}}^l = \mathrm{diag}\left\{\left((\mathbf{C}^l)^H\mathbf{S}\right)_{1,1},\ldots\left((\mathbf{C}^l)^H\mathbf{S}\right)_{K,K}\right\}\mathbf{a} + (\mathbf{C}^l)^H\mathbf{n} = \mathbf{Ga} + \mathbf{z} \tag{12}$$

where the diagonal matrix $\mathbf{G}$ can be viewed as the gains of the equivalent symbol rate channels and $\mathbf{z}$ is the additive noise component after the linear equalization stage. Now, it can be noted that even if we have the correct decision for **a** at iteration l–1, we might obtain a poor decision for **a** at iteration l due to the effect of **G.** In fact, PIC has a worse performance than MMSE without compensation of **G** for higher order QAM modulations to the extent that PIC cannot work properly for higher QAM modulation without scaling. Notice that since **G** is real valued, it does not have any effect on PSK modulations. One simple possibility to undo the effect of **G** is to multiply the input of the slicers $\tilde{\mathbf{a}}^l$ by

$$\mathbf{G}^{-1} = [\mathrm{diag}\{((\mathbf{C}^l)^H\mathbf{S})_{1,1}, \ \cdots, \ ((\mathbf{C}^l)^H\mathbf{S})_{K,K}\}]^{-1} \tag{13}$$

Notice that since the scaling stage is diagonal and thus operates component-wise, it does not change the slicer input SINR in anyway; it only restores the signal levels to their original "range" being thus crucial for high order QAM.

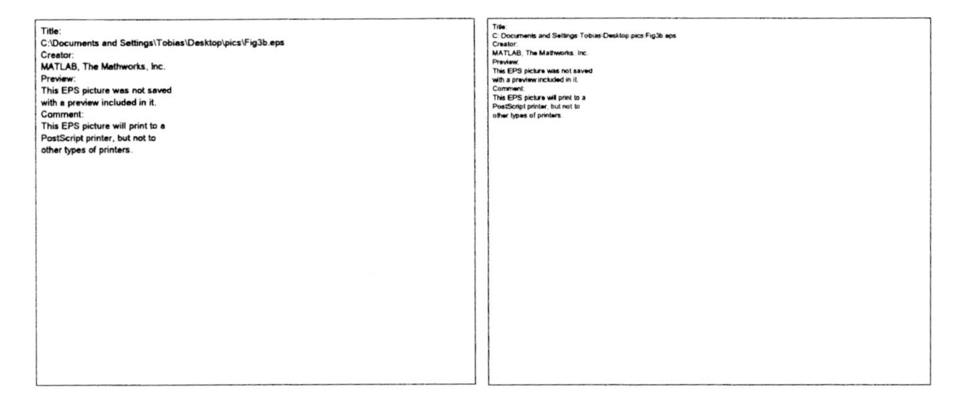

Figure 3. Theoretical and simulated performance of linear MMSE and PIC detectors with 256 and 1024 subchannels with 16-QAM modulation. Downlink.

Figure 3 shows similar results as in Figure 2, but with 16-QAM modulation. Note that the required number of subchannels to get close to the theoretical performance is much higher than in case of QPSK. In this example, we show the curves for 256 and 1024 subchannels. Also the error propagation problem persists for 256 subchannels but for 1024 subchannels the simulation results match the theory with 16-QAM. Thus, in general, there is a clear tendency of needing higher number of subchannels with higher order modulations to approach the theoretical performance.

Finally we show some results for the uplink case. The performance of the uplink is clearly worse when using the linear MMSE receiver compared to the downlink, as can be seen from Figure 4. However, the PIC performance is very similar in the uplink and the downlink for the QPSK modulation. In the 16-QAM case, similar observations are suggested by theory but with up to 256 subchannels, this is not realized in practice (simulations). In the theoretical case, the performances end up being very similar, although in the downlink, the convergence in the iterations is achieved faster and the first PIC iteration performs better than in the uplink.

Nevertheless, the improvement in performance compared to the linear MMSE receiver is still substantial.

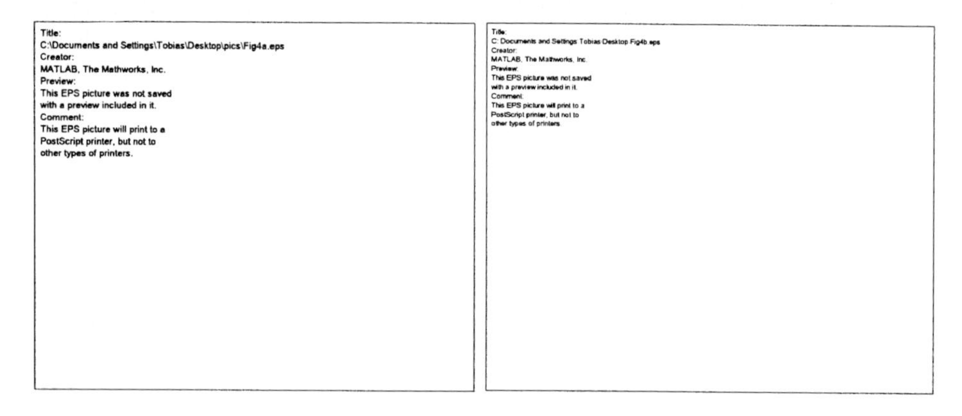

Figure 4. Theoretical and simulated performance of linear MMSE and PIC detectors with 64 subchannels for QPSK and 256 subchannels for 16-QAM modulation. Uplink.

Zhanyun Duan, Tobias Hidalgo Stitz, Mikko Valkama & Markku Renfors
Institute of Communications Engineering
Tampere University of Technology
P.O. Box 553, FIN-33101 Tampere
Finland

REFERENCES

[1] S. Hara and R. Prasad, "Overview of multicarrier CDMA," *IEEE Commun. Mag.*, vol. 35, pp. 126-133, Dec. 1997.

[2] S. Tsumura and S. Hara, "Design and performance of quasi-synchronous multi-carrier CDMA system," in *Proc. IEEE Veh. Technol Conf.*, Atlantic City, NJ, Oct. 2001, pp. 843-847.

[3] M. Debbah, "Linear Precoders for OFDM Wireless Communications." PhD thesis, http://www-syscom.univ-mlv.fr/~debbah/indexeng.html.

[4] R. van Nee and R. Prasad, *OFDM for Wireless Multimedia Communications*. Boston, MA: Artech House, 2000.

[5] J. G. Proakis, *Digital Communications*, 4th ed. New York, NY: McGraw-Hill, 2001.

[6] S. Beheshti, S. H. Isabelle and G. W. Wornell, "Joint intersymbol and multiple-access interference suppression algorithms for CDMA systems," *European Trans. Telecommun.*, vol. 9, pp. 403-418, Sept./Oct. 1998.

[7] S. M. Kay, *Fundamentals of Statistical Signal Processing: Estimation Theory*. Englewood Cliffs, NJ: Prentice-Hall, 1993.

[8] M. Valkama, T. Hidalgo Stitz, and M. Renfors, "Enhanced per-carrier processing for MC-CDMA downlink," Asilomar Conference on Signals, Systems, and Computers, Pacific Grove, CA, Nov. 2003, to appear.

OFDM, DS-CDMA, AND MC-CDMA SYSTEMS WITH PHASE NOISE AND FREQUENCY OFFSET EFFECTS

Nizar Hicheri, Michel Terré, Bernard Fino

Abstract. This paper analyses the effects of either a frequency offset or a phase noise process on the performances of different waveforms. Waveforms taken into consideration are candidates for different wireless access networks. We focus our work on DS-CDMA (IS95, UMTS, Cdma2000) and OFDM (IEEE802.11a, HiperlanII, DVB-T). The case of MC-CDMA is analysed. Different results, that give directly the SNR degradation due to one of the two defaults analysed, are presented. These formula are verified through simulations.

1. INTRODUCTION

In this paper, we consider a multi-users downlink through an AWGN channel. We investigate the influence of frequency offset and Wiener phase noise on OFDM, DS-CDMA and MC-CDMA systems. OFDM, and MC-CDMA have several advantages, as the possibility of efficiently solving multipath channel by performing a simple equalization of the received signal. Nevertheless, multi-carrier systems, are known to be very sensitive to frequency offset and phase noise introduced by local oscillators. Many papers treats about the influence of such impairments, each one with its own approach. In [1-8], the OFDM system was considered. In [1], Jinwen Shentu and al considers a Gaussian phase noise impacting a PCC-OFDM system (Polynomial Cancellation Coded Orthogonal Frequency Division Multipleing) which is a modified OFDM system to combat phase noise effects and compare it to a classic OFDM, results show a better performances for PCC-OFDM. In [2] the influence of Wiener-Levy phase noise on the Cut-off rate and capacity of OFDM systems was studied. In [3] Lorenzo Piazzo and al, analysed the phase noise effects on OFDM system for three the coherent, the CPEC and the differential receivers. These authors have shown that the SNR depends only on the power of the phase noise and not on the shape of the spectrum. Ana Gacia Armada analysed in [4-5] SNR degradation in presence of (little) phase noise and shows when the bandwidth of the phase noise is less bigger than the inter-carrier spacing then a correction of the Common Phase Error (CPE) can contribute to a deacrease of the BER. Otherwise there is no improvement in BER when inter-carrier spacing is quite equal to phase noise bandwidth. In [6], Thierry Pollet and al, have written the expression of SNR degradation in presence of phase noise and frequency offset both for OFDM and SC (Single Carrier) systems, they have concluded that OFDM system is more sensible than SC system. In [7], K. Sathananthan and al, have calculated by numerical tools

K. Fazel and S. Kaiser (eds.), Multi-Carrier Spread-Spectrum, 489–500.

the error probability of OFDM system with frequency offset. In [8], Luciano Tomba analysed the effect of Wiener phase noise in OFDM system and proposed to calculate the corresponding BER by the moment generating function (MGF). In [9-12] the MC-CDMA system is considered. In [9], Jiho Jan and al analysed the performances of MC-CDMA system with frequency offset in the case of selective channel and for EGC and MRC receivers. In [10] Luciano Tomba used a semi-analytic approach to study the sensitivity of the MC-CDMA to carrier phase noise and frequency offset. In [11] Heidi Steendam and Marc Moeneclay analysed the effect of carrier phase jitter on MC-CDMA by using matrix notations, and they give expression and curves for SNR degradation. Finally, in [12] Younsum Kim and al, calculated the effect of frequency offset on MC-CDMA system, they also proposed a particular shape of the pulse to combat ICI caused by the offset. Our paper will complete these previous studies and, using a global formalisation for these different waveforms OFDM, DS-CDMA and MC-CDMA, we will analyse their immunity to these frequency offsets and phase noise defaults, we will derive BER expressions due to theses defaults.

2. NOTATIONS

2.1 Spreading description

We introduce a complex vector $X(nT_s) = [x_0(nT_s) \;\; \dots \;\; x_{N-1}(nT_s)]^T$ containing a set of N independent communication symbols. These symbols will be transmitted during the same time slot, of duration T_s, from a base station or an access point to terminals. DS-CDMA, OFDM, MC-CDMA solutions could be formalised through a same vector transformation :

$$Y(nT_s) = \boldsymbol{Z}X(nT_s) \tag{1.}$$

where $\boldsymbol{Z}$ stands for an (NxN) spreading matrix whose each column represents a particular spreading sequence. The kth column of the matrix $\boldsymbol{Z}$ is the sequence used for the $x_k(nT_s)$ symbol. The $Y(nT_s)$ vectors represents signal samples that will be transmitted through the air interface. For DS-CDMA, $\boldsymbol{Z}$ is the product of a diagonal matrix $\boldsymbol{D}$ whose elements would typically come from a long scrambling Gold sequence times a square Hadamard matrix $\boldsymbol{H}$.

$$\boldsymbol{Z} = \boldsymbol{DH} \tag{2.}$$

for OFDM, $\boldsymbol{Z}$ is a Fourier matrix $\boldsymbol{F}$, $f_{n,m} = e^{j2\pi\, n\, m/N}$

$$\boldsymbol{Z} = \boldsymbol{F} \tag{3.}$$

and for MC-CDMA, we have directly :

$$Z = FDH \tag{4.}$$

2.2 Channel description

We make the hypothesis that the fading is constant over the time T_s, we can then write the expression of the channel as : $c(nT_s) = \sum_{k=0}^{L} \alpha_k (nT_s) \delta(nT_s - \tau_k)$

Coefficients α_k are independent random gaussian variables. Thanks to the insertion of a cyclic prefix for OFDM or even for CDMA, and with its suppression at the receiver stage, the propagation channel has a bounded effect and the receiver can consider only symbols coming from the $X(nT_s)$ vector. Let L be the number of paths in the channel, we can then formalise the channel with a matrix notation :

$$\boldsymbol{C}(nT_s) = \sum_{k=0}^{L} \alpha_k (nT_s) \boldsymbol{J}^k \tag{5.}$$

In this expression the (NxN) J matrix is defined as :

$$\boldsymbol{J} = \begin{pmatrix} 0 & 1 \\ \boldsymbol{I}_{(N-1xN-1)} & 0 \end{pmatrix} \tag{6.}$$

With these notations, the received signal can directly be written as :

$$R(nT_s) = \boldsymbol{C}(nT_s)\boldsymbol{Z}(nT_s)X(nT_s) + B(nT_s) \tag{7.}$$

$B(nT_s)$ stands for a complex gaussian white noise vector with a power spectral density N_0 and with $E\left[B(nT_s)B^H(nT_s)\right] = \sigma^2 \boldsymbol{I}$. For the sake of simplicity, we will forget to specify the temporal interval $\left[nT_s, (n+1)T_s\right[$, and just write the received signal as : $R = \boldsymbol{CZ}X + B$.

2.3 Receivers description

With a minimum mean squared errors receiver (***MMSE***), we must identify a matrix $\boldsymbol{W}$ of despreading sequences that leads to a minimum squared error between $\hat{X}$ and X :

$$W \;/\; Min\; E\left[\left|W^H R - X\right|^2\right] \tag{8.}$$

The well known solution of this equation is given by :

$$W_{mmse} = \left(CZPZ^H C^H + \sigma^2 I\right)^{-1} C\,ZP \tag{9.}$$

where P is a diagonal matrix with powers of transmitted symbols. In the case of ***OFDM***, the receiver usually used is constituted by a FFT and a frequency equalization. With our matrix notations we can write this receiver as :

$$W_{ofdm} = C^{-H} Z \tag{10.}$$

In the case of ***CDMA***, we usually use a rake receiver (or MRC receiver) which can be written as (using all paths) :

$$W_{rake} = CZ \tag{11.}$$

3. FREQUENCY OFFSET IN AWGN CHANNEL

In this paragraph we will consider an additive white gaussian noise channel (AWGN). Having orthogonal spreading sequences, we will use Z^H for despreading. We consider a frequency offset Δf between the carrier of the transmitter and the receiver, we have then to introduce a Φ diagonal matrix with elements defined by : $\Phi_{nn} = e^{j\varphi_n}$, with $\varphi_n = 2\pi\Delta f\, T_s \frac{n}{N}$.

Without any frequency correction, the estimated signal $\hat{X}$ is given by :

$$\hat{X} = Z^H \Phi\left(ZX + B\right) \tag{12.}$$

Introducing a new matrix $M = diag(Z^H \Phi\, Z)$, we could propose the following decomposition for $\hat{X}$:

$$\hat{X} = M\,X + \left\lfloor Z^H \Phi\, Z - M\,\right\rfloor X + Z^H \Phi\, B \tag{13.}$$

In this equation the first term represents the useful component, the second term the inter-user or inter-carrier interference (ICI), and the last term is the contribution of the gaussian noise. We introduce the energy per symbol E_s and we suppose independent symbols with the same energy such that : $E\left\lfloor XX^H\right\rfloor = \left(E_s / T_s\right) I$.

Let P_u be the useful power, we have :

$$P_u = \frac{1}{N} E\left[|MX|^2 \right] = \frac{1}{N} E\left[trace\left\{ M\, X\, X^H\, M^H \right\} \right] \quad (14.)$$

After some derivation we obtain :

$$P_u = \frac{E_s}{NT_s} \sum_{i=0}^{N-1} \left| \sum_{n=0}^{N-1} |Z_{ni}|^2\, e^{j\varphi_n} \right|^2 \quad (15.)$$

The power of the inter-channel interference P_{ici} is given by :

$$P_{ici} = \frac{1}{N} E\left[\left| \left[Z^H \Phi\, Z - M \right] X(nT_s) \right|^2 \right] \quad (16.)$$

To calculate this power, we could remark that $Z^H \Phi Z$ is a unitary matrix, then we have :

$$E\left[|X|^2 \right] = E\left[\left| Z^H \Phi Z X \right|^2 \right] = N \frac{E_s}{T_s} \quad (17.)$$

using orthogonality between vectors $M\, X$ and $\left[Z^H \Phi Z - M \right] X$, we have :

$$P_{ici} = \frac{E_s}{T_s} - P_u \quad (18.)$$

The power of the Gaussian noise is given by :

$$P_b = \frac{1}{N} E\left[\left| Z^H \Phi B \right|^2 \right] = \frac{N_0}{T_s} \quad (19.)$$

Finally we can write the signal to noise ratio as :

$$SNR = \frac{\sum_{i=0}^{N-1} \left| \sum_{n=0}^{N-1} |Z_{ni}|^2\, e^{j\varphi_n} \right|^2}{N - \sum_{i=0}^{N-1} \left| \sum_{n=0}^{N-1} |Z_{ni}|^2\, e^{j\varphi_n} \right|^2 + N \frac{N_0}{E_s}} \quad (20.)$$

In the case of OFDM and DS-CDMA, we have: $\left|Z_{ij}\right|^2 = \frac{1}{N}$, developing equation (23), we obtain directly the SNR with respect to the frequency offset :

$$SNR = \frac{\sin^2(\pi\,\Delta f\,T_s)}{N^2\left(1+\frac{N_0}{E_s}\right)\sin^2(\pi\Delta f\,T_s/N) - \sin^2(\pi\,\Delta f\,T_s)} \quad (21.)$$

4. FREQUENCY OFFSET IN MULTIPATH CHANNEL

In the general case, we can write the estimated vector as :

$$\hat{X} = \boldsymbol{W}^H \boldsymbol{\Phi}(\boldsymbol{CZ}X + B) \quad (22.)$$

Using the same decomposition as before, we obtain :

$$\hat{X} = diag\left(\boldsymbol{W}^H \boldsymbol{\Phi}\,\boldsymbol{CZ}\right)X + \left[\boldsymbol{W}^H \boldsymbol{\Phi}\,\boldsymbol{CZ} - diag\left(\boldsymbol{W}^H \boldsymbol{\Phi}\,\boldsymbol{CZ}\right)\right]X + \boldsymbol{W}^H \boldsymbol{\Phi} B \quad (23.)$$

Introducing the matrix $\boldsymbol{K} = diag\left(\boldsymbol{W}^H \boldsymbol{\Phi}\,\boldsymbol{CZ}\right)$, the desired power can be written as :

$$P_u = \frac{1}{N} E\left[\left|\boldsymbol{K}X\right|^2\right] = \frac{E_s}{NT_s} trace\left\{\boldsymbol{K}\,\boldsymbol{K}^H\right\} \quad (24.)$$

The second term of (eq 23) is the ICI component, its power is given by :

$$P_{ici} = \frac{1}{N} E\left[\left|\left[\boldsymbol{W}^H \boldsymbol{\Phi}\,\boldsymbol{C}\,\boldsymbol{Z} - \boldsymbol{K}\right]X\right|^2\right] \quad (25.)$$

The last term of (eq 23) is supposed to be gaussian with power given by :

$$P_b = \frac{1}{N} E\left[\left|\boldsymbol{W}^H \boldsymbol{\Phi} B\right|^2\right] = \frac{N_0}{NT_s} trace\left\{\boldsymbol{W}^H \boldsymbol{W}\right\} \quad (26.)$$

Finally we can write the signal to noise ratio as :

$$SNR = \frac{trace\left\{\boldsymbol{K}\,\boldsymbol{K}^H\right\}}{trace\left\{\left(\boldsymbol{W}^H \boldsymbol{\Phi}\,\boldsymbol{CZ} - \boldsymbol{K}\right)\left(\boldsymbol{W}^H \boldsymbol{\Phi}\,\boldsymbol{CZ} - \boldsymbol{K}\right)^H\right\} + \frac{N_0\, trace\left\{\boldsymbol{W}^H \boldsymbol{W}\right\}}{E_s}} \quad (27.)$$

5. PHASE NOISE

The phase noise is the difference between the phase of the carrier and the phase of the LO. Theoretically to account for phase noise or frequency offset the output of a microwave oscillator is given, in complex form, by : $e^{j\{2\pi f_c t+\varphi(t)\}}$, where f_c is the nominal frequency of the oscillator and $\varphi(t)$ is for the phase noise case a continuous-path Brownian motion (or Wiener-Levy) process [13] with zero mean and variance $2\pi\beta t$. If we consider a little variation of the phase noise, we can approximate $\mathbf{\Phi}$ by $(\boldsymbol{I_d}+j\mathbf{\Psi})$ where $\mathbf{\Psi}$ is a diagonal matrix defined by : $\Psi_{nn}=\varphi_n$. By considering the discrete model of Wiener-Levy process, we can write the phase φ_n as : $\varphi_n=\sum_{l=0}^{n}\theta_l$, where θ_l is a zero-mean Gaussian variable, with $\sigma_\theta^2=2\pi\beta$. β represents the two-sided 3-dB bandwidth of the spectrum [14]. The demodulated data is finally given by :

$$\hat{X}=\boldsymbol{Z}^{\boldsymbol{H}}(\boldsymbol{I_d}+j\mathbf{\Psi})\boldsymbol{Z}X+\boldsymbol{Z}^{\boldsymbol{H}}(\boldsymbol{I_d}+j\mathbf{\Psi})B \tag{28.}$$

By introducing $\mathbf{\Gamma}=diag(\boldsymbol{Z}^{\boldsymbol{H}}\mathbf{\Psi}\boldsymbol{Z})$, we can write :

$$\hat{X}=X+j\mathbf{\Gamma}X+j\left(\boldsymbol{Z}^{\boldsymbol{H}}\mathbf{\Psi}\boldsymbol{Z}-\mathbf{\Gamma}\right)X+\boldsymbol{Z}^{\boldsymbol{H}}(\boldsymbol{I_d}+j\mathbf{\Psi})B \tag{29.}$$

The first term is the desired signal, the second term corresponds to the *common phase error* (CPE), the third term is the result of inter-symbol interference (ICI) due of phase noise. The last term is the contribution of the gaussian noise. In the case of OFDM and DS-CDMA systems, we can write $|Z_{ni}|^2=1/N$, the CPE term shows that for OFDM, all the sub-carriers undergo a phase rotation with the same angle $\Gamma_{ii}=\frac{1}{N}\sum_{n=0}^{N-1}\varphi_n$. The power P_{cpe} of the CPE term is given by :

$$P_{cpe}=\frac{1}{N}E\left[|\mathbf{\Gamma}X|^2\right]=\frac{E_s}{NT_s}E\left[trace\left\{\mathbf{\Gamma}\,\mathbf{\Gamma}^H\right\}\right] \tag{30.}$$

Developing this equation yields to :

$$P_{cpe}=\frac{E_s}{NT_s}E\left[\sum_{i=0}^{N-1}\left|\sum_{n=0}^{N-1}|Z_{ni}|^2\varphi_n\right|^2\right] \tag{31.}$$

After some derivations we obtain :

$$P_{cpe} = \frac{E_s}{NT_s} \frac{\sigma_\theta^2}{6} (2N^2 + 3N - 5) \tag{32.}$$

The power of the ICI contribution is given by :

$$P_{ici} = \frac{1}{N} E\left[\left| \left(\mathbf{Z}^H \mathbf{\Psi} \mathbf{Z} - \mathbf{\Gamma} \right) X \right|^2 \right] \tag{33.}$$

Developing (36) yields to :

$$P_{ici} = \frac{E_s}{NT_s} E\left[\sum_{i=0}^{N-1} \sum_{\substack{n=0 \\ n \neq i}}^{N-1} \left| \sum_{m=0}^{N-1} Z_{mi}^* Z_{mn} \varphi_m \right|^2 \right] \tag{34.}$$

Finally we will obtain :

$$P_{ici} = \frac{E_s}{NT_s} \sigma_\theta^2 \frac{N^2 + N - 2}{2} - P_{cpe} \tag{35.}$$

The signal to noise ratio (SNR) is defined as : $SNR = \dfrac{E_s / T_s}{P_{cpe} + P_{ici} + N_0 / T_s}$

If we replace each term by its expression we obtain finally :

$$\boxed{SNR = \frac{E_s}{N_0} \left(\sigma_\theta^2 \left(\frac{N^2 + N - 2}{2N} \right) (E_s / N_0) + 1 \right)^{-1}} \tag{36.}$$

6. SIMULATION RESULTS

Figure 1 and 2 illustrate equations (21) and (36). Simulations results are very close to theoretical curves. As expected by theoretical results performances of OFDM, DS-CDMA and MC-CDMA are similar.

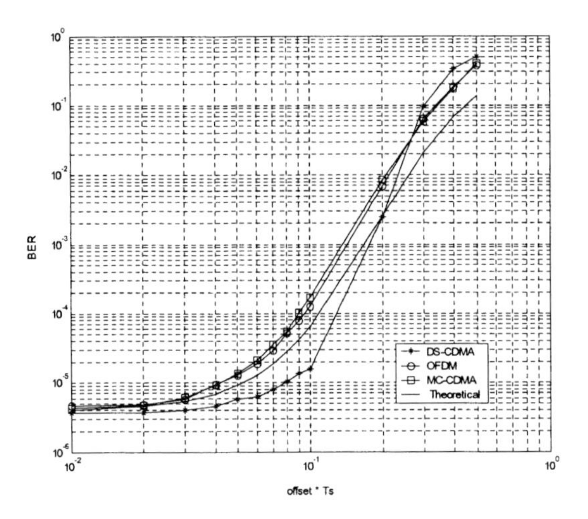

Figure 2. Theoretical and simulated BER in presence of frequency offset with $E_b/N_0 = +10\ dB$ *and for AWGN channel*

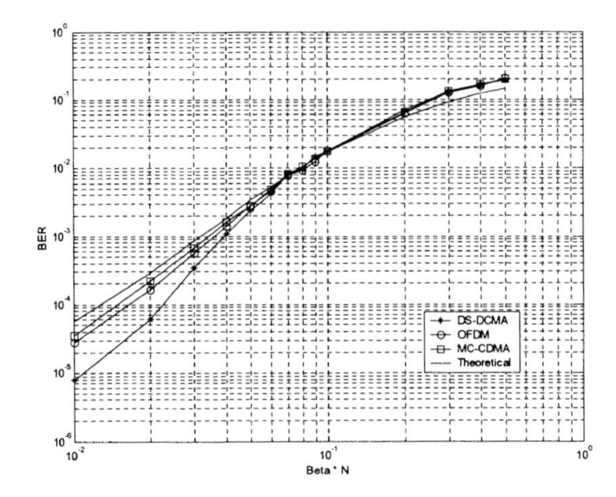

Figure 3. Theoretical and simulated BER function of βN *in presence of phase noise with* $E_b/N_0 = +10\ dB$.

7. CONCLUSION

In this paper the carried out work enabled us to develop a direct estimation of the BER and SNR degradation due either to frequency offset or to phase noise effect. These results are valid for DS-CDMA, OFDM and MC-CDMA waveforms.

8. REFERENCES

[1] J. Shentu, and J. Armstrong, "Effects of Phase Noise on Performance of PCC-OFDM," internet, Telecommunications & Signal Processing Workshop (WITSP 2002), Wollongong, Australia, pp. 50-54, 9-11 December 2002.

[2] D. Petrovic, W. Rave, and G. Ferrweis, "Phase Noise Influence en Bit Error Rate, Cut-off Rate and Capacity of M-QAM OFDM Signaling," In Proc. International OFDM Workshop (InOWo)'02, pp. 188-193, Hamburg, Germany, September 2002.

[3] L. Piazzo, P. Mandarini, "Analysis of Phase Noise Effects in OFDM Modems," IEEE Trans. Commun, vol 50, pp. 1696 -1705, Oct. 2002.

[4] A. G. Armada, "Understanding the effects of Phase Noise in Orthogonal Frequency Division Multiplexing," IEEE Trans on Braodcasting, vol. 47, no2, pp. 153-159, June. 2001.

[5] A. G. Armada and M. Calvo, "Phase noise and sub-carrier spacing effects on the performance of an OFDM communication systems," IEEE commun, Lett., vol. 2, no. 1, pp. 11-13, Jan. 1998.

[6] T. Pollet, M. Van Bladel, and M. Moeneclaey, "BER sensitivity of OFDM systems to carrier frequency offset and wiener phase noise," IEEE Trans. Commun, vol. 43, pp. 191-193, Feb./Mar./Apr. 1995.

[7] K. Sathananthan and C. Tellambura, "Probability of Error Calculation of OFDM Systems With Frequency Offset," IEEE Trans. Commun, vol. 49, no11, pp. 1884-1888, Nov 2001.

[8] L. Tomba, "On the effect of Wiener phase noise in OFDM systems," IEEE Trans. Comun., vol 46, no5, pp. 580-583, May 1998.

[9] J. Jang, and K. B. Lee, "Effects of frequency offset on MC/CDMA system performance," IEEE commun, Lett., vol. 3, no. 7, pp. 196-198, July. 1999.

[10] L. Tomba, and W. A. Krzymien, "Sensitivity of the MC-CDMA Access Scheme to Carrier Phase Noise and Frequency Offset," IEEE Transactions on Vehicular Technology, vol. 48, no. 5, pp. 1657-1665, September 1999.

[11] H. Steendam, M. Moeneclaey, "Sensitivity of OFDM and MC-CDMA to Carrier Phase Errors", ICC 1999, Vancouver, June 99, pp. 1510-1514.

[12] Younsum Kim, Sooyong Choi, Chulwoo You and Daesik Hong, "Effect of carrier frequency offset on performance of MC-CDMA systems and its countermeasure using pulse shaping,", ICC '99, vol. 1, pp. 167-171, June 1999.

[13] A. Papoulis, Probability, Random Variables and Stochastic Process, 3rd ed. New York: McGraw-Hill, 1991.

[14] D.KreB, O. Ziemann, and R. Dietzel, "Electronic simulation of phase noise," Eur. Trans. Telecommun., vol. 6, pp. 671-674, Nov./Dec. 1995.

Nizar Hicheri
e-mail: hicheri_nizar@yahoo.fr

Michel Terré
tel: (33) 1 40272767
fax: (33) 1 40 27 24 81
e-mail: terre@cnam.fr

Bernard Fino
(33) 1 40272551
fax: (33) 1 40 27 24 81
e-mail: fino@cnam.fr

Conservatoire National des Arts et Métiers
Chaire de RadioCommuinications
292 rue Saint Martin
75003 Paris
FRANCE

A New Phase Noise Mitigation Method in OFDM Systems with Simultaneous CPE and ICI Correction*

Songping Wu and Yeheskel Bar-Ness
Center for Communications and Signal Processing Research
ECE Dept., New Jersey Institute of Technology
Newark, NJ 07102

Abstract

OFDM suffers severe performance degradation in the presence of phase noise. Different methods have been proposed to mitigate phase noise effects either in time or in frequency domain. While the time-domain approach is impractical in most applications due to the need of ideal assumptions, such as special pilot pattern or AWGN channel, the alternative frequency-domain approach appears to be a better choice in correcting common phase error (CPE) and intercarrier interference (ICI), which are the results from phase noise. To the best of the authors' knowledge, ICI has never been considered simultaneously with CPE correction except in [1].

In this paper, a new simultaneous CPE and ICI correction method is proposed for OFDM systems with phase noise. In this method, CPE, as well as ICI, is derived as a function of a set of weighting coefficients which can be estimated by means of maximum likelihood estimation (MLE) method. These coefficients are then used for ICI cancellation followed by CPE correction. In order to reduce computational complexity, the proposed method is further simplified by using only a subset of those weighting coefficients. Numerical results are provided to show the effectiveness of the proposed scheme with very good performance as well as high spectral efficiency.

1 Introduction

Orthogonal frequency division multiplexing (OFDM) has raised a lot of interest both in wire and wireless communications. It has been deployed in a varieties of

*This work was supported by NSF under Grant CCR-0085846.

K. Fazel and S. Kaiser (eds.), Multi-Carrier Spread-Spectrum, 501–508.

applications. In comparison to single carrier transmission, OFDM is quite effective to combat channel multipath fading hostility while providing high spectral efficiency. In fact, OFDM receiver can be implemented with one-tap channel equalizer and use simple hardware.

The disadvantage of OFDM, however, is its sensitivity to phase noise caused by the random phase difference between the transmitter and receiver oscillators.

To compensate for phase noise, several methods have been proposed in [8]-[1] and references therein. These methods can be categorized into time-domain [8][9] and frequency-domain approaches [10]-[1].

Time domain approach aims to eliminate the multiplicative phase noise before the digital Fourier transform (DFT) unit of receiver. This can be achieved by extracting a single pilot subcarrier signal from each OFDM symbol to drive a phase lock loop (PLL) for phase noise correction [8]. Despite its low cost, this method requires ideal pilot pattern which is not feasible in practice. Another time-domain method proposed in [9] interpretes time domain phase noise using orthogonal transforms to turn phase noise correction into the recovery of DCT-based real-value waveforms. This method may not be spectrally efficient in case of random phase noise and is not applicable to fading channel.

Frequency domain methods correct the effects of phase noise after DFT, i.e., the resultant CPE and ICI, and prove to be feasible in practice regardless of pilot patterns or channel environments. A conventional frequency domain method, as introduced in [10] and [5], directly compensates for CPE or its phase. Later, in [1], a new phase noise suppression (PNS) algorithm was proposed, which first obtains the estimates of CPE and the energy of random ICI, and then applies them to a minimum mean square error (MMSE) equalizer. By correcting both CPE and ICI , this method achieves a better performance than conventional CPE correction.

We notice that, in the presence of phase noise, each received subcarrier signal (excluding AWGN noise) is actually the weighted sum of all transmitted signals multiplied by the corresponding channel response in frequency domain. Moreover, CPE and ICI are both functions of those weighting coefficients, so that, once we obtain those weighting coefficients, CPE and ICI can be eliminated simultaneously. Based on this observation, we propose in this paper an alternative approach to mitigate phase noise over fading channels.

This paper is organized as follows. Section II gives the OFDM system model in the presence of phase noise. Conventional CPE correction method is introduced in Section III. A new method is presented in Section IV with simultaneous CPE and ICI correction; and numerical results are given in Section V to demonstrate the effectiveness of the proposed method. This paper is concluded in Section VI.

2 System Model

The principle of OFDM is to transform the incoming data symbol into N low-rate parallel signals, which modulates a set of subcarriers using inverse digital

Fourier transform (IDFT) so as to obtain time-domain signals. A cyclic prefix is then added to these time-domain signals to combat inter-symbol interference (ISI) caused by channel multipath fading effects and enables simple channel equalization at the receiver.

At the receiver side, after removing the cyclic prefix and taking the length-N DFT at the receiver, the received kth subcarrier signal of the mth symbol is expressed by [1]

$$\begin{aligned} y_m(k) &= x_m(k)h(k)c_m(0) \\ &\quad + \sum_{\substack{l=0 \\ l \neq k}}^{N-1} x_m(l)h(l)c_m(l-k) + n_m(k) \end{aligned} \tag{1}$$

where $x_m(k)$ and $h(k)$ are the subcarrier data signal and the channel fading gain in frequency domain, respectively; $n_m(k)$ denotes the AWGN noise with zero mean and variance σ^2. The transmitted signals $x_m(k)$ are assumed to be mutually independent with $E\left[|x_m(k)|^2\right] = E_x$, while the energy of the channel gain $h(k)$ is normalized to unity, namely $E\left[|h_m(k)|^2\right] = 1$. $c_m(p)$ is given by $c_m(p) = \frac{1}{N}\sum_{n=0}^{N-1} e^{j2\pi np/N + j\phi_m(n)}$, with $\phi_m(n)$ denoting the random phase noise. Note that $c_m(p)$ is actually the IDFT of $e^{j\phi_m(n)}$ with $c_m(p) = c_m(p \bmod N)$.

Equation (1) suggests that the received signal $y_m(k)$ is the weighted sum of $x_m(k)h(k)$ plus AWGN noise with the weighting coefficients denoted by $\{c_m(p)\}_{p=0}^{N-1}$ .It's well known that phase noise has strong effects on OFDM systems, destroying orthogonalities among subcarrier signals. As a result, two detrimental effects occur: common phase error (CPE) denoted by $c_m(0)$; and intercarrier interference (ICI) represented by the second term of (1) as a function of $\{c_m(p)\}_{p=1}^{N-1}$.

3 Conventional CPE Correction (CPEC)

Conventional CPE correction (CPEC) method was introduced in [10] where the phase of CPE was considered. This is somewhat sufficient since for small phase noise, the amplitude of CPE, or $c_m(0)$, is approximately unity and can thus be neglected. Instead, as reported in [1], directly estimating CPE saves extra computational complexity needed for extracting its phase from pilot signals, and results in the improved estimation accuracy and therefore better receiver performance. Therefore, $c_m(0)$ can be obtained by the LS estimator [1]

$$\hat{c}_m(0) = \frac{\sum_{k \in P} y_m(k)x_m^*(k)h^*(k)}{\sum_{k \in P} |x_m(k)h(k)|^2} \tag{2}$$

where P represents the set of N_p pilot signals.

Note that, the negative effect of ICI has not been considered in CPEC. With the conclusion that ICI energy is approximately constant for large N [1], i.e., ICI energy can be estimated together with AWGN noise energy using guard band which is, in general, available at both sides of OFDM signal spectrum. This result, combined with CPEC, achieves a better performance with minimum mean square error (MMSE) equalizer as in [1].

It is of interest to estimate the value of ICI term and cancel it directly. And if possible, we do not have to estimate the energy of ICI as it will be removed from the desired signal by interference cancellation. As such, we can correct CPE and cancel ICI simultaneously to achieve an extra performance gain. For comparison, we would also like to see the performance of the proposed method in case of channel estimation errors.

4 Simultaneous CPE and ICI Correction

4.1 Maximum-Likelihood Estimation (MLE) of $c_m(p)$

Motivated by the parallel interference cancellation (PIC) method in DS-CDMA systems, we propose a new method to evaluate CPE and ICI simultaneously and remove them. The basic idea is to obtain the estimates of $\{c_m(p)\}_{p=0}^{N-1}$, which are the corresponding DFT output of time-domain phase noise $\{e^{j\phi_m(n)}\}_{n=0}^{N-1}$, serving as the weighting coefficients of CPE and ICI as in (1).

We start by rewriting (1) as

$$\underbrace{\begin{pmatrix} y_m(0) \\ y_m(1) \\ \vdots \\ y_m(N-1) \end{pmatrix}}_{\mathbf{y}} = \underbrace{\begin{pmatrix} \mathrm{w}_0 & \mathrm{w}_1 & \cdots & \mathrm{w}_{N-1} \end{pmatrix}}_{\mathbf{W}} \cdot \underbrace{\begin{pmatrix} c_m(0) \\ c_m(1) \\ \vdots \\ c_m(N-1) \end{pmatrix}}_{\mathbf{c}} + \underbrace{\begin{pmatrix} n_m(0) \\ n_m(1) \\ \vdots \\ n_m(N-1) \end{pmatrix}}_{\mathbf{n}} \tag{3}$$

where $\mathrm{w}_k = \begin{pmatrix} a_k & \cdots & a_{N-1} & a_0 & \cdots & a_{k-1} \end{pmatrix}^T$ with $a_k = x_m(k)h(k)$, $k = 0, 1, ..., N-1$. Note that any w_k with $k \neq 0$ is a left circular shift of w_0.

We need to recover the weighting coefficient vector $\mathbf{c}$ from the received signal vector $\mathbf{y}$. Given the knowledge of channel and transmitted signals, $\mathbf{W}$ is deterministic and known. Moreover, conditioned on $\mathbf{c}$, $\mathbf{y}$ is an N-dimensional Gaussian random variable vector with mean $\mathbf{Wc}$ and covariance matrix $\sigma^2\mathbf{I}$ with $\mathbf{I}$ denoting identity matrix. In order to estimate vector $\mathbf{c}$, we choose the

MLE method, which asymptotically achieves Cramer-Rao lower bound (CRLB) by minimizing the conditional probability distribution function [13]. In additive white Gaussian noise **n**, this is equivalent to minimizing the squared Euclidean distance[1]

$$\check{c}=\arg\min_{c} \|\mathbf{y}-\mathbf{W}\mathbf{c}\|^2$$

which leads to

$$\begin{aligned} \check{c} &= (\mathbf{W}^H\mathbf{W})^{-1}\mathbf{W}^H\mathbf{y} \\ &= \mathbf{W}^{-1}\mathbf{y} \end{aligned} \tag{4}$$

To solve (4), we have to know the matrix **W**, which requires the knowledge of channel and transmitted signals. The former can be estimated by means of the preamble of each block, as discussed in Section 3. On the other hand, since it is impractical to use all transmitted subcarrier signals as pilots, we can directly take the decision output of CPEC as the approximation of transmitted signals towards the solution of (4). Consequently, we obtain the estimate of **W**, whose entries are replaced by $a_k = \hat{x}_m(k)\hat{h}(k)$.

Equation (3) can be further rewritten as

$$\mathbf{y} = c_m(0)\mathbf{H}\mathbf{x}+\mathbf{W}^1\mathbf{c}^1 + \mathbf{n} \tag{5}$$

where $\mathbf{W}^1\mathbf{c}^1$ indicates the ICI vector produced by random phase noise, with $\mathbf{W}^1$ and $\mathbf{c}^1$ denoting matrix **W** without the first column and vector **c** without the first entry, respectively; $\mathbf{H} = diag(h(0), ..., h(N-1))$ and $\mathbf{x} = \left(x_m(0) \quad \cdots \quad x_m(N-1)\right)^T$. Therefore, subtracting ICI from the received signal leads to

$$\mathbf{y}^s = \mathbf{y}-\mathbf{W}^1\check{\mathbf{c}}^1 \tag{6}$$

The ICI-cancelled signal vector y^s is then compensated for its CPE error and yields

$$\begin{aligned} \mathbf{y}^{sc} &= \check{c}_m^{-1}(0)\mathbf{y}^s \\ &= \check{c}_m^{-1}(0)\left(\mathbf{y}-\hat{\mathbf{W}}^1\check{\mathbf{c}}^1\right) \end{aligned} \tag{7}$$

where $\check{c}_m(0)$, the first entry of vector $\check{\mathbf{c}}$, denotes the estimate of CPE. The phase noise corrected signals are then used for channel equalization and detection.

5 Numerical Results

Numerical results given in this section demonstrate the effectiveness of our new approach. An uncoded OFDM system is chosen with symbol size $N = 64$.

[1] The estimation result is indicated by the superscript $\check{c}$ in order to distinguish it from that of CPEC, where the estimation result is denoted by $\hat{\mathbf{c}}$.

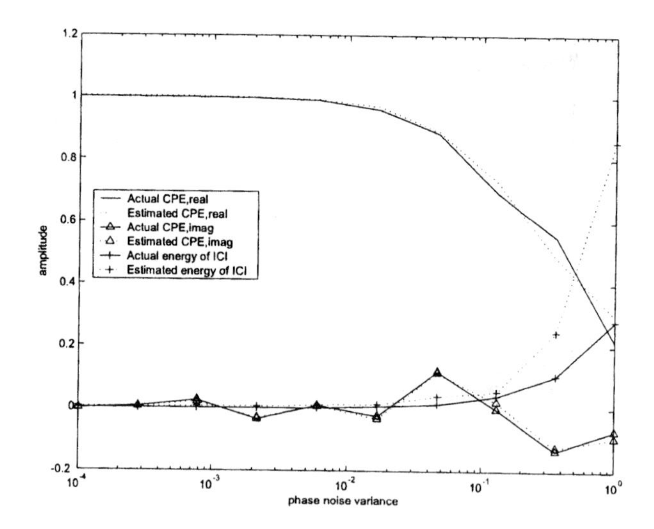

Figure 1: Actual and estimated values of CPE and ICI versus phase noise variance, SNR=10dB, $N_p = 4$

A Rayleigh fading channel consisting of six multiple paths is assumed in the simulation. Each subcarrier signal is 16-QAM modulated. N_p pilot subcarrier signals are evenly distributed within a symbol for conventional CPEC in (2).

Simulation results are presented in Fig. 1-2, which provide an clear insight into the role of phase noise on SER performance and the effectiveness of different phase noise mitigation approaches.

We show in Fig. 1 the actual versus estimated values of parameters, i.e., CPE and ICI. It shows that, for different phase noise levels, both real and imaginary parts of the estimated CPE consistently match the actual values. The energy of the estimated ICI, based on the weighting coefficients $\{c_m(p)\}_{p=1}^{N-1}$, approaches that of the actual value for phase noise variance less than 10^{-1}, which holds in practice. This suggests the robustness of the proposed method to phase noise levels.

Fig. 2 shows that, when number of pilots N_p increases, the performance gap decreases, indicating some performance improvement. Whereas, this improvement is never significant even if we use half of bandwidth as pilots, i.e., $N_p = 32$. Hence, $N_p = 4$ (occupying 6.25% bandwidth) is enough in all cases, implying the high spectral efficiency of our new approach with sufficient performance.

6 Conclusions and Discussions

OFDM suffers severe performance degradation in the presence of phase noise. Different methods have been proposed in the literature to correct phase noise either in time domain or in frequency domain. Time-domain approach requires

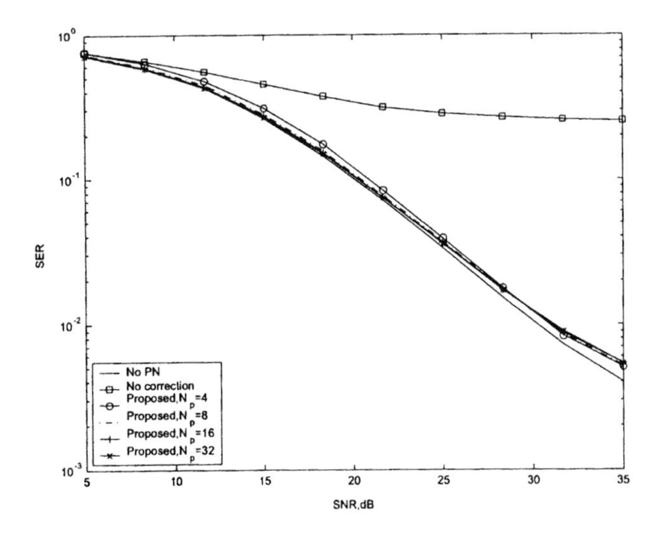

Figure 2: SER performance versus SNR with different number of pilots N_p, where $L = N$

some impractical assumptions on pilot patterns or channel environments, making frequency-domain approach preferable in practice. Conventional frequency-domain methods which emphasize CPE correction was further enhanced by the PNS algorithm proposed in [1] using the estimate of ICI plus noise energy to minimize the mean square error.

To complement the work in [1], we have proposed, in the presence of channel estimation errors, an alternative approach for simultaneous CPE and ICI correction, which first estimates a set of weighting coefficients that are the DFT output of time-domain phase noise; then evaluate and correct both CPE and ICI as a function of those coefficients. In order to reduce computational complexity, the proposed method has further been simplified by estimating a subset of those coefficients. Numerical results have been provided to demonstrate the potential of our new approach and its simplified version with the performance close to PNS.

References

[1] S. Wu and Y. Bar-Ness, "A phase noise suppression algorithm for OFDM based WLANs," *IEEE Commun. Lett.*, vol. 6, pp. 535 –537, Dec. 2002.

[2] T. Pollet, M. Bladel, and M. Moeneclaey, "BER sensitivity of OFDM systems to carrier frequency offset and Wiener phase noise," *IEEE Trans. Commun.*, vol. 43, pp. 191 –193, Feb. 1995.

[3] L. Tomba, "On the effect of wiener phase noise in OFDM systems," *IEEE Trans. Commun.*, vol. 46, pp. 580 –583, May 1998.

[4] J. Scott, "The effects of phase noise in COFDM," *EBU Technical Review*, Summer 1998.

[5] A. G. Armada and M. Calvo, "Phase noise and sub-carrier spacing effects on the performance of an OFDM communication system," *IEEE Commun. Lett.*, vol. 2, pp. 11 –13, Jan. 1998.

[6] A. G. Armada, "Understanding the effects of phase noise in orthogonal frequency division multiplexing (OFDM)," *IEEE Trans. Broadcast.*, vol. 47, pp. 153 –159, Jun. 2001.

[7] S. Wu and Y. Bar-Ness, "Performance analysis on the effect of phase noise in OFDM systems," in *Proc. ISSSTA'02*, (Prague, Czech), pp. 133 –138, Sep. 2002.

[8] M. S. El-Tanany, Y. Wu, and L. Hazy, "Analytical modeling and simulation of phase noise interference in OFDM-based digital television terrestrial broadcasting systems," *IEEE Trans. Broadcast.*, vol. 47, pp. 20 –31, Mar. 2001.

[9] R. A. Casas, S. L. Biracree, and A. E. Youtz, "Time domain phase noise correction for OFDM signals," *IEEE Trans. Broadcast.*, vol. 48, pp. 230 –236, Sep. 2002.

[10] P. Robertson and S. Kaiser, "Analysis of the effects of phase noise in orthogonal frequency division multiplexing (OFDM) systems," in *Proc. ICC'95*, (Seattle, WA), pp. 1652 –1657, 1995.

[11] IEEE Std 802.11a-1999, *Supplement to IEEE standard for information technology - telecommunications and information exchange between systems - local and metroplitan area networks - specific requirements. Part 11: wireless LAN medium access control (MAC) and physical layer (PHY) specifications: high-speed physcial layer in the 5GHz band.* http://www.ieee.org, Dec. 1999.

[12] O. Edfors, M. Sandell, J. V. D. Beek, S. K. Wilson, and P. O. Borjesson, "OFDM channel estimation by singular value decomposition," *IEEE Trans. Commun.*, vol. 46, pp. 931 –939, Jul. 1998.

[13] S. C. Kay, *Fundamentals of statistical signal processing: Estimation Theory.* NJ, Prentice Hall, 1993.

WOLFGANG RAVE, DENIS PETROVIC AND GERHARD FETTWEIS

PERFORMANCE COMPARISON OF OFDM TRANSMISSION AFFECTED BY PHASE NOISE WITH AND WITHOUT PLL

Vodafone Chair Mobile Communications Systems
Dresden University of Technology, D-01062 Dresden, Germany
email {rave, petrovic ,fettweis}@ifn.et.tu-dresden.de

Abstract. Oscillator phase noise present in OFDM systems limits the performance by producing a common phase error and additional intercarrier interference. While phase noise is most often modelled as a Wiener process corresponding to free running oscillators, we use a recently developed approach [1] to describe the stochastical behaviour of phase locked loops (PLL) to compare the performance degradation of OFDM-transmission for a receiver operating with a free running oscillator to one using a first order charge-pump PLL having a VCO with the same linewidth. In addition to the relative linewidth of the oscillator with respect to subcarrier spacing, a characteristic quantity determining the performance is the time constant of the PLL with respect to the OFDM symbol duration. Comparing the performance for different QAM constellations it is found, that a PLL will improve bit and symbol error rate, if its gain can be made high enough to obtain time constants smaller than the OFDM symbol duration.

1. INTRODUCTION

The application of OFDM has become widespread in wireless communication systems, because of its flexibility to adapt transmission rates, its high spectral efficiency and its robust behaviour with respect to multipath effects. The price to be paid is increased sensitivity with respect to nonlinearities in the transmission channel arising from a high peak to average power ratio, non-ideal synchronization and phase noise. Phase noise in OFDM has been studied by several authors (see e.g. [2]) predominantly considering free-running oscillators, thus modelling phase noise as a Wiener process. Analysis of the demodulated signal shows that generally phase noise leads to a common symbol rotation on all subcarriers (common phase error, CPE) which can be corrected using known pilot symbols. In addition residual intercarrier interference (ICI) occurs which is approximately Gaussian and remains uncorrected. It limits the performance, because an effective SNR ultimately determined by phase noise (when all other noise contributions become small) is left which explains the occurrence of a bit error floor.
A question one might ask in this context is, to which degree a PLL used for tracking the carrier phase will change the performance observed in OFDM transmission and under which condition the above-mentioned Wiener process is or is *not* an appropriate model for phase noise in OFDM. We address this problem within the framework of the phase noise theory which has recently been developed by *Demir* [3] et al. and *Mehrotra* [1]. In that work a thorough characterization of noise processes in

K. Fazel and S. Kaiser (eds.), Multi-Carrier Spread-Spectrum, 509–516.

free-running oscillators is provided as well as an extension to the behaviour of PLLs which we use to generate a well-defined stochastical phase noise process for a PLL in an OFDM transmission chain.
In the following we consider the combined effect of phase noise introduced in the analog frontend and the post-FFT correction of the common phase error as depicted in Fig. 1. To concentrate on these two effects, we make the assumption, that the time-discrete OFDM signal is directly downconverted, received and sampled with perfect symbol timing, i.e. further sampling clock or frequency synchronization errors are excluded. To compare bit and symbol error rate performance with a free oscillator to that of an OFDM receiver with a PLL, a free-running oscillator replaces the PLL in the model sketched in Fig. 1.

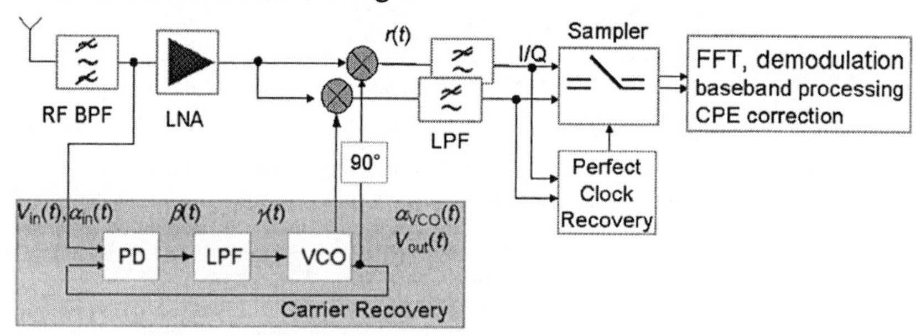

Fig. 1: OFDM receiver model with a PLL for carrier phase tracking.

The outline of the paper is as follows: In section 2 we introduce the baseband model for the OFDM system under investigation and specify how phase noise affects the received signal. Section 3 describes the stochastic models used for a free oscillator and a charge-pump PLL (which serves as an example PLL) together with asymptotic expressions for their spectra. Symbol and bit error rate curves are presented in section 4 for uncoded and coded transmission using different QAM-constellations. Given the same oscillator linewidths of free-running oscillator and VCO we illustrate, that a PLL only makes a difference, if its tracking speed is high enough to suppress phase noise on the time scale of an OFDM symbol.

2. OFDM TRANSMISSION MODEL

We study the properties of OFDM transmission. First the input bit stream is mapped to QAM-symbols. Performing an inverse Fourier transform on a group of N such QAM-symbols the time domain signal which occupies the system bandwidth $W = 1/T_u$ is obtained. The complex envelope of the transmitted signal within one OFDM symbol interval $t \in [0, T_u]$ can be written as $s(t) = \sum_{k=0}^{N-1} S(k) e^{j2\pi f_k t}$, where $f_k = k\Delta f_{car} = kW/N$ denotes the kth carrier frequency and $S(k)$ the data symbol modulating the kth carrier. The corresponding discrete time sequence reads

$$s(n) = \sum_{k=0}^{N-1} S(k) e^{j2\pi kn/N}, \tag{1}$$

to which a cyclic prefix of length T_{GI} is added. The received sampled OFDM signal $r(n) = s(n)e^{j\phi(n)} + \xi(n)$ is disturbed by AWGN samples $\xi(n)$ (a non-dispersive channel is assumed to concentrate on the influences of PLL and CPE correction) and the phase noise process $\phi(n)$ that causes a random phase modulation.
Due to phase noise up- and down conversion oscillator signals cannot be described by a perfectly harmonic carrier signal $x_c(t) = e^{j2\pi f_c t}$ (with carrier frequency f_c). Instead a random phase shift $\phi(t)$ which will be described in terms of a random time shift $\alpha(t)$ (with respect to an ideal oscillator) has to be taken into account producing the imperfect oscillator signal $x_c(t+\alpha(t)) = e^{j2\pi f_c(t+\alpha(t))}$. The demodulated complex carrier amplitude received on the lth subcarrier in an OFDM symbol becomes:

$$R_l = \frac{1}{N}\sum_{n=0}^{N-1} r(n)e^{-j2\pi nl/N} = \frac{1}{N}\sum_{n=0}^{N-1}\left[\sum_{k=0}^{N-1} S(k)e^{j2\pi kn/N}\right]e^{j\phi(n)}e^{-j2\pi nl/N} + N_l \ . \quad (2)$$

N_l still represents the additive gaussian noise, the properties of which are unaffected by the demodulation. The sum in eq. (2) can be split into interference created by the common phase error term and a remaining sum of intercarrier interference:

$$R_l = \underbrace{S(l)\frac{1}{N}\sum_{n=0}^{N-1} e^{j\phi(n)}}_{CPE} + \underbrace{\frac{1}{N}\sum_{n=0}^{N-1}\sum_{k=0,k\neq l}^{N-1} S(k)e^{-j2\pi(k-l)n/N}e^{j\phi(n)}}_{ICI} + N_l \ . \quad (3)$$

We assume ideal CPE correction which is equivalent to filtering the phase noise spectrum $S_{osc}(f)$ by a low pass filter with transfer function $H(f) = \mathrm{sinc}^2(f/\Delta f_{car})$. In that case the remaining ratio between intercarrier interference and signal power is [4]

$$P_{ICI}/S = \sum_{k\neq l}\int_0^{\infty} \mathrm{sinc}^2(f/\Delta f_{car} - (k-l))\, S_{osc}(f)\, df \ . \quad (4)$$

Using this result an effective SNR (with respect to an AWGN channel) can be calculated approximately as:

$$\left(\frac{S}{P_{ICI}+P_{AWGN}}\right)_l \cong \frac{1}{\int_{-\infty}^{\infty}\left[1\text{-}\mathrm{sinc}^2(f/\Delta f_{car})\right]S_{osc}(f)df + 1/SNR_{AWGN}} \ . \quad (5)$$

In section 3 we will apply this argument to study the conditions under which a PLL improves the performance with respect to a free-running oscillator.

3. PHASE NOISE MODEL

To quantify the influence of phase noise we use the baseband representations of stochastic phase processes for a free-running oscillator and a charge-pump PLL (CP-PLL) described in the two following subsections.

Power Spectral Density and Phase Process of a free-running Oscillator

Demir et al. presented in [3] a phase noise theory for free-running oscillators which are asymptotically described by a Wiener process. The time shift $\alpha(t)$ of the oscillating signal with respect to an ideal reference is described as $\alpha(t) = \sqrt{c}B(t)$, where $B(t)$ represents a standard Wiener process [5]. The variance of the process $\alpha(t)$ thus grows linearly in time proportional to a constant c which describes the oscillator quality. The Lorentzian spectrum associated with a Wiener process can be specified by a single parameter, its 3 dB bandwidth (which is related to c by $\Delta f_{3dB} = \pi c f_c^2$):

$$S^{free}(f) = \frac{1}{\pi} \cdot \frac{\Delta f_{3dB}}{\Delta f_{3dB}^2 + f^2} . \tag{6}$$

The phase process of the free-running oscillator $\phi(t) = 2\pi f_c \alpha(t)$ is described as the integral of a zero-mean unit variance gaussian R.V. $\xi \sim N(0,1)$:

$$\phi(t) = 2\pi f_c \sqrt{c_{free}} \int_0^{B(t)} dB(\tau) = 2\pi f_c \sqrt{c_{free}} \int_0^t \xi(\tau) d\tau . \tag{7}$$

Power Spectral Density and Phase Process of a Charge-Pump PLL

The basic structure of a PLL is shown in Fig. 1, where the reference signal V_{in} with time shift $\alpha_{in}(t)$ is to be tracked. The reference signal exhibits the noise spectrum of the oscillator by which it was generated which we assume has good quality (at least as good as the receiving LO).

The spectrum of the PLL output signal V_{out} (with time shift $\alpha_{VCO}(t)$) will be determined at high offset frequencies (relative to the fundamental oscillation frequency) by the properties of the VCO. The characteristics of the VCO can be obtained as the open loop spectrum of the PLL. If the PLL is locked, it follows at low frequencies the spectrum of the reference signal with a transition frequency defined by the loop bandwidth. For our investigation the spectrum of interest, is the spectrum of the signal at the phase detector output, denoted $\beta(t)$, which affects the signal after down-conversion.

To obtain a description of the stochastical process at the output of a PLL, we follow the approach outlined in [1] which starts from the definition of $\beta(t)$:

$$\beta(t) = \alpha_{VCO}(t) - \alpha_{in}(t) . \tag{8}$$

PLL analysis proceeds by solving the associated stochastical differential equation (SDE) for $\beta(t)$ coupled to the time shift at the low pass filter output $\gamma(t)$. The spectrum is derived from the autocorrelation function of this output signal which is asymptotically independent of t, because the PLL output corresponds to a wide-sense stationary stochastic process.

For further details the reader is referred to [1]. Here we only want to outline, how we generated stochastical processes for typical PLLs. As an example we use the generation of such a process for the particular case of a charge-pump PLL. For a CP-PLL the system of coupled SDEs for $\beta(t)$ and $\gamma(t)$ can be found in [1]:

$$\begin{bmatrix} \dot{\beta} \\ \dot{\gamma} \end{bmatrix} = -\underbrace{\begin{bmatrix} 0 & -\sqrt{c_{PLL}} \\ \omega_1 & \sqrt{c_{PLL}} \end{bmatrix}}_{A} \cdot \begin{bmatrix} \beta \\ \gamma \end{bmatrix} + \underbrace{\begin{bmatrix} \sqrt{c_{VCO}} & -\sqrt{c_{in}} \\ -\sqrt{c_{VCO}} & \sqrt{c_{in}} \end{bmatrix}}_{D} \begin{bmatrix} \xi_{VCO} \\ \xi_{in} \end{bmatrix} . \tag{9}$$

The quantities in eq. (9) have the following meaning: $\sqrt{c_{PLL}}$ with dimension [s^{-1}] is defined as $\sqrt{c_{PLL}} = k_{PD}\sqrt{c_{control}}$, where k_{pd} is the phase detector gain, $c_{control}$ characterizes noise sources in the control node of the VCO (phase detector plus low pass filter) and $\bar{\gamma} = \gamma / k_{PD}$. The parameters c_{VCO} and c_{in} characterize the strength of noise sources of VCO and input signal, respectively. Finally, ξ_{VCO} and ξ_{in} are uncorrelated white noise sources and ω_1 specifies the corner frequency of the loop filter.
The asymptotic spectrum is governed by the eigenvalues of the matrix A which are determined to be $\lambda_{1,2} = \left(\sqrt{c_{PLL}} \pm \sqrt{c_{PLL} - 4\omega_1\sqrt{c_{PLL}}}\right)/2 = \left(\sqrt{c_{PLL}} \pm \sqrt{\varepsilon}\right)/2$.

Similarly to ordinary first order differential equations these eigenvalues play the role of time constants which determine the speed of the PLL response. In addition certain coefficients $\mu_{1,2}$, $\nu_{1,2}$ are used in [1][1] to express the asymptotic spectrum (eq.(13) in [1]). If one is only interested in checking the correctness of the numerically generated spectrum of the stochastical process, the first order approximation (which is already very precise) can be expressed also by the circuit parameters themselves. The power spectral density of the VCO output of the charge-pump PLL reads

$$S_{\alpha\alpha}(\Delta\omega) = \omega_c^2 \exp\left\{-\omega_c^2(c_{VCO} + c_{in})\frac{\sqrt{c_{PLL}} - 2\omega_1}{2\varepsilon}\right\}\left[\frac{c_{in}}{\Delta\omega^2} + \dots \right. \\ \left. + \frac{c_{VCO}(\lambda_1 - \omega_1) - c_{in}(\lambda_1 + \omega_1)}{\sqrt{\varepsilon}(\lambda_1^2 + \Delta\omega_1^2)} + \frac{c_{VCO}(-\lambda_2 + \omega_1) - c_{in}(\lambda_2 + \omega_1)}{\sqrt{\varepsilon}(\lambda_2^2 + \Delta\omega_1^2)}\right]. \quad (10)$$

The relevant spectrum for the reception of OFDM signals at the output of the mixer is obtained from the AKF of $\beta(t)$ and equals the sum of two Lorentzians:

$$S_{\beta\beta}(\Delta\omega) = \omega_c^2 \frac{c_{VCO} + c_{in}}{\sqrt{c_{PLL} - 4\omega_1\sqrt{c_{PLL}}}} \cdot \left[\frac{\lambda_1 - \omega_1}{\lambda_1^2 + \Delta\omega^2} - \frac{\lambda_2 - \omega_1}{\lambda_2^2 + \Delta\omega^2}\right]. \quad (11)$$

Influence of CPE Correction on the Spectra of free-running Oscillator and CP-PLL

Typical spectra for a CP-PLL are shown on the left hand side in Fig. 2, where the asymptotic behaviour at the VCO output of the PLL and mixer outputs (equivalent to phase detector output) according to eqs. (10, 11) are shown as thick lines and compared to numerically calculated '*noisy*' looking spectra obtained from integrating eq. (9) with the Euler-Maruyama method [5] averaging over several realisations of the power spectrum of the stochastical process. The PLL output spectrum $S_{\alpha\alpha}(\Delta\omega)$ follows at low frequencies the spectrum of the high-quality input signal and coincides after a bump with the open-loop VCO output spectrum at high frequencies which falls off with -20 dB/decade. The spectrum $S_{\beta\beta}(\Delta\omega)$ of the difference signal

[1] A factor of 1/2 which appears erroneously in the second term for ν_1 in the appendix of [1] has been omitted.

$\beta(t)$ is a high-pass filtered version of the VCO output spectrum $S_{\alpha\alpha}(\Delta\omega)$ filtering out the phase noise at low frequencies.

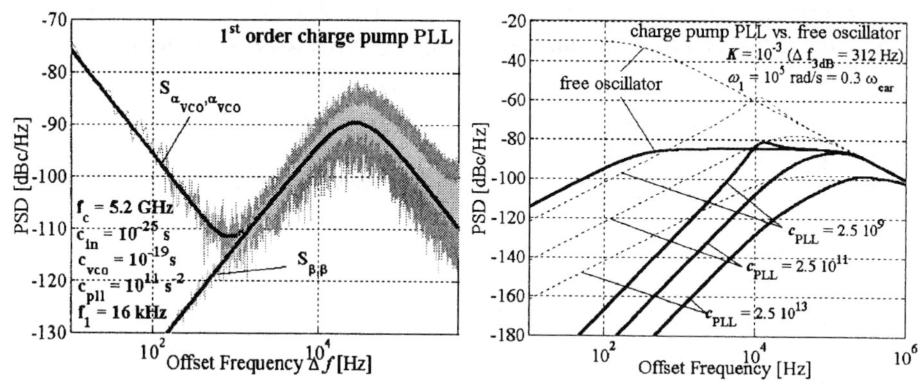

Fig. 2: Comparison of numerically generated power spectral densities at the PLL and phase detector outputs with the analytical expressions (left). Comparison between phase noise spectra of a free oscillator and a CP-PLL before (dashed lines) and after (full lines) multiplication with the ICI weighting function (CPE correction).

To understand under which circumstances use of a PLL will alter OFDM performance, one has to note, that a similar effect is obtained with CPE correction, as shown on the right hand side of Fig. 2. We compare weighted (thick lines) and unweighted (dashed lines) spectra of a free oscillator and a PLL. The 3dB frequencies of VCO and free oscillator are set in both cases to $\Delta f_{3\text{dB}} = 312$ Hz ($K = 10^{-3}$). Weighting the phase noise spectrum with $1-\text{sinc}^2(f/\Delta f_{car})$ removes the low-frequency part of the spectrum which masks the differences between free oscillators with CPE correction and PLLs, if the gain in the phase locked loop is not sufficiently high (its time constant sufficiently small). Setting the parameter Δf_{car}, to 312.5 kHz, appropriate for WLANs, implicitly introduces a time scale, which corresponds to $T_u = 3.2$ µs. Therefore three different gain values of the phase detector are shown which is equivalent to changing the time constant of the phase noise processes generated at the phase detector output of the PLL. For the particular values here, the time constants are given approximately by $1/\lambda = 40, 2.76$ and 0.20 µs (defining a relative time constant $\tau = 1/(\lambda \cdot T_u)$ this corresponds to $\tau = 12.5, 0.86$ and 0.064). Only the two smaller ones improve the performance with respect to the free-running oscillator.

4. PERFORMANCE EXAMPLE FOR WLAN-RELATED PARAMETERS

To study quantitatively the influence of different local oscillators with and without PLL on OFDM transmission we simulated a typical WLAN system with 64 carriers per OFDM symbol. The transmitted signal was disturbed by additive white Gaussian noise plus phase noise. Stochastical sample processes for free running oscillators and charge-pump PLLs were generated according to the method described in section 3. The effect of varying relative oscillator linewidth $K = \Delta f_{3\text{dB}}/\Delta f_{\text{car}}$ is seen in Fig. 3.

We changed the parameter c_{free} defined in eq.(7) for the free oscillator and used the same c-values for c_{VCO} in the charge-pump PLL with fixed values of $\omega_1 = 10^5$ rad/s comparing a relatively low gain of the PLL with $c_{PLL} = 2.5 \cdot 10^{10}$ s^{-2} ($1/\lambda = 12.65$ μs, $\tau = 3.95$) to one that is 3 orders of magnitude larger ($1/\lambda = 0.2$ μs, $\tau = 0.064$).

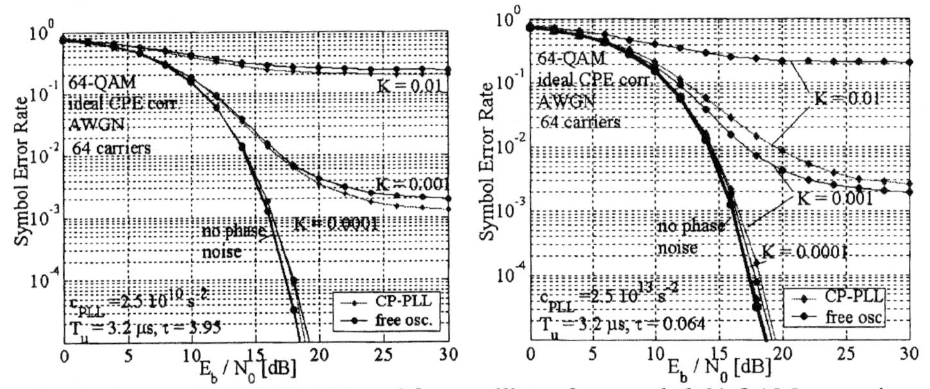

Fig. 3: Comparison of CP-PLL and free oscillator for uncoded 64-QAM transmission for relative linewidths $K = 10^{-2} \ldots 10^{-4}$ for slow (left) and fast (right) tracking.

Input phase noise, determined by c_{in}, was kept fixed at 10^{-21} s. For a PLL time constant longer than the OFDM symbol duration of 3.2 μs (Fig. 3, left) the improvement with a PLL is negligible. For a relative time constant of 0.064 a PLL markedly improves performance with respect to a free running oscillator (right).

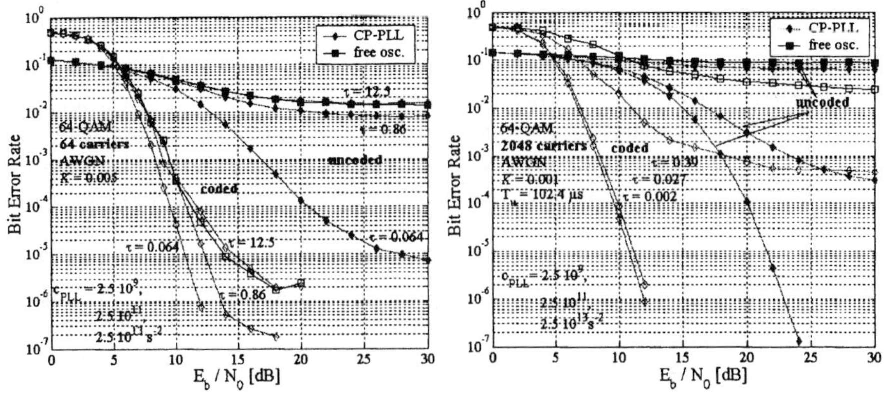

Fig. 4: Comparison of coded (open symbols) and uncoded 64-QAM transmission for several time constants of the PLL with 64 (left) and 2048 carriers (right).

The influence of the PLL time constant is further illustrated in Fig.4 for uncoded and coded 64-QAM transmission using bit-interleaved coded modulation with a convolutional code of memory 6 and rate 1/2. Coding significantly improves the performance reducing the error floor. In addition we compared the influence of OFDM symbol duration by increasing the number of carriers from 64 to 2048, applying ideal CPE correction in all cases. Although a better quality oscillator ($K = 10^{-3}$) was used for 2048 carriers, ideal CPE correction even with coding is insufficient to achieve a BER$<10^{-2}$, if the oscillator is free running. Using the same phase detector gains (c_{PLL}-values) as for 64 carriers the influence of the PLL becomes stronger, because the relative time constant is shorter, showing that for longer OFDM-symbols a carrier recovery circuit is as well more feasible and helpful. The influence of constellation size is shown in Fig. 5 for uncoded transmission. Note that the tracking capability of PLLs should benefit from the high SNR which is inherently necessary to use large constellations. Again long and short time constants ($\tau = 3.95 / 0.064$) of the PLL are compared as in Fig. 3, demonstrating, that the tracking capability of the PLL has to be brought into the time range of one OFDM symbol.

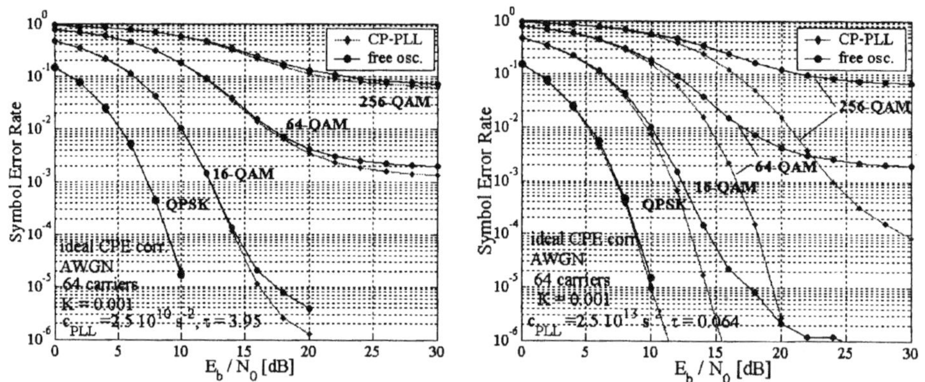

Fig. 5: Influence of constellation size for slow (left) and fast (right) tracking.

CONCLUSIONS

A stochastical phase noise model to describe the behaviour of a PLL was applied to characterize the performance of OFDM taking into account the effect of carrier tracking in an OFDM receiver. Apart from the relative linewidth of the oscillator the time constant of the phase process at the phase detector output of the PLL with respect to OFDM symbol duration becomes important. To model the performance with a free-running oscillator and CPE correction appears to be valid, as long as the time constant remains long with respect to one OFDM symbol duration. For short symbols (WLANs) and/or time variant mobile channels, where the gain in the PLL can not be made very large due to stability considerations, this is a reasonable approximation. For broadcasting standards such as DVB this might be different.

REFERENCES

[1] A. Mehrotra, "Noise Analysis of Phase-Locked Loops", IEEE Trans. Circuits & Systems-I, vol. 49, no.9., pp. 1309-1316, Sept. 2002

[2] E. Costa, S. Pupolin, "*M*-QAM-OFDM System Performance in the Presence of a Nonlinear Amplifier and Phase Noise", IEEE Trans. Comm., vol.50, no.3., March 2002.

[3] A. Demir, A. Mehrotra, J. Roychowdhury, "Phase Noise in Oscillators: A Unifying Theory and Numerical Methods for Characterisation", IEEE Trans. Circuits & Systems-I, vol.47, no.5., pp. 655-674, May 2000

[4] J. Stott, "The Effects of Phase Noise in COFDM", BBC Res.&Dev. Techn. Rev., pp.1-10, 1998

[5] D. J. Higham, An Algorithmic Introduction to Numerical Simulation of Stochastic Differential Equations", SIAM Rev. vol. 43, no 3, pp. 525-546, 2001